Physics and Chemistry in Space Vol. 12

Edited by L. J. Lanzerotti, Murray Hill

A. V. Gurevich E. E. Tsedilina

Long Distance Propagation of HF Radio Waves

With 126 Figures

Springer-Verlag
Berlin Heidelberg New York Tokyo

Dr. ALEXSANDR V. GUREVICH

Academy of Sciences of the USSR
P. N. Lebedev Physical Institute
Leninsky Prospect, 53
117924 Moscow, USSR

Dr. ELENA E. TSEDILINA

Institute of Terrestrial Magnetism
Ionosphere and Radio Wave Propagation
Academy of Sciences (IZMIRAN)
Troitsk
142092 Moscow Region, USSR

Translated from Russian by:

Ya. S. DIMANT, A. G. ISRAETEL

IZMIRAN
Troitsk
142092 Moscow Region, USSR

ISBN-13:978-3-642-70251-8 e-ISBN-13:978-3-642-70249-5
DOI: 10.1007/978-3-642-70249-5

Library of Congress Cataloging in Publication Data. Gurevich, A. V. (Aleksandr V.) Long distance propagation of HF radio waves. (Physics and chemistry in space ; v. 12) Bibliography: p. 1. Ionospheric radio wave propagation. I. Tsedilina, E. E. (Elena E.), 1935- . II. Title, III. Title: Long distance propagation of H. F. radio waves. IV. Series. QC801.P46 vol. 12 [TK6553] 530'.0919 s 85-4713 [621.3841'1]

Softcover reprint of the hardcover 1st edition 1985

Typesetting: Fotosatz GmbH, Beerfelden.

2131/3130-543210

Preface

The study of very long-distance and around-the-world propagation of HF radio waves becomes more urgent in connection with the problems of long-distance ground-based radio communications, communications with spacecrafts and satellites, satellite-to-satellite communications, around-the-world radar scanning, around-the-world sounding of the ionosphere, etc. At present, these investigations have acquired particular interest because transmitters which make it possible to intentionally modify the ionospheric properties by powerful radio waves have become available.

In the case of radio wave propagation over comparatively small distances (about 3000 – 5000 km, one – two hops), in a first approximation, the ionosphere can be considered homogeneous in the horizontal direction. The radio wave propagation theory in a horizontally-homogeneous, i. e., spherically-symmetric ionosphere was developed with sufficient completeness as early as in the 1920 – 1940's by Appleton, Ratcliffe, Beynon, Booker, Martyn, and others. This theory is presented in detail in the well-known monographs by Ginzburg (1967), Bremmer (1946), and Budden (1961). Based on this theory, detailed methods for the calculation of radio paths, determination of field amplitude, and interpretation of vertical and oblique ionograms have been developed. All these methods are well-known and widely used in practice, see monographs by Al'pert (1974), Shchukin (1940), and Davies (1969).

An altogether different situation takes place in the case of very long-distance multihop and around-the-world propagation. In this case radio waves pass ionospheric regions, which greatly differ from one another as, for instance, the daylight and night hemispheres or the polar, midlatitude and equatorial regions. In these regions propagation conditions are quite different. That is why the ionospheric horizontal inhomogeneity defining the variation of the ionospheric properties along a radio path proves to be of great importance for very long-range propagation.

Due to the Earth's sphericity and stratified structure of the ionosphere, waveguide channels (ducts) appear in which radio waves propagate over long distances. These ducts have a complicated structure, they are inhomogeneous, interrelated, and vary with time variations in the ionosphere. To understand the main features of long-distance propagation, it is necessary to study the properties of the waveguide channels. The interlayer ionospheric duct is of particular interest. The absorption in this duct is less than in the case of conventional hop propagation in the Earth – ionosphere waveguide. Thus, the interlayer duct is the most convenient for very long-distance and around-the-world propagation. Of importance are not only the properties of the interlayer duct, but also the possibility of its excitation by ground-based sources, that is, the possibility of capture into the duct of radio waves transmitted from the Earth and their escape back to the Earth. In the ray approximation, in the horizontally-homogeneous ionosphere, direct capture of radio waves from the Earth into the interlayer duct is impossible. Therefore, the capture is defined by such fine effects as the horizontal inhomogeneity of the ionosphere, radio wave scattering and tunneling due to wave effects. The artificial ionospheric irregularities produced during the modification of the ionosphere by powerful radio waves can be of particular importance for the capture into the duct of radio waves and their escape from the duct.

In addition to the smooth regular structure of the ionosphere, irregular ionospheric inhomogeneities also significantly influence long-distance propagation. These irregularities are characterized by a wide range of scales: from tens of meters to hundreds of kilometers. Scattering by these irregularities results in random oscillations of the wave field. The role of scattering increases rapidly with the distance covered by a wave in the ionosphere. This significantly affects the long-distance signal characteristics, such as fluctuations of the amplitude and phase, fluctuations and distortions of the bearing, and fluctuations of the time delay, resulting in a great broadening and even complete spreading of the mode characteristics in the ionograms of long-distance oblique sounding. Besides, scattering results in variations of the average wave-field amplitude, a broadening of the receiving frequency range, and variations of the diurnal characteristics of long-distance radio communication, etc. Nonlinear scattering by artificial irregularities can also play a marked role.

Thus, long-distance radio wave propagation is defined to a great extent by the horizontal inhomogeneity and irregular structure of the ionosphere. This complicates significantly a theoretical description of radio wave propagation. If, for instance, only the horizontal ionospheric inhomogeneity is taken into account, this will lead to a fundamental mathematic complication of the problem. This complication is connected with the transition from the scalar wave equation in one-dimensional inhomogeneous medium (methods for the solution of this equation are well-developed), to much more complicated and not so well-studied vector wave equation in three-dimensional inhomogeneous medium. As a consequence, the theory of long-distance HF radio wave propagation was being developed for a long time in the approximation of a spherically-symmetric, i. e., horizontally-homogeneous ionosphere. This made it possible to obtain only a qualitative idea of the effects actually taking place in the real ionosphere.

In the 1960's, numerical calculation techniques were developed for ray trajectories in the horizontally-inhomogeneous ionosphere in the geometrical-optics approximation. These methods, developed in detail later on (see, for example Kerblay and Kovalevskaya 1974) are successfully applied for the determination of important characteristics of long-distance propagation such as, for instance, the maximum usable frequency of radio path. However, many of the finer problems have proved to be too complicated and awkward for the ray-tracing technique. These problems involve the capture of radio waves into the interlayer ducts and their escape from the ducts, the identification of the propagation modes on a radio path, the determination of the field amplitude for various modes, and also the construction of ionograms of the long-distance oblique sounding. Due to the fact that the problems were too complicated they were practically not solved by means of the ray-tracing methods.

The situation has changed during the last decade. The use of the features of inhomogeneous structure of the actual ionosphere has allowed to significantly simplify the description of long-distance propagation. This theory has received the name of adiabatic approximation. An intense development of the adiabatic approximation theory allowed the global characteristics of long-distance propagation to be investigated, and to work out in detail the techniques for radio path calculation, including the identification of the modes, the determination of the wave-field amplitudes, the construction of ionograms, and so on. The next important step was the transition to the statistical methods of wave-field description. By means of these methods the effects of irregular ionospheric inhomogeneities on radio wave propagation have been systematically described. The adiabatic approximation has also played an important role in solving the above mentioned problem. Thus, a considerable contribution to the development of the theory of long-distance HF propagation in the ionosphere has been recently made.

The present work is devoted to the description of the current state of the theory. The work deals with a systematic description of the theory of radio wave propagation in a three-dimensional inhomogeneous ionosphere. Consideration of the features of the actual horizontally-inhomogeneous ionosphere makes it possible to significantly simplify the problems of radio wave propagation and to derive the main relations of the adiabatic approximation theory. Its rigorous substantiation within the framework of the wave theory is given in Appendix 3, written by one of the authors in collaboration with N. D. Borisov.

Considerable attention in the present work is paid to the investigation of the general features and global characteristics of very long-distance radio wave propagation as well as to the description of the specific methods for radio path calculation. The radio path calculation by the adiabatic approximation method proves to be especially simple for the parabolic model of an ionospheric layer. The calculation can be performed quickly and easily used in practice. When more complicated ionospheric models are used, the theory takes into account the possibility of wave propagation in various ducts as well as the transition of waves from duct to duct due to the regular inhomogeneous structure. To carry out these investigations it was necessary to develop a new model for the global distribution of electron density and electron collision frequency (see Appendix 2). This model, specially constructed for the calculation of radio wave propagation, appears to be most convenient. It describes quite accurately the main features of the ionospheric structure and is characterized by a minimum of parameters.

When an irregular inhomogeneous structure of the ionosphere is taken into account, the classical formulation of the problems of the radio wave propagation theory changes fundamentally. In this case the ionosphere is considered to be a randomly inhomogeneous medium, so that methods of statistical radiophysics (Rytov, Kravtsov, and Tatarskii 1978) should be applied for a theoretical description of radio wave propagation. In the present monograph the equation of radiation transfer is used. Its considerable simplification is achieved by the use of the small angle and adiabatic approximation. Thus, a wide range of the phenomena connected with the effect of multiple scattering on the processes of capture and radio wave propagation in ionospheric waveguide channels could be described.

The solution of the inverse problem is of particular interest. The problem involves the determination of regular and irregular three-dimensional inhomogeneous structures of the ionosphere according to the results of radio investigations. The general principles of the theory required for the problem to be solved are set forth in the present work. However, it is important to carry out extensive investigations based on the data of oblique sounding over long-distance and round-the-world paths. Using the data of oblique sounding one can study the radio wave scattering. Thus, global distribution models for the ionospheric irregularities could be constructed. Round-the-world sounding combined with a network of oblique and vertical sounding stations should in future lead to the creation of a system allowing to determine the three-dimensional global distribution of both electron concentration in the ionosphere and the ionospheric irregularities as well as their time variations. The creation of such a system would be of fundamental importance for investigations of solar-terrestrial relationships, the ionosphere, and radio wave propagation.

Similar questions arise when radio wave propagation in the magnetosphere is studied. The magnetospheric plasma density is small, so, generally speaking, it slightly affects the HF radio wave propagation. But under certain conditions a wave can be trapped by the magnetospheric inhomogeneities. In this case a kind of radio wave guiding along the geomagnetic lines of force similar to propagation in inhomogeneous ionospheric waveguides is observed. The same guiding takes place for the VLF radio waves as well. When describing the magnetospheric guiding the methods developed in the theory of very long-distance HF ionospheric propagation and, in particular, the adiabatic approximation prove to be rather effective. The closing chapter of the present work is devoted to the investigation of the problems of magnetospheric guiding.

In conclusion, the authors express their gratitude to V. L. Ginzburg and V. V. Migulin for their support, to N. D. Borisov, A. G. Israetel, V. Yu. Kim, N. I. Manaenkova, T. N. Soboleva, I. A. Tushentsova and D. I. Fishchuk for active participation in the investigation of a number of problems considered in the monograph, to V. V. Vas'kov, S. F. Golyan, Yu. K. Kalinin, L. A. Lobachevskii, V. Yu. Trakhtengerts, and Yu. V. Chugunov for their helpful discussions and valuable advice and also to T. A. Ivanova and N. V. Kukhareva for their help in the preparation of the manuscript and the drawings.

Moscow, 1985

A. V. GUREVICH
E. E. TSEDILINA

Contents

1 Introduction

High frequency radio wave propagation over very long-range and around-the-world paths is of great interest for radio communications. In particular, round-the-world signals are often observed at frequencies exceeding the maximum usable frequency of the *F*-layer (MUF-*F*) [1]. This means that the waves at frequencies higher than the MUF-*F* can be confined by the ionosphere. A possible use of such frequencies for stable ground-based radio communications is of great practical importance.

A number of other problems connected with very long-distance and around-the-world propagation also arises, e.g., the communication with spacecrafts and satellites, and satellite-to-satellite communications. Further, around-the-world propagation can provide, in principle, an around-the-world radar scanning of the ionosphere. Using the pulse train at different frequencies one can carry out an around-the-world sounding of the ionosphere.

It should be especially noted that recently radio facilities have become available allowing the production of HF radio emission fluxes of high power and directivity. Ionospheric properties are significantly modified by the high-power radio waves (Ginzburg and Gurevich 1960, Utlaut and Cohen 1971, Shlyuger 1974, Getmantsev et al. 1973). Thus, a possibility of using artificial ionospheric modifications for the improvement of very long-range radio communications arises.

1.1 Very Long-Distance and Magnetospheric Propagation

1.1.1 Round-the-world Signals

It has been known for a long time that short waves can propagate over very long and round-the-world distances. It was Quaeck (1927) who first reported the phenomenon of round-the-world echo in 1927. Later, the phenomenon was observed by Eckersley (1927). Quaeck and Möegel (1929), Hess (1948, 1949, 1952), and others (Isted 1958, Luscombe 1957, Taylor and Young 1928,

[1] The maximum usable frequency of the F-layer is the highest frequency on which radio waves transmitted from the Earth can still be reflected from the ionosphere. Radio waves with a frequency higher than the MUF-*F* penetrate through the ionosphere without reflection.

De Voogt 1950, Ortner 1959, Bailey 1958, Al'pert 1974). It was Fenwick and Villard (1963a,b,c) of Stanford University (USA) who made the most systematic observations of round-the-world (RTW) signals, namely, they studied experimentally their main characteristics. At present, RTW observations on various paths continue to be intensively carried out. The work by Barker and Grossi (1970) should be especially noted. They studied RTW signals with two satellites, OV4-1R and OV4-1T, and with a number of ground-based stations.

Let us consider in brief the main experimental characteristics of RTW signals.

The Observed Frequency Range for RTW Signals. As a rule for RTW-signal observations on the Earth, frequencies within the range of 10 – 27 MHz are used. The dominant frequency range in which RTW signals are detected is 12 – 22 MHz (Fenwick 1963, 1966; Fenwick and Villard 1963a,b,c). The dependence of RTW-propagation frequencies on solar activity appears to be weak. When solar activity increases significantly, the frequencies increase by 2 – 3 MHz (Golyan and Shlionskii 1971, Golyan 1972, 1973, 1974). Note, however, that during the solar maximum years, the RTW signals were observed at higher frequencies as well (up to 40 MHz) (Isted 1958). The ranges of observed frequencies for RTW signals from the maximum (MUF of RTW propagation) up to the minimum (LUF of RTW propagation) for March – April 1962 and for March 1964 are shown in Fig. 1.1. The abbreviation LUF means the lowest usable frequency for a given RTW path. The value of the LUF depends on the transmitter power, absorption, and path length.

At night MUF of RTW propagation appears to be lower than in the daytime and is equal to 16 – 18 MHz (Fenwick 1963, Kiyanovskii and Ostrovskii 1972, Bubenik et al. 1971). However, the RTW signals are usually not observed during all the 24 h. That is why here the winter – day observations are compared with the summer – night ones. Some increase of the RTW frequency in the twilight zone and especially in the post-sunset zone is observed (Fenwick 1963).

Diurnal, Seasonal, and Sunspot-Cycle Variations of RTW-Signal Occurrence. During the equinox periods and in winter, the RTW signals are mainly observed in the daytime, while in summer they are observed at night (Fenwick 1963, 1966; Kiyanovskii and Ostrovskii 1972, Bubenik et al. 1971). In this case the frequency of occurrence of RTW signals is significantly higher in the daytime than at night. According to Fenwick (1963), in the equinox transition period and on some summer days, the morning onset of RTW propagation began very near the time of ground sunrise. In summer, RTW propagation was found to often begin considerably in advance of ground sunrise and it was occasionally possible throughout the entire night period (Fenwick 1963) (over a very limited frequency range, however). The diurnal and seasonal dependencies of RTW occurrence are closely related to the azimuthal ones. Near the terminator, the RTW signals are observed during all the seasons. The percent

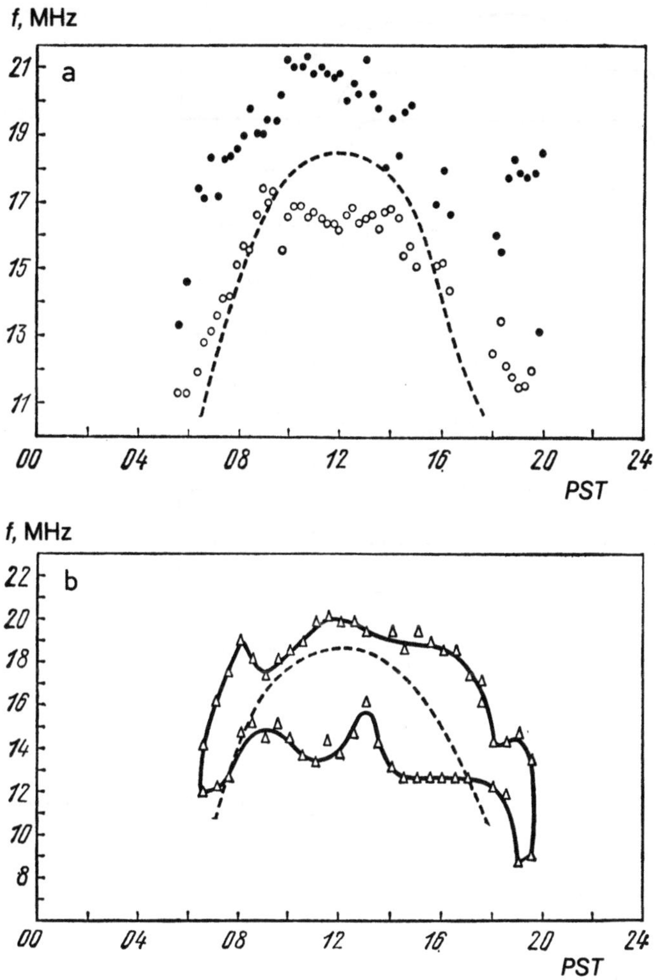

Fig. 1.1 a, b. Range of observed frequencies for RTW signals vs time of day (PST) at Stanford (USA): **a** March – April 1962 (Fenwick 1963); **b** March 1964 (Fenwick 1966); *dashed line* shows the results of theoretical calculations (see Sect. 3.4)

of time of the RTW propagation observed for different seasons versus time of day is shown in Fig. 1.2. The effect of solar activity on the frequency of RTW propagation occurrence is not yet clear. For instance, in 1958 during the solar activity maximum, a number of papers dealing with the RTW observations were published. This, however, can be associated with an increase of the whole number of observations during the IGY. According to Golyan and Shlionskii (1971) and Ostrovskii (1972), when solar activity increased in 1970 to 1971, the frequency of RTW propagation decreased.

The Effect of Magnetic Activity on the RTW Propagation Characteristics. This effect has been studied by Fenwick (1963), Fenwick and Villard (1963),

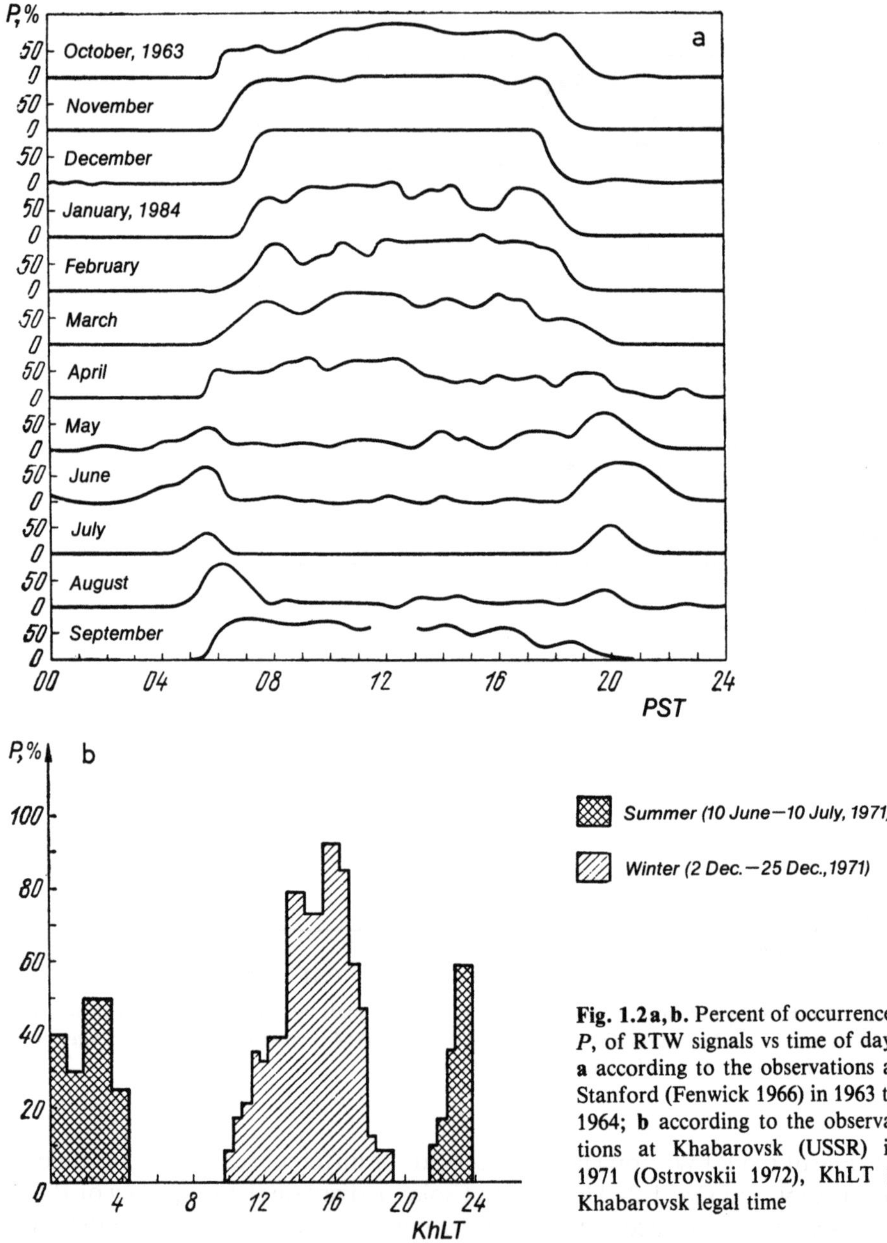

Fig. 1.2a, b. Percent of occurrence, *P*, of RTW signals vs time of day: **a** according to the observations at Stanford (Fenwick 1966) in 1963 to 1964; **b** according to the observations at Khabarovsk (USSR) in 1971 (Ostrovskii 1972), KhLT is Khabarovsk legal time

and Bubenik (1975). The main results show that the MUF decreases with the increase of the magnetic index ΣK_P, while the LUF increases. The amplitude of RTW signals and the duration of their reception also decreases. In some cases when the index ΣK_P (or the index A_P) is great, the RTW signals are absent at all the frequencies altogether.

Latitude–Longitude Dependence of RTW Propagation Occurrence. The experiments of Fenwick (1963, 1966) and Ostrovskii (1972) show a very high percent of occurrence of RTW signals (up to 100%) on midlatitude paths in day time (see Fig. 1.2). On the contrary, the RTW propagation over meriodional paths is hindered. According to Fenwick (1963), Ostrovskii (1972), and Gerson et al. (1969), the percent of RTW propagation occurrence over the polar region becomes lower. In the equinox periods in the twilight belt, the percent of RTW propagation over a meridional path increases (Golyan and Shlionskii 1971, Bubenik 1975).

Time Delay of RTW Signals. This is an important characteristic. Figure 1.3 shows the shape of RTW pulses received. According to Fenwick (1963, 1966) and Fenwick and Villard (1963a, b, c) the time delay of RTW signals t_1 depends on frequency and time of day and varies within 137–139 ms. The time of maximum amplitude propagation t_2 is more stable ($t_2 \sim 139.5$ to 140 ms). The time delay of RTW signals measured with the OV4 satellites is within 138–142 ms. The pulse dispersion is small at the frequencies close to the MUF. At lower frequencies it is about 2–5 ms. It should be emphasized that on the whole, the dispersion of RTW signals is rather significant. This can be seen from subtracting the principal "geometrical part" of time delay. This part is determinated by wave circulation around the Earth ($\Delta t_0 = 2\pi R_0/c \approx 134$ ms).

Splitting of RTW Signals. The phenomenon of RTW signal splitting was observed in the RTW ionograms (Bubenik et al. 1971) obtained by means of

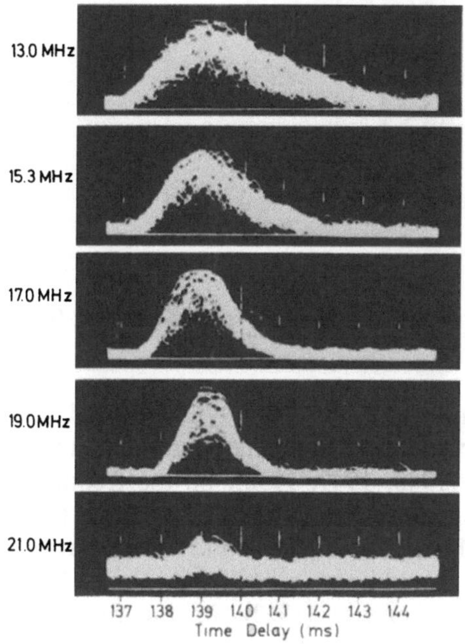

Fig. 1.3. Pulse shape of RTW signals at various frequencies vs time delay (in ms) according to Fenwick and Villard (1963c). The time marks are shown spaced by 1 ms. The transmitted pulse length was 1 ms. Records show film integrations of 80 received pulses for varying frequency. The pulse dispersion is seen to be greater at lower frequencies

the sweep frequency continuous wave (SFCW) sounding technique. At lower frequencies the filling of a signal is uniform. At frequencies approaching the MUF a continuous signal is split into several separate signals. At frequencies close to the MUF one can observe 2 – 3 split signals with a width of about 1 ms, being apart from each other by 1 – 2 ms (Fig. 1.4).

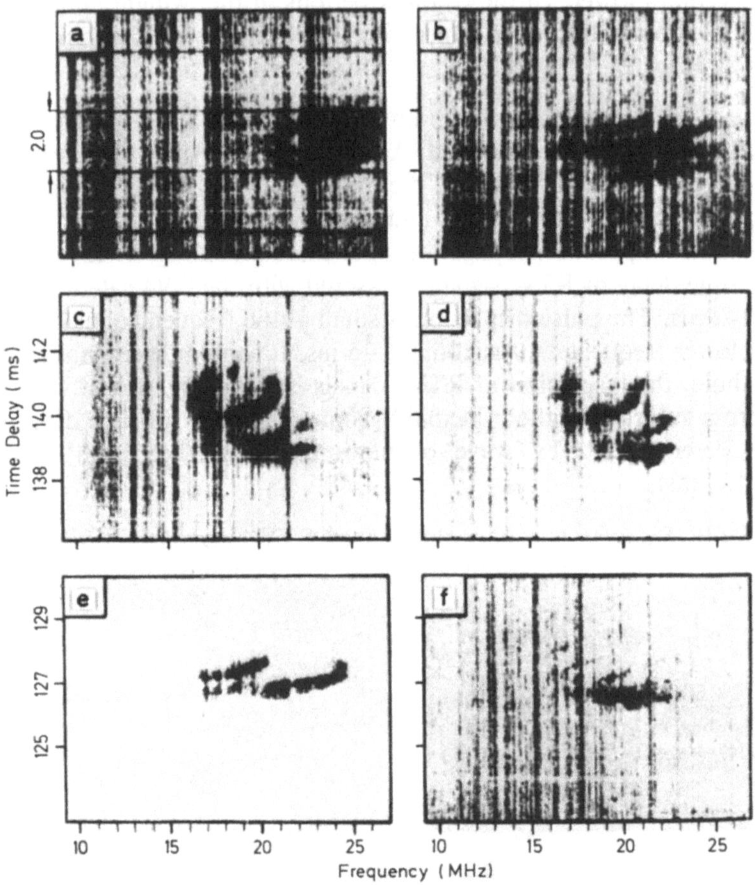

Fig. 1.4a – f. Ionograms of RTW signals **(a – d)** and the signals close to RTW ones **(e – f)** obtained by the SFCW sounding technique

Absorption. The observed RTW signals are not strongly attenuated and signals covering short distances often come to be much more attenuated than the RTW signals. According to Hess (1948, 1949, 1952), on the frequency 20 MHz, the absorption of the RTW signals circling the Earth two and three times is about 5 – 10 dB. In this case the signals of four and five circulations have been often observed. The same data obtained on the frequency 37 MHz is given by Isted (1958). According to Fenwick (1963), on the frequency 15.1 MHz from 13.10 to 17.15 PST, the RTW-signal absorption varies within

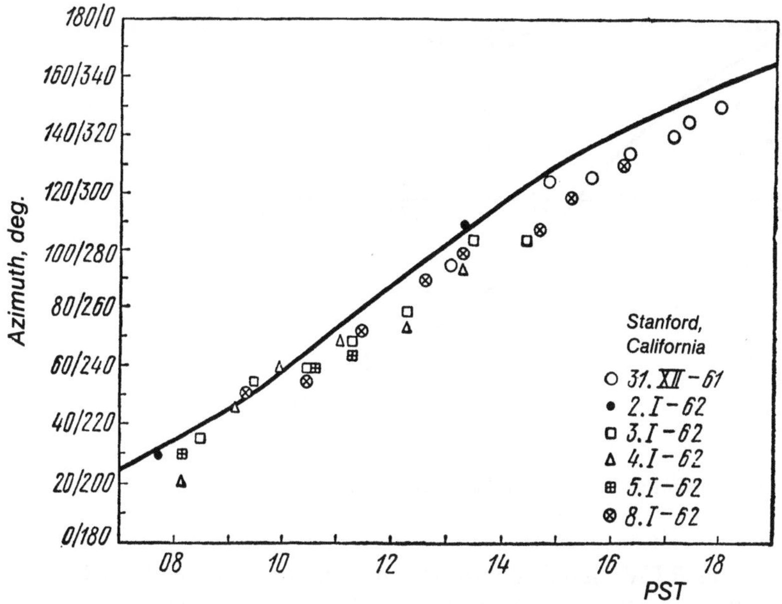

Fig. 1.5. Experimental points for the optimum RTW propagation azimuth vs time of day at Stanford, winter 1961 – 1962 (Fenwick 1963). Azimuth is in degrees and read off from the northern direction. The *solid line* shows the direction orthogonal to the direction of the subsolar point (the curve of predicted optimum azimuths)

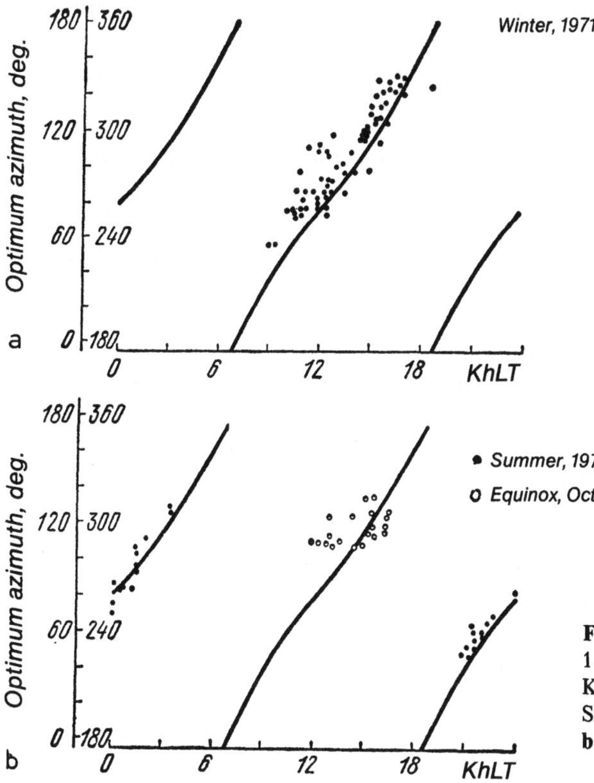

Fig. 1.6 a,b. The same as in Fig. 1.5. The results were obtained at Khabarovsk (Ostrovskii and Shlionskii 1974): **a** winter; **b** summer and autumn

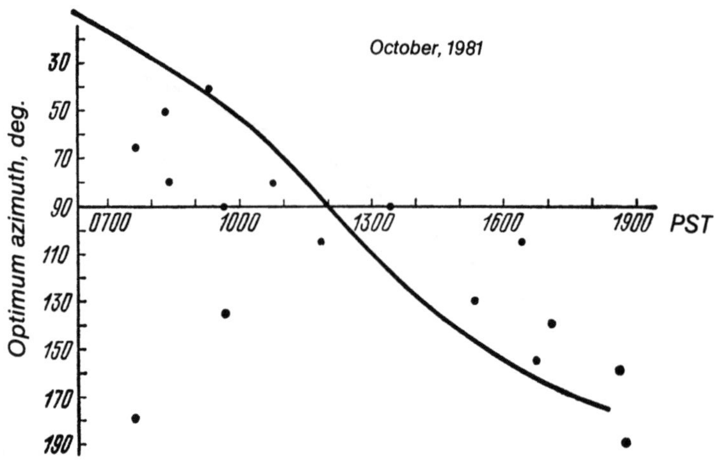

Fig. 1.7. The same as in Fig. 1.5, for October 1961 (Fenwick 1963)

15 – 20 dB. The RTW signal absorption measured with the OV4-1R and OV4-1T satellites on the frequency 34.3 MHz in an ionospheric duct lies within 3 – 18 dB for daytime conditions of reception (Barker and Grossi 1970). All the given values of absorption are measured by comparison of the amplitudes of the two-circulation RTW signals with the amplitudes of the one-circulation RTW signals. These values characterize the total absorption in RTW duct in the ionosphere. These values, however, do not involve the additional absorption occurring at the capture into an ionospheric wave duct (IWD) and at the descent from the IWD to the Earth.

The Angles of Arrival in the Vertical Plane and the Angles of Deviation from the Great Circle Plane. There is a small number of measurement data on vertical angles of RTW-signal arrival (Agaryshev and Unuchkov 1975, Neyer and Rawer 1951, Agaryshev 1981). According to Agaryshev and Unuchkov (1975), and Neyer and Rawer (1951), the vertical angles of arrival for the frequencies of 15 – 20 MHz are within 4 – 10°.

Ray deviations from the great circle plane were studied by Fenwick and Villard (1963a) on the Standford – Urbana path, 37000 km long. Ninety percent of the measurements made on the frequency of 15.4 MHz in January to March 1963 showed that the ray deviations from the great circle plane were within the range of ± 3° (400 km). On the paths located in the twilight zone, the ray deviation is much greater (Bubnov and Rumyantsev 1975).

The Optimum Directions of RTW-Signal Propagation as a *Function of Time of Day and Season.* These were studied by Fenwick (1966) and Ostrovskii (1972). The optimum azimuth of RTW-signal propagation in the summer and winter periods is close to the direction orthogonal to the direction of the subsolar point from a transmission point. Curves of the optimum azimuths are shown in Figs. 1.5 – 1.7. In the summer and winter period the experimental data are grouped near the curves of the optimum azimuths with an average spread of about ± 20° (Fenwick 1966, Ostrovskii and Shlionskii 1974).

Note that for the conditions close to the spring and autumn equinoxes, the optimum directions, according to Fenwick (1963), can differ greatly from the above mentioned optimum azimuths (Fig. 1.7).

As it is seen from the observational data on RTW signals, these signals are characterized by a number of stable regular properties (time delay, frequency range, lateral deviations, optimum azimuths, seasonal and diurnal variations of frequency of occurrence, etc.). At the same time it should be emphasized that the frequency of RTW occurrence is rather irregular. For instance, at one and the same path under the same conditions, the RTW signals can or cannot be observed. Besides, the frequency of RTW occurrence significantly depends on the transmitter power and on the sensitivity of reception facilities. The frequency of RTW occurrence becomes regular only after being averaged with respect to a certain time interval of observations.

1.1.2 Long-Distance Oblique Sounding

Radio wave propagation over long-distance paths was studied in many papers (see, for example, Golyan and Shlionskii 1971, Golyan 1972, 1973, 1974, Khodzha-Akhmetov and Koshurko 1975, Lobachevskii et al. 1981, Vovk et al. 1981, Kunsucker and Bates 1969, Martines Brunet and Popov 1980, Smirnov 1972, Golyan et al. 1981, Bates and Albee 1966).

Here, we shall consider the results of long-distance oblique sounding (LDOS), enabling one to obtain important information about the character of radio wave propagation. This sounding is like the usual vertical and oblique one.

A series of short pulses covering a broad frequency range for a short period of time is emitted from a transmission point. The reception of signals is carried out a long distance from the transmitter. The ionograms of LDOS are obtained by measuring the time delay of pulses versus the wave frequency at the reception point.

The Structure of Oblique Sounding (OS) Ionograms. To find out the features of LDOS ionograms, first let us consider a typical ionogram of a usual OS obtained for a distance of 1000 – 3000 km (Fig. 1.8). The OS ionogram consists of several bands. Each of them is formed by pulses whose character of propagation is similar at different frequencies. The bands are called modes. In Fig. 1.8, for instance, the lowest band is formed by the mode reflected from the E-layer. This mode has the least time delay and depends slightly on wave frequency f. The modes reflected from the F-layer (the bands $1F$ and $2F$ in Fig. 1.8) have greater time delay. In this case the $1F$ mode is formed by one-hop signals, i.e., one reflection from the F-layer takes place between transmission and reception points, while the $2F$ mode is formed by two-hop signals. The upper part of the $1F$ mode (and often of some other F modes) varies

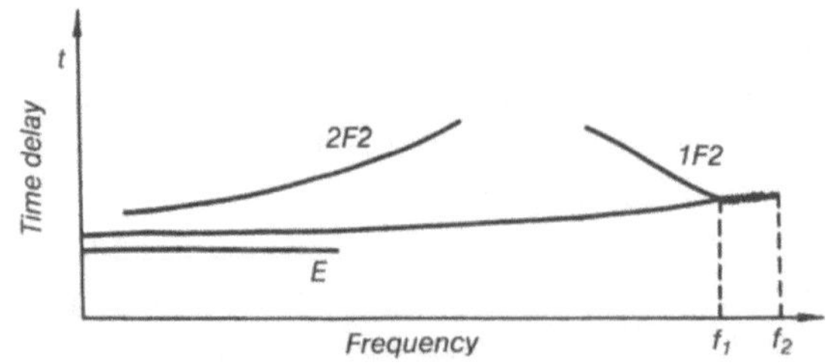

Fig. 1.8. A typical OS ionogram on a short path

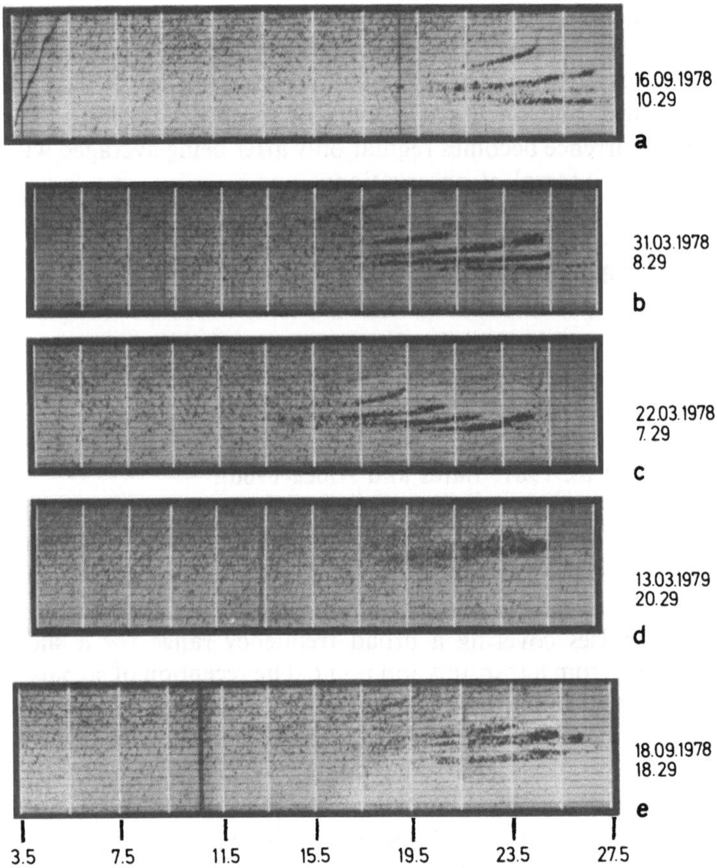

Fig. 1.9. LDOS ionograms on the Moscow – Molodiozhnaya (Antarctic Continent) path (Lobachevskii et al. (1981). The equipment with the continuous frequency sweep in the range of 3.5 to 27.5 MHz allowed to obtain an ionogram during 2 min. Pulses of length 0.1 ms were transmitted with the repetition frequency 50 Hz. The receiver band width is equal to 14 kHz. Date, month, year and Moscow legal time (MLT) is indicated

inversely with frequency. This part is formed by the signals, reflected in the vicinity of the *F*-layer maximum (the Pedersen ray). The frequency at which the upper and the lower parts of the $1F$ mode join is the maximum propagation frequency for a path in question (the frequency f_1 in Fig. 1.8). It

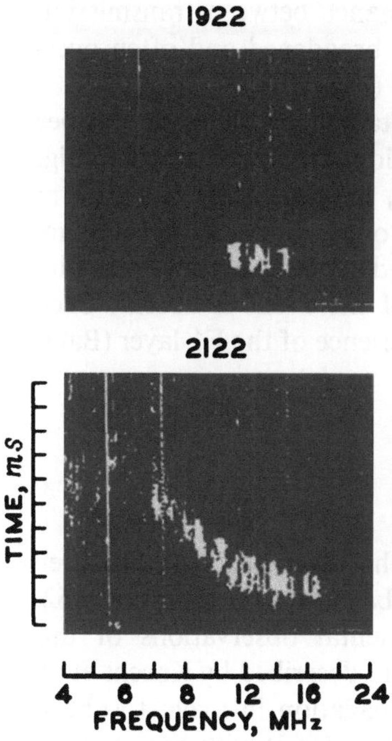

Fig. 1.10. Long-tailed trace sequence recorded in July 18, 1964 on the Okinawa to College path (Bates and Albee 1966). Times are in UT. The pulse width used was 1 ms

corresponds to the MUF of the path. Reflections may practically be often observed at frequencies higher than f_1. Such a signal is of a diffuse character. The signal determines the maximum observed frequency (MOF) (the frequency f_2 in Fig. 1.8).

Features of LDOS *Ionograms.* Figure 1.9 shows typical LDOS ionograms. The following main features of LDOS ionograms can be noted.

1. The observed modes in LDOS ionograms are mainly related to the reflection from the *F*-layer.
2. The upper part of the modes formed by the Pedersen ray (see Fig. 1.8) is usually absent in LDOS ionograms.
3. The diffuse character of the bands in the ionograms is essentially intensified. The width of the bands appears to be much greater than the length of the emitted pulses. The broadening of the ionograms becomes more intensified when approaching the MOF (see Fig. 1.9b, c).
4. The mode broadening is much greater in the morning – evening time and especially at night than in the daytime. For instance, on the Moscow – Molodiozhnaya path (the path length S = 13720 km), the modes at night are often broadened five times or more (Khodzha-Akhmetov and Koshurko 1975, Lobachevskii et al. 1981, Smirnov 1972, Golyan 1981). As a result, the modes can be overlapped, so it is difficult to separate them from one another (Fig. 1.9d).

5. The mode broadening increases with the distance between transmission and reception points. The modes are greatly broadened and often overlapped for RTW or close to the RTW signals (Fig. 1.4).
6. The ionograms of unusual structure are obtained by Bates and Albee (1966) and Kunsucker and Bates (1969). Such ionograms are shown in Fig. 1.10. They vary inversely with frequency, i.e., the time delay of signals at lower frequencies is greater than at higher ones. These ionograms are similar to the upper part of the F1-band formed by the Pedersen ray in Fig. 1.8, in this case, however, the lower part of the mode is absent. Such ionograms were observed in summer in the presence of the F1-layer (Bates and Albee 1966).

1.1.3 Magnetospheric Propagation

The possibility for HF radio wave guiding in the magnetosphere has been discussed in literature since the early 1960's (Obayashi 1959, Booker 1962, Pitteway and Cohen 1961). The first experimental observations of this phenomenon carried out by means of rockets were described by Knecht et al. in 1961. A systematic study of radio wave propagation in magnetospheric ducts was carried out by Muldrew (Muldrew 1963, 1967) with the satellites Alouette 1, Alouette 2, and also by Loftus et al. with the satellite Explorer 20 (Loftus et al. 1966).

Magnetospheric propagation of HF radio waves between geomagnetic conjugate points on the Earth was observed and investigated by Ben'kova et al. (1980), Bukin (1978), Ben'kova et al. (1978), Belikovich et al. (1976), and Benediktov et al. (1971).

Let us consider the main observational characteristics of magnetospheric HF signals.

The Observed Frequency Range. Magnetospheric wave guiding was observed with the satellites in the frequency range of 0.2 – 5.6 MHz (Muldrew 1963, 1967; Loftus et al. 1966). In this case the signals were most often observed in the lower frequency range (as low as 1.6 MHz) (Muldrew 1967).

When a radio wave propagates in the magnetosphere between the conjugate points on the Earth (Earth – Earth path), the situation is quite different. Here, the observed frequency range is 5 – 15 MHz, the frequencies of 8 – 12 MHz being observed most often (Bukin 1978).

The Diurnal Dependence. This is quite pronounced. Low-frequency magnetospheric signals (as low as 1.6 MHz) are observed mainly at night (Muldrew 1967) (Fig. 1.11). At higher frequencies (1.6 MHz $< f \lesssim 2$ MHz) the signals have a tendency to group in the morning period (Muldrew 1967, Loftus et al. 1966). Magnetospheric signals are observed on the Earth – Earth path mainly in the periods close to sunrise and sunset. The signals are observed most often when one of the geomagnetic conjugate regions of the ionosphere is already

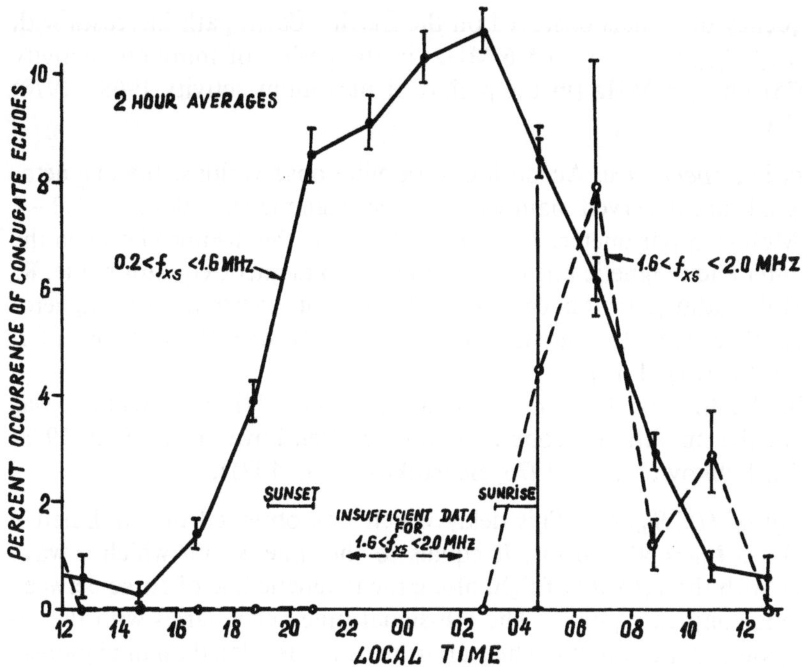

Fig. 1.11. Percent of occurrence of the radio signals trapped in magnetospheric ducts vs time of day, according to the Alouette data. *Solid curve* corresponds to the frequency of extraordinary wave $0.2 < f_{xs} < 1.6$ MHz, *broken curve* corresponds to the frequencies $1.6 < f_{xs} < 2.0$ MHz

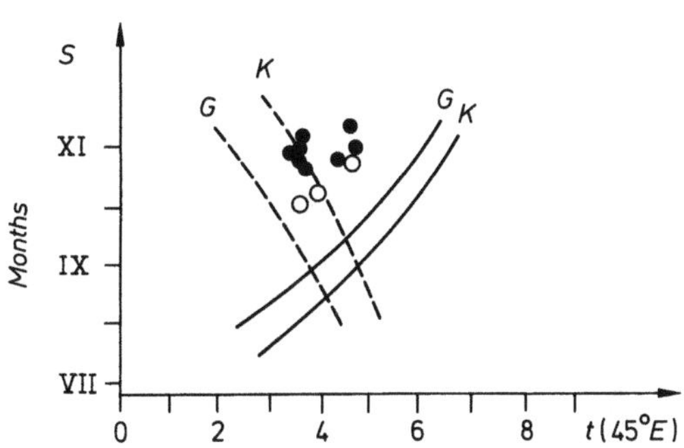

Fig. 1.12. Times of observation of the magnetospheric signals on the Earth–Earth path. The transmitter was situated near Gorki in the northern hemisphere. The capture of signals into the magnetospheric duct must take place near Kharkov. Observations were carried out at a geomagnetic conjugate point in the southern hemisphere in October to November 1968 and 1970. The days when signals were observed are shown by *circles* (1968) and *dots* (1970) vs local time. *Solid lines* show the time of sunrise at the height of 200 km above Gorki (*G*) and Kharkov (*K*), *broken lines* show the time of sunrise at geomagnetic conjugate points

lighted by the Sun, and the other region is not (Ben'kova et al. 1980) (Fig. 1.12).

The Effect of Magnetic Activity. This effect on frequency of occurrence of magnetospheric signals is slight (Muldrew 1963, 1967; Bukin 1978). The minimum frequency of signals observed on the Earth – Earth path increases with solar activity: f_{min} varies from 5.6 MHz (in the period of minimum activity 1974 – 1975) up to 8 MHz (in the period of maximum activity 1968 – 1970) (Bukin 1978).

The Latitude Dependence. According to satellite observations, the magnetospheric signals are observed mainly at low geomagnetic latitudes $L \sim 1.3-2$ (L is the McIlwain parameter: $L = r/R_0$, where r is the distance between the Earth's center and magnetic line of force in the geomagnetic equator plane, R_0 is the Earth's radius). In this case the frequency of occurrence of magnetospheric signals aboard the satellite increases sharply with the decrease of L (Muldrew 1967) (Fig. 1.13).

On the Earth – Earth path the magnetospheric HF propagation was observed in the midlatitude zone $L \sim 2-2.5$ (Ben'kova et al. 1980, 1978; Bukin 1978, Belikovich et al. 1976, Benediktov et al. 1971).

Time Delay of HF Signals. This delay which was observed on the Earth – Earth path, is $t_{HF} = 60-80$ ms. It equals to the time within which a wave propagates with the velocity of light along the magnetic line of force between the magnetic conjugate points. The HF-signal time delay varies with the increase of geomagnetic activity. This variation of t_{HF} is related unambiguously

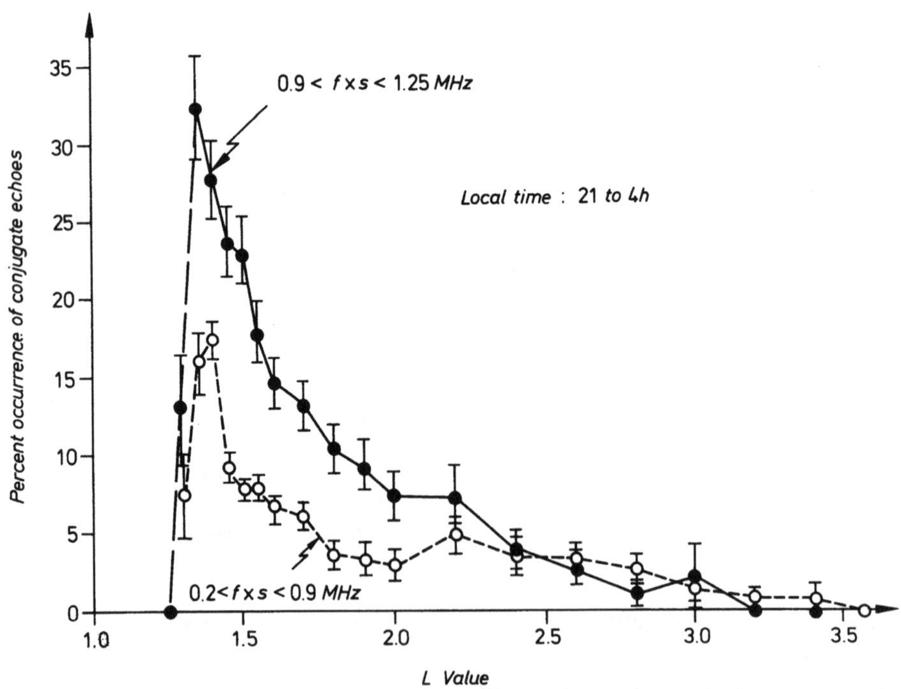

Fig. 1.13. Percent of occurrence of radio signals trapped into magnetospheric ducts vs the McIlwain parameter L, according to the results of Alouette II

to the variation of time delay for VLF signal propagating between the same points on the frequency $f_{VLF} = 15$ kHz (Fig. 1.14) (Bukin 1978, Bukin and Molchanov 1976). This shows that the two HF and VLF signals propagate in the magnetosphere along one and the same trajectory, i.e., along one and the same magnetic line of force (Ben'kova et al. 1980).

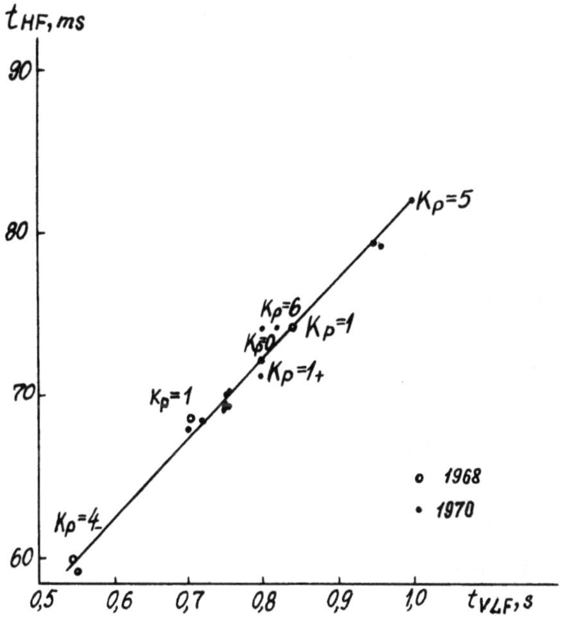

Fig. 1.14. Propagation time of HF signal between the geomagnetic conjugate points vs the propagation time of VLF signal between the same points. The propagation time changes due to variation of the length of magnetic line of force with magnetic activity (values of K_p are indicated in the figure)

Low-frequency signals ($f \lesssim 2-3$ MHz) transmitted from a satellite at low geomagnetic latitudes ($L < 2$) are often recorded by the same satellite, several times with different time delay. This corresponds to multiple reflections from the ionosphere of a wave trapped into magnetospheric duct (Muldrew 1963) (Fig. 1.15).

Magnetoionic Splitting of Signal. This is recorded both on the Earth – Earth path and on board the satellites. On the Earth – Earth path, the difference in time delay of the ordinary and extraordinary components is $\Delta t_{0x} \sim 1-3$ ms. According to satellite observation, $\Delta t_{0x} < 1$ ms. According to Muldrew (1963, 1967), the extraordinary wave is confined in a magnetospheric duct on higher frequencies than the ordinary one. The observed difference between the maximum recorded frequency of the extraordinary wave, and the ordinary one is $\Delta f_m = f_{x_{max}} - f_{0_{max}} \approx 0.3 - 0.5$ MHz at $f_{max} \sim 5.5$ MHz.

Thus, for the time being the experimental data available prove beyond doubt the existence of magnetospheric guiding for HF signals and establish a number of its quite stable properties.

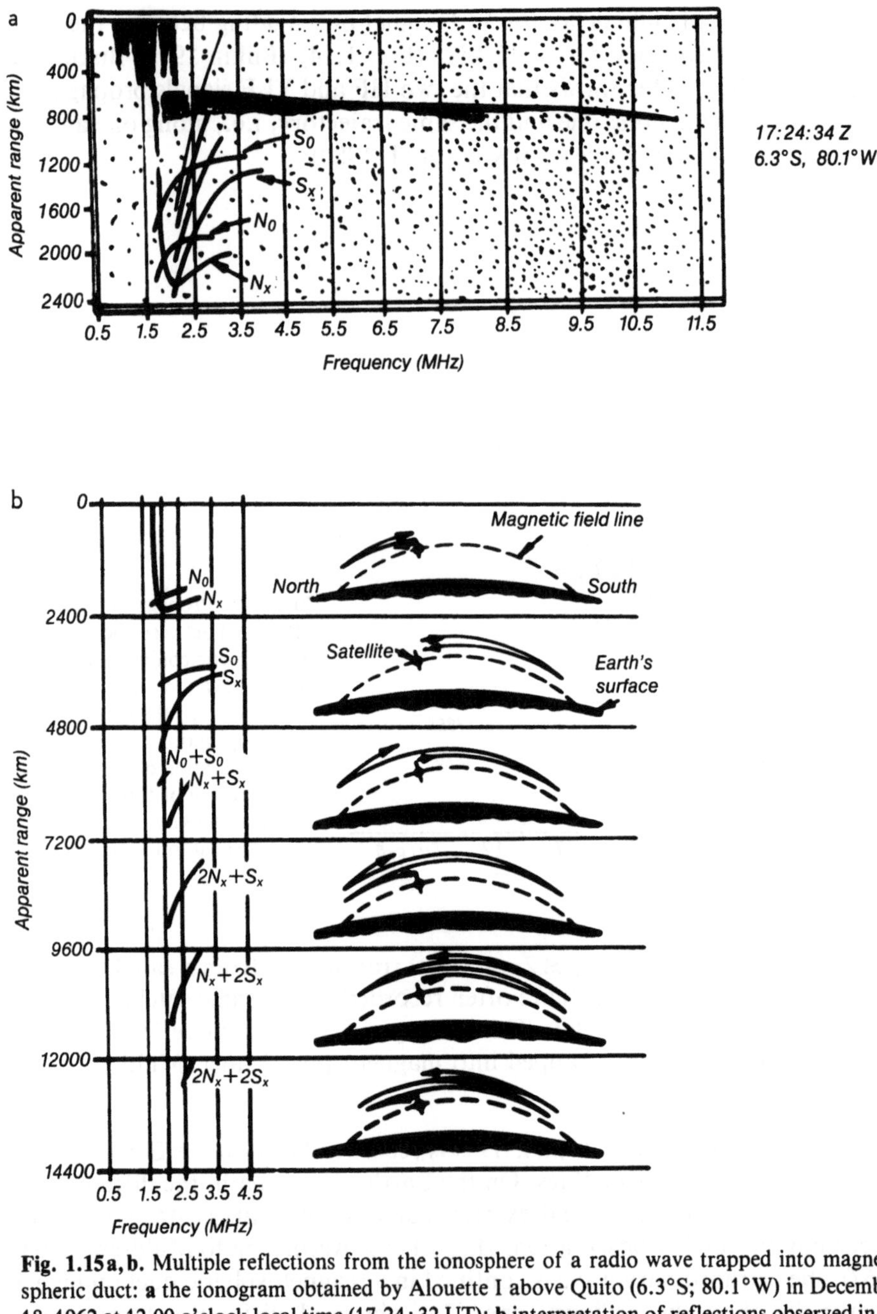

Fig. 1.15a, b. Multiple reflections from the ionosphere of a radio wave trapped into magnetospheric duct: **a** the ionogram obtained by Alouette I above Quito (6.3°S; 80.1°W) in December, 18, 1962 at 12.00 o'clock local time (17.24:32 UT); **b** interpretation of reflections observed in the ionogram.

The pulse repetition frequency is 125 s^{-1}, i.e., the distance between two neighboring pulses is 2400 km. The ionogram exhibits the tracks of six successive pulses. N_o and N_x are the ordinary and extraordinary components reflected once in the northern hemisphere, S_o and S_x are those in the southern hemisphere, N_o+S_o and N_x+S_x are those reflected twice in the northern hemisphere and in the southern one and so on

1.2 Features of Radio Wave Propagation Over Long Distances

Ionospheric Wave Ducts. Long-distance radio wave propagation is a propagation in ionospheric waveguide channels that defines to a great extent its main features.

Short radio waves with a frequency less than the MUF-*F* are reflected from the ionosphere and come back to Earth. In the case of oblique incidence on an ionospheric layer, they cover a certain distance along the Earth's surface. This distance depends on the angle of ray elevation. The maximum distance covered by a wave at single reflection (one hop) is equal to 2000 – 3000 km. When propagating over very long distances, radio waves are reflected repeatedly from the ionosphere and Earth, making many hops (Fig. 1.16a, trajectory 1). In this case radio waves propagate in the Earth – ionosphere waveguides, namely: in the *F*-duct if a wave is reflected from the *F*-layer, in the *E*-duct when reflected from the *E*-layer. There is considerable absorption in these ducts, especially in the daytime. The absorption region is situated in the ionosphere at heights of 50 – 130 km (Fig. 1.16a, shaded area). As a consequence, in the daytime a radio signal becomes considerably attenuated already after two – three hops.

Due to the stratified structure of the ionosphere, a wave can propagate not only in ducts between the Earth and the ionosphere, but also in the waveguide duct *FE* between the *F*- and *E*-layers at heights of 120 – 130 km (Fig. 1.16a, trajectory 2). The absorption in the duct is comparatively small, consequently, the interlayer ionospheric duct is convenient for long-distance propagation.

The Earth sphericity is of importance in the formation of waveguide channels. A ray propagates in an ionospheric duct along the ricocheting

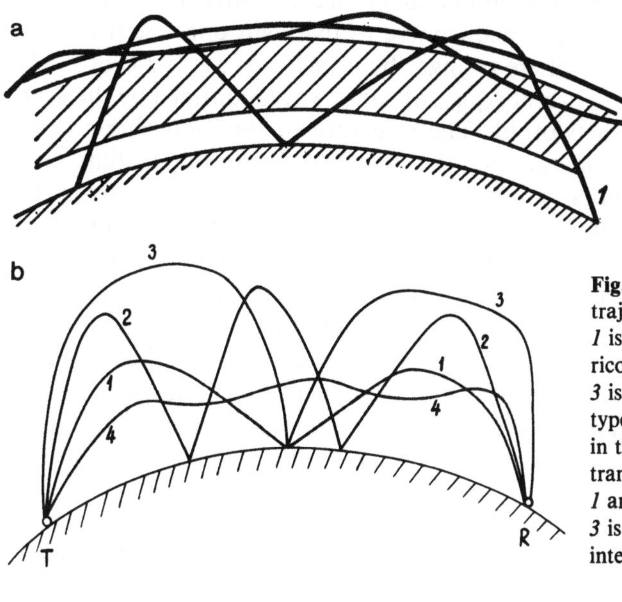

Fig. 1.16. a Types of radio wave trajectories in the ionosphere: *1* is hop trajectory, *2* is ricocheting (chordal) trajectory, *3* is sliding trajectory; **b** various types of radio wave trajectories in the ionosphere between transmitter (*T*) and receiver (*R*): *1* and *2* are the hop trajectories, *3* is the Pedersen ray, *4* is the interlayer trajectory

trajectory, being reflected from the "inner surface" of the ionosphere (trajectory 2). In the limiting case such a ray is sliding at one and the same height (trajectory 3). It is essential that the ricocheting and the more so sliding rays are incident on reflecting surface of the ionosphere at a greater angle than the rays emitted from the Earth. That is why they can have higher frequency. The maximum frequency of the sliding ray, f_{sm}, is the maximum frequency of radio waves confined by the ionosphere. The frequency f_{sm} is higher than the MUF-*F*. The ratio of $f_{sm}/f_{\text{MUF-}F}$ is about 1.8 – 2.5. Thus, the sliding mode frequency can greatly exceed the MUF-*F*.

One should bear in mind the following feature of ionospheric waveguides. The Earth's surface is only a rigid waveguide wall, while the reflection of a wave from the ionosphere significantly depends on the height distribution of electron density, on the angle of elevation, and wave frequency. Therefore, even at a given electron density distribution, the form of an ionospheric waveguide varies greatly with wave frequency f. For instance, at high frequencies f close to f_{sm}, ionospheric waveguides are narrow ducts in the vicinity of the height of sliding. They widen rapidly with the decrease of frequency f, and when $f \lesssim f_{\text{MUF-}F}$ they have covered the whole space between the Earth and the maximum of the ionosphere. In this case the duct width is three – four orders greater than the wavelength. This means that the wave volume of ionospheric waveguides is very large. The number of wave modes confined by waveguides is $n \sim 10^3 - 10^4$.

Influence of the Horizontal Inhomogeneity of the Ionosphere. The ionosphere is inhomogeneous in the horizontal direction, i.e., along the Earth's surface. At shorter distances up to 2000 – 5000 km, the influence of the inhomogeneity is not strong, while at long distances, about 10000 km and longer, the variations of the ionosphere in the horizontal direction are greater. For example, in the transition from the daylight hemisphere to the night one, the maximum electron density in the ionosphere changes by a factor of 3 – 5. Accordingly, the maximum frequencies f_{sm} and $f_{\text{MUF-}F}$ of radio waves confined by the ionosphere change more than twofold. The structure of the ionospheric layers changes more strongly. For example, the *D*-layer exists in the daytime only. The ionosphere undergoes similar changes in the transition from the equatorial to the midlatitude and further to the polar region. The most sharp horizontal changes of the ionosphere are observed in the vicinity of the terminator separating the sunlit and dark sides of the ionosphere. The inhomogeneous structure of the ionosphere also undergoes considerable time variations, for instance, seasonal, sunspot cycle, and so on. It is clear from the above that over the long-distance paths, the properties of ionospheric waveguides vary rather strongly. That is why the horizontal inhomogeneity of the ionosphere proves to be of primary effect on long-distance radio wave propagation.

So, as it was pointed out above, the confinement of a wave in an ionospheric waveguide depends on the height distribution of electron density, the angle of elevation, and wave frequency. The height distribution, in turn, varies greatly over long distances. Will the wave of a given frequency be confined in a waveguide? At what angles of incidence on a layer will the wave of a given frequency be confined by the waveguide? Will the wave reach the reception point? What frequencies can be used to establish radio communications between given transmission and reception points? And so on. The answers on these main questions of long-distance radio communications completely depend on the character of the horizontal inhomogeneity of the ionosphere.

The horizontal inhomogeneity affects significantly *the field amplitude, the angles of arrival,* and *the time delay* of a signal at the reception point. The inhomogeneity completely defines *the lateral deviations* and *bearing variations* of rays, because in the case of spherical symmetry, a ray always propagates in a fixed direction, i.e., along a great circle arc.

Over long distances, waveguide channels vary, transform, sometimes completely disappear and appear again. This, of course, strongly affects radio wave trajectories in the ionosphere. The complex trajectories related to the transitions between ducts are defined completely by the horizontal inhomogeneity of the ionosphere (or by irregular ionospheric inhomogeneities). In particular, the inhomogeneity is of primary importance for radio wave *capture* into the interlayer duct *FE* and for escape from the duct.

Let us also note the phenomenon of the focusing of radio wave field at the antipode and round-the-world points. The region of focusing would have been very small in a spherically – symmetric ionosphere, and so the effect of field intensification within this region would have been very strong. But, actually, the focusing region is spread and the field intensity is lower by several orders. In this case the horizontal ionospheric inhomogeneity plays the principal part.

An important feature of the inhomogeneous structure of the ionosphere is that it varies rapidly enough along the height z (over a distance of about the height of the ionospheric layer maximum, z_m), varying much slower in the horizontal direction (over a distance of about the Earth's radius, R_0). As a consequence, the characteristic scale of oscillations of a radio wave propagating in ionospheric waveguide channels (i.e., the hop scale, $x_0 \sim \sqrt{R_0 z_m}$) is less than R_0. Therefore, on an average, the horizontal ionospheric variation over a distance of about one oscillation is of a slow adiabatic character. Due to this, *the adiabatic invariant I* (Gurevich 1971) of short radio waves propagating in the horizontally – inhomogeneous ionosphere is approximately conserved. The adiabatic invariant conservation enables one to describe radio wave propagation in the horizontally – inhomogeneous ionosphere in a rather simple and consistent way. This method is named *the adiabatic approximation.*

Multimode Structure of Signals. One of the typical properties of signals, received over long distances, is their multimode or multiray structure. As an example, let us consider a radio path between the transmission and reception points, being a long distance from each other. The two points can be situated both on the Earth's surface and above the Earth below the *F*-layer maximum. A radio wave of a given frequency transmitted from the transmission point can arrive at the reception point by different ways. Therefore, the wave field at the reception point is a sum of several rays (modes). Usually $\sim 1-10$ modes are observed on a long-distance path. This is illustrated by Fig. 1.16b. In the Earth – ionosphere duct a ray can make various number of hops (trajectory 1, 2) between the transmission and reception points, which results in appearance of various modes (bands) in the LDOS and OS ionograms (Figs. 1.8, 1.9). The occurrence at the reception point of two rays with the same number of hops is possible. It is due to the sharp increase of the hop length when reflecting near the ionospheric layer maximum. This ray is known as the Pedersen ray (trajectory 3 in Fig. 1.16b). It is well seen in the ionograms for short, single, hop paths (Fig. 1.8). One cannot see the Pedersen ray on the long-distance paths (Fig. 1.9). The disappearance of the Pedersen ray on long-distance paths is a typical manifestation of the ionospheric horizontal inhomogeneity. At long distance paths the complex trajectories can exist as well. The trajectories are connected with the wave capture into the interlayer duct *FE*, and with propagation in the duct, and, finally, with the departure to the Earth (trajectory 4). The complex trajectories are also connected with the transitions from one duct to another (for instance, a hop trajectory changes to a ricocheting one, and vice versa).

Radio Wave Scattering. There is always a complex structure of irregular inhomogeneities in the real ionosphere. They are characterized by a wide range of scales: from tens of meters to hundreds of kilometers. The relative perturbation of electron density $\Delta N_e/N_{e0}$ in large-scale inhomogeneities can amount to several percent. The perturbation $\Delta N_e/N_{e0}$ usually drops by power law with decreasing inhomogeneity scale. The inhomogeneous ionospheric structure intensifies significantly in the polar and auroral regions, as well as in the equatorial zone. At night the perturbations $\Delta N_e/N_{e0}$ are usually greater than in the daytime. The extent of inhomogeneity generation varies significantly with geomagnetic and solar activity.

Radio waves are scattered by the irregular inhomogeneities. As a result, multiple propagation paths appear. When being scattered, the wave field can acquire random character. Therefore, the role of scattering increases with the distance, covered by a wave in the ionosphere. The scattering is of great importance for long-distance and, moreover, for round-the-world propagation. It is seen, for example, from the comparison of Figs. 1.8, 1.9, and 1.3. The spreading in the ionograms increases rapidly with distance due to the scattering.

It is fluctuations of the wave amplitude and phase at the reception point, fluctuations of lateral deviations and bearing, bearing distortions, and fluctuations of time delay that are caused by the scattering. The fluctuations mentioned above result in great widening and even in full spreading of the mode characteristics in the LDOS ionograms. Due to scattering, redistributions of wave energy both in the volume of a given wave duct and between various ducts take place. Scattering is one of the reasons of radio wave capture into the interlayer duct and escape from it. Scattering results in significant horizontal and vertical widening of narrow radio beams and in an increase of the duration of pulse signals. Due to scattering, the propagation is possible at frequencies higher than the MUF for a given path, and so on.

The spectrum of ionospheric irregularities varies significantly when modifying the ionosphere by powerful radio waves. A nonlinear scattering by artificial irregularities produced in such a way can significantly affect the HF propagation. In particular, the nonlinear scattering is the origin of very strong, aspect-sensitive scattering of HF and VHF radio waves in the disturbed F-region.

Magnetospheric Propagation. Electron density in the magnetosphere is small, and usually does not exceed $(1-5) \cdot 10^{+3}$ cm^{-3}. It decreases monotonically when moving away from the Earth and varies significantly only over a distance of about the Earth's radius R_0. The magnetospheric plasma density changes quite sharply only near the boundary of the plasmasphere, $L \sim 3-5$ (Banks 1972) (Fig. 1.17).

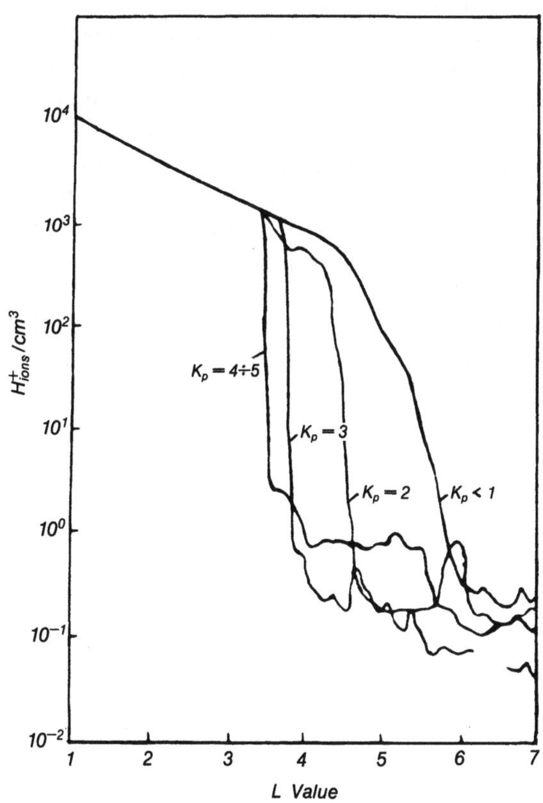

Fig. 1.17. Plasma density in the magnetosphere vs the McIlwain parameter L at different values of the magnetic activity index K_p. The plasmospheric boundary is shifted to smaller values of L with increasing magnetic activity

Plasma with electron density $N_e < 5 \cdot 10^3 \cdot \mathrm{cm}^{-3}$ is transparent for HF radio waves. It results only in a small change of the refractive index. That is why on the average the magnetospheric plasma weakly influences the HF propagation.

This simple situation, however, changes when the magnetospheric irregularities are taken into account. Quasistationary magnetospheric irregularities are always elongated in the direction of the geomagnetic field, because here the longitudinal diffusion is by many orders of magnitude greater than the transverse one. The longitudinal size of magnetospheric irregularities can be equal to the length of line of force between the geomagnetic conjugate points. The transverse size ρ has a wide range of scales from tens of meters to hundreds of kilometers. According to Al'pert (1971), Booker (1962), Muldrew (1963), the optimum transverse scale is $\rho_0 \approx 200-1000$ m. The perturbations of electron density in the irregularities vary from several percent to nearly unity. However, the probability of great values of $\Delta N_e / N_{e0}$ is essentially lower. The frequency of occurrence of magnetospheric irregularities sharply increases in the periods of sunrise and sunset (Al'pert 1971).

The presence of highly elongated irregularities can strongly affect the trajectories of HF radio signals in the magnetosphere although the dielectric constant perturbations remain small. Indeed, let us consider a radio wave propagating in the direction close to the irregularity axis. Such a wave can be successively reflected from the irregularity boundaries (see Fig. 1.18). The wave appears to be trapped by the irregularity and, on the average, propagates along its axis. The ray trajectory is bent following the bending of the irregularity which forms a kind of wave duct in the magnetosphere. If the irregularity covers the whole line of force, the radio waves trapped into the duct can propagate between the conjugate points on the Earth. It is important that when a magnetospheric duct traverses the ionosphere, the electron density in the duct sharply increases in the vicinity of the F-layer maximum. As a result, a barrier for radio waves is formed. If the wave frequency f is less than the barrier critical frequency, f_0, then a radio wave penetrates slightly through the ionospheric barrier. Hence, the duct is cut off for radio waves with $f < f_0$. It is opened for radio waves with $f > f_0$, and a wave is free to leave it for the Earth.

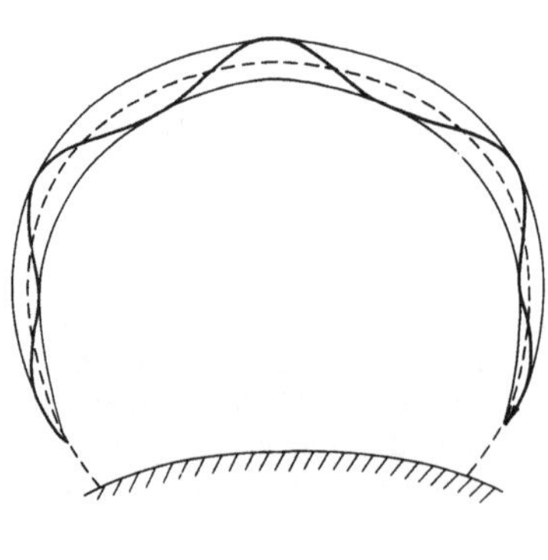

Fig. 1.18. A trajectory of a radio wave trapped into the waveguide channel formed by magnetospheric irregularity

Of importance is the problem of the wave duct excitation. If a transmitter is situated on board a satellite somewhere above the F-layer maximum, then the radio waves trapped in the duct can be both with the frequency $f>f_0$ and with the frequency $f<f_0$. In this case mainly the radio waves with $f>f_0$ come out to the Earth. The radio waves with $f<f_0$ trapped into the duct can be several times successively reflected from the ionosphere at the conjugate points (see Fig. 1.15). Naturally, between the conjugate points only radio waves with the frequency $f>f_0$ can propagate.

Theory Problems. The qualitative analysis carried out above shows that for theoretical description of long-range radio wave propagation, it is necessary, first of all, to take into account the waveguide propagation in the horizontally-inhomogeneous, i.e., three-dimensional inhomogeneous ionosphere. Radio wave scattering by the irregular ionospheric inhomogeneities is of principal importance. The nonlinearity can play a significant role. The present monograph is devoted to the up-to-date state of the long-range propagation theory taking into account all the main factors. An important problem of the theory is to develop the calculation methods for long-range radio paths. This problem is covered by a special chapter of the monograph. The simplest method taking into account the horizontal ionospheric inhomogeneity involves the adiabatic approximation and parabolic model of an ionospheric layer. This method is presented in detail in Sec. 5.1. The computations by the method are rather simple, they can be performed quickly and successfully used in practice.

For theoretical description of magnetospheric HF propagation, it is important to study the wave guiding in magnetospheric ducts and their excitation by the sources located in the magnetosphere and on the Earth.

These problems, in theoretical aspect, are close to problems discussed for HF radio waves in ionospheric ducts. That is why the common approach developed in the monograph for the theory of HF propagation in the ionosphere and magnetosphere seems to be quite natural.

It goes without saying that at present the theory is not yet complete, many problems are far from being fully investigated. In particular, among these are such problems as the effects of magnetic field and radio wave polarization, the role of caustics, wave effects, and so on. Further development of global distribution models of ionospheric electron density taking into account seasonal sunspot cycle and other variations in the ionosphere will be of great importance for investigations of the very long-range propagation. The same can be said about global distribution models of the ionospheric irregularities.

In this context we should point out an important role of the experimental and theoretical investigations based on the data of LDOS and RTW sounding. Using the LDOS data one can study radio wave scattering. It will enable one, by using the satellite data, to create the reliable models of global distribution of the ionospheric irregularities. The RTW sounding carried out at several points combined with a network of oblique and vertical sounding stations can be used as a system allowing to determine the three-dimensional global distribution of both electron density in the ionosphere and ionospheric irregularities, as well as their time variations. The creation of such a common system would be of fundamental importance for HF radio communications and also for investigations of solar – terrestrial relationships, the ionosphere and radio wave propagation.

Small Parameters. Before setting forth the subject matter in detail, let us consider the main small parameters of the problem. These parameters allow to simplify significantly the problem of radio wave propagation for the conditions of interest.

For very long-distance and RTW propagation, the HF waves are important with the frequency $f \gtrsim 10$ MHz. In this case

$$\omega_H^2 \ll \omega^2 , \qquad \omega_H = \frac{eH}{mc} , \tag{1.1}$$

$$\omega_0^2 \ll \omega^2 , \qquad \omega_0^2 = \frac{4\pi e^2 N_e}{m} . \tag{1.2}$$

Here $\omega = 2\pi f$, ω_H is the gyrofrequency, ω_0 is the plasma frequency.

Due to Eq. (1.1), in a first approximation we can consider an ionospheric plasma as an isotropic one. The effects of the magnetic field (corrections of the order of ω_H/ω and ω_H^2/ω^2) can be taken into account in the next approximations.

Due to Eq. (1.2), the dielectric constant of a plasma ε is always nearly unity:

$$\varepsilon \approx 1 - \frac{\omega_0^2}{\omega^2} , \qquad \varepsilon \approx 1 . \tag{1.3}$$

This means that we can neglect the resonance effects, arising when $\varepsilon \approx 0$. It also follows from Eqs. (1.2) and (1.3) that a wave can propagate in ducts only in the direction close to the horizontal one:

$$\frac{k_z}{k_\theta} \ll 1 , \qquad k_\theta \approx k_0 = \frac{\omega}{c} \sqrt{\varepsilon} \approx \frac{\omega}{c} . \tag{1.4}$$

Here, z is the vertical direction, θ is the ray direction in the horizontal plane, k_z and k_θ are the components of the wave vector $\boldsymbol{k}$.

According to Eqs. (1.3) and (1.4), one should consider only radio waves emitted from the Earth at a small angle to the horizon,

$$\alpha_0 \ll 1 , \tag{1.5}$$

where α_0 is the ray emission (take off) angle. Besides, for HF waves always

$$\frac{\nu_e}{\omega} \ll 1 , \tag{1.6}$$

where ν_e is the electron collision frequency. According to Eq. (1.6), one can ignore the effect of collisions on the dielectric constant and, to a first approximation, neglect the wave absorption. The following conditions describe the main features of electron density, N_e distribution in the ionosphere. First,

$$\Delta z \ll R_0 . \tag{1.7}$$

Here, Δz is the thickness of an ionospheric layer, R_0 is the Earth's radius. Equation (1.7) means that the ionosphere is a thin spherical layer. Next,

$$\frac{\lambda}{N_e}\frac{\partial N_e}{\partial z} \ll 1 , \tag{1.8}$$

where λ is the wavelength. Due to Eq. (1.8), one can use with a good accuracy the geometric – optics approximation.

The ionosphere is slightly inhomogeneous in the horizontal direction

$$N_e = N_e(z, \delta_1 R_0 \theta, \delta_2 R_0 \varphi) ; \qquad \delta_1 \ll 1 , \qquad \delta_2 \ll 1 , \tag{1.9a}$$

or, more precisely,

$$\delta_1 \approx \frac{\partial N_e}{R_0 \partial \theta} \bigg/ \frac{\partial N_e}{\partial z} \sim 10^{-2}, \qquad \delta_2 \approx \frac{\partial N_e}{R_0 \partial \varphi} \bigg/ \frac{\partial N_e}{\partial z} \sim 10^{-2} . \tag{1.9b}$$

Due to Eq. (1.9), the approximation of a spherical ionosphere often appears to be good.

Finally, the condition

$$\frac{1}{2\pi}\frac{\Theta}{P_i}\frac{dP_i}{d\theta} \ll 1 \tag{1.10}$$

defines the possibility for using the adiabatic approximation. In Eq. (1.10) Θ is the average period of ray oscillations, P_i is a parameter of the ionospheric duct (for example, its effective width or height).

2 Spherically – Symmetric Ionosphere

As indicated in Eq. (1.9), the ionosphere, at first approximation, is a thin spherically-symmetric plasma layer. The study of radio wave propagation in a simple case of a spherically-symmetric layer enables one to make clear many important properties of long-distance propagation. We shall discuss them briefly in the present chapter, concentrating mainly on the problems associated with ionospheric wave ducts (IWD).

2.1 Wave Equation in the Spherically – Symmetric Ionosphere; Ionospheric Wave Ducts

An electromagnetic field is described by the Maxwell equations. Assuming the field cyclic frequency ω to be fixed, the Maxwell equations have the form

$$\begin{aligned} \operatorname{rot}\boldsymbol{H} &= \frac{i\omega}{c}\tilde{\varepsilon}\,\mathscr{E} + \frac{4\pi}{c}\boldsymbol{j}\,, \qquad \operatorname{div}\boldsymbol{H} = 0\,, \\ \operatorname{rot}\mathscr{E} &= -\frac{i\omega}{c}\boldsymbol{H}\,, \qquad \operatorname{div}(\tilde{\varepsilon}\,\mathscr{E}) = 4\pi\rho\,. \end{aligned} \tag{2.1}$$

Here, $\tilde{\varepsilon}$ is the complex dielectric constant of the medium

$$\tilde{\varepsilon} = \varepsilon + i\frac{4\pi\sigma}{\omega}\,. \tag{2.2}$$

The magnetic permeability has been set equal to unity. The sources for electric and magnetic fields in Eq. (2.1) are electric currents $\boldsymbol{j}$ and charges ρ. At a given frequency ω, the relation between their densities is given by

$$\rho = \frac{i}{\omega}\operatorname{div}\boldsymbol{j}\,. \tag{2.3}$$

In the absence of external currents and charges, the system of equations (2.1) reduces to a single vectorial wave equation (Ginzburg 1967, Sec. 2)

$$\Delta\mathscr{E} - \operatorname{grad}\operatorname{div}\mathscr{E} + k_0^2\tilde{\varepsilon}\mathscr{E} = 0\,, \qquad k_0 = \omega/c\,. \tag{2.4}$$

The system of equations [Eq. (2.1)] or the single vectorial wave Eq. (2.4) describes electromagnetic wave propagation in an arbitrary inhomogeneous medium $\tilde{\varepsilon} = \tilde{\varepsilon}(\mathbf{r})$. In general, the investigation of these equations is a rather difficult problem. However, in a spherically-symmetric medium, where

$$\tilde{\varepsilon} = \tilde{\varepsilon}(r)\,, \tag{2.5}$$

the vectorial Eqs. (2.1) and (2.4) are essentially simplified owing to the fact that they can be decomposed into independent scalar components, each of them being further decomposed into the radial $Z(r)$ and angular $Y(\theta, \varphi)$ parts. The decomposition, which is similar to the well-known Debye decomposition for homogeneous media, is carried out in detail in Appendix 1. The angular part $Y(\theta, \varphi)$ is described in terms of the spherical functions. As a result, the vectorial equations [Eq. (2.1)] reduce to two independent equations for the radial parts of the scalar components $Z(r)$ [Eqs. (A1.6) and (A1.7), see Appendix 1]. Further simplifications are possible due to the assumption that the ionosphere is a thin spherical layer, see Eq. (1.7). Making a transition from $Z(r)$ to a function $\Psi(z)$ which describes the height distribution of the field within the layer (here, $z = r - R_0$ is the altitude, $z \ll R_0$) we reduce Eqs. (A1.6) and (A1.7) to a single scalar Eq. (A1.8) for the function $\Psi(z)$

$$\frac{d^2\Psi}{dz^2} + k_0^2(E - U(z))\,\Psi = 0\,. \tag{2.6}$$

Here, E is a constant (A1.8), and an effective potential $U(z)$ is given by

$$U(z) = -\varepsilon(z) - \frac{2z}{R_0} \approx -1 + \frac{4\pi e^2 N_e(z)}{m\omega^2} - \frac{2z}{R_0}\,. \tag{2.7}$$

The form of $U(z)$, according to Eq. (2.7), is determined by the altitude profile of the electron density in the ionosphere. When $N_e(z) = 0$, i.e., in the absence of the electrons, the function $U(z)$ decreases monotonically with height. In the presence of pronounced ionospheric layers, the function becomes nonmonotonic. The height dependence of U for typical profiles of daytime and nighttime ionospheres, listed in Table 2.1, is shown in Fig. 2.1. The curves $U(z)$ at rather low frequencies ω are seen to be nonmonotonic, with only one or two minima within the altitude range 100 – 250 km, i.e., in the E and F regions. The effect of the ionosphere weakens as the wave frequency ω increases.

Let us emphasize the fact that Eq. (2.6) is equivalent to the Schrödinger equation for a particle in the potential field $U(z)$. The constant E plays here the role of the energy level. The mechanical analogy for Eq. (2.6) makes it possible to clear up the properties of IWD. Indeed, insofar as the function $U(z)$ plays the role of the potential in the Schrödinger equation, the minima of $U(z)$ form potential wells for a particle to be trapped in. In terms of wave concept, it means that in the presence of minima of $U(z)$, wave ducts appear in which electromagnetic waves can propagate. Of importance is also the fact that the Earth's surface (due to its great conductance) reflects radio waves well, so that a kind of infinite potential barrier is formed at $z = 0$ (see Fig. 2.2).

Waves can propagate in the waveguides: between the Earth and E-layer (ground-based duct I, or duct E), between the E and F-layers (ionospheric duct II, or duct FE), and between the Earth and F-layer (ground-based duct III, or duct F). Different propagation paths of the wave at a given frequency ω are characterized by different values of the constant E. For the transmitter located on the Earth's surface, the constant E is always within the range $-1 \leqslant E \leqslant 0$, being connected with the angle of emission α_0 (i.e., the angle between the wave normal and horizontal on the Earth's surface) by

$$\cos^2\alpha_0 = -E\,, \qquad \cos^2\alpha_0 \approx 1-\alpha_0^2\,. \tag{2.8}$$

Table 2.1. Electron density N_e, plasma (Langmuir) frequency ω_0, and angular frequency of the sliding ray ω_s for the daytime and nighttime ionosphere

z, km	Day			Night		
	N_e, cm^{-3}	ω_0, Hz	ω_s, Hz	N_e, cm^{-3}	ω_0, Hz	ω_s, Hz
50	5	$1.2 \cdot 10^5$	–	–	–	–
60	80	$5 \cdot 10^5$	–	–	–	–
70	$2 \cdot 10^2$	$8 \cdot 10^5$	–	–	–	–
80	$1 \cdot 10^3$	$1.8 \cdot 10^6$	–	10	$1.8 \cdot 10^5$	–
90	$8 \cdot 10^3$	$5 \cdot 10^6$	$5 \cdot 10^7$ } I	60	$4.4 \cdot 10^5$	$5 \cdot 10^6$ } I
100	$8 \cdot 10^4$	$1.6 \cdot 10^7$	$2.2 \cdot 10^8$	$1.2 \cdot 10^3$	$2.0 \cdot 10^6$	$2 \cdot 10^7$
110	$1.2 \cdot 10^5$	$2.0 \cdot 10^7$	–	$1.8 \cdot 10^3$	$2.4 \cdot 10^6$	–
120	$1.3 \cdot 10^5$	$2.1 \cdot 10^7$	$1.0 \cdot 10^8$	$2.1 \cdot 10^3$	$2.6 \cdot 10^6$	–
130	$1.5 \cdot 10^5$	$2.2 \cdot 10^7$	$1.37 \cdot 10^8$	$2.2 \cdot 10^3$	$2.65 \cdot 10^6$	–
150	$2.1 \cdot 10^5$	$2.6 \cdot 10^7$	$1.86 \cdot 10^8$	$2.4 \cdot 10^3$	$2.8 \cdot 10^6$	–
175	$3.5 \cdot 10^5$	$3.3 \cdot 10^7$	$2.35 \cdot 10^8$	$2.6 \cdot 10^3$	$2.9 \cdot 10^6$	–
200	$5 \cdot 10^5$	$4 \cdot 10^7$	$2.75 \cdot 10^8$ } II	$3 \cdot 10^3$	$3.1 \cdot 10^6$	$1.3 \cdot 10^7$
225	$7.2 \cdot 10^5$	$4.8 \cdot 10^7$	$3.1 \cdot 10^8$	$4.3 \cdot 10^3$	$3.7 \cdot 10^6$	$3 \cdot 10^7$
250	$1 \cdot 10^6$	$5.7 \cdot 10^7$	$3.32 \cdot 10^8$	$1 \cdot 10^4$	$5.7 \cdot 10^6$	$5.8 \cdot 10^7$
275	$1.35 \cdot 10^6$	$6.6 \cdot 10^7$	$4 \cdot 10^8$	$3.8 \cdot 10^4$	$1.1 \cdot 10^7$	$1.26 \cdot 10^8$ } II
300	$1.6 \cdot 10^6$	$7.2 \cdot 10^7$	–	$1 \cdot 10^5$	$1.8 \cdot 10^7$	$1.72 \cdot 10^8$
325	$1.64 \cdot 10^6$	$7.25 \cdot 10^7$	–	$1.8 \cdot 10^5$	$2.4 \cdot 10^7$	$1.75 \cdot 10^8$
350	$1.63 \cdot 10^6$	$7.2 \cdot 10^7$	–	$2 \cdot 10^5$	$2.8 \cdot 10^7$	–
400	$1.5 \cdot 10^6$	$7 \cdot 10^7$	–	$3 \cdot 10^5$	$3.1 \cdot 10^7$	–
450	$1.2 \cdot 10^6$	$6.3 \cdot 10^7$	–	$2.6 \cdot 10^5$	$3.0 \cdot 10^7$	–
500	$9 \cdot 10^5$	$5.4 \cdot 10^7$	–	$2 \cdot 10^5$	$2.3 \cdot 10^7$	–

The dashed and dash-dot lines in Fig. 2.2 show the situations when the wave frequency is equal to the maximum usable frequencies (MUF) for the E and F regions: $f = f_{\text{MUF-}E}$ and $f = f_{\text{MUF-}F}$. In these cases, the wave emitted horizontally from the Earth's surface ($\alpha_0 = 0$, i.e., $E = -1$) is reflected from the peak of the barrier E or barrier F. In Fig. 2.1 in the case of $f = f_{\text{MUF-}E}$, we have $\omega \approx 1 \cdot 10^8$ (daytime), and in the case of $f = f_{\text{MUF-}F}$, we have $\omega \approx 2.3 \cdot 10^8$ (daytime), $\omega \approx 8.9 \cdot 10^7$ (nighttime). A more detailed classification of IWD and the analysis of radio wave trajectories will be performed in Sec. 4.1.

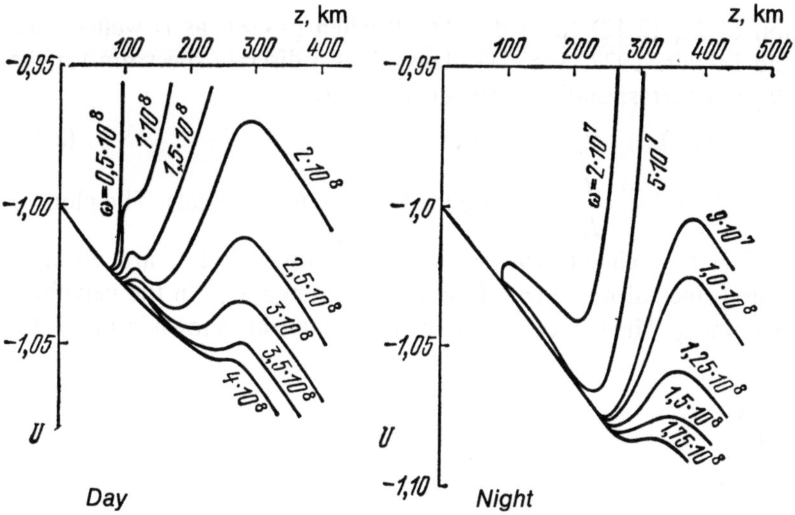

Fig. 2.1. Function $U(z)$ for various frequencies (shown at the *curves*)

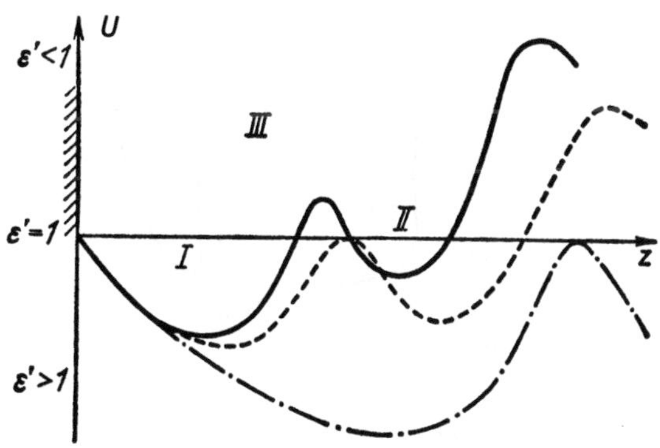

Fig. 2.2. Various types of characteristic potentials, $U=-\varepsilon'$: *solid line* corresponds to $f<f_{\text{MUF-}E}$, *dashed line* corresponds to $f=f_{\text{MUF-}E}$, *dash-dotted line* corresponds to $f=f_{\text{MUF-}F}$

2.2 Radio Wave Field in an Ionospheric Duct

Wave ducts arise in the neighborhood of the $U(z)$-minima. The Schrödinger Eq. (2.6) possesses here a discrete spectrum of stationary energy levels which refer to different modes of the duct.

Let us consider the region near one of the minima, $z \approx z_s$. In this region

$$U(z)=U(z_s)+\frac{1}{2}U''(z_s)(z-z_s)^2+\dots, \tag{2.9}$$

so Eq. (2.6) takes the form

$$\frac{d^2\Psi}{d\zeta^2}+(R-\zeta^2)\Psi=0\,,$$

$$\zeta=\frac{z-z_s}{\rho_s},\quad \rho_s=\left(\frac{2}{U''(z_s)k_0^2}\right)^{1/4},\quad R=\frac{\sqrt{2}\,k_0(E-U(z_s))}{[U''(z_s)]^{1/2}}. \tag{2.10}$$

The solution of Eq. (2.10) (provided $\Psi \to 0$ when $\zeta \to \infty$), as is well-known (Landau and Lifshitz 1977, Sec. 21), determines a discrete spectrum for the constants R_n and corresponding eigenfunctions Ψ_n

$$\Psi_n = e^{-\zeta^2/2} H_n(\zeta) ; \qquad R_n = 2n+1 , \tag{2.11}$$

where $H_n(\zeta) = (-1)^n e^{\zeta^2} \frac{d^n}{d\zeta^n} e^{-\zeta^2}$ are the Hermite polynomials. The plots of $\Psi(\zeta)$ for a few first eigenfunctions, i.e., first wave modes, are given in Fig. 2.3a. The principal mode ($n = 0$) covers the range $\sim \rho_s$ in the neighborhood of the point z_s. The values of ρ_s in the ionosphere are shown in Fig. 2.3b.

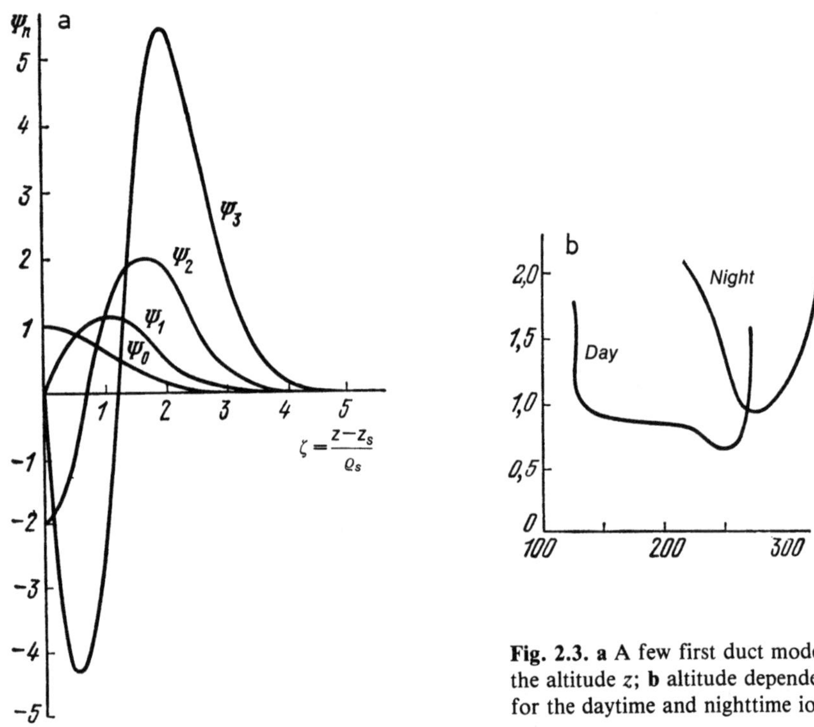

Fig. 2.3. a A few first duct modes, Ψ_n, vs the altitude z; **b** altitude dependence of ρ_s for the daytime and nighttime ionosphere

It is seen that usually $\rho_s \sim 1$ km, i.e., the principal mode essentially slides in the vicinity of the constant altitude z_s. The frequency ω_s of the "sliding" wave for different z_s is listed in Table 2.1. The frequency ω_s is 6–10 times more than the corresponding Langmuir plasma frequency $\omega_0(z_s)$. The maximum frequency for the sliding wave is $\omega_{sm} \approx 4 \cdot 10^8$ by day and $\omega_{sm} \approx 1.8 \cdot 10^8$ by night. It exceeds ω_{MUF} by 1.7 – 2 times [for the model considered (Table 2.1), $\omega_{MUF} \approx 2.3 \cdot 10^8$ by day, and $\omega_{MUF} \approx 8.9 \cdot 10^7$ by night]. Thus, radio waves whose frequencies significantly exceed those for hop propagation can propagate in the ionospheric ducts. The maximum frequency of a wave confined by the ionosphere is ω_{sm}.

Subsequent modes are of an oscillatory type and near z_s they cover regions greater than ρ_s, the region size increasing with the mode number n. It is quite natural, since the modes with $n \neq 0$ refer to the rays which intersect the sliding line, $z = z_s$, at a certain angle, the latter grows with the mode number. Simultaneously, the number of field oscillations produced by the interference between the incident and reflected rays also increases. For large $n (n \gg 1)$ we can make use of the geometrical – optics approximation, quite similar to the quasi-classical (WKB) approximation for the solutions of the Schrödinger equation (Landau and Lifshitz 1977).

In the geometrical-optics approximation for a wave travelling in one direction along z, the solution of Eq. (2.6) is given by Landau and Lifshitz (1977, Sec. 19)

$$\Psi = \frac{C}{\sqrt{k(z)}} e^{i\int k(z)dz}, \qquad k(z) = \pm k_0 \sqrt{E - U(z)}\,. \tag{2.12}$$

Noting that the wave in the duct is always composed of the incident and reflected waves, we obtain (Landau and Lifshitz 1953, Sec. 47)

$$\Psi = \frac{C}{\sqrt{k(z)}} \sin\left(\int_{z_{min}}^{z} k(z)dz + \chi\right) = \frac{C_1}{\sqrt{k(z)}} \cos\left(\int_{z}^{z_{max}} k(z)dz + \chi\right), \tag{2.13}$$

where χ is the phase shift which is dependent on the type of the reflecting barrier; z_{min} and z_{max} are the reflection points or points of the minimum and maximum trajectory heights, the points are defined by the condition $k(z) = 0$ or

$$E - U(z) = 0\,; \qquad z_{min} = z_{min}(E)\,, \qquad z_{max} = z_{max}(E)\,. \tag{2.13a}$$

In a smooth wall duct [i.e., when $U(z)$ increases smoothly] we have (Landau and Lifshitz 1977, Sec. 47)

$$\chi = \frac{\pi}{4}\,. \tag{2.14}$$

In this case, bearing in mind that both expressions for Ψ in Eq. (2.13) should coincide within $z_{min} \leqslant z \leqslant z_{max}$, we obtain from Eqs. (2.13) and (2.14) the condition for determination of duct modes or quantum constants $E = E_n$ (the Bohr's quantization rule) (Landau and Lifshitz 1977, Sec. 48)

$$k_0 \int_{z_{min}}^{z_{max}} \sqrt{E_n - U(z)}\,dz = \pi\left(n + \frac{1}{2}\right). \tag{2.15}$$

In particular, for the parabolic duct profile, i.e., near the minimum of $U(z)$, we have

$$E_n = U(z_s) + \left(n + \frac{1}{2}\right) \cdot \frac{(2U''(z_s))^{1/2}}{k_0}\,,$$

$$\Psi_n(x) = \frac{C}{\sqrt[4]{1-x^2}} \sin\left\{\left(n+\frac{1}{2}\right)[x\sqrt{1-x^2}+\arcsin x] + \frac{\pi}{2}(n+1)\right\},$$

$$x = \frac{\zeta}{\sqrt{2n+1}} = \frac{z-z_s}{\rho_s\sqrt{2n+1}}. \tag{2.16}$$

The constants E_n in the geometrical-optics approximation are seen to coincide identically with those obtained in the precise theory [Eq. (2.11)]. The functions $\Psi(\zeta)$ for $n=3$, $n=4$ are shown in Fig. 2.4. The normalizing constant C in Eq. (2.16) has been chosen $C=(-1)^n 2^{n/2} n!!$ in order that both expressions for Ψ [Eqs. (2.11) and (2.16)] coincide at $\zeta=0$. It is seen that for the mode number as small as $n=3$, the geometrical-optics solution (the dashed line) differs strongly from the precise one only in the vicinity of the reflection points $k(z)=0$, or $x=\pm 1$, $\zeta=\pm\sqrt{2n+1}$. The reflection points are marked in Fig. 2.4 by thin vertical lines. Note that in the vicinity of the reflection points, geometrical-optics approximation can be significantly refined by matching it in this region with the precise solution of the wave equation in terms of Airy's function (Ginzburg 1970, Landau and Lifshitz 1977).

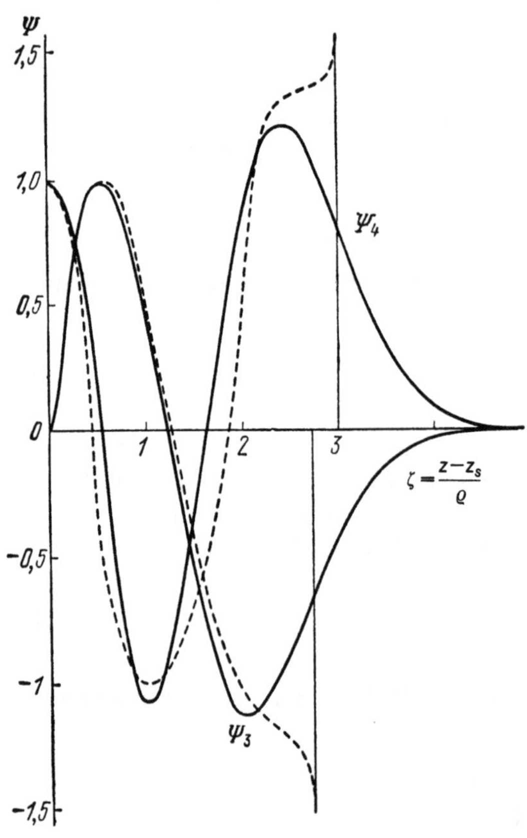

Fig. 2.4. Comparison of the exact solution (*solid curve*) with the geometrical – optics one (*dashed line*)

The expressions obtained define essentially the field structure in the ionospheric ducts. For the first modes, i.e., those in the neighborhood of the duct minimum, the parabolic approximation [Eq. (2.9)] is valid, so the wave field is described by Eqs. (2.10) and (2.11). For high-order modes ($n \gg 1$) equations of the geometrical-optics approximation are valid. We emphasize that in IWD the number of modes is great enough $\sim 10^3 - 10^4$. Therefore, here the geometrical optics is quite a good approximation. The only exception is a number of special regions in the vicinity of the duct boundary, they are considered in detail in Appendix 1.

It is the field of only a separate mode that was considered above. The total field is given by adding up all the modes which are generated by the source current $\boldsymbol{j}$ (i.e., by the current in the transmitter antennas). The generation is described by an inhomogeneous wave equation [Eqs. (A1.6) and (A1.7); Appendix 1]. The latter can be solved by a decomposition of the source currents $\boldsymbol{j}$ into the eigenmodes of the homogeneous equation. This procedure is described in detail in Krasnushkin (1948), Landau and Lifshitz (1977), and Bremmer (1946), so we shall not consider it here.

2.3 Radio Wave Tunneling Through the Interduct Barriers; Capture Coefficient

It is the radio wave field in an isolated wave duct that was considered above. The real structure of IWD is of a more complicated type: two or even more mutually connected ducts commonly exist (Figs. 2.1 and 2.2). In this case, the wave tunneling, penetration from one duct to another becomes possible. It is important that the transmitter is located, as a rule, inside one of the ducts. Then the transmitter excites eigenmodes of the duct. For instance, the antenna placed directly on the Earth's surface generates wave modes of the ground-based wave ducts I, III only (see Fig. 2.2). However, due to wave tunneling a portion of wave energy transfers from duct I to duct II. The ratio between the energy penetrated into duct II and the total energy of the emitted wave is called the capture coefficient.

A detailed study of radio wave field distribution in coupled ducts is made in Appendix 1. There the capture coefficient is determined as well which characterizes the penetration through the interduct barrier. By using Appendix 1, let us evaluate now the intensity of tunneling under the Earth's ionosphere conditions.

In the neighborhood of its peak, the barrier can be approximated as parabolic

$$U(z) = U_m\left[1+\left(\frac{z-z_m}{a}\right)^2\right]; \tag{2.17}$$

Here, z_m is the point of the barrier maximum, a is the peak width, and $U_m = U(z_m)$ is the peak "height" ($U_m < 0$). The coefficient of a capture into duct II for a wave generated in duct I at a given constant E [or at a given emission angle α_0, see Eq. (2.8)] is given by Eq. (A1.39), see (Appendix 1)

$$p \approx \exp\left\{\frac{2\pi^2 a}{\lambda}(E - U_m)\right\}. \tag{2.18}$$

Here, as usual, $E < 0$ and the unequality $-E > -U_m$ must be fulfilled, since only in this case the wave is reflected from the barrier and remains in the duct I; $\lambda = 2\pi c/\omega = c/f$ is the wave length. For the ionospheric E-layer, $a \approx 200-300$ km; for F-layer, $a \approx 400-500$ km. The wave length, $\lambda \sim 15$ to 30 m. Hence, the penetration is not negligibly small only if $|E - U_m| \lesssim 10^{-4}$, the penetration is significant when $|E - U_m| \lesssim 10^{-5}$. The barrier "height" (i.e., the duct depth) in the real ionosphere is $(U_{\max} - U_{\min}) \sim 10^{-2} - 10^{-3}$. Hence, any wave tunneling through the barrier is noticeable for highest levels only, which occupy $\sim 10^{-2}$ of the whole duct volume.

The total energy penetrating into duct II (or FE) (see Sec. 4.1) is given by integration all over the excited levels of the constant E. Let a function $D(\alpha_0)$ be the energy density distribution with respect to angles of emission α_0, which is determined by the transmitter directivity diagram. Then the total energy tunneling into the channel is given by

$$W = W_0 \int_0^{\alpha_{0m}} d\alpha_0 D(\alpha_0) e^{-2\Gamma(\alpha_0)} p(\alpha_0), \tag{2.19}$$

where W_0 is the transmitter power, α_{0m} is the takeoff angle referred to the peak of U_m, $\Gamma(\alpha_0)$ is the wave absorption [Eq. (2.33)], p is the capture coefficient at given α_0. According to Eq. (A1.39), it should be put that

$$D(\alpha_0) \approx D_m = D(\alpha_{0m}), \qquad \Gamma \approx \Gamma_m = \Gamma(\alpha_{0m}).$$

We have allowed for the fact that the penetration is localized essentially within a narrow band near $\alpha_0 = \alpha_{0m}$. Putting also that by Eq. (A1.39), $E = -(1 - \alpha_0^2)$, $U_m = -1 + \alpha_{0m}^2$, we obtain

$$W = W_0 D_m e^{-2\Gamma_m}(\pi k_0 a)^{-1/2} F(x), \qquad F(x) = e^{-x^2}\int_0^x e^{t^2} dt,$$
$$x = (\pi k_0 a)^{1/2}\alpha_{0m}, \qquad U_m \approx -1. \tag{2.20}$$

The total capture coefficient, i.e., the fraction of the total emitted energy which penetrates into the duct, is given by

$$G = \frac{W}{W_0} = (\pi k_0 a)^{-1/2}\frac{e^{-2\Gamma_m}}{\bar{\alpha}} F(x). \tag{2.21}$$

Here, $\bar{\alpha}$ is the effective angular width of the emitted beam, $\bar{\alpha} = 1/D_m$. If $x \gg 1$, then

$$G \approx \frac{e^{-2\Gamma_m}}{2\pi k_0 a\, \alpha_{0m}\bar{\alpha}}. \tag{2.22}$$

If $x \ll 1$, then $G \to 0$, since $F(x) \to 0$ if $x \to 0$. The coefficient has the maximum value $G_{\max}$ at $x = x_m \approx 0.9$

$$G_{\max} \approx 0.43 \frac{e^{-2\Gamma_m}}{(k_0 a)^{1/2} \bar{\alpha}} . \quad (2.23)$$

In the ionosphere a capture into the *FE*-duct through wave tunneling is possible provided $\alpha_{0m} > 0$, i.e., at frequencies $f < f_{\text{MUF-}E}$ (as is the case of Fig. A1.1a; see Appendix 1). Putting $a \sim 200-300$ km (the *FE*-duct's barrier), $\lambda \sim 15-30$ m, $\bar{\alpha} \sim 0.3$, $\Gamma_m < 1$, we obtain that for a frequency f noticeably different from $f_{\text{MUF-}E}$ (i.e., at $x \gg 1$, or $\alpha_{0m} \gg (k_0 a)^{1/2}$), $G \sim 10^{-3}$ to 10^{-4}. The maximum value of G, $G_{\max} \approx 5 \cdot 10^{-3}$ is reached close by $f_{\text{MUF-}E}$ (or $f_{\text{MUF-}F_1}$) at $\alpha_{0m} \approx 0.6 \cdot (k_0 a)^{1/2} \approx 3 \cdot 10^{-3} \approx 0.16°$.

When evaluating the capture coefficient, the penetration in the vicinity of the barrier peak was assumed to be almost the same for adjacent levels. Generally speaking, this does not always take place, so in this case only the upper levels of the *FE*-duct are mainly filled up. Nevertheless, this fact only slightly changes the total capture coefficient. Note also that insofar as the total absorption in the ionosphere is given by

$$\int_{z_1}^{z_2} k_2 dz \sim 0.01 - 1 ,$$

the resonant effects can be important for $p \sim 0.001 - 1$ only, i.e., for the highest levels, $|E - U_m| < 10^{-4}$ (see Appendix 1).

2.4 Ray Trajectories; Pulse Propagation

To deal with specific problems such as the study of ray trajectories and radio wave absorption in the ionosphere, time delay and pulse dispersion, etc., it is convenient to apply the geometrical-optics approximation.

Ray Trajectories. In the spherically-symmetric ionosphere radio waves propagate around-the-world in a great circle plane. Let θ be the azimuthal angle in this plane. Then a wave vector has two components: k_z and k_θ

$$k^2 = k_\theta^2 + k_z^2 = \frac{\omega^2}{c^2} \varepsilon , \quad k_\theta = \frac{q}{r} , \quad r = R_0 + z , \quad (2.24)$$

in the spherically-symmetric case [$\varepsilon = \varepsilon(r)$], the quantity q being constant (this follows from Snell's law; Al'pert 1974, Davies 1969). Ray trajectory in the great circle plane is determined by the dependence of z on the horizontal coordinate $r\theta$

$$\frac{dz}{r\,d\theta} = \frac{k_z}{k_\theta} = \pm \sqrt{\frac{\omega^2}{c^2} \frac{\varepsilon}{k_\theta^2} - 1} = \pm \frac{1}{\sqrt{-E}} \sqrt{E + \frac{\varepsilon r^2}{R_0^2}} . \quad (2.25)$$

In the latter expression we have introduced a constant

$$-E=\frac{c^2q^2}{\omega^2R_0^2}\,. \tag{2.26}$$

Let us take into account now that by Eq. (1.7), the altitude $z \ll R_0$. Then Eq. (2.25) can be rewritten in the following form:

$$\frac{dz}{d\theta}=\pm R_0\sqrt{E-U(z)}\,,\qquad U(z)=-1+\frac{\omega_0^2}{\omega^2}-\frac{2z}{R_0}\,. \tag{2.27}$$

Differentiating it with respect to θ, we obtain

$$\frac{d^2z}{d\theta^2}=-\frac{R_0^2}{2}\frac{\partial U}{\partial z}\,. \tag{2.28}$$

Equation (2.28) describes the motion of a classical particle in the potential field $U(z)$, i.e., it is a classical analog for the Schrödinger Eq. (2.6). The mechanical analog for a "sliding" ray is a particle which locates in the state of stable equilibrium on the bottom of the potential well.

$$z=z_s\,,\qquad \left(\frac{dz}{d\theta}\right)_{z=z_s}\equiv 0\,.$$

The second equilibrium point is the potential maximum $z=z_m$. It is an unstable equilibrium point and corresponds to the Pedersen ray (Bremmer 1946).

The constant E is unambiguously connected with the angle α formed by a given ray trajectory with the horizontal at the level z (elevation angle) by the following relation

$$\alpha=\frac{1}{R_0}\frac{dz}{d\theta}=[E-U(z)]^{1/2}\,. \tag{2.29}$$

In particular, at the sliding level z_s (Fig. 2.5), we have

$$\alpha_s=[E-U(z_s)]^{1/2}=\left[\varepsilon(z_s)+\frac{2z_s}{R_0}+E\right]^{1/2},$$

$$E=\alpha_s^2-\varepsilon(z_s)-2z_s/R_0\,,\qquad \alpha_s\ll 1\,,\qquad \varepsilon\approx -E\approx 1\,. \tag{2.29a}$$

The minimum and maximum trajectory heights $z_{min}(E)$, $z_{max}(E)$ are defined by the reflection condition, $E-U(z)=0$ (see Sec. 2.2).

The solution of Eq. (2.28) determining the angle θ as a function of z is obvious to have the form

$$\theta=\frac{1}{R_0}\int\limits_{z_0}^{z}\frac{dz}{\sqrt{E-U(z)}}+\theta_0\,. \tag{2.30}$$

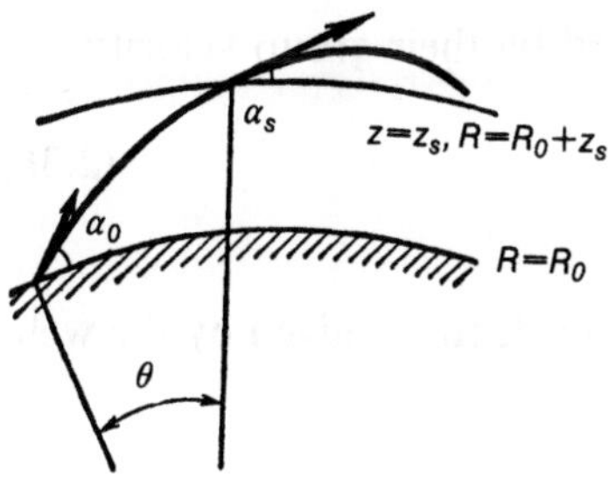

Fig. 2.5. Wave trajectory pattern. The ray takeoff angle, α_0, and the angle of ray elevation at the height of sliding, α_s, are shown

Here, z_0 is the ray altitude at $\theta = \theta_0$. Hence, the ray oscillation period with respect to θ is given by

$$\Theta = \frac{2}{R_0} \int_{z_{min}}^{z_{max}} \frac{dz}{\sqrt{E - U(z)}}. \tag{2.31}$$

In particular, for E close in value to $U(z_s)$, Eq. (2.9) for $U(z)$ may be used, which yields

$$\Theta = \frac{2\sqrt{2}\,\pi}{[R_0^2 U''(z_s)]^{1/2}} = \frac{2\pi k_0 \rho_s^2}{R_0}. \tag{2.32}$$

In the real ionosphere, $\Theta \sim 0.2 - 0.7$, and, thus, all over the round-the-world path, i.e., within the range of $\theta = 2\pi$, the ray undergoes $10 - 30$ oscillations.

Wave Absorption. The total absorption factor Γ of HF radio waves ($\omega \gg \omega_H$, $\omega \gg \nu_e$) is given by

$$\Gamma = \frac{\omega}{c} \int \varkappa ds = \frac{2\pi e^2}{m c \omega^2} \int \frac{N_e(z)\, \nu_e(z)\, dz}{[E - U(z)]^{1/2}}. \tag{2.33}$$

Here $\varkappa$ is the absorption coefficient, ν_e is the electron collision frequency. In Eq. (2.33) the specific expression for $\varkappa$ under conditions given in Eqs. (1.1) and (1.6) is taken into account; from the integration with respect to the wave path S we have proceeded to the integration with respect to z by using Eq. (2.27).

To deal with trapped waves or a multihop propagation, it is convenient to average the total absorption factor Γ over the oscillation period Θ. In this case

$$K = \left(\frac{d\Gamma}{d\theta}\right) = \frac{4\pi e^2}{m c \omega^2 \Theta} \int_{z_{min}}^{z_{max}} \frac{N_e(z)\, \nu_e(z)\, dz}{\sqrt{E - U(z)}}. \tag{2.34}$$

Hence, for the sliding ray we obtain

$$K_s = \left(\frac{d\Gamma}{d\theta}\right)_s = \frac{R_0 \omega_0^2(z_s)\, \nu_e(z_s)}{2 c \omega^2}. \tag{2.35}$$

Group Time Delay of Pulses. This is determined by their group velocity

$$v_{\mathrm{gr}} = \frac{d\omega}{dk}. \tag{2.36}$$

For the case in question, $k = \frac{\omega}{c}\sqrt{\varepsilon}$, the group velocity is given by the well-known equation

$$v_{\mathrm{gr}} = c\sqrt{\varepsilon} = c\sqrt{1 - \frac{\omega_0^2}{\omega^2}}. \tag{2.37}$$

The time delay is given by

$$dt = \frac{ds}{v_{\mathrm{gr}}} = \frac{1}{v_{\mathrm{gr}}}\sqrt{r^2 d\theta^2 + dz^2} = \frac{r d\theta}{v_{\mathrm{gr}}}\sqrt{1 + \left(\frac{dz}{r d\theta}\right)^2}. \tag{2.38}$$

Now, taking into account Eqs. (2.25) – (2.27), we obtain

$$\sqrt{1 + \left(\frac{dz}{r d\theta}\right)^2} = \sqrt{\frac{\varepsilon r^2}{-ER_0^2}} = \sqrt{\frac{U(z)}{E}}, \tag{2.39}$$

so the final expression for the pulse group time delay is given by

$$t = \frac{1}{c}\int \frac{r}{\sqrt{\varepsilon}}\sqrt{\frac{U(z)}{E}}\, d\theta \approx \frac{R_0}{c}\int d\theta \frac{(1 + 2z/R_0)}{\sqrt{-E}}. \tag{2.40}$$

In the latter expression, using Eq. (2.27), one can go over from the integration with respect to θ to the integration with respect to z. The group time delay, as it is seen from Eq. (2.40), depends on the constant E and altitude z, i.e., on the wave frequency and mode number. It is quite natural, since rays corresponding to different values of E travel in different trajectories. In particular, for a parabolic duct, according to Eqs. (2.10) and (2.16), we have

$$-E = -E_n = -U(z_{\mathrm{s}}) - \frac{2n+1}{k_0^2 \rho_{\mathrm{s}}^2}.$$

Hence, the pulse propagation time, for instance, over an RTW path is given by

$$t_{\mathrm{RTW}} = \frac{2\pi R_0}{c} \frac{1 + 2z_{\mathrm{s}}/R_0}{\sqrt{-U(z_{\mathrm{s}}) - \frac{2n+1}{k_0^2 \rho_{\mathrm{s}}^2}}}. \tag{2.41}$$

Here, the fact has been taken into account that the mean altitude z for a parabolic duct is z_{s}. For instance, by Eq. (2.41), for the sliding mode ($n = 0$) in the model presented in Table 2.1 for the frequency $f_{\mathrm{s}} = 20$ MHz, the time $t_{\mathrm{RTW}} = 138.8$ ms and for $f_{\mathrm{s}} = 27$ MHz, $t_{\mathrm{RTW}} = 139.4$ ms. When calculating

t_{RTW}, we have supposed that the ray propagates one-half of the path in the daytime ionosphere, and the other in the nighttime one.

The pulse distortion is determined by the frequency dependence of the group velocity. The dependence is very complicated and is studied numerically (Kim 1974, 1975). We emphasize, however, that both the time delay and dispersion of pulses, which propagate long distances in the real ionosphere, are determined essentially by the horizontal inhomogeneity. Therefore, these questions will be considered in detail below (Sec. 4.1).

Radio Path Modes. Now let us consider a radio path. Let us assume, for simplicity, that both the emission point $\theta_0 = 0$ and reception point $\theta = \theta_1$ are situated on the Earth's surface, $z = 0$. Only a wave which makes an integer number of oscillations (or hops) between the emission and reception points can arrive at the latter one. This requires the following condition

$$n = \frac{\theta_1}{\Theta(E)} = \frac{\theta_1 R_0}{2 \int\limits_{z_{min}}^{z_{max}} \frac{dz}{\sqrt{E - U(z)}}} . \tag{2.42}$$

Here, n is an integer number of oscillations. Equation (2.42) yields the fixed $E = E_n$ or the fixed emission angles $\alpha_0 = \alpha_{0n} = \sqrt{E_n + 1}$ (2.8). The condition [Eq. (2.42)] singles out from all the emitted waves, only those which arrive at the reception point. They are called the modes for a given radio path.

The integer n has both the upper and lower limits

$$n_{min} \leqslant n \leqslant n_{max} , \qquad n_{min} = \frac{\theta_1}{\Theta_1} , \qquad n_{max} = \frac{\theta_1}{\Theta_{min}} . \tag{2.43}$$

Here, Θ_1 is the maximum oscillation period related to the wave emitted at the smallest angle with respect to the horizon: $\alpha_0 = 0$, $E \approx -1$; Θ_{min} is the minimum period related to the ray which is reflected in the vicinity of the F-region maximum. According to Eq. (2.43), the number of modes, m, on the path is given by

$$m = n_{max} - n_{min} + 1 . \tag{2.44}$$

The number m depends on the wave frequency f and in a spherically-symmetric ionosphere increases with distance between the emission and reception points. This, naturally, pertains to frequencies $f > f_0 = (e^2 N_m / \pi m)^{1/2}$ only, where N_m is the maximum electron density in the ionosphere. When $f < f_0$, the number of modes is great (infinite, if formally), being determined by the emitted signal power and receiver sensitivity rather than by propagation conditions. Of importance is the fact that for each f the mode reflected near F-region maximum is bifurcated, since at $n = n_{max}$ Eq. (2.42) has two solutions (Pedersen 1929). One of them corresponds to the standard ray and the other – to the Pedersen one, which is reflected from the F-region peak. It causes the occurrence of an extra mode in Eq. (2.44), which is natural to be called the Pedersen mode.

Field Amplitude. The mean field amplitude of a signal received is determined by the radio wave energy flux at the reception point. Let the transmitter be situated on the Earth's surface at the point $\theta_0 = 0$, $z = 0$. The emission power is

$$W = W_0 D(\alpha_0, \varphi_0) . \tag{2.45}$$

Here, W_0 is the total transmitter power and $D(\alpha_0, \varphi_0)$ is the antenna directivity diagram with respect to the vertical (α_0) and azimuthal (φ_0) angles

$$\int D(\alpha_0, \varphi_0) \cos\alpha_0 \, d\alpha_0 \, d\varphi_0 = 1 . \tag{2.45a}$$

The received signal energy density on the Earth at one hop distance from the transmitter, $d = R_0 \Theta_{\alpha_0}$, is given by (Davies 1969)

$$w = \frac{W_0 D(\alpha_0, \varphi_0) \cos\alpha_0 \exp(-2K_{\alpha_0} \Theta_{\alpha_0})}{c R_0^2 \sin\Theta_{\alpha_0} \sin\alpha_0 \left(\dfrac{d\Theta}{d\alpha}\right)_{\alpha_0}} . \tag{2.46}$$

Here, Θ_{α_0} is the oscillation period of a wave emitted at an angle of α_0; K_{α_0} is the average absorption coefficient in the ionosphere [Eq. (2.34)]. The received signal energy density [Eq. (2.46)] has the well-known singularities: $w \to \infty$ at $\alpha_0 \to 0$, which is a so-called horizontal focusing, and at $(d\Theta/d\alpha)_{\alpha_0} \to 0$, which is a so-called focusing in the vicinity of the shadow region (Bremmer 1946).

Equation (2.46) can be easily generalized in the case of long-distance multihop propagation in the spherically-symmetric ionosphere. The total wave energy is then represented in the form

$$w = \sum_{n_{\min}}^{n_{\max}} w_n + w_p \tag{2.47}$$

where w_n is the energy density for each mode

$$w_n = \frac{W_0 D(\alpha_{0n}, \varphi_0) \cos\alpha_{0n} \cdot \exp(-2K_{\alpha_{0n}} n \Theta_{\alpha_{0n}})}{c R_0^2 n \sin(n\Theta_{\alpha_{0n}}) \cdot \sin\alpha_{0n} \left(\dfrac{d\Theta}{d\alpha}\right)_{\alpha_{0n}}} \cdot p_n . \tag{2.48}$$

The summation in Eq. (2.47) is carried out with respect to all the modes from $n_{\min}$ to $n_{\max}$ [Eq. (2.43)], w_p refers to the Pedersen mode. The factor p_n determines a change of the wave energy density after reflections from the Earth

$$p_n = \frac{1 + F_n}{2} \cdot F_1 \cdot F_2 \cdot \ldots F_{n-1} . \tag{2.48a}$$

Here, F_k is the factor of k-th single radio wave reflection from the Earth. When $F_1 = F_2 = \ldots = F$, Eq. (2.48a) reduces to the Kazantsev formula (1956).

Note that in order to characterize the signal amplitude, logarithmic loss coefficients are commonly applied which give in dB the energy losses caused by the ray divergence (L_d), focusing (L_f), and ionospheric absorption (L_a) (see Davies 1969). In our case, we have for n-th mode [Eq. (2.48)]

$$L_{d_n} = 20 \log (R_0 \, n \, \Theta_{\alpha_{0n}}) \, ,$$

$$L_{f_n} = 10 \log \left[\frac{\sin(n \, \Theta_{\alpha_{0n}})}{n \, \Theta_{\alpha_{0n}}} \cdot \frac{1}{\Theta_{\alpha_{0n}}} \left(\frac{d\Theta}{d\alpha_0} \right)_{\alpha_{0n}} \cdot p_n^{-1} \cdot \tan \alpha_{0n} \right] \tag{2.49}$$

$$L_{a_n} = 8.68 \, K_{\alpha_{0n}} \cdot n \Theta_{\alpha_{0n}} \, .$$

Note that logarithmic coefficients are well applicable for a separate mode rather than for the total field amplitude [Eq. (2.47)].

Maximum Frequencies. Radio waves propagating in ricochet trajectories can have the highest frequencies. The maximum frequency for a wave confined in the ionosphere is the sliding wave frequency, $f_{sm} = \omega_{sm}/2\pi$. It is defined by the condition for the disappearance of the duct, i.e., when the maximum and minimum of the potential $U(z)$ merge, forming an inflection point $z = z_s$

$$\left(\frac{d^2U}{dz^2} \right)_{z = z_s} = 0 \, , \qquad \left(\frac{dU}{dz} \right)_{z = z_s} = 0 \, . \tag{2.50}$$

Taking Eq. (2.7) into account, we have

$$\omega_{sm} = \left[\frac{2\pi e^2 R_0}{m} \left(\frac{dN}{dz} \right)_{z_s} \right]^{1/2} , \qquad f_{sm} = \left[\frac{e^2 R_0}{2\pi m} \left(\frac{dN}{dz} \right)_{z_s} \right]^{1/2} \tag{2.51}$$

where z_s is defined by the conditions in Eq. (2.50).

The maximum frequency for waves emitted from the Earth and reflected from the ionosphere is the maximum frequency for the communication between given points on the Earth's surface, i.e., the maximum usable frequency, $f_{MUF} = \omega_{MUF}/2\pi$ (Chap. 1). The value of ω_{MUF} is determined by the reflection condition for the horizontal ray $\alpha = 0°$ [Eq. (2.8)] at the maximum of $U(z)$ (see Fig. (2.2)

$$\left(\frac{dU}{dz} \right)_{z = z_k} = 0 \, , \qquad \alpha_{0k} = \sqrt{1 + E} = 0 \, . \tag{2.52}$$

Hence, we obtain (see also Davies 1969, Al'pert 1974)

$$\omega_{MUF} = \omega_{0m} \sqrt{\frac{R_0}{2 z_k}} \, , \qquad \omega_{0m} = \sqrt{\frac{4\pi e^2 N_m}{m}} \tag{2.53}$$

where the altitude z_k is defined by

$$z_k = N_m \Bigg/ \left(\frac{dN}{dz} \right)_{z_k} . \tag{2.54}$$

Here, N_m is the maximum electron density in an ionospheric layer.

Comparing ω_{sm} with ω_{MUF}, we obtain

$$\frac{\omega_{sm}}{\omega_{MUF}} = \left[\frac{(dN/dz)_{z_s}}{(dN/dz)_{z_k}}\right]^{1/2} . \tag{2.55}$$

For real height distributions of the electron density in the ionosphere, the ratio ω_{sm}/ω_{MUF} is about 1.8 – 2.2.

For any given emission and reception points on the Earth, there is the maximum or limiting frequency ω_M for the communications between these points (Davies 1969). The frequency ω_M is defined as the maximum frequency for the first (lowest) mode (always $\omega_M \leqslant \omega_{MUF}$). For long-distance paths, as the path length increases, the frequency ω_M makes damping oscillations, tending to ω_{MUF}.

3 Horizontally – Inhomogeneous Ionosphere

The spherically-symmetric ionosphere has been examined above, i.e., ionospheric plasma distribution was supposed to vary only with height z. However, a significant asymmetry is inherent to the real ionosphere. Charged particle concentration at a fixed height z changes strongly when passing from the daylight hemisphere to the night one or from the polar zone to the midlatitude and equatorial ones. Though this variation in the horizontal direction x is rather slow in comparison with the height variation, Eq. (1.9)

$$\varepsilon = \varepsilon(z, x), \qquad \delta = \left| \frac{\partial\varepsilon/\partial x}{\partial\varepsilon/\partial z} \right| \sim 10^{-2},$$

nevertheless, the horizontal variation proves to be dominant in the case of very long-distance propagation. The point is that over very long-range paths a wave passes through quite different regions of the ionosphere, for example, through its day and night sides. The propagation conditions, the character and properties of the waveguide channels in these regions are quite different (see, for example, Fig. 2.1 and Chap. 4). That is why the horizontal inhomogeneity which determines a gradual variation of ionospheric properties along the ray path, plays an important role.

The present chapter deals with the study of radio wave propagation in the inhomogeneous ionosphere. In this chapter the problems are considered only within the framework of the geometrical optics, while Appendix 3 deals with exact solutions of the wave equation in a horizontally-inhomogeneous ionosphere.

3.1 Geometrical Optics

When describing radio wave propagation in the inhomogeneous ionosphere $\varepsilon = \varepsilon(\boldsymbol{r})$, it is natural to use the geometrical-optics approximation, Eq. (1.8). In this approximation, the solution of wave equation is sought as follows (Landau and Lifshitz 1975)

$$\mathscr{E} = \mathscr{E}_0(\boldsymbol{r}) \exp\{iS(\boldsymbol{r})\}, \tag{3.1}$$

where $\mathscr{E}_0(\boldsymbol{r})$ is the field amplitude which varies comparatively slow, and $S(\boldsymbol{r})$ is the eikonal which varies rapidly. The latter defines the wave vector (i.e., ray propagation direction) at any point

$$\boldsymbol{k} = \operatorname{grad} S. \tag{3.2}$$

As follows from Eq. (2.4) and (3.2), $\boldsymbol{k}$ is described by the eikonal equation

$$(\boldsymbol{k})^2 = (\operatorname{grad} S)^2 = k_0^2 \varepsilon(\boldsymbol{r}) , \qquad k_0 = \frac{\omega}{c} , \tag{3.3}$$

where $\varepsilon(\boldsymbol{r})$ is the dielectric constant. Here, we neglect a small imaginary part of ε, see Eq. (1.6). By differentiating Eq. (3.3) and taking into account the fact that $\operatorname{rot}\boldsymbol{k} = 0$ [Eq. (3.2)], Eq. (3.4) can be rewritten in the form

$$(\boldsymbol{k}\,\nabla)\boldsymbol{k} = \frac{k_0^2}{2} \nabla \varepsilon . \tag{3.4}$$

Let us write down Eq. (3.4) in the spherical coordinates r, θ, φ:

$$\boldsymbol{k} = \frac{\partial S}{\partial \boldsymbol{r}} = \left\{ \frac{\partial S}{\partial r} , \frac{1}{r} \frac{\partial S}{\partial \theta} , \frac{1}{r \sin\theta} \frac{\partial S}{\partial \varphi} \right\} = \{ k_r, k_\theta, k_\varphi \} .$$

For the radial component of Eq. (3.4), by differentiating Eq. (3.3), we have

$$\frac{\partial}{\partial r} \left[\left(\frac{\partial S}{\partial r} \right)^2 + \frac{1}{r^2} \left(\frac{\partial S}{\partial \theta} \right)^2 + \frac{1}{r^2 \sin^2\theta} \left(\frac{\partial S}{\partial \varphi} \right)^2 \right]$$

$$= 2 \left[\frac{\partial S}{\partial r} \frac{\partial^2 S}{\partial r^2} - \frac{1}{r} \left(\frac{1}{r^2} \left(\frac{\partial S}{\partial \theta} \right)^2 + \frac{1}{r^2 \sin^2\theta} \left(\frac{\partial S}{\partial \varphi} \right)^2 \right) \right.$$

$$\left. + \frac{1}{r^2} \frac{\partial S}{\partial \theta} \frac{\partial^2 S}{\partial \theta \partial r} + \frac{1}{r^2 \sin^2\theta} \frac{\partial S}{\partial \varphi} \frac{\partial^2 S}{\partial r \partial \varphi} \right] .$$

Taking into account Eq. (3.3), let us replace the term in the parentheses

$$\frac{1}{r^2} \left(\frac{\partial S}{\partial \theta} \right)^2 + \frac{1}{r^2 \sin^2\theta} \left(\frac{\partial S}{\partial \varphi} \right)^2 = k_0^2 \varepsilon - \left(\frac{\partial S}{\partial r} \right)^2 = k_0^2 \varepsilon - k_r^2$$

and rewrite Eq. (3.4) for the radial component as follows

$$\left[k_r \frac{\partial}{\partial r} + \frac{k_r}{r} + \frac{k_\theta}{r} \frac{\partial}{\partial \theta} + \frac{k_\varphi}{r \sin\theta} \frac{\partial}{\partial \varphi} \right] k_r = \frac{k_0^2}{2 r^2} \frac{\partial}{\partial r} (r^2 \varepsilon) . \tag{3.4a}$$

Analogously we have for the horizontal components

$$\left[k_r \frac{\partial}{\partial r} + \frac{k_r}{r} + \frac{k_\theta}{r} \frac{\partial}{\partial \theta} + \frac{k_\varphi}{r \sin\theta} \frac{\partial}{\partial \varphi} \right] k_\theta - \frac{\cot\theta}{r} k_\varphi^2 = \frac{k_0^2}{2r} \frac{\partial \varepsilon}{\partial \theta} , \tag{3.4b}$$

$$\left[k_r \frac{\partial}{\partial r} + \frac{k_r}{r} + \frac{k_\theta}{r} \frac{\partial}{\partial \theta} + \frac{k_\theta}{r} \cot\theta + \frac{k_\varphi}{r \sin\theta} \frac{\partial}{\partial \varphi} \right] k_\varphi = \frac{k_0^2}{2 r \sin\theta} \frac{\partial \varepsilon}{\partial \varphi} . \tag{3.4c}$$

The characteristic equations for the set of Eqs. (3.4) are given by

$$\frac{dr}{k_r}=\frac{rd\theta}{k_\theta}=\frac{r\sin\theta d\varphi}{k_\varphi}=\frac{dk_r}{\dfrac{k_0^2}{2r^2}\dfrac{\partial}{\partial r}(r^2\varepsilon)-\dfrac{k_r^2}{r}}$$

$$=\frac{dk_\theta}{\dfrac{\cot\theta k_\varphi^2}{r}+\dfrac{k_0^2}{2r}\dfrac{\partial\varepsilon}{\partial\theta}-\dfrac{k_rk_\theta}{r}}=\frac{dk_\varphi}{\dfrac{k_0^2}{2r\sin\theta}\dfrac{\partial\varepsilon}{\partial\varphi}-\dfrac{k_\theta}{r}\cot\theta\cdot k_\varphi-\dfrac{k_rk_\varphi}{r}}. \tag{3.5}$$

Hence, it follows that

$$\frac{dr}{d\theta}=\frac{rk_r}{k_\theta},\qquad \frac{d\varphi}{d\theta}=\frac{k_\varphi}{k_\theta\sin\theta}. \tag{3.5a}$$

By differentiating Eq. (3.5a) with respect to θ and substituting the derivatives $dk_r/d\theta$, $dk_\theta/d\theta$, $dk_\varphi/d\theta$ from Eq. (3.5), we obtain the following second-order equations

$$\frac{d^2r}{d\theta^2}=\frac{k_0^2}{2k_\theta^2}\frac{\partial}{\partial r}(r^2\varepsilon)-\frac{rk_rk_0^2}{2k_\theta^3}\frac{\partial\varepsilon}{\partial\theta}-\frac{rk_rk_\varphi^2}{k_\theta^3}\cot\theta+\frac{rk_r^2}{k_\theta^2}, \tag{3.6}$$

$$\frac{d^2\varphi}{d\theta^2}+2\cot\theta\frac{d\varphi}{d\theta}+\cos\theta\sin\theta\left(\frac{d\varphi}{d\theta}\right)^3=\frac{k_0^2}{2k_\theta^2\sin^2\theta}\frac{\partial\varepsilon}{\partial\varphi}-\frac{k_0^2}{2k_\theta^2}\frac{d\varphi}{d\theta}\frac{\partial\varepsilon}{\partial\theta}. \tag{3.7}$$

These equations will be used further for the determination of radio wave trajectories in the ionosphere.

Now let us go to the solution of the geometrical-optics equation using the main features of ionospheric structure and very long-distance HF propagation. Note that the ionosphere is a thin plasma layer [Eq. (1.7)] which is slightly inhomogeneous in the horizontal direction [Eq. (1.9)]. The radio waves propagate at a small angle to the horizon [Eqs. (1.4) and (1.5)]. Accordingly, let us introduce two small parameters δ and γ. According to Eqs. (1.4), (1.5), (1.7), and (1.9), we have

$$\frac{\Delta z}{N_eR_0}\frac{\partial N_e}{\partial\theta},\ \frac{\Delta z}{N_eR_0}\frac{\partial N_e}{\partial\varphi}\sim\delta\ll 1\,,\quad \frac{k_r}{k_0},\ \alpha_0,\ \frac{\Delta z}{R_0}\sim\gamma\ll 1\,. \tag{3.8}$$

Solution of Eqs. (3.4) – (3.7) is naturally to be sought in the form of a series with respect to these small parameters. For a zero approximation with respect to δ, taking into account Eq. (3.8), in terms of Eq. (3.7), we obtain the following equation for the ray trajectory in the horizontal direction:

$$\frac{d^2\varphi}{d\theta^2}+2\cot\theta\frac{d\varphi}{d\theta}+\cos\theta\sin\theta\left(\frac{d\varphi}{d\theta}\right)^3=0\,. \tag{3.9}$$

The general integral of Eq. (3.9) is given by

$$\frac{d\varphi}{d\theta} = -\frac{C}{\sin\theta\sqrt{\sin^2\theta - C^2\cos^2\theta}}, \qquad \tan\theta\sin(\varphi-\beta) = C. \tag{3.10}$$

Here β and C are the integration constants. It is easy to be sure that the two-parametrical family of curves [Eq. (3.10)] represents the intersection of spherical surface by the planes, passing through the center of the sphere, i.e., by the great circle planes. Thus, for a zero approximation with respect to δ, the ray trajectory lies in the great circle plane as in the case of spherical symmetry. In particular, according to Eq. (3.10), all the trajectories passing through a given point θ_0, φ_0, are described by the equation

$$\tan\theta\sin(\varphi-\beta) = \tan\theta_0\sin(\varphi_0-\beta). \tag{3.11}$$

The constant β specifies the ray azimuth at the point θ_0, φ_0. The constant varies within $\varphi_0 \ll \beta < \varphi_0 + \pi$. At $\beta = \varphi_0$, the ray is directed along the meridian $\varphi = \varphi_0$. At $\beta = \varphi_0 + \pi/2$ the ray at the point θ_0, φ_0 is directed along the parallel.

The trajectory form in the great circle plane is described by Eq. (3.6). Choosing the axis $\boldsymbol{e}_\theta$ coinciding with the direction of $\boldsymbol{k}_0$ and changing r for the height $z = r - R_0$, and also taking into account that $z/R_0 \sim k_z/k_0 \sim \gamma$, $k_\varphi \sim \delta$, we rewrite Eq. (3.6) for a first approximation with respect to γ as follows [Gurevich (1971), Gurevich and Tsedilina (1973)]

$$\frac{d^2 z}{d\theta^2} = \frac{R_0^2}{2}\frac{\partial}{\partial z}\left(\varepsilon + \frac{2z}{R_0}\right) = -\frac{R_0^2}{2}\frac{\partial U}{\partial z}. \tag{3.12}$$

This trajectory equation coincides in fact with Eqs. (2.37) and (2.38). The only difference is that in the horizontally-inhomogeneous ionosphere ε varies not only with z, but with θ as well. Therefore, the potential

$$U = U(z,\theta) = -\varepsilon(z,\theta) - \frac{2z}{R_0}. \tag{3.13}$$

From Eqs. (3.4b) and (3.3) for the same approximation the equation $k_0 r = \text{const}$ follows. This corresponds to Snell's law which has that very form in the chosen coordinate system, $\boldsymbol{e}_\theta \| \boldsymbol{k}_0$ [see Eq. (2.34)].

Ray deviation from the great circle plane is described by Eq. (3.7) taking into account the correction of the order of δ. For a first approximation with respect to δ, we have

$$\frac{d^2\varphi}{d\theta^2} + 2\cot\theta\frac{d\varphi}{d\theta} + \cos\theta\sin\theta\left(\frac{d\varphi}{d\theta}\right)^3$$
$$= \frac{1+\sin^2\theta\left(\frac{d\varphi}{d\theta}\right)_0^2}{2}\left[\frac{1}{\sin^2\theta}\frac{\partial\varepsilon}{\partial\varphi} - \left(\frac{d\varphi}{\partial\theta}\right)_0\frac{\partial\varepsilon}{\partial\theta}\right]. \tag{3.14}$$

Here, $(d\varphi/d\theta)_0$ is determined for the unperturbed trajectory according to Eq. (3.10). It is taken into account that $k_{\theta_0} = k_0[1+(d\varphi/d\theta)_0^2 \sin^2\theta]^{-1/2}$ [it follows from Eq. (3.5a) taking into account Eq. (3.3): $k_{\theta_0}^2 + k_{\varphi_0}^2 = k_0^2$]. So far as the corrections to the unperturbed trajectory are small, the left part of Eq. (3.14) permits the linearization:

$$\frac{d\varphi}{d\theta} = \left(\frac{d\varphi}{d\theta}\right)_0 + \left(\frac{d\varphi}{d\theta}\right)_1, \quad \left|\left(\frac{d\varphi}{d\theta}\right)_1\right| \ll \left|\left(\frac{d\varphi}{d\theta}\right)_0\right| .$$

Besides, Eq. (3.14) is simplified in the coordinate system with the axis $\boldsymbol{e}_\theta \| \boldsymbol{k}_0$. In this system the great circle plane is given by $\varphi = \beta = \varphi_0$, $(d\varphi/d\theta)_0 = 0$ [see Eqs. (3.10) and (3.11)]. Consequently, here $d\varphi/d\theta$ is small (of the order of δ). Then Eq. (3.14) takes the simple form

$$\frac{d^2\varphi}{d\theta^2} + 2\cot\theta \frac{d\varphi}{d\theta} = \frac{1}{2\sin^2\theta} \frac{\partial\varepsilon}{\partial\varphi} . \tag{3.14a}$$

Let us note, however, that the condition for smallness of $d\varphi/d\theta$ fails in the vicinity of the antipode and round-the-world points $\theta \to \pi$, $\theta \to 2\pi$, where the coefficients of Eq. (3.14a) have a singularity related to the special choice of coordinate system. The same singularity appears in the case of linearization of the general Eq. (3.14) in the vicinity of the turning points θ_m of unperturbed trajectory $\sin\theta_m = C\cos\theta_m$ [see Eq. (3.10)]. The analysis of the solution in the vicinity of singular points is given in (Gurevich and Tsedilina 1979), see also Sec. 3.3.

3.2 Adiabatic Approximation

The Adiabatic Invariant. In the case of very long-distance radio wave propagation, ray trajectory oscillates along the height in the waveguide channel between the ionosphere maximum and the Earth or in still more narrow wave ducts. These ducts do not remain constant. They vary along the ray path because of variation of height profiles of ionospheric electron density. For instance, the ducts can rise or descend, converge or widen and change their form. However, it is important that usually in the ionosphere, the waveguide channels vary slowly, as they say, adiabatically. The latter means that the ducts vary slightly over one ray oscillation

$$\mu = \frac{\Theta}{2\pi P_i} \frac{dP_i}{d\theta} \ll 1 . \tag{3.15}$$

Here, Θ is the period of ray oscillations [Eq. (2.31)], P_i is a certain characteristic of a waveguide channel, for example, its effective width or height, μ is the adiabaticity parameter.

In the case of adiabatic variations of waveguide channels one can find the approximate integrals defining the ray trajectory with rather high accuracy. These integrals are called adiabatic invariants.

Consider Eq. (3.12) describing the ray trajectory oscillations along the height z. Let us assume that the potential U [Eq. (3.13)], i.e., the plasma dielectric constant ε, varies quite slowly versus θ [Eq. (3.15)]. Note that if the potential U did not depend on θ, then the ray would make the constant oscillations in the duct (the case of spherical symmetry). The level of ray oscillations would have been characterized then by the constant E [Eqs. (2.26) and (2.8)]. In our case, when $U = U(\theta)$, the quantity E is no longer constant. It varies along the ray path, $E = E(\theta)$. Indeed, defining E according to Eq. (2.27), we have

$$E = \frac{1}{R_0^2}\left(\frac{dz}{d\theta}\right)^2 + U, \qquad \frac{dz}{d\theta} = \pm R_0\sqrt{E-U}\,. \tag{3.16}$$

Then

$$\frac{dE}{d\theta} = \frac{2}{R_0^2}\,\frac{dz}{d\theta}\,\frac{d^2z}{d\theta^2} + \frac{\partial U}{\partial z}\,\frac{dz}{d\theta} + \frac{\partial U}{\partial \theta}$$

and taking into account Eq. (3.12), we obtain

$$\frac{dE}{d\theta} = \left.\frac{\partial U}{\partial \theta}\right|_{z=z(\theta)}. \tag{3.16a}$$

The quantity E defined according to Eq. (3.16), involves the part smoothly varying with θ and another one rapidly oscillating (oscillation period Θ). The amplitude and oscillation phase of E depend on the specific ray trajectory. When the conditions for the adiabaticity [Eq. (3.15)] are satisfied, the amplitude of E oscillations is very small: $|E_{\sim}| \ll |E|$. In this case it is natural to separate the quantity $\bar{E}$ averaged with respect to oscillations and smoothly varying with θ. Now let us consider the adiabatic invariant I (Gurevich 1971):

$$I = \frac{4}{R_0}\int_{z_{min}}^{z_{max}} \sqrt{\bar{E}-U}\,dz\,. \tag{3.17}$$

Here, z_{min} and z_{max} are the wave reflection points defined according to Eqs. (2.29) and (2.13a) with the substitution of $\bar{E}$ for E. Though the quantities $\bar{E}$, U, z_{min}, and z_{max} vary significantly with θ, the invariant I, when Eq. (3.15) is satisfied, remains almost constant. Indeed, as it follows from Eq. (3.17)

$$\frac{dI}{d\theta} = \frac{4}{R_0}\left\{\left(\sqrt{\bar{E}-U}\right)_{z_{max}}\frac{dz_{max}}{d\theta} + \left(\sqrt{\bar{E}-U}\right)_{z_{min}}\frac{dz_{min}}{d\theta}\right\}$$
$$+\frac{2}{R_0}\int_{z_{min}}^{z_{max}} \frac{\dfrac{d\bar{E}}{d\theta} - \dfrac{\partial U}{\partial\theta}}{\sqrt{\bar{E}-U}}\,dz \approx 0\,. \tag{3.18}$$

Here, Eqs. (2.13a) and (3.16a) have been used. They are quite accurate when substituting $\bar{E}$ for E in the case of adiabatic variations of parameters.

Thus, the adiabatic invariant I [Eq. (3.17)] is the approximate integral of trajectory Eq. (3.12). The invariant is conserved with high accuracy in the slowly varying ionospheric waveguide [Eq. (3.15)].

For example, in the vicinity of the point of sliding, the ionospheric waveguide channel is a parabolic one [Eq. (2.9)]

$$U(z) = U(z_s) + \frac{1}{2} U''(z_s)(z - z_s)^2 + \dots .$$

In this case the adiabatic invariant, according to Eq. (3.17) is equal to

$$I = \Theta(\bar{E} - U(z_s)), \qquad \Theta = \frac{2\sqrt{2}\pi}{[R_0^2 U''(z_s)]^{1/2}} . \tag{3.19}$$

Here, Θ is the period of ray oscillations [Eq. (2.32)]. The potential [Eq. (2.9)] can vary along the ray path due to variation of the height of sliding $z_s = z_s(\theta)$, i.e., due to variation of waveguide channel height or due to variation of $U''(z_s) = U''(\theta)$, i.e., because of the broadening or narrowing of the duct. For example, let the sliding height remain fixed, while the value of $U''(\theta)$ characterizing the duct width varies in the vicinity of the point θ_0 from U_0'', when $\theta < \theta_0 - \frac{1}{\delta}$, to U_1'', when $\theta > \theta_0 + \frac{1}{\delta}$

$$U''(\theta) = \frac{U_0'' + U_1'' \exp[\delta(\theta - \theta_0)]}{1 + \exp[\delta(\theta - \theta_0)]} . \tag{3.20}$$

Here, the parameter δ characterizes the rapidity of the $U''(\theta)$ variation. According to Eq. (3.20), we have for the maximum value

$$\left| \frac{1}{U''} \frac{dU''}{d\theta} \right|_{\max} \leqslant \delta .$$

When changing $U''(\theta)$ according to Eq. (3.20), the adiabatic invariant is conserved with an accuracy of ΔI (see Landau and Lifshitz 1976, Sec. 51)

$$\Delta I \sim \exp(-4\pi^2/\delta\Theta_0) \quad \text{when} \quad U_0'' > U_1'' ,$$

$$\Delta I \sim \exp(-4\pi^2/\delta\Theta_1) \quad \text{when} \quad U_0'' < U_1'' .$$

Here, Θ_0 and Θ_1 are the oscillation periods defined according to Eq. (3.19) for $U''(z_s) = U_0''$ and $U(z_s)'' = U_1''$, respectively. Bearing in mind that the parameter of adiabaticity in the case considered is

$$\mu = \frac{\Theta}{2\pi} \left| \frac{1}{U''} \frac{dU''}{d\theta} \right|_{\max} \leqslant \frac{\delta\Theta}{2\pi} ,$$

we find that the change of I is

$$\Delta I \sim \exp\left(-\frac{C}{\mu}\right), \quad C \approx 2\pi. \tag{3.21}$$

When Eq. (3.15) is satisfied, the change of the invariant is vary small. Note that the analysis of solutions of wave equation shows that the meaning of the adiabatic invariant introduced here within the framework of the geometrical optics is closely connected with eigenmodes of the horizontally-inhomogeneous ionospheric waveguide. The adiabatic invariant conservation means to a certain extent the conservation of the wave field distribution with respect to wave modes of the duct. A detailed analysis of these problems and, in particular, the problem of wave mode conservation and adiabatic approximation accuracy in a horizontally-inhomogeneous duct is carried out in Appendix 3.

Let us also emphasize the mechanical analogy considered above in Secs. 2.1 and 2.4. Geometrical-optics Eq. (3.12) is similar to the equation of classical particle motion $z(\theta)$, oscillating in the potential field $U(z, \theta)$. The height z plays the role of particle coordinate, θ plays the role of "time". Equation (3.15) means that the potential $U(z)$ varies adiabatically slow in "time" θ. In this case, according to (Landau and Lifshitz 1976), the adiabatic invariant $I = \oint p\,dz$ (p is the particle momentum) is conserved. This invariant is equivalent to the invariant described by Eq. (3.17).

Main Characteristics of Ray Trajectory. Using the adiabatic invariant conservation, one can construct a ray trajectory in a horizontally-inhomogeneous ionosphere. One can also determine the trajectory main characteristics, i.e., the main parameters characterizing long-range radio wave propagation. Such a solution is called the adiabatic approximation.

The ray trajectory both in the case of adiabatic approximation and in the case of spherical symmetry is characterized by the quantity E which varies now along the ray path: $E = E(\theta)$ (here the quantity $\bar{E}$ is meant, the symbol of averaging will be omitted below). The dependence of E on θ is defined from the condition for adiabatic invariant conservation

$$I(E, \theta) = I_0, \tag{3.22}$$

wherefrom we have

$$E = E(\theta, I_0). \tag{3.23}$$

Equations (3.22) and (3.23) define the E variation along the ray path for a given initial invariant I_0. The other parameters are defined in terms of E by the same formulas as in the case of spherical symmetry. Let us list here the main relations.

The reflection points $z_{min}(\theta)$ and $z_{max}(\theta)$ are determined from Eq. (2.13a):

$$E(\theta) - U(z_m, \theta) = 0. \tag{3.24}$$

The ray trajectory in the vertical plane, i.e., the dependence of z on θ is defined for every oscillation in an implicit form by Eq. (2.30)

$$\theta = \theta_0 + \frac{1}{R_0} \int\limits_{z_0}^{z} \frac{dz}{\sqrt{E(\theta) - U(z,\theta)}} . \tag{3.25}$$

The period of ray oscillations

$$\Theta = \frac{2}{R_0} \int\limits_{z_{\min}}^{z_{\max}} \frac{dz}{\sqrt{E - U(z,\theta)}} = \frac{dI}{dE} . \tag{3.26}$$

In the vicinity of the sliding point z_s, the oscillation period is related to the adiabatic invariant by the simple Eq. (3.19).

The number of ray oscillations in a duct over the path between the points θ_0 and θ is given by

$$\mathscr{P} = \int\limits_{\theta_0}^{\theta} \frac{d\theta}{\Theta} , \tag{3.27}$$

where the period $\Theta(\theta)$ is defined according to Eq. (3.26). Generally, the quantity $\mathscr{P}$ is not integer. It describes not only the number of full oscillations made by a ray, but the phase of oscillations at a given point as well. The oscillation phase

$$\Phi = 2\pi(\mathscr{P} - [\mathscr{P}]) \tag{3.28}$$

shows what part of the oscillation period the ray has passed up to a given point θ (here $[\mathscr{P}]$ is an integer part of $\mathscr{P}$).

Consider a radio path with the emission point $\theta = \theta_0$ and the reception point $\theta = \theta_1$, both situated on the Earth. To arrive at the reception point, the ray is bound to make an integer number of oscillations [see Eq. (2.42)]

$$n = [\mathscr{P}] = \int\limits_{\theta_0}^{\theta_1} \frac{d\theta}{\Theta(E)} . \tag{3.29}$$

Equation (3.29) defines the modes of a given path θ_0, θ_1 in the adiabatic approximation and it is equivalent to the condition $\Phi(\theta_1) = 0$. Equations (3.29) and (3.23) define the specific value of the invariant $I_0 = I_{0n}$ for every n. If I_{0n} is known, according to Eqs. (3.22) and (3.23), one can find $E = E_n(\theta)$ related to a given mode. At the emission point, $E = E_n(0) = -1 + \alpha_{0n}^2$ [see Eq. (2.8)]. Hence, as in the case of spherical symmetry, a definite angle of emission, $\alpha_0 = \alpha_{0n}$, corresponds to every mode.

The total number of modes on the path is defined as before by Eq. (2.44).

The total absorption A of a ray over the path from θ_0 to θ, and the group time delay are defined analogously in the adiabatic approximation. The absorption K averaged with respect to the oscillation period Θ is equal to [see Eq. (2.34)]

$$K = \overline{\left(\frac{d\Gamma}{d\theta}\right)} = \frac{4\pi e^2}{mc\omega^2\Theta} \int\limits_{z_{\min}}^{z_{\max}} \frac{N_e(z)\,\nu_e(z)\,dz}{\sqrt{E - U(z)}} . \tag{3.30}$$

Accordingly, the total absorption is given by

$$A=\int\limits_{\theta_0}^{\theta} K d\theta=\frac{4\pi e^2}{mc\omega^2}\int\limits_{\theta_0}^{\theta}\frac{d\theta}{\Theta}\int\limits_{z_{\min}}^{z_{\max}}\frac{N_e(z)\,\nu_e(z)\,dz}{\sqrt{E-U(z)}}. \tag{3.31}$$

The group time delay of pulse signals [see Eq. (2.40)] is equal to

$$t=\frac{R_0}{c}\int\limits_{\theta_0}^{\theta} d\theta\frac{1}{\sqrt{-E}}\left(1+\frac{2}{R_0}\bar{z}(\theta)\right), \tag{3.32}$$

where

$$\bar{z}(\theta)=\frac{2}{R_0\Theta}\int\limits_{z_{\min}}^{z_{\max}}\frac{z\,dz}{\sqrt{E-U(z)}}. \tag{3.33}$$

In deriving Eq. (3.32), the averaging in Eq. (2.40) with respect to the ray oscillations was carried out.

The Field Amplitude. The average field amplitude or energy density w of a radio wave emitted from the point $\theta_0=0$ and propagating in one wave duct is defined at the reception point $\theta=\theta_1$ analogously to Eqs. (2.46) – (2.48)

$$w=\sum_{n_{\min}}^{n_{\max}} w_n+w_p \tag{3.34}$$

$$w_n=\frac{W_0 D(\alpha_{0n},\varphi_0)\cos\alpha_{0n}\exp\left(-2\int\limits_0^{\theta_1}K_n d\theta\right)}{cR_0^2\sin\theta_1\cdot\sin\alpha_{1n}\cdot Q_n}\cdot p_n. \tag{3.35}$$

Here, $K_n=K(E_n)$ [see Eq. (3.30)], and $E_n(\theta)$ is the value of E for n-th mode of the radio path. The mode is defined according to Eqs. (3.29), (3.22), and (3.23) by means of the invariant I_{0n}. At the emission point $\theta_0=0$, we have

$$E_n(0)=E_{0n}=-1+\alpha_{0n}^2. \tag{3.36}$$

As it is seen from Eq. (3.36), the angle α_{0n} is related unambiguously to E_{0n} and, consequently, to the invariant I_{0n} of the n-th mode. The angle α_{1n} in the denominator of Eq. (3.35) is taken as usual at the reception point, i.e., at $\theta=\theta_1$.

Now, the quantity Q_n is defined similarly to Eqs. (2.46) and (2.48). We have

$$Q_n=\left[\frac{d}{d\alpha_0}\theta_n(\alpha_0)\right]_{\alpha_0=\alpha_{0n}}. \tag{3.37}$$

Here, $\theta_n(\alpha_n)$ is the distance (in radians) from the emission point $\theta_0=0$ to the point of arrival to the Earth after n reflections from the ionosphere. The distance depends on the emission angle α_0 or on the invariant I_0 and is defined by the equation

$$n=\int_0^{\theta_n(\alpha_0)}\frac{d\theta}{\Theta(\alpha_0,\theta)}=\int_0^{\theta_n(I_0(\alpha_0))}\frac{d\theta}{\Theta(I_0(\alpha_0),\theta)}.$$

Clearly, at $\alpha_0=\alpha_{0n}$

$$\theta_n(\alpha_{0n})=\theta_1 .$$

In Eq. (3.37) the differentiation is carried out at a fixed $\alpha_0=\alpha_{0n}$, i.e., at $I_0=I_{0n}$

$$\left(\frac{d\theta_n}{d\alpha_0}\right)_{\alpha_{0n}}=\left(\frac{d\theta_n}{dI_0}\right)_{I_{0n}}\cdot\left(\frac{dI_0}{dE_0}\right)_{I_{0n}}\cdot\left(\frac{dE_0}{d\alpha_0}\right)_{I_{0n}}=\left(\frac{d\theta_n}{dI_0}\right)_{I_{0n}}\Theta_{0n}\cdot 2\alpha_{0n} .$$

Thus,

$$Q_n=2\,\alpha_{0n}\Theta_{0n}\left(\frac{d\theta_n}{dI_0}\right)_{I_{0n}} . \tag{3.38}$$

Here, $\Theta_{0n}=\Theta_{\theta=0}(E_{0n})$ is the initial oscillation period for the n-th mode.

In Eq. (3.34) the summation is carried out with respect to all the modes from $n_{\min}$ to $n_{\max}$ [Eq. (2.43)]. Possibility for the Pedersen mode presence is taken into account in Eq. (3.34) by the term w_p which is calculated by the same Eqs. (3.35) and (3.38). The factor p_n determines as above the influence of radio wave reflection from the Earth [Eq. (2.48a)].

As the comparison of Eqs. (3.35) and (3.36) with Eqs. (2.48) and (2.49) shows, the logarithmic coefficient of losses L_d for the n-th mode due to the ray divergence in the horizontally-inhomogeneous ionosphere does not change its form [Eq. (2.49)]. The absorption factor L_a changes in an obvious manner, and only the coefficient of focusing, L_f, undergoes a nontrivial variation

$$\begin{aligned}&L_d=20\log(R_0\theta_1)\,,\qquad L_a=8.68\int_0^{\theta_1}K_n\,d\theta\\&L_f=10\log\left(\frac{\sin\theta_1}{\theta_1}\cdot\frac{\sin\alpha_{1n}}{\cos\alpha_{0n}}\,\frac{Q_n}{\theta_1}\,p_n^{-1}\right).\end{aligned}\tag{3.39}$$

At the caustic points where Q_n [Eq. (3.38)] vanishes, the field amplitude sharply increases just like in the neighborhood of the shadow region in the spherically-symmetric ionosphere [Eq. (2.46)]. In the horizontally-inhomogeneous ionosphere the number of caustics can increase. Note that at the antipodal ($\theta_1\to\pi$) and round-the-world ($\theta_1\to 2\pi$) points the field amplitude also increases as in the case of spherical symmetry. However, in a horizontally-inhomogeneous ionosphere in the vicinity of these singular points, Eqs. (3.34) – (3.39) are inapplicable. Here, a more precise analysis is necessary taking into account the lateral deviations of rays (see Secs. 3.4, 5.1, 5.3).

The quantity Q_n is proportional to the derivative $(d\theta_n/dI_0)_{I_{0n}}$, hence, the derivative appears in the denominator of Eq. (3.35). In the case of specific calculations of Q_n, much more higher accuracy is needed than when determining the other terms in Eq. (3.35). On the other hand, in the horizontally-inhomogeneous ionosphere due to the caustic effect, the spatial field distribution has a sharply fluctuating character. Besides, it should be taken into account that in the real ionosphere usually there are irregular large-scale inhomogeneities (with the scale ~500 – 1000 km) due to which the caustic distribution becomes still more complicated and intricate. At the same time, in the ionosphere there is a spectrum of small-scale irregularities, scattering by which results in dispersal of the caustics and in appearance of strong spatial fluctuations in the field amplitude. That is why when calculating the amplitude, instead of the exact value of Q_n, it may be sufficient to use the averaged quantity $\bar{Q}_n$

$$\bar{Q}_n = n\left(\frac{d\Theta}{d\alpha_0}\right)_{\alpha_{0n}} = \int_0^{\theta_1} \frac{d\theta}{\Theta}\left(\frac{d\Theta}{d\alpha_0}\right)_{\alpha_{0n}} .$$

Taking into account that

$$\left(\frac{d\Theta}{d\alpha_0}\right)_{\alpha_{0n}} = \left(\frac{d\Theta}{dI_0}\right)_{I_{0n}} \cdot \left(\frac{dI_0}{d\alpha_0}\right)_{I_{0n}} = \left(\frac{d\Theta}{dE}\right)_{I_{0n}} \cdot \left(\frac{dE}{dI}\right)_{I_{0n}}$$

$$\cdot \left(\frac{dI}{dE_0}\right)_{I_{0n}} \cdot \left(\frac{dE_0}{d\alpha_{0n}}\right)_{I_{0n}} = \left(\frac{d\Theta}{dE}\right)_{I_{0n}} \cdot \frac{2\,\alpha_{0n}\Theta_{0n}}{\Theta}$$

we obtain

$$\bar{Q}_n = 2\,\alpha_{0n}\Theta_{0n}\left|\int_0^{\theta_1}\left(\frac{1}{\Theta^2}\,\frac{d\Theta}{dE}\right)_{I_{0n}} d\theta\right| . \tag{3.39a}$$

In the latter expression, the absolute value of the integral is taken, because in the vicinity of the layer maximum the derivative $(d\Theta/dE)$ can change its sign. The calculation of the quantity $\bar{Q}_n$ does not require a higher accuracy. To what extent Eq. (3.39a) is sufficiently accurate for the long-distance paths, can be established only by a detailed comparison with the observational data.

The Maximum Adiabatic Invariant. Transition through the Separatrix. Figure 3.1a shows a typical form of the potential $U(z)$ in the ionosphere. Special trajectories called the separatrices separating the different capture regions are shown by the broken line.

The separatrix 1 separates the rays trapped in the ducts II and III; the separatrix 2, in the ducts I and III; the separatrix 3 separates all the radio waves trapped in the duct III, i.e., the waves confined by the ionosphere. The separatrices are characterized by the quantity $E = E_m = U_m$, where U_m is the

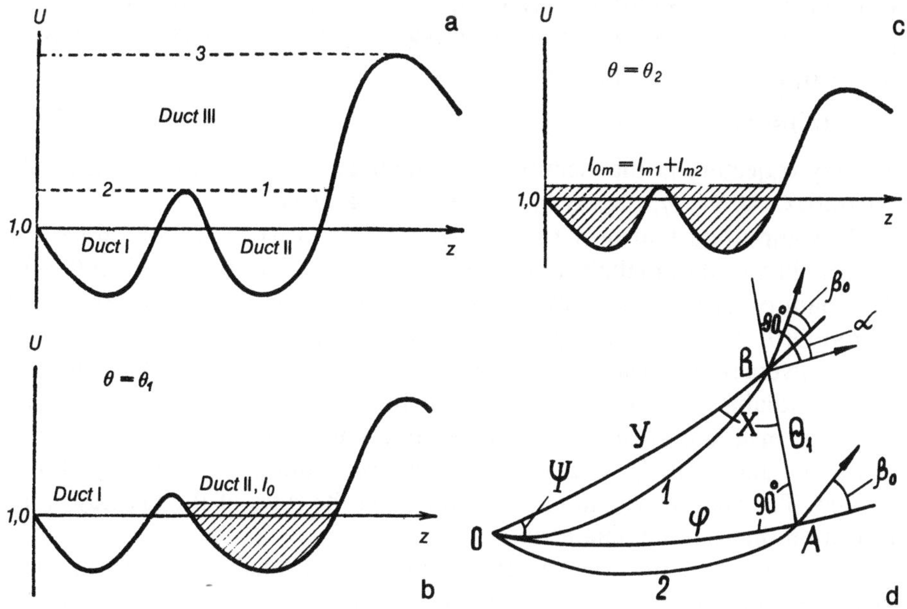

Fig. 3.1a – d. A typical shape of the potential U for the daytime ionosphere (**a**). Change of the ray invariant at the departure from the duct is shown: **b** before the departure; **c** after the departure; **d** the determination of azimuthal bearing deviation β_0 at the reception point B or A, and also of the initial emission azimuth at the point O

value of potential at the maximum defining a given special trajectory. The adiabatic invariants corresponding to these trajectories are equal to

$$I_m = \frac{4}{R_0} \int_{z_{min}}^{z_{max}} [U_m - U(z)]^{1/2} dz\,. \tag{3.40}$$

Here, either z_{min} (duct II) or z_{max} (duct I, III) is the point at which the potential $U(z)$ has the maximum. The adiabatic invariant of any wave trapped in a given duct is less than I_m. Therefore, the quantity I_m is called maximum adiabatic invariant of a duct.

The maximum adiabatic invariant which is determined only by the form of the potential $U(z)$ is the most important characteristic of a given duct. In a horizontally-inhomogeneous ionosphere, the quantity I_m varies depending on the horizontal coordinates θ and φ.

Let us consider a ray trajectory which is characterized by the invariant I_0. As usual, let us choose the axis $\boldsymbol{e}_\theta$ in the direction of the ray propagation. The maximum adiabatic invariant I_m varies along the ray path due to horizontal inhomogeneity of the ionosphere, $I_m = I_m(\theta)$, the invariant I_0 remaining constant. Clearly, the ray will be confined in a given duct as long as the condition

$$I_0 < I_m(\theta) \tag{3.41}$$

is fulfilled. If this condition is not fulfilled, the ray leaves the duct. The critical point of the trajectory, $\theta = \theta_s$ at which the condition [Eq. (3.41)] is no longer satisfied, is defined by the equation

$$I_m(\theta_s) = I_0 . \tag{3.42}$$

The ray trajectory in the vicinity of the singular point θ_s is close to the duct separatrix. At $\theta = \theta_s$, the ray intersects the separatrix.

It is important that in the vicinity of the separatrix the condition for the applicability of the adiabatic approximation is not valid. Indeed, when $\theta \to \theta_s$, the quantity $E(\theta)$ for the ray considered tends to U_m. But the oscillation period $\Theta(\theta)$, according to Eq. (3.26), increases when $E \to U_m$. The oscillation period diverges in a logarithmic manner at $E = U_m$. In this case the condition for the adiabatic invariant conservation [Eq. (3.15)] is broken. Hence, the adiabatic invariant of the ray can undergo a significant change in the vicinity of the singular point θ_s. The special analysis of solution in the vicinity of θ_s is needed. The analysis would allow establishing the rules of the ray trajectory transition through the separatrix.

It is important that the region of violation of adiabatic approximation near the separatrix, $\delta\theta = \theta_s - \theta_0$, is exponentially small with respect to the adiabaticity parameter, as one can easily make sure from Eqs. (3.15), (3.25), and (3.26) (see Appendix 3, Sec. A3.5). Hence,

$$\delta\theta \ll \Theta . \tag{3.43}$$

Thus, the region of adiabaticity violation can be considered to be infinitely small, and the point of discontinuity, $\theta = \theta_s$, can be introduced. At this point the ray adiabatic invariant changes by a jump.

The value of the jump of the adiabatic invariant is easy to find. When a ray leaves duct II for duct III, the invariant I_0 changes from $I_0 = I_{m2}$ (Fig. 3.1b) to $I_{0m} = I_{m2} + I_{m1}$ (Fig. 3.1c). Here, I_{m2} and I_{m1} are the maximum adiabatic invariants of the ducts II and I, I_{0m} is the minimum adiabatic invariant of the duct III. On the contrary, when a ray is trapped into the duct II, namely, when a ray leaves duct III for duct II, I_0 decreases from $I_0 = I_{m1} + I_{m2}$ to $I_0 = I_{m2}$. When the ray leaves duct I or when it is trapped into this duct, the same relations take place. Furthermore, transitions from duct I to duct II and vice versa are possible. Thus, the boundary condition for the invariant at the point of discontinuity, $\theta = \theta_s$, in the general case can be written as follows

$$[I_0]_{\theta_s} = \pm I_m^* . \tag{3.44a}$$

Here, $[I_0]_{\theta_s} = I_0(\theta_s + 0) - I_0(\theta_s - 0)$ is the jump of the invariant in the transition through the separatrix. The sign "+" is taken when the ray leaves a given duct, the sign "−" when the ray is trapped into the duct, I_m^* is the difference of the maximum adiabatic invariants (for instance, in Fig. 3.1 when the wave leaves duct III and is trapped into duct II or vice versa, the quantity I_0 changes for $\pm I_{m1}$). Naturally, the quantity E does not undergo a jump at $\theta = \theta_s$

$$E(\theta_s - 0) = E(\theta_s + 0) . \tag{3.44b}$$

The ray trajectory is also continuous at

$$z(\theta_s-0) = z(\theta_s+0)\,. \tag{3.44c}$$

The boundary conditions [Eq. (3.44)] define the rules of transition through the singular point θ_s. The adiabatic approximation solution together with Eq. (3.44) at the discontinuity point proves to be completely closed and allows the ray trajectories in the ionosphere to be determined in the case of arbitrary slow variations of waveguide channels.

The phase of ray oscillations also undergoes a jump at the discontinuity point θ_s. This is connected with the sharp change in the character of ray oscillations. The magnitude of the phase jump can be determined by constructing the trajectory from θ_s to the nearest turning point $z(\theta_M)$, where $\theta_M > \theta_s$. The following relation must be satisfied

$$\Phi(\theta_M) - \Phi(\theta_m) = \pi\,.$$

Here, $z(\theta_m)$ is another nearest to θ_s turning point, $\theta_m < \theta_s$. This very relation defines the phase jump $[\Phi]_{\theta_s}$ at the discontinuity point.

Note that the presence of sharp jumps at the singular point θ_s is connected with the use of the geometrical – optics approximation. The precise solution of the wave equations in the vicinity of the separatrix shows that here the sharp transition is somewhat spread due to the wave effects. Comparison with the results of the wave theory allows the region of applicability of the boundary conditions [Eq. (3.44)] to be established (see Appendix 3, Sec. A3.5).

Let us also note that such boundary conditions as Eq. (3.44) are true for regions of very rapid change of duct parameters as well. These regions are those of sharp nonadiabaticity where the duct parameters vary significantly over the distances of $\delta\theta$ satisfying Eq. (3.43).

Thus, in the slightly inhomogeneous ionosphere the geometrical – optics equations for the ray trajectories in the adiabatic approximation are integrated by quadratures. This approximation determines all the main trajectory parameters as well as the trajectory itself and the oscillation phase. That is why the approximation proves to be rather efficient when analyzing the very long-range and round-the-world propagation. The adiabatic approximation has an exponential accuracy with respect to the adiabaticity parameter μ [Eq. (3.15)]; at sufficiently small values of μ, the approximation proves to be perfectly rigorous. Actually, in the ionosphere the value of the parameter μ is usually small. However, it is different in various ionospheric regions and significantly depends on the ionospheric model used.

Besides, when calculating the phase of ray oscillation Φ [Eq. (3.28)] and the value of $z(\theta)$ [Eq. (3.25)], the error of adiabatic approximation increases proportionally to the number of oscillations $\mathscr{P}$ [Eq. (3.27)]. That is why the adiabatic approximation provides a higher accuracy when calculating the parameters independent of the phase Φ: Θ, $z_{\min}$, $z_{\max}$, K, A, t, $\bar{z}$, and so on. The comparison of calculation results in the adiabatic approximation and exact numerical solutions of the geometrical – optics equations for the specific ionospheric model is carried out in Sec. 5.3.

3.3 Lateral Deviations of the Radio Wave Trajectories; Bearing Variation

For a zero approximation with respect to the parameter of horizontal inhomogeneity, δ, a ray travels along the great circle arc defined by the initial direction of the wave vector $\boldsymbol{k}_0$. The trajectory departure from the great circle plane, i.e., the trajectory lateral deviation, is described by a first-order perturbation with respect to δ [Eqs. (3.14) and (3.14a)]. Linearizing Eq. (3.14) we obtain

$$\frac{d^2\varphi_1}{d\theta^2}+\frac{\cot\theta}{\sin^2\theta-C^2\cos^2\theta}[2(1+C^2)\sin^2\theta+C^2]\frac{d\varphi_1}{d\theta}$$
$$=\frac{(1+C^2)}{\sin^2\theta-C^2\cos^2\theta}F_\varphi+\frac{C(1+C^2)\sin\theta}{(\sin^2\theta-C^2\cos^2\theta)^{3/2}}F_\theta. \tag{3.45}$$

Here, $\varphi=\varphi_0(\theta)+\varphi_1(\theta)$, where $\varphi_1(\theta)$ is the small deviation from the great circle arc defined by Eq. (3.10). The constant C is also defined by Eqs. (3.10) and (3.11). In a particular case, $C=0$, the direction of the wave vector $\boldsymbol{k}_0$ coincides with the axis $\boldsymbol{e}_\theta$ [in this case $\beta=\varphi_0$, see Eqs. (3.10) and (3.11)], and from Eq. (3.45) we obtain

$$\frac{d^2\varphi_1}{d\theta^2}+2\cot\theta\frac{d\varphi_1}{d\theta}=\frac{1}{\sin^2\theta}F_\varphi. \tag{3.46}$$

Naturally, the same equation follows directly from Eq. (3.14a).

The functions $F_\varphi(\theta,\varphi)$ and $F_\varphi(\theta,\varphi)$ in Eqs. (3.45) and (3.46) are defined by the equations

$$F_\varphi=\frac{1}{2}\frac{\partial\varepsilon}{\partial\varphi}(z,\theta,\varphi)\,,\qquad F_\theta=\frac{1}{2}\frac{\partial\varepsilon}{\partial\theta}(z,\theta,\varphi)\quad\text{at}\quad z=z(\theta,\varphi)\,. \tag{3.47}$$

Here, the function $z(\theta,\varphi)$ describes the ray trajectory in the vertical plane. Taking into account the smallness of δ, one can average the functions F_φ and F_θ with respect to the period of ray oscillations. In the adiabatic approximation this averaging has a simple form

$$\bar{F}_\varphi=\frac{1}{R_0\Theta}\int\limits_{z_{\min}}^{z_{\max}}\frac{\partial\varepsilon}{\partial\varphi}\frac{dz}{(E-U)^{1/2}}\,,$$
$$\bar{F}_\theta=\frac{1}{R_0\Theta}\int\limits_{z_{\min}}^{z_{\max}}\frac{\partial\varepsilon}{\partial\theta}\frac{dz}{(E-U)^{1/2}}\,. \tag{3.48}$$

Further, we shall omit the sign of averaging. It is important that on the unperturbed trajectory along the great circle arc, $\varphi = \varphi_0(\theta)$, the functions F_φ and F_θ depend only on θ

$$F_\varphi(\theta,\varphi)\,|_{\varphi_0(\theta)} = F_\varphi[\theta,\varphi_0(\theta)] = F_{\varphi_0}(\theta)\,, \qquad F_\theta = F_{\theta_0}(\theta)\,. \tag{3.49}$$

Insofar as the actual trajectory deviation from the great circle arc is small (of the order of δ), then with this degree of accuracy the dependence of F on φ in Eq. (3.45) is insignificant and one can substitute the functions F_φ and F_θ by $F_{\varphi_0}(\theta)$ and $F_{\theta_0}(\theta)$ in the right part of Eq. (3.45). Due to this, Eq. (3.45) can be integrated. We have

$$\frac{d\varphi_1}{d\theta} = \frac{\sin\theta(1+C^2)}{[\sin^2\theta - C^2\cos^2\theta]^{3/2}} \int_{\theta_0}^{\theta} \left\{ F_{\varphi_0}(\theta') \frac{[\sin^2\theta' - C\cos^2\theta']^{1/2}}{\sin\theta'} + CF_{\theta_0}(\theta') \right\} d\theta'\,,$$

$$\varphi_1(\theta) = (1+C^2) \int_{\theta_0}^{\theta} \left\{ F_{\varphi_0}(\theta') \frac{[\sin^2\theta' - C^2\cos^2\theta']^{1/2}}{\sin\theta'} + CF_{\theta_0}(\theta') \right\} \times \left[\frac{-\cos\theta}{(\sin^2\theta - C^2\cos^2\theta)^{1/2}} + \frac{\cos\theta'}{(\sin^2\theta' - C^2\cos^2\theta')^{1/2}} \right] d\theta'\,. \tag{3.50}$$

Here, it has been taken into account that at the initial point, (φ_0, θ_0), the perturbation of the great circle trajectory, $\varphi_0(\theta_0)$, is absent, that is, $\varphi_1(\theta_0) = d\varphi_1/d\theta(\theta_0) = 0$. In particular, at $C = 0$ (i.e., when $\boldsymbol{k}_0 \,\|\, \boldsymbol{e}_\theta$), we obtain

$$\frac{d\varphi_1}{d\theta} = \frac{1}{\sin^2\theta} \int_0^{\theta} F_{\varphi_0}(\theta')\, d\theta'\,, \tag{3.51}$$

$$\varphi_1 = \int_0^{\theta} (\cot\theta' - \cot\theta) F_{\varphi_0}(\theta')\, d\theta'\,. \tag{3.52}$$

Here, we also have put $\theta_0 = 0$.

The solution obtained has the divergence at the turning points of the trajectory, $\theta = \theta_m = \arctan C$. This is due to the presence of singularity in coefficients of linearized equations. The singularity occurrence is quite clear: at the turning point, θ reaches its extreme value (maximum or minimum), that is, $d\theta/d\varphi = 0$ at $\theta = \theta_m$. Consequently, here the derivative $d\varphi/d\theta$ tends to infinity, so even small shift, $\Delta\theta$, in the vicinity of the turning point significantly changes the value of φ. To eliminate the divergence, it is necessary to take into account the nonlinear term $(d\varphi/d\theta)^3$ in the vicinity of the singular point. On the other hand, in the narrow region near the singular point, the small right part of Eqs. (3.54) and (3.46) is negligible, so one can directly use the solution of Eqs. (3.10) and (3.11) taking into account only the shift of the turning point (for detail see Gurevich and Tsedilina 1979).

Note that appearance of the singularity in Eqs. (3.45) and (3.46) is connected with the choice of variables: for the main trajectory [in the great circle plane, Eqs. (3.10) and (3.11)] the angle φ is a double – valued function of θ. Consequently, it is convenient sometimes to change the variables and to seek θ as a function of φ rather than φ as a function of θ. The dependence of θ on φ for the main trajectory is unambiguous, and singularities do not appear in this case. Let us take briefly the corresponding solution (Gurevich and Tsedilina 1977, Baranov et al. 1977). According to Eq. (3.5a), $\frac{d\theta}{d\varphi} = \frac{k_\theta \sin\theta}{k_\varphi}$. Differentiating this equation with respect to φ, and taking into account Eq. (3.5), we obtain

$$\frac{d^2\theta}{d\varphi^2} - 2\cot\theta\left(\frac{d\theta}{d\varphi}\right)^2 - \sin\theta\cos\theta = \frac{k_0^2\sin^2\theta}{2k_\varphi^2}\,\frac{\partial\varepsilon}{\partial\theta} - \frac{k_0^2}{2k_\varphi^2}\,\frac{d\theta}{d\varphi}\,\frac{\partial\varepsilon}{\partial\varphi}\,. \tag{3.53}$$

This equation is similar to Eq. (3.7). Neglecting the right part of Eq. (3.53) for a zero approximation with respect to δ, we find that its solution in this approximation is the great circle trajectory [Eqs. (3.10) and (3.11)]: $\theta = \theta_0(\varphi)$. The trajectory departure from the great circle plane is described by first-order perturbations with respect to δ. Putting in Eq. (3.53) $\theta = \theta_0(\varphi) + \theta_1(\varphi)$, and linearizing it with respect to θ_1, we find

$$\begin{aligned} &\frac{d^2\theta_1}{d\varphi^2} + \left[-\cos 2\theta_0 + 2\frac{\sin^2\theta_0 - C^2\cos^2\theta_0}{C^2}\right]\theta_1 \\ &\quad + \frac{4}{C}\cos\theta_0\sqrt{\sin^2\theta_0 - C^2\cos^2\theta_0}\,\frac{d\theta_1}{d\varphi} \\ &= \frac{C^2}{1+C^2}F_\theta + \frac{C\sqrt{\sin^2\theta_0 - C^2\cos^2\theta_0}}{(1+C^2)\sin\theta_0}F_\varphi\,. \end{aligned} \tag{3.54}$$

Here, $\theta_0 = \theta_0(\varphi)$ is defined according to Eq. (3.10), while the functions F_θ and F_φ are described as earlier by Eq. (3.47), with respect to the needed accuracy that $F_\theta = F_\theta[\theta_0(\varphi),\ \varphi]$, $F_\varphi = F_{\varphi_0}(\varphi)$. In particular, if the unperturbed ray travels along the equator $\boldsymbol{k}_0 \,\|\, \boldsymbol{e}_\varphi$, $\theta_0 = \pi/2$, $d\theta_0/d\varphi = 0$, then Eq. (3.54) takes the simple form

$$\frac{d^2\theta_1}{d\varphi^2} + \theta_1 = F_{\theta_0}(\varphi)\,, \qquad F_{\theta_0} = \left(\frac{1}{2}\,\frac{\overline{\partial\varepsilon}}{\partial\theta}\right)_{\theta=\pi/2}. \tag{3.55}$$

Its solution is

$$\theta_1 = \int_0^\varphi F_{\theta_0}(\varphi')\sin(\varphi - \varphi')\,d\varphi'\,. \tag{3.56}$$

One can also obtain the general solution of the linear Eq. (3.54) which has, however, a rather complicated form, so we shall not present it here (see Gurevich and Tsedilina 1977, Baranov et al. 1977).

Equation (3.56) can be obtained directly from Eq. (3.52) by means of the double rotation of the coordinate system r, θ, φ; first, with respect to the axis x ($x_{\theta,\varphi} = r\sin\theta\cos\varphi$) through an angle $\psi_1 = \pi/2$, and then the second rotation of the system x, y, z through an angle $\psi_2 = \pi/2$ with respect to the axis z (for detail see Manaenkova and Tsedilina 1981, 1982). Note that Eqs. (3.51) and (3.52) are derived for the ray, travelling along a "meridian" of the system r, θ, φ, while Eqs. (3.55) and (3.56) are derived for the "equator".

Using Eq. (3.56), it is easy to derive the expression for the bearing deviation, β_0, (at the reception point) of a wave coming from a given direction, for instance, from the direction of the transmitting station (see Fig. 3.1d). Let OA be the great circle arc on a sphere between the points O and A, the spherical coordinate system r, θ, φ being situated so that its equator ($\theta = \pi/2$) be coincident with the arc OA. Then a ray transmitted from the point O along the arc OA (the ray 1 in Fig. 3.1d) due to the lateral deviation will come into the point B, but not into the point A. In this case the ray will have the lateral deviation θ_1 with respect to point A, and with respect to the great circle arc OB, the ray will have the azimuthal angle (or azimuth deviation) β_0. The initial azimuth of the ray will change by an angle of ψ with respect to the unperturbed trajectory OA. According to Fig. 3.1d

$$\beta_0 = x + \alpha - \pi/2$$
$$\alpha = \frac{d\theta_1}{d\varphi} \tag{3.57}$$

where φ is the length of the arc OA. Next, considering the spherical triangle ΔOAB, we have

$$\sin x = \frac{\sin\varphi}{\sin y} = \frac{\sin\varphi}{\sqrt{1-\cos^2\varphi\cdot\cos^2\theta_1}} = \frac{1}{\sqrt{1+\theta_1^2\cot^2\varphi}}\,. \tag{3.58}$$

In Eq. (3.58) it has been taken into account that $\theta_1 \ll 1$. From Eqs. (3.57) and (3.58), we have

$$\beta_0 = \frac{d\theta_1}{d\varphi} - \frac{\pi}{2} + \arcsin\left(\frac{1}{\sqrt{1+\theta_1^2\cot^2\varphi}}\right). \tag{3.59}$$

Taking into account Eq. (3.56), we rewrite Eq. (3.59) in the form

$$\beta_0 = \int_0^\varphi \cos(\varphi-\varphi')F_{\theta_0}(\varphi')\,d\varphi' - \frac{\pi}{2} + \arcsin\left(\frac{1}{\sqrt{1+\theta_1^2\cot^2\varphi}}\right). \tag{3.60}$$

In the vicinity of the antipode, $\varphi \approx \pi$, and round-the-world point, $\varphi \approx 2\pi$, the latter term in Eq. (3.60) is equal to π or 0, hence, $\beta_0 \approx \pm\pi/2$. At all the other values of φ, when $\theta_1^2 \cot^2\varphi \ll 1$, from Eq. (3.60) we have

$$\beta_0 = \int_0^{\varphi} \frac{\sin\varphi'}{\sin\varphi} F_{\theta_0}(\varphi')\, d\varphi' \, . \tag{3.61}$$

From the triangle OAB one can also obtain the expression for deviation of the initial azimuth at the transmission point ψ (Baranov et al. 1977). Taking into account that $\theta_1 \ll 1$, we have

$$\sin\psi = \theta_1/\sin y$$

or

$$\psi = \arcsin\left(\frac{\theta_1}{\sin\varphi\sqrt{1+\theta_1^2\cot^2\varphi}}\right) . \tag{3.62}$$

In the vicinity of the antipode, $\varphi \approx \pi$, we have $\psi \approx \pm\pi/2$, i.e., here the quantity ψ has a discontinuity.

3.4 Capture of Radio Waves into the Interlayer Ionospheric Duct and Their Escape from the Duct

The interlayer ionospheric wave duct is of great interest for very long-distance propagation (for detail see Chaps. 4 and 5). Let us consider the capture of energy emitted from the Earth into such a channel. The capture takes place due to horizontal variation of the duct, i.e., due to horizontal refraction of radio waves. First, let us consider the simplest case, when the capture occurs immediately after the first ray reflection from the ionosphere (Gurevich and Tsedilina 1973).

The Conditions for Capture into the Interlayer Channel FE. The capture into the interlayer channel is impossible in the spherically – symmetric ionosphere in the geometrical – optics approximation. However, the actual ionosphere is inhomogeneous in the horizontal direction. In other words, the dielectric constant ε (or the potential U) varies along the ray path, $U = U(z, \theta)$. In this case, if the potential well "deepens" effectively in the propagation direction, then rays, transmitted from the Earth and close to the critical ray (Fig. 3.2) can be trapped.

For the critical ray, $E \approx E_m = U_m$. The value $E = E_m$ for the critical ray is shown by the broken line in Fig. 3.2a. The well deepened effectively along the propagation path (Fig. 3.2b, $\theta_2 > \theta_1$) and the ray appeared to be trapped in the interlayer duct.

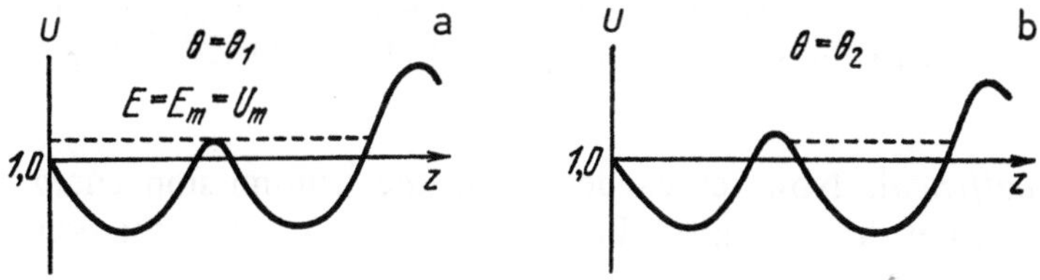

Fig. 3.2. Capture of the critical ray ($E = E_m = U_m$) into the duct when the well is deepened along the path

Hence, the conditions for capture are as follows

$$\alpha_0 \approx \alpha_{0m}, \qquad \frac{dI_m}{d\theta} > 0 . \tag{3.63}$$

Here, α_{0m} is the emission angle of the critical ray (at a given frequency f), reflecting from the E-layer (when $\alpha > \alpha_{0m}$, the ray is reflected from the F-layer). The angle α_{0m} is defined by Eq. (2.8)

$$\cos^2\alpha_{0m} \approx 1 - \alpha_{0m}^2 = -E_m = -U_m = \left[\varepsilon(z, f) + \frac{2z}{R_0}\right]_{\min} . \tag{3.64}$$

Here, U_m is the maximum value of the function $U(z)$. U_m defines the boundary of an ionospheric duct for a given frequency f.

If the emission angle, $\alpha_0 = \alpha_{0m}$, is fixed, while the frequency f varies, then Eq. (3.64) defines the wave frequency f_m, in the vicinity of which the capture takes place ($f \approx f_m$). Note that Eq. $U_m = E_m = -1$, according to Eq. (3.64), determines the maximum usable frequency for the E-layer, $f_{\text{MUF-}E}$, i.e., the frequency at which only the horizontal ray, $\alpha_{0m} = 0$, is reflected from the E-layer. From Eq. (3.64) it is seen that direct capture into the interlayer channel is possible only at $E_m = U_m \geqslant -1$, i.e., for frequencies $f \lesssim f_{\text{MUF-}E}$. The critical frequency of capture, f_m, decreases with increasing angle α_{0m}.

The physical meaning of the second Eq. (3.63) is also quite clear. In fact, it has been explained already in Sec. 3.2 when analyzing the conditions of transition through the separatrix. So far as the maximum adiabatic invariant of a duct varies along the path, $I_m = I_m(\theta)$, while the adiabatic invariant of a given ray is conserved, then an increase of I_m with θ is necessary for capture. Only in this case the ray which was critical at $\theta = \theta_0$, i.e., which had the invariant $I_m(\theta_0)$ after the transition through the separatrix would occur at $\theta > \theta_0$ inside the duct, $I_m(\theta) > I_m(\theta_0)$. Clearly, not only the critical ray is trapped, but the rays in a certain region of the angles $\Delta\alpha$ as well. The width of the capture region increases with $dI_m/d\theta$. Note that at sufficiently rapid variation of a duct, i.e., when the value of $dI_m/d\theta$ is quite great, the width of the capture region, $\Delta\alpha$, in the vicinity of critical angle α_{0m}, as well as the width of frequency range, Δf, in the vicinity of critical frequency f_m can be significant.

Specific characteristics of conditions for radio wave capture into the *FE* duct for actual models of the inhomogeneous ionosphere will be studied in Chapter 4.

The Capture Coefficient. Now let us determine the transmission energy trapped in the duct. Let the transmitter be situated on the Earth at the point $\theta = 0$, $z = 0$

$$W = W_0 D(\alpha_0, \varphi_0) . \tag{3.65}$$

Here, W_0 is the total power of the transmitter, $D(\alpha_0, \varphi_0)$ is the directivity diagram of transmission with respect to vertical, α_0, and azimuthal, φ_0, angles

$$\int D(\alpha_0, \varphi_0) \cos \alpha_0 \, d\alpha_0 \, d\varphi_0 \equiv 1 .$$

The emission angle of the critical ray for a given azimuth, φ_0, on the Earth is $\alpha_{0m}(\varphi)$. Accordingly, $E_m = -1 + \alpha_{0m}^2$. When $\alpha < \alpha_{0m}$, the rays are reflected from the *E*-layer and miss the duct. When $\alpha > \alpha_{0m}$, they pass at $\theta \approx \theta_m$ through the region of the barrier maximum $U_m(\theta_m) = E_m$ and reach the duct. Further, the rays are reflected from the *F*-layer and again come back to the barrier maximum at the point $\theta \approx \theta_m + \Theta$ (Fig. 3.3). In this case, if $E(\theta_m + \Theta) < U_m(\theta_m + \Theta)$, then the ray is reflected from the barrier and appears to be trapped in the duct. Thus, the conditions for capture are as follows

$$E(\theta_m) \geqslant U_m(\theta_m) ; \qquad E(\theta_m + \Theta) \leqslant U_m(\theta_m + \Theta) , \tag{3.66}$$

i.e., all the rays between the critical ray 1, for which $E_{k_1}(\theta_m) = U_m(\theta_m) = E_m$, and the critical ray 2, for which $E_{k_1}(\theta_m + \Theta) = U_m(\theta_m + \Theta)$, are trapped in the duct.

Let us take into account that

$$E(\theta_m + \Theta) = E(\theta_m) + \int_{\theta_m}^{\theta_m + \Theta} \frac{dE}{d\theta} \, d\theta ;$$

$$U_m(\theta_m + \Theta) = U_m(\theta_m) + \int_{\theta_m}^{\theta_m + \Theta} \frac{dU_m}{d\theta} \, d\theta ,$$

and, consequently,

$$U_m(\theta_m + \Theta) - E(\theta_m + \Theta) = U_m(\theta_m) - E(\theta_m) + \int_{\theta_m}^{\theta_m + \Theta} \left(\frac{dU_m}{d\theta} - \frac{dE}{d\theta} \right) d\theta .$$

Hence, taking into account that for the critical ray 2, $E_{k_1}(\theta_m + \Theta) = U_m(\theta_m + \Theta)$, we find that

$$E_{k_2}(\theta_m) = U_m(\theta_m) + \Delta = E_{k_1}(\theta_m) + \Delta ,$$
$$\Delta = \int_{\theta_m}^{\theta_m + \Theta} \left(\frac{dU_m}{d\theta} - \frac{dE}{d\theta} \right) d\theta . \tag{3.67}$$

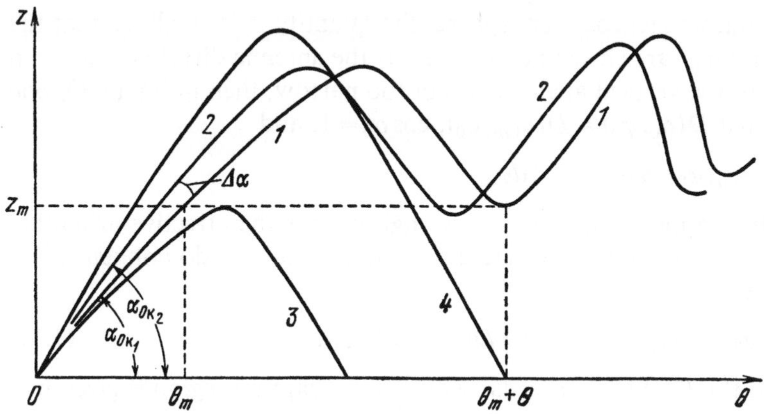

Fig. 3.3. Capture of the critical rays 1 and 2 into the duct *FE*

Next, assuming that the variation of $U(\theta)$ and $U_m(\theta)$ is quite smooth, we go to integration over z from integration over θ [Eq. (2.27) is used: $d\theta = dz/R_0\sqrt{E-U}$)]. Then, instead of Eq. (3.67), we obtain

$$\Delta = \frac{2}{R_0}\int_{z_{\min}}^{z_{\max}} \frac{\dfrac{dU_m}{d\theta} - \dfrac{dE}{d\theta}}{\sqrt{E-U}}\,dz\,.$$

Next, according to Eq. (3.16a), $\dfrac{dE}{d\theta} = \dfrac{\partial U}{\partial\theta}$ and $E \approx U_m$, and we finally obtain

$$\Delta = \frac{2}{R_0}\int_{z_{\min}}^{z_{\max}} \frac{\dfrac{dU_m}{d\theta} - \dfrac{\partial U}{\partial\theta}}{\sqrt{U_m-U}}\,dz = \frac{dI_m}{d\theta}\,. \tag{3.68}$$

Here, $I_m(\theta)$ is the maximum adiabatic invariant of the duct [Eq. (3.40)]. Let us emphasize that the integral in Eq. (3.68) always converges by virtue of $\dfrac{dU_m}{d\theta} - \dfrac{\partial U}{\partial\theta} \to 0$ when $U \to U_m$, i.e., $z \to z_{\min}$.

Thus, we have found the quantities $E_{k_1}(\theta_m)$ and $E_{k_2}(\theta_m)$ for the critical rays 1 and 2 defining the boundaries of the capture region. The parameters E_{k_1} and E_{k_2} are related unambiguously to the emission angles α_{0k_1} and α_{0k_2} [Eq. (2.8); Fig. 3.3]. Taking into account that $\alpha_{0k_1} = \alpha_{0m}$, we have

$$\alpha_{0k_2} = \sqrt{\alpha_{0m}^2 + E_{k_2} - E_{k_1}} = \sqrt{\alpha_{0m}^2 + \Delta}\,. \tag{3.69}$$

This equation determines the angular width of the capture region with respect to the vertical angles α_0 in a given direction of the horizontal azimuth φ_0.

The capture coefficient, i.e., the ratio of the transmission power trapped into the duct to the total transmitting power is given by

$$G = \int d\varphi_0 \int_{\alpha_{0m}}^{\alpha_{0k_2}} D(\alpha_0, \varphi_0)\cos\alpha_0 d\alpha_0\,. \tag{3.70}$$

In a slightly inhomogeneous ionosphere, the quantity Δ is small, so that the angles α_{0k_2} and α_{0k_1} are close to each other. If the antenna directivity diagram with respect to the vertical angles α_0 is not too narrow, then in Eq. (3.70) one can assume that $D(\alpha_0, \varphi_0) \approx D(\alpha_{0m}, \varphi_0)$, $\cos\alpha_0 \approx 1$, and

$$G = \int D(\alpha_{0m}, \varphi_0)(\alpha_{0k_2} - \alpha_{0m})\, d\varphi_0 .$$

Finally, if the trapping conditions vary slightly over the effective azimuthal width of the antenna directivity diagram, then α_{0k_2} and α_{0m} do not depend on φ_0, and hence,

$$G = D_m(\alpha_{0k_2} - \alpha_{0m}) , \qquad D_m = \int D(\alpha_{0m}, \varphi_0)\, d\varphi_0 . \tag{3.71}$$

According to Eq. (3.69), the angular width of the capture region is given by

$$\Delta\alpha = \alpha_{0k_2} - \alpha_{0m} = \sqrt{\alpha_{0m}^2 + \Delta} - \alpha_{0m} .$$

Hence, when $\alpha_{0m}^2 \gg \Delta = dI_m/d\theta$, we have

$$\Delta\alpha = \frac{\Delta}{2\,\alpha_{0m}} , \qquad \Delta = \frac{dI_m}{d\theta} . \tag{3.72}$$

When $\alpha_{0m} \to 0$, the angular width increases. It has its maximum at $\alpha_{0m} \approx 0$, i.e., when $f \to f_{\text{MUF-}E}$ or $f \to f_{\text{MUF-}F_1}$. In this case

$$(\Delta\alpha)_{\max} = \sqrt{\Delta} = \sqrt{dI_m/d\theta} . \tag{3.73}$$

The capture coefficient [Eq. (3.71)] is given by

$$G = \Delta\alpha/\bar{\alpha} , \tag{3.74}$$

where $\bar{\alpha} = 1/D_m$ is the characteristic width of the antenna directivity diagram with respect to the vertical angles α. In the ionosphere, $\Delta\alpha$ can be equal to $0.1 - 5°$ (see Secs. 4.1 and 5.3). Accordingly, G can reach the value ~ 1 under the optimum conditions, when the directivity diagram is sufficiently narrow. On the average, assuming that $\bar{\alpha} \approx 10°$, we have $G \sim 0.01 - 0.5$.

Let us note the specific case when the MUF for the F-layer is close to the MUF for the E-layer (see Fig. 3.4). In this case, the capture coefficient can increase in comparison with that in Eqs. (3.71) – (3.74), since it is important that the turning point of a ray is situated in the vicinity of the F-layer maximum (the Pedersen ray). The role of the Pedersen ray in long-distance HF propagation has been discussed in the papers (Sazhin and Tinin 1975a, b; Tinin 1973).

The Capture Induced by Large-Scale Inhomogeneities. Radio wave capture into the interlayer wave duct due to the regular horizontal gradient of the ionosphere was considered above. In this case the ionosphere varies smoothly within the period Θ of ray oscillations. Under real conditions, large-scale ionospheric inhomogeneities can play a significant role in addition to the smooth regular gradient. Such inhomogeneities are usually formed because of the gravity or tidal waves, and have the horizontal scale L of about 300 to

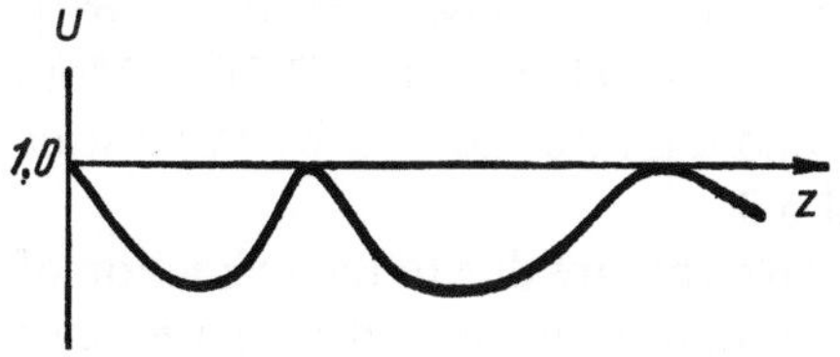

Fig. 3.4. Form of the potential U when MUF-E = MUF-F

600 km (Gershman 1974, Drobzhev et al. 1975, Francis 1975). The angular size of the inhomogeneities, $\Delta\theta \approx L/R_0$, is less than the period Θ of ray oscillation.

Consider the radio wave capture into the interlayer duct, caused by refraction on the large-scale inhomogeneities. Let a given ray reach the barrier maximum, U_m at the point z_m; θ_m, being characterized at this point by the level $E > U_m$. The ray will be trapped in the duct if, after reflection from the F-layer and return to the barrier, the level $E(\theta_m + \Theta)$ appears to be below the barrier maximum $U_m(\theta_m + \Theta)$. Consequently, the capture conditions, as before, have the form of Eq. (3.66):

$$E(\theta_m, z_m) > U_m(\theta_m)\,, \qquad E(\theta_m + \Theta, z_{m1}) \leqslant U_m(\theta_m + \Theta)\,.$$

Here, z_{m1} is the point of the barrier maximum at $\theta = \theta_m + \Theta$. Let us take into account that when the ray propagates in the duct, according to Eq. (3.16a), we have

$$\frac{\partial E}{\partial \theta} = \frac{\partial U}{\partial \theta}(\theta, z(\theta, E))\,. \tag{3.75}$$

Here, $z(\theta, E)$ is the trajectory of the ray which had the level $E(\theta_m)$ at the point (z_m, θ_m). Usually the trajectory perturbation due to the inhomogeneity is weak. This allows to neglect the variation of E at the right-hand side of Eq. (3.75), i.e., here we assume that $E = E(\theta_m) = E_m$. Then

$$E(\theta_m + \Theta, z_{m1}) = E(\theta_m, z_m) + \int_{\theta_m}^{\theta_m + \Theta} \frac{\partial U}{\partial \theta}(\theta, z(\theta, z_m))\, d\theta\,.$$

Next, repeating the reasoning of the previous section, we find that all the rays between the critical rays 1 and 2 for which

$$\begin{aligned} \Delta &= E_2(z_m, \theta_m) - E_1(z_m, \theta_m) \\ &= U_m(\theta_m + \Theta) - U_m(\theta_m) - \int_{\theta_m}^{\theta_m + \Theta} \frac{\partial U}{\partial \theta}(\theta, z(\theta, E_2))\, d\theta \end{aligned} \tag{3.76}$$

are trapped in the duct.

In the case of the smooth horizontal gradient, taking into account Eqs. (2.27) and (3.67), we obtain Eq. (3.76) from Eq. (3.68).

Thus, radio wave capture into the duct due to refraction by large-scale inhomogeneities is described, as above, by Eqs. (3.69)–(3.71), but now the parameter Δ is not only related to variation of the maximum invariant [Eq. (3.68)]. This parameter is expressed by a more general relation [Eq. (3.76)]. More detailed consideration of the refraction and capture induced by large-scale inhomogeneities is given in Secs. 6.2 and 6.3.

The just considered process of radio wave capture due to the refraction of radio beams by the smooth regular horizontal gradient and by large-scale ionospheric inhomogeneities appears to be the most effective mechanism of capture into the duct. However, other capture mechanisms can be also important. They are: radio wave tunneling (see Appendix 1.3), scattering by irregularities (Chap. 6), and nonlinear effects (Chap. 7).

Capture into the Interlayer Duct F_1F_2. If the F_1-layer is quite pronounced, then the second interlayer duct F_1F_2 situated above the maximum of F_1-layer (see Sec. 4.1) appears. The conditions for capture into the duct F_1F_2 are the same [Eq. (3.63)] as in the case of capture into the FE duct. In this case, however, α_{0m} is the critical angle for the maximum of the F_1-layer (when $\alpha_0 < \alpha_{0m}$, the ray is reflected from the F_1-layer, when $\alpha > \alpha_{0m}$ it is reflected from the F_2-layer) and I_m is the maximum adiabatic invariant of the F_1F_2 duct. The coefficient of capture into the F_1F_2 duct is also described by Eqs. (3.70) to (3.74) and (3.76) with respective substitution of the angle α_{0m} and the adiabatic invariant I_m. The capture is most effective for the frequencies $f \approx f_{\text{MUF-}F_1}$, where $f_{\text{MUF-}F_1}$ is the maximum usable frequency for the F_1-layer.

Capture after Several Hops. In the interlayer duct, radio waves undergo comparatively small absorption. Therefore, in the presence of strong absorption in the lower layers of the ionosphere (for example, in the daylight zone), the most advantageous is the capture immediately after the first reflection of the ray from the ionosphere. On the other hand, the capture itself is most effective in the regions of sharp horizontal variation of the ionosphere in the direction of more rapid growth of the adiabatic invariant of the duct. That is why in the case of weak absorption in lower ionospheric layers (the night zone), a radio wave can be trapped after several hops as well, i.e., the transmitter in the night zone is separated by a distance of several hops from the region of effective capture (the latter, usually, is the terminator region). In this case, the coefficient of capture into the duct is described as it was earlier by Eqs. (3.70) to (3.74) and (3.76), but now the critical angle α_{0m} is defined with allowance for several successive reflections of a ray from the Earth and the F-layer.

The Departure of Radio Waves from the Duct. The process of departure to the Earth of radio waves propagating in the interlayer duct is similar to the process of capture considered above. The radio waves trapped in the upper levels of the duct, $(E \sim U_m)$, come out from the duct. The observer on the Earth must see the radio waves arriving at an angle of α_{0m} to the horizon [or somewhat greater than α_{0m}, Eq. (3.69)], where α_{0m} is defined according to Eq. (3.64). If the arrival angle of radio waves, α_{0m}, is fixed, then their frequency must be close to f_m as defined by Eq. (3.64) (or somewhat higher than f_m). The condition for radio wave departure from the duct due to the regular refraction is given by

$$\frac{dI_m}{d\theta} < 0 . \tag{3.77}$$

The amount of energy leaving the duct is proportional to $|dI_m/d\theta|$ and to the energy density on the upper levels of the duct. Refraction by large-scale inhomogeneities can also significantly affect the departure of radio waves.

Note that Eq. (3.69) together with Eq. (3.77) show that optimum conditions for observation of RTW signals, propagating in the interlayer ionospheric duct, take place when the transmitter and receiver are situated in the region of $I_m(\theta)$ minimum.

In conclusion, let us discuss the problem of long-distance radio communications through the interlayer duct. Let the transmission point, θ_0, and the reception point, θ_1, be situated on the Earth. An important parameter characterizing the possibility for reception of long-distance radio signals is the phase of ray oscillations [Eq. (3.28)]. A radio wave must make an integer number of oscillations between the transmission and reception points [Eq. (3.29)]. The fact that a ray beam is emitted at various vertical angles, α_0, from the transmitter favors the fulfillment of the latter condition [Eq. (3.29)] when propagating in one duct Earth – ionosphere. Only the rays for which the relation, Eq. (3.29), is satisfied, come to the reception point. The relation defines several modes, i.e., several rays, emitted from the transmission point at various angles $\alpha_{01}, \alpha_{02}, \ldots \alpha_{0k}$ and coming to the reception point.

If a wave propagates in the interlayer duct, then the additional conditions, $\alpha_0 \approx \alpha_{0m}$ [Eq. (3.63)] must also be satisfied at the points of entrance and departure from the interlayer duct. The fact that the conditions are true for a certain angular range, $\Delta\alpha$ [Eq. (3.69)] in the vicinity of α_{0m}, favors the fulfillment of these conditions. If among the angles $\alpha_{01}, \ldots \alpha_{0k}$ there are angles lying in the interval α_{0m}, $\alpha_{0m} + \Delta\alpha$, then the corresponding modes can propagate in the interlayer duct. Rapid spreading of the oscillation phase over long-distance paths due to scattering (see Sec. 6.2) can favor radio wave departure from the duct. The scattering can also help the capture into the duct (Sec. 6.1). The azimuthal focusing of rays at round-the-world and antipodal points favors round-the-world and antipodal propagation. On the whole, Eq. (3.29) together with the Eqs. (3.63) and (3.77) are, of course, far from being always fulfilled. That is the reason for irregularity of signals, propagating over the long-distance Earth – Earth paths in the interlayer duct.

The propagation in the interlayer duct is more regular over the Earth – ionosphere and ionosphere – Earth paths, so this duct is used for HF communications with satellites and for long-distance HF radar. Note that artificial irregularities created intentionally when modifying the ionosphere by intense radio waves (see Chap. 7) can significantly favor the improvement of

HF wave capture into the interlayer duct and their departure from the duct. The conditions of propagation and long-distance radio communications on various paths are discussed in more detail in Chaps. 4 and 5, where the calculations are carried out using specific ionospheric models.

4 Regularities of Very Long-Distance Radio Wave Propagation in the Ionosphere

Simple adiabatic approximation enabling one to describe long-distance radio wave propagation in the horizontally – inhomogeneous ionosphere has been constructed above in a general form.

Within the framework of the approximation one can establish the presence of ionospheric wave ducts (IWD) making long-distance or around-the-world propagation possible. One can also determine the optimum conditions for radio wave capture and departure from ducts, and to find the absorption and the main characteristics of ray trajectories on various radio paths. When determining them we are usually not concerned with the precise ray trajectory, although we can determine all its major features. The number of parameters defining the radio wave propagation is reduced in the adiabatic approximation, hence, the approximation proves to be rather efficient when analyzing the global properties of very long-distance and round-the-world propagation.

In this chapter we shall describe the application of the adiabatic approximation method (the method of the adiabatic invariant) for investigation of the global properties of HF propagation. All the specific calculations are performed with the available analytical model of electron density and electron collision frequency distribution in a three-dimensional inhomogeneous ionosphere (see Appendix 2).

This model is constructed for the quiet ionosphere under equinoctial conditions, thus, the specific results of calculations are related to these conditions.

4.1 Investigations of the Global Characteristics of Long-Distance Propagation

Classification of the Ionospheric Wave Ducts. Ray trajectories in the inhomogeneous ionosphere are similar to trajectories of particles, oscillating along the height z in the waveguide channels which are formed at the minima of the potential $U(z) = -\varepsilon'(z, \theta, \varphi)$. The value of effective dielectric constant ε', according to Eq. (2.7), is equal to

$$\varepsilon'(z, \theta, \varphi, f) = \left[1 - 8.06 \cdot 10^{-5} \frac{N_e(z, \theta, \varphi)}{f^2}\right]\left(1 + \frac{2z}{R_0}\right). \tag{4.1}$$

Here, f is in MHz, $N_e(z, \theta, \varphi)$ is the electron number in cm^{-3}. The value of ε' for wave frequency f depends significantly on the variation of the electron density N_e along the path, i.e., not only on the ionospheric profile along the height z, but also on its variation in the horizontal direction (i.e., versus the latitude θ and the longitude φ).

Figure 4.1a shows a typical shape of the potential $U = -\varepsilon'$ for the smoothed model of the daytime $N_e(z)$ profile (Fig. 4.1b). There are two peaks on the distribution curve of N_e defining E and F layers of the ionosphere. There is also a "step" in the D-layer, namely, a sharp curve bend in $N_e(z)$ distribution at the height of 75 km. In Fig. 4.1a the curve $U = -\varepsilon'(z)$ can be seen to have two maxima, A and B, and two minima, P and Q, forming three characteristic regions of $\varepsilon'(z)$ variation – I, II, III. We shall term them E, FE, and F ducts, respectively.

The *duct F* is situated below the maximum of the F-layer of the ionosphere. Radio waves trapped in the waveguide usually propagate being reflected successively from the F-layer and from the Earth (trajectories 1a, 1b, and 2a in Fig. 4.1a). In this case hop propagation is said to take place through the F-layer. The emission angle α_0, read off from the horizontal plane on the Earth, is related to the $\varepsilon'(z)$ at the reflection point by Snell's law [Eq. (2.8)].

$$\cos^2\alpha_0 = \varepsilon', \qquad \alpha_0 \approx \sqrt{1-\varepsilon'}\,. \tag{4.2}$$

It is evident that radio waves emitted from the Earth at the angles

$$\alpha_E^m \leqslant \alpha_0 \leqslant \alpha_F^m, \qquad \alpha_E^m = \arccos\sqrt{\varepsilon'_{\min E}}, \qquad \alpha_F^m = \arccos\sqrt{\varepsilon'_{\min F}} \tag{4.3}$$

propagate in the duct F in hops.

Here, $\varepsilon'_{\min E}$ is the value of $\varepsilon'(z)$ at the point A, $\varepsilon'_{\min F}$ is the value of $\varepsilon'(z)$ at the point B (see Fig. 4.1a). Thus, the greatest angle α_F for capture into the duct F is given by

$$\alpha_{\max F} = \alpha_F^m = \arccos\sqrt{\varepsilon'_{\min F}}\,. \tag{4.4}$$

Note that the propagation in the duct F is of a hop type, if $\varepsilon'_{\min E} < 1$, i.e., if the wave frequency $f \geqslant f_{\text{MUF-}E}$.

Otherwise, some ricochet trajectories, separated from the Earth, appear in the duct F. The region of such trajectories is shaded in Fig. 4.1c.

If, in addition, $\varepsilon'_{\min F} > 1$, i.e., $f > f_{\text{MUF-}F}$, then the duct F is completely isolated and none of its trajectories touch the surface of the Earth (Curves 4 and 5 in Fig. 4.1c).

The *duct E* is located between the Earth and the height $z_{\max E}$ (the latter is approximately equal to the height of the E-layer maximum). The radio waves trapped in the duct E can propagate in hops, being reflected from the E-layer and from the Earth (this is the hop propagation through the E-layer, see trajectory 2b in Fig. 4.1a). It is true for $f < f_{\text{MUF-}E}$, i.e., $\varepsilon'_{\min E} < 1$ (see Fig. 4.1c, curve 1). In this case the emission angle α_0 satisfies the unequality

$$0 \leqslant \alpha_0 \leqslant \arccos\sqrt{\varepsilon'_{\min E}}\,. \tag{4.5}$$

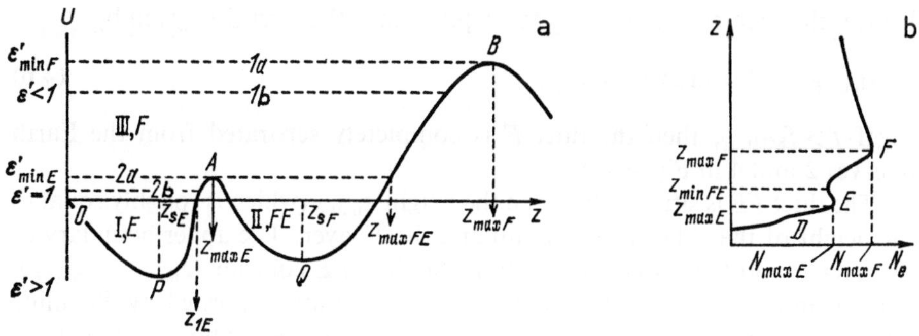

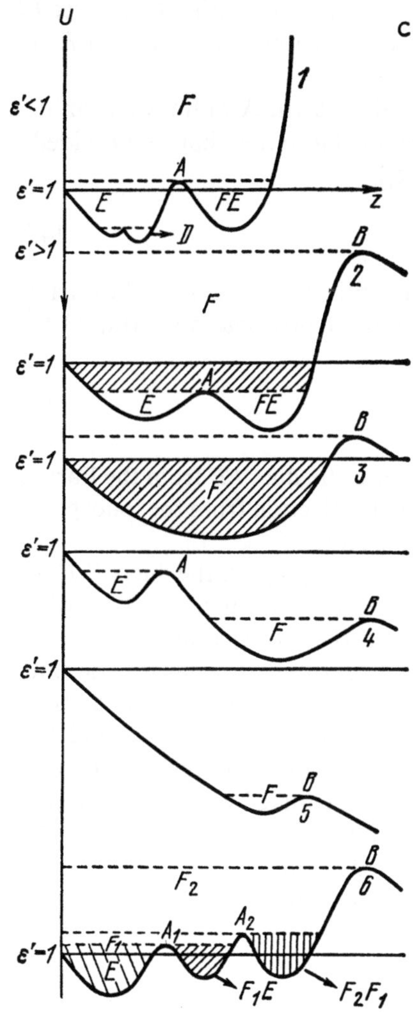

Fig. 4.1 a – c. A typical shape of the potential $U = -\varepsilon'(z)$ (**a**) and $N_e(z)$ profile in daytime (**b**). **c**. Form of the potential $U = -\varepsilon'(z)$ for different ionospheric conditions

Hence, the greatest value of α_E for capture into the duct E is given by

$$\alpha_{\max E} = \alpha_E^m = \arccos \sqrt{\varepsilon'_{\min E}} \,. \tag{4.6}$$

If $f > f_{\text{MUF-}E}$, then the duct E is completely separated from the Earth (Curves 2 and 4 in Fig. 4.1c).

The duct FE is situated between the height $z_{\max E}$ (which is roughly equal to the height of the E-layer maximum) and the F-layer. The upper boundary of the duct E and the lower boundary of the duct FE coincide ($z_{\max E} = z_{\min FE}$). Bear in mind that we shall follow the classification suggested by Tsedilina (1974). According to this classification, we shall speak about the duct FE coupled with the duct E only. But the duct E can exist without the duct FE (Curve 4 in Fig. 4.1c). In this case, the channel, existing above the duct E and well separated from it, will be called the duct F.

The interlayer duct FE is isolated from the Earth. A radio wave emitted from the Earth on the frequency f can be trapped into this channel provided it is emitted at an angle close to α_E^m [Eq. (3.64)]

$$\alpha_{FE} = \alpha_E^m = \arccos \sqrt{\varepsilon'_{\min E}} \,. \tag{4.7}$$

For a radio wave to be trapped, the total adiabatic invariant I_{FE} of the duct FE is bound to increase in the direction of radio wave propagation, θ [Eq. (3.63)]

$$\frac{\Delta I_{FE}}{\Delta \theta} > 0 \,. \tag{4.8}$$

In this case the invariant I_0 of ray entry into the duct FE from the Earth is equal to the total duct invariant I_{FE} at the first reflection from the ionosphere, i.e., 700 – 1000 km apart from the emission point.

When the F-layer is bifurcated into the F_2 and F_1 layers, the ducts F_1E, F_1, and F_2F_1 appear in addition to the ducts E, F_2 (see Curve 6 in Fig. 4.1c, where the ducts F_1E and F_2F_1 are shown in different types of shading).

The duct F_1 is situated below the maximum of the F_1 – layer. The interlayer ducts F_2F_1 and F_1E are situated between the F_1 and F_2 layers and also F_1 and E layers, respectively. As a rule, the volumes of the ducts F_1E and F_2F_1 are small. Note that some other shallow channels appear in the duct E at low frequencies in the daytime in the region of the "step-like" variation of electron density in the D-layer at the heights $z \sim 50 - 80$ km (see Curve 1 in Fig. 4.1c). Further, we shall mainly analyze the ducts E, FE, and F.

The fundamental characteristic of a wave duct is its total adiabatic invariant I_m [Eq. (3.40)]

$$\begin{aligned} I_m(f, \theta, \varphi) &= \frac{4}{R_0} \int_{z_{\min}}^{z_{\max}} [U_{\max} - U(z)]^{1/2} dz \\ &= \frac{4}{R_0} \int_{z_{\min}}^{z_{\max}} [\varepsilon'(z, \theta, \varphi, f) - \varepsilon'_{\min}(z_{\max}, \theta, \varphi, f)]^{1/2} dz \,, \end{aligned} \tag{4.9}$$

where ε'_{min} is the minimum value of ε' in the duct. It should be remembered that $\varepsilon' = -U$, consequently, the minima of ε' correspond to the maxima of the potential U. For the duct F we have $\varepsilon'_{min} = \varepsilon'_{min F}(z_{max F})$, and for the ducts E and FE we have $\varepsilon'_{min} = \varepsilon'_{min E}(z_{max E}) = \varepsilon'_{min E}(z_{max FE})$ (see Fig. 4.1a). We shall denote by I_E, I_{FE}, I_F the total adiabatic invariants of the ducts E, FE, F, respectively. When propagating in the duct a wave conserves the value of the adiabatic invariant I (Sec. 3.2). The wave is confined in the duct provided the total adiabatic invariant I_m of the duct is equal to or exceeds I along the path. But if the total duct invariant becomes less than the wave invariant ($I_m < I$), then the wave leaves the duct for another one or leaves the ionosphere altogether. The adiabatic invariant undergoes a jump, according to Eq. (3.44a), at the points of ray departure from the duct or its entry into the duct. For example, at the point A (Fig. 4.1 a, c) when the wave leaves the duct FE, the adiabatic invariant varies from $I = I_{FE}$ to $I = I_{FE} + I_E$ or $I = I_E$, and when the wave is trapped into the duct FE, the invariant varies from $I = I_{FE} + I_E$ or $I = I_E$ to $I = I_{FE}$. Thus, due to the variation of duct volumes along the path, a wave can leave one channel for another.

The adiabatic invariant I_1 of a horizontal ray, i.e., a ray emitted from the Earth at an angle $\alpha_0 = 0$, is also of importance. It is the level $\varepsilon' = 1$ that corresponds to the ray (see Fig. 4.1a), i.e.,

$$I_1 = \frac{4}{R_0} \int_0^{z_1} (\varepsilon' - 1)^{1/2} dz\,, \qquad \varepsilon'(z_1) = 1\,, \tag{4.10}$$

where z_1 is the reflection height for the horizontal ray. The horizontal ray can be reflected from both the E-region and F-region. In the first case $I_1 = I_{1E}(z_{1E})$, (Fig. 4.1a, and Curves 1, 6 in Fig. 4.1c), and in the latter one, $I_1 = I_{1F}(z_{1F})$ (Fig. 4.1c, Curves 2 and 3). A wave emitted from the Earth, or reaching it, is bound to have the invariant $I > I_1$. In particular, the capture into the duct FE of the ray emitted from the Earth is possible, provided $I_1 \leqslant I_E$.

Equations (4.1)–(4.10) for a given model of the electron density $N_e(z, \theta, \varphi)$ distribution can be applied when analyzing the general properties of the IWD's F, E, and FE. The equations can also be applied for the specific calculations of the radio wave propagation over long-distance and round-the-world paths in the ionosphere (Sec. 5.2). To determine the main characteristics of the IWD for a given frequency f, it is necessary to calculate the values of $\varepsilon'(z, f)$ at different θ, φ and to determine all the specific points on the curves $U(z) = -\varepsilon'(z)$. Such points (see Fig. 4.1a) include, first, P and Q – the minima of $U(z) = -\varepsilon'(z)$ in the ducts E and F, i.e., the maxima of the curve $\varepsilon'(z)$. (Their heights z_{SE} and z_{SF} are the ones for the sliding mode propagation, see Sec. 2.2.) Second, these are points A and B, i.e., the maxima of $U(z) = -\varepsilon'(z)$ (the minima of the curve $\varepsilon'(z)$). By determining ε' at these points we thereby determine the constants ε'_{min} and the respective values of $z_{min}(\varepsilon'_{min})$ and $z_{max}(\varepsilon'_{min})$ in every duct. These values are necessary to calculate

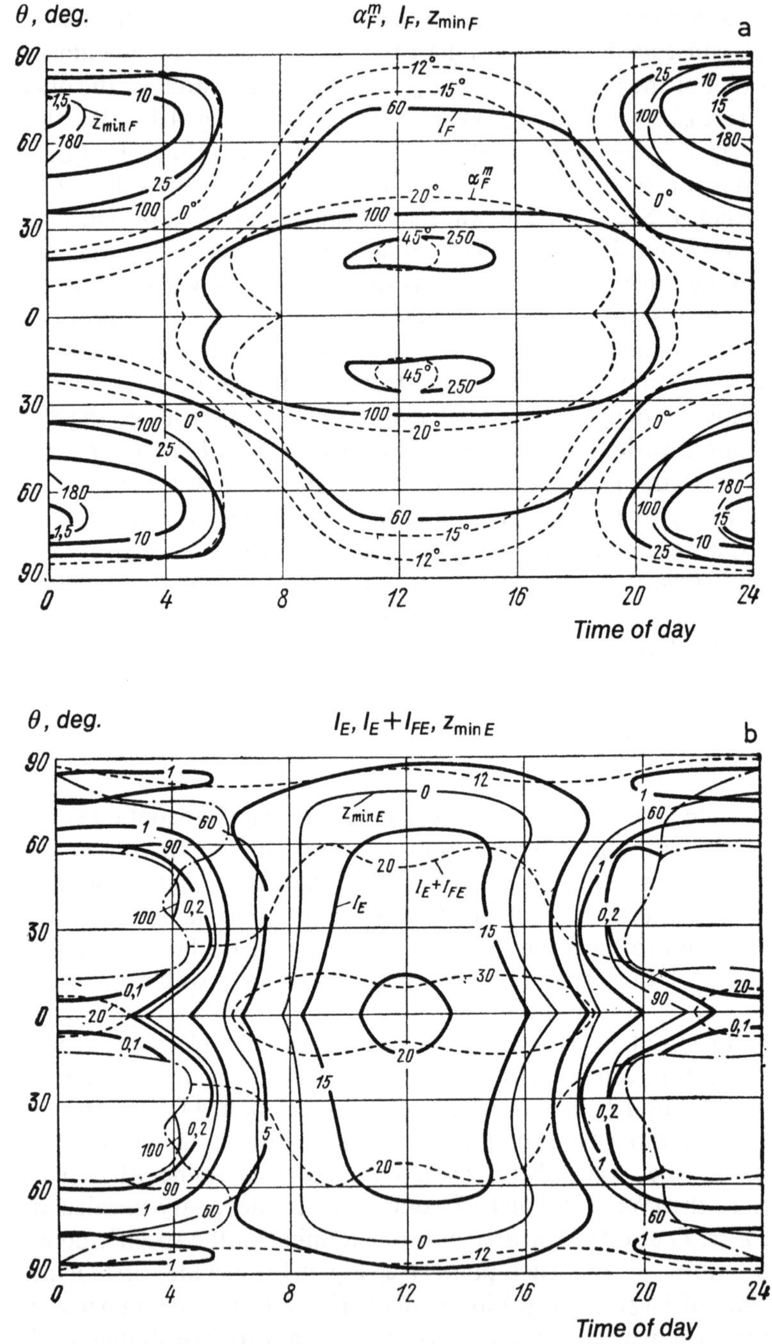

Fig. 4.2 a, b. Maps of the adiabatic invariants, angles of capture, and heights of the ducts, $f = 15$ MHz (Tushentsova et al. 1977). **c, d** Maps of the adiabatic invariants, angles of capture, and heights of the ducts, $f = 15$ MHz (Tushentsova et al. 1977). **e – h** Potential forms illustrating the change of the horizontal ray invariant I_1 for the frequency $f \leqslant f_{MUF}$. The reflection heights of the horizontal ray z_{1E} or z_{1F} are indicated

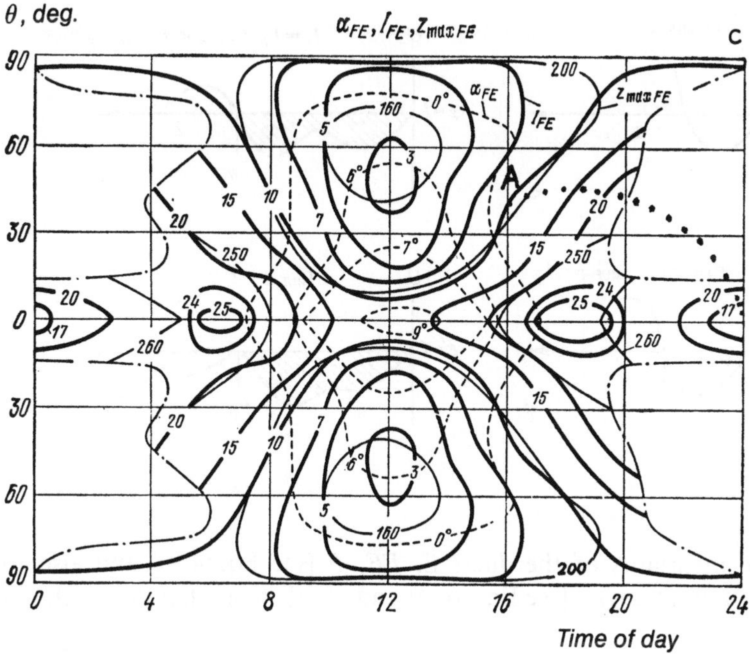

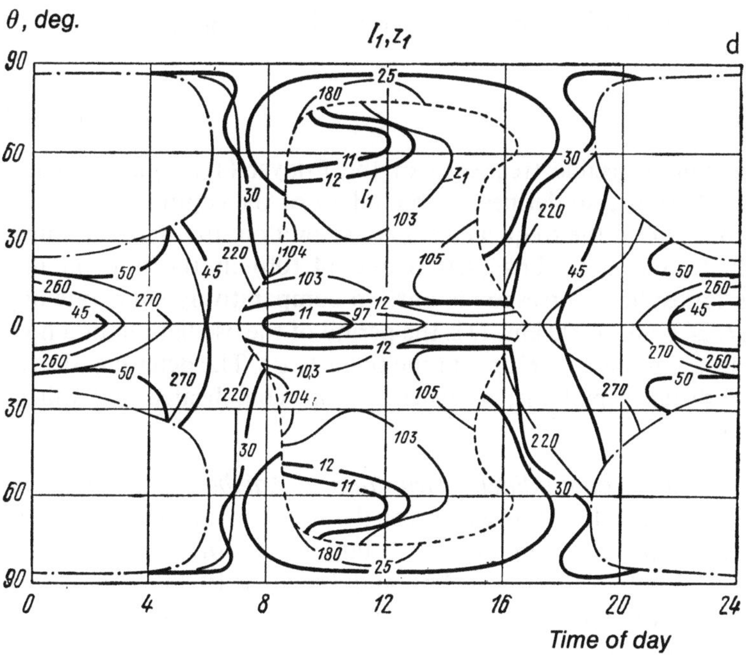

Fig. 4.2 c, d (Fig. 4.2e – h see next page)

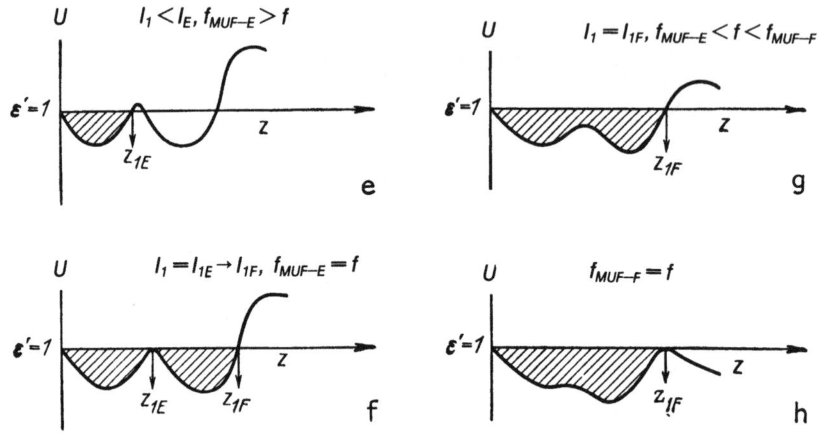

Fig. 4.2e – h

the adiabatic invariants of the ducts *E, FE, F* [see Eq. (4.9)] and also to calculate the capture angles [see Eqs. (4.4) and (4.5)]. Note that the condition

$$\alpha_F^m(\theta, \varphi, f) \to 0 \tag{4.10a}$$

defines the maximum usable frequency of the *F*-layer ($f_{\text{MUF-}F}$), whereas the condition

$$\alpha_E^m(\theta, \varphi, f) \to 0 \tag{4.10b}$$

defines the maximum usable frequency of the *E*-layer ($f_{\text{MUF-}E}$). The latter condition [Eq. (4.10b)] also defines the boundary of the capture region of a wave into the duct *FE* at the frequency *f*. Let us point out that the determination of MUF according to Eqs. (4.10a) and (4.10b) in the horizontally-homogeneous ionosphere corresponds only to comparatively short one-hop paths ($S = R_0\Theta \lesssim 3000-4000$ km, Θ is the period of the oscillation (or a hop) of a horizontal ray in the duct *F* or *E* at the point θ, φ). The determination of MUF for long-distance paths is, in general, more complicated and will be discussed below.

Maps of Latitudinal – Longitudinal Distribution of Adiabatic Invariants and Other IWD Parameters. It is convenient for the study of long-distance propagation to use maps of the global distribution of adiabatic invariants (Tushentsova et al. 1974, 1977; Fishchuk and Tsedilina 1975). Examples of such maps plotted for radio waves of frequency $f = 15$ MHz, and for the model EMI-81, W = 10 (Appendix 2) are shown in Fig. 4.2 a – d. The isopleths of the total adiabatic invariants I_m of the ducts *F, E, FE* [see Eq. (4.9)], those of the horizontal ray invariant I_1 [Eq. (4.10)], and of the invariant $I = I_E + I_{EF}$ are shown on the maps (in fact, the quantities $R_0I/4$ are shown everywhere). In Fig. 4.2 a – c the duct boundaries ($I_m \to 0$) are shown by the dash – dot line. The region of transition of the invariant $I_{1E} \rightarrow I_{1F}$ in Fig. 4.2b is shown by the broken line. When using the maps, one should bear in mind that, according to the accepted classification, the duct *FE* transforms into the duct *F* when the duct *E* disappears ($I_E \rightarrow 0$). Besides, when the duct *FE* is present, the total invariant I_F of the duct *F* involves the volumes of the ducts *E* and *FE* (for detail see Tushentsova et al. 1974, 1977; Fishchuk and Tsedilina 1975).

From Fig. 4.2a the duct *F* is seen to have the maximum volume in the daytime. At night, the values of I_F sharply decrease. They are especially small at night in the subpolar latitudes. The volume of duct *E* (Fig. 4.2b) is by one order less than that of *F*.

At night in the middle latitudes, the duct *E* vanishes altogether. The inter-layer duct *FE* (Fig. 4.2c) is of an opposite type than the ducts *E* and *F*: it has the minimum volume in the daytime in the middle latitudes. The duct *FE* volume increases towards the equatorial and polar latitudes and the night zone.

The horizontal ray adiabatic invariant I_1 (Fig. 4.2d) also has a minimum in the daytime, since at this moment the ray is reflected from the *E*-layer, i.e., $I_1 = I_{1E}$ (see Fig. 4.2e). When the considered frequency $f = 15$ MHz exceeds the MUF-*E*, the horizontal ray begins to be reflected from the *F*-layer, $I_1 = I_{1F}$, ant its adiabatic invariant increases by a jump (Fig. 4.2 f, g). The discontinuity curve of I_1 is shown in Fig. 4.2d by the dashed line. The invariant I_1 becomes meaningless when the frequency exceeds the MUF-*F* (see Fig. 4.2g). The region of $f = 15 \text{ MHz} > f_{\text{MUF-}F}$ is indicated in Fig. 4.2d by the dash – dot curve.

The angles of capture into IWD and duct heights are also of importance for the analysis of the propagation conditions. The distribution curves of these quantities for $f = 15$ MHz are shown in Fig. 4.2 a – d.

Radio Path Analysis Using the Maps of Adiabatic Invariants. Using maps of the IWD adiabatic invariants one can find out if it is possible for the signals to propagate on any radio path, including the round-the-world path, too. The MUF is determined as well. The maps allow the most favorable conditions of wave capture into IWDs and departure from the ducts to be determined.

As an example, let us consider the radio path shown in Fig. 4.2c by the dashed line. The emission point *A* is located at the latitude $\theta = 40°$, local time LT = 16 h. The emission azimuth read off eastward from the north direction is equal to 67.5°. The receiver is located at the point *B*. Using maps in Fig. 4.2, we determine the adiabatic invariant variation along the ray path. The variation is shown in Fig. 4.3a versus the central angle Ψ along the path[2], i.e., the distance $S = R_0 \Psi$ from the emission point according to Fig. 4.2c. The

---

[2] In Chap. 3 the variable θ was mainly used for the designation of an angular coordinate along the propagation path. In the present chapter the letter θ designates the geographical latitude, thus, in the specific path calculations the angular distance along the paths is designated by Ψ.

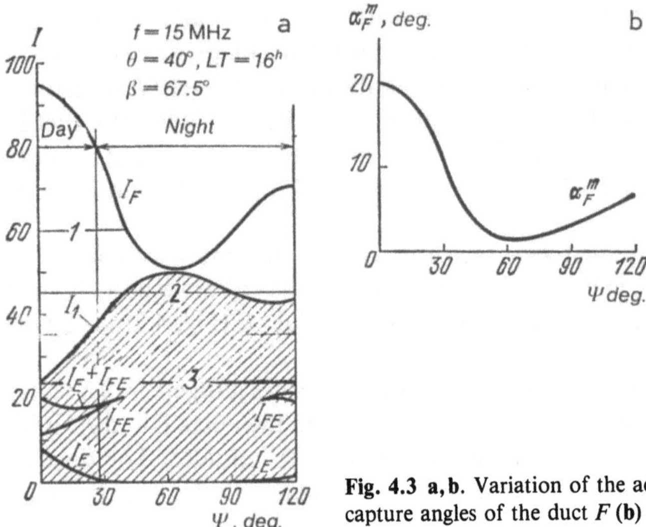

Fig. 4.3 a, b. Variation of the adiabatic invariants (**a**) and capture angles of the duct F (**b**) along the path shown by *broken line* in Fig. 4.2 c

minimum value of the invariant I_F in the path is seen to be equal to 50. This value gives the upper limit for ray propagation over the path: the ray with $I > 50$ is not confined in the ionosphere. Thus, the ray with $I = 60$ (trajectory 1 in Fig. 4.3) leaves the ionosphere at $\Psi = 40°$.

The quantity I_1 gives the lower limit for permissible invariants of a wave emitted from the Earth's surface, thus, for the emission point, $I_1 = 24$ (Fig. 4.3a). Therefore, the radio waves emitted from the Earth at an angle of $\alpha_0 \geqslant 0$ (see Fig. 4.3b) have at the initial point of the path, $\Psi = 0$, the invariants $I \geqslant 24$. If $24 \leqslant I \leqslant 50$, they reach the terminal point of the path (trajectory 2). In this case, however, if $I < 42$, the rays pass to ricochet trajectories in the duct F (the shaded area) and at the terminal point they fail to come down to the Earth. Thus, only the rays with $42 < I < 50$ can reach the terminal point of the path $\Psi = 120°$. The emission angles of those rays, according to Fig. 4.3b, are within $5-8°$, the angles of arrival $\sim 0-2°$. In this case, according to Fig. 4.3a, the rays cover a part of the path without reflecting from the Earth. As it is seen from Fig. 4.3a, the invariants $I_1 > I_E$. Thus, the duct E for the path in question is separated from the Earth's surface and there are no conditions for a direct ray capture into the duct FE on the frequency $f = 15$ MHz.

Now let us consider around-the-world meridional midday-midnight path. The adiabatic invariant variation of all the ducts along the path is shown in Fig. 4.4a. Figure 4.4b shows the respective variation of maximum and minimum heights of the ducts F, FE, E, and the height of the sliding mode, z_{SF}, in the duct F. Figure 4.4c shows the variation of the capture angles α_F^m, $\alpha_E^m = \alpha_{FE}$ for the path. From Fig. 4.4a the horizontal ray adiabatic invariant I_1 is seen to be small in daytime ($I_1 = 11$). At this moment in the duct F and E radio waves propagate in hops (trajectories 1 and 2), if $I_1 = 11$. All of them,

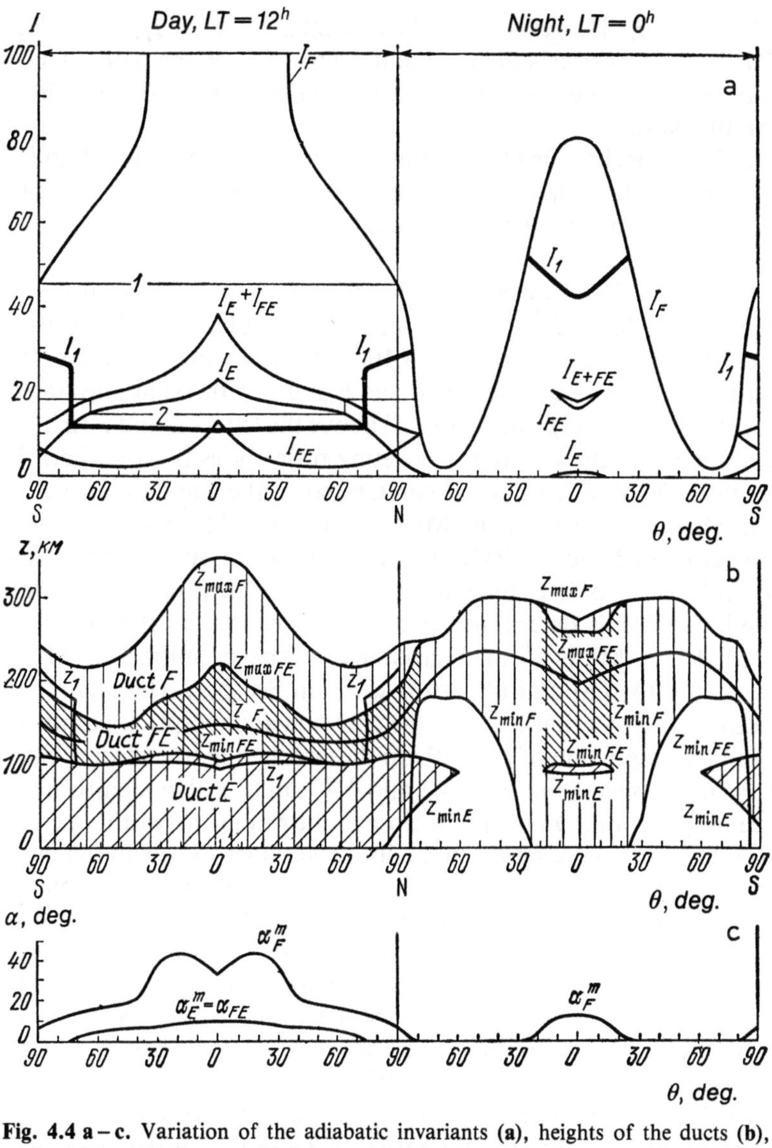

Fig. 4.4 a–c. Variation of the adiabatic invariants (**a**), heights of the ducts (**b**), and capture angles (**c**) for RTW noon-midnight meridional path on the frequency $f = 15$ MHz. The ducts *F*, *FE*, and *E* are shown by different types of shading

however, leave the ionosphere at night at latitudes $\theta \approx 60-90°$. This is caused by a sharp minimum of the total invariant ($I \approx 2$) at $\theta \sim 70°$ in the night part of the path. It means that on the path only the rays with fairly small invariants, $I \lesssim 2$, can be confined. Figure 4.4a shows that propagation in the interlayer duct *FE* is also possible. In this case capture into the duct *FE* is possible practically all over the day part of the path, provided the emission of energy is directed towards an increase of the invariant I_{FE} [Eq. (4.8)]. It can be seen from Fig. 4.4a ($I_1 < I_E$) and Fig. 4.4c (in the day part of the path, the capture angle, $\alpha_{FE} > 0$). However, even the minimum value of I_{FE} at the latitudes $30-70°$ is ~ 3. It exceeds the minimum value of I_F on the path. Hence, there are no regular conditions for a wave emitted from the Earth to propagate on the path.

Some additional mechanisms (for example, scattering, see Chaps. 6 and 7) are needed. They lead to a change of the ray adiabatic invariant I both along the path and at the instant of wave capture into the duct.

Determination of the Radio Path MUF. Figure 4.5 illustrates the curves of global distribution of MUF-E^h, MUF-F^h in MHz, calculated by the EMI-81 model at $W = 10$ by Eqs. (4.10a) and (4.10b). Henceforth, the index "h" above the letters E and F designates the ducts E and F with usual hop trajectories (h – hops). Respectively, the index "r" above the letters E and F (E^r, F^r) will designate the channels with the ricocheting trajectories (r – ricochet). In Fig. 4.5 the curves MUF-E^h, MUF-F^h correspond to the standard curves of MUF – 4000, which are given in the MUF predictions (Monthly prediction 1975, Ionospheric predictions 1982). When analyzing a specific path, the predictions can be used to determine the hop MUF for the given time or time interval. The hop MUF is determined in a regular manner by the minimum value of MUF-F^h or MUF-E^h at a given path. The definition of MUF, different from Eqs. (4.10a) and (4.10b) is given in Sec. 5.1. This definition takes into account the fact that between the emission and reception points the radio wave can pass a portion of the path on the ricocheting trajectories. Such trajectories, as it was seen from Figs. 4.3 and 4.4, can be rather significant in long-distance propagation. On the paths close to the round-the-world ones, of great importance is the propagation in the interlayer duct *FE* (or in the night duct F^r). The presence of such trajectories contributes to enlarged possibilities of long-distance propagation and to an increase of MUF in the path against MUF calculated by standard simplified techniques with only hop trajectories taken into account (Kerblay and Lyakhova 1969, Monthly prediction 1975).

Figure 4.6a illustrates the analysis of the propagation conditions at various frequencies on the path close to the round-the-world one (Tushentsova et al. 1977). The transmitter is located at the latitude $\theta_1 = 45°$, the transmission azimuth is $\beta = 315°$, the receiver is at the latitude $\theta_2 = 55°$ (the Krasnodar – Kaliningrad path). The data refer to the period of 8 – 12 LT for the location of $\theta_2 = 55°$. Curve 1 shows the variation of the minimum value of MUF-F_{min} for hop propagation over the round-the-world path at the given time, obtained with the help of Fig. 4.5b. Curve 2 shows MUF-$E_{\alpha_0 = 0}$ and Curve 3 shows MUF-$F_{\alpha_0 = 0}$ for the location of $\theta_2 = 55°$. On the RTW path, the MUF-F_{min} is seen to be essentially lower than the MUF-E and MUF-F on the direct path. Curve 4 indicates the maximum frequencies MUF of RTW_1 when round-the-world propagation in the interlayer duct *FE* and duct F^h is possible. Quantities MUF of RTW exceed MUF-F_{min} by 4 – 10 MHz. Capture into the round-the-world duct occurs at frequencies about the MUF-E and lower. Taking into account Curve 4 which defines the maximum propagation fre-

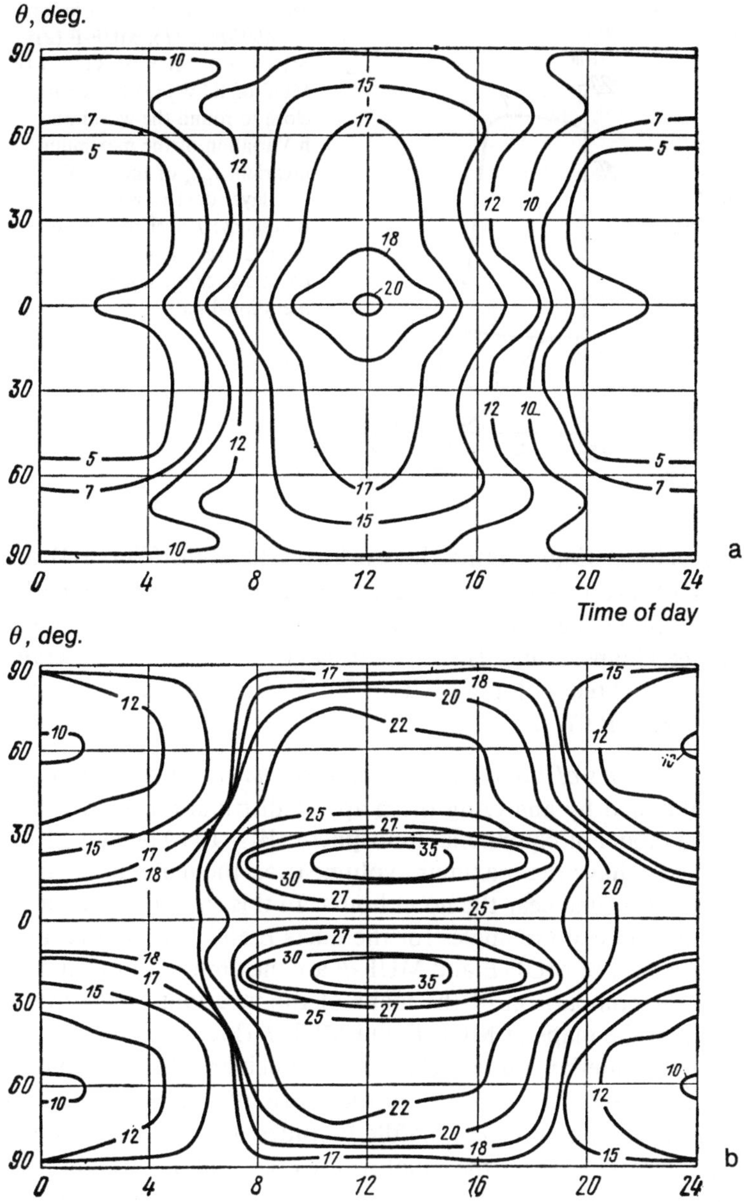

Fig. 4.5 a, b. Curves of the MUF-E^{h} in MHz (**a**); the same for the MUF-F^{h} (**b**)

quencies for the path and Curve 2 which defines the maximum frequency of capture, we obtain the resultant Curve 5. This curve defines the maximum propagation frequencies MUF of RTW_2 on the path allowing for direct capture. The MUF of RTW_2 is seen from Fig. 4.5 to be significantly greater than MUF-F_{min} (Curve 1) on the path. Note that in the interval 8 – 11 LT the

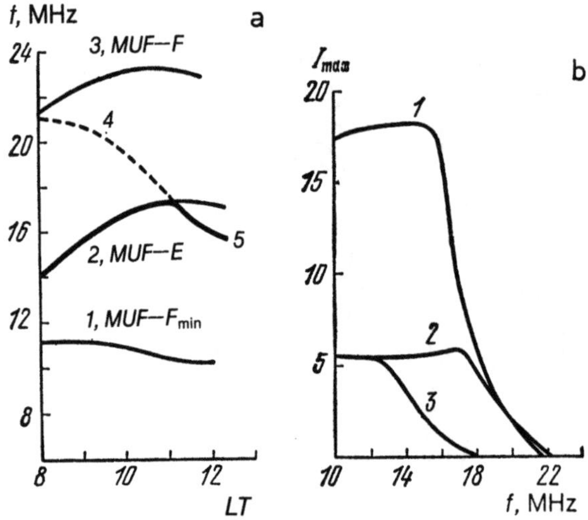

Fig. 4.6. a Curves of variation of MUF-F_{min} (*1*), MUF-*E* (*2*); MUF-*F* (*3*); MUF of RTW_1 (*4*); MUF of RTW_2 (*5*) for a path close to round-the-world one. **b** Variation of the maximum invariant I_{max} of the RTW interlayer duct vs wave frequency: *1* over the equatorial path; *2* over the midlatitude path; *3* over the noon–midnight meridional path

maximum frequencies defining round-the-world propagation (Curve 4) are greater than MUF of direct capture into the duct *FE* (Curve 2). Consequently, the presence of additional mechanisms resulting in radio wave capture into the interlayer duct (see Chaps. 6 and 7) can significantly increase the observed MUF of RTW as compared with the MUF of RTW_2 (the first, however, can not exceed the MUF of RTW_2).

Some General Regularities of Long-Distance and Round-the-World Propagation. These are obtained by analyzing the maps of distribution of the IWD parameters for the model of equinoctial ionosphere, $W = 10$ (Appendix 2).

The ducts *E* and *F* have the maximum volume (i.e., the maximum values of invariants) in the daytime near the noon hours. In this case, the maximum values of invariants and capture angles for the ducts *E* and *F* are in the same regions where the values of MUF-*E* and MUF-*F* are maximum. For example, this is seen from the comparison of the maps of the invariants and capture angles (Fig. 4.2), and the maps of MUF-*E* and MUF-*F* (Fig. 4.5).

The volume of the interlayer duct *FE* is always much less than the volume of the duct *F*. The characteristic property of the distribution curves I_{FE} is that the values of I_{FE} are minimum at the middle latitudes near the noon hours. The values of I_{FE} increase towards the equator, the pole, and the twilight belt. In the same region, the angles of capture into the duct *FE* exist as well ($\alpha_{FE} > 0$, i.e., in the duct *E* the height $z_{minE} = 0$). Consequently, there are favorable conditions for the capture of energy from the Earth into the interlayer duct *FE*. Simultaneously, as it follows from Eq. (3.77), this region is the most favorable for the output of energy from the duct *FE* to the Earth. That is why in this region, under the equinoctial conditions RTW signals can be expected to appear most frequently.

The conditions in the twilight belt are rather uniform during about an hour after sunrise and sunset (Tushentsova et al. 1977). Here, the RTW duct exists up to the frequencies ~22 MHz, which corresponds to the experimental data (Golyan and Shlionskii 1971, Golyan 1974). The propagation by the ricocheting trajectories failing to descend to the Earth over a greater part of the path is possible in this belt for lower frequencies ($f \leqslant 17$ MHz). At higher frequencies, the duct volume is small in the subpolar region. That is why, only the rays trapped at the middle latitudes into the duct *FE*, having a small volume, can pass through the subpolar region. At the frequencies $f \geqslant 12$ MHz, there are no conditions for the direct capture into the duct *FE*. An additional turn of the ray in the ionosphere is needed, for instance, in the case of scattering by the irregularities (see Chaps. 6 and 7).

The minimum values of the invariant I_F take place in the dark hemisphere in the subpolar region. In this region the most rapid disappearance of the ionospheric duct is observed. It is accounted for by the presence of a "trough" in the characteristics of the *F*-layer. Consequently, the maximum frequencies at which round-the-world propagation is still possible, have their minima on the paths, passing through the night polar region. The frequencies are about 16–18 MHz, which follows, for example, from Fig. 4.6, where the maximum values of the invariant I_{max} on the three different paths are given. This corresponds to the experimental results (Fenwick 1963, 1966; Fenwick and Villard 1963a, b, c; Harnischmacher 1960, Nevolin and Shlionskii 1973, Banks 1965, Golyan and Shlionskii 1971, Golyan 1972, 1973, 1974; Benediktov et al. 1974, Kiyanovskii and Ostrovskii 1972, Bubenik et al. 1971, Ostrovskii and Shlionskii 1974, Ostrovskii 1972, Gerson et al. 1969).

The maximum frequencies of propagation over midlatitude, round-the-world, and subequatorial paths are about 20–22 MHz (Fig. 4.5). These results agree with the RTW observations at Standford ($\varphi = 120°$W, $\theta = 37°25'$) in the years close to the solar activity minimum (see Fig. 1.1). Figure 1.1 a, b shows the comparison of the experimental results with the theoretical ones. These figures give the theoretical curve (broken line) determined by the maps (Tushentsova et al. 1974) (EMI-72, Appendix 2) as the maximum frequency of the direct capture into the duct *FE* (MUF of RTW_2, Sec. 3.4). The maximum frequency of the round-the-world propagation in the duct (MUF of RTW) for every hour of LT by day is more than MUF of RTW_2 by 2–3 MHz, and in the evening and morning hours by 5–8 MHz. The maximum frequencies of signal passage in the channel depend significantly on the azimuthal direction of the path. They depend markedly on the choice of a model. For instance, according to EMI-81, $W = 10$, the MUF of the interlayer duct is much higher than for the model EMI-72. As it is seen from the figures, the RTW signals are observed in the daytime at the frequencies close to the calculated curve. However, the theoretical MUF of the direct capture MUF of RTW_2 is by 2–3 MHz lower than the experimental MUF of RTW in the daytime, i.e., the experimental MUF of RTW is apparently close to the MUF of RTW_1. Besides, the RTW signals are observed experimentally during the longer period of time. The broadening of the time and frequency range of RTW reception as compared with the MUF of RTW_2 indicates a significant role of the additional mechanisms of radio wave capture (for example, due to the scattering by the irregularities of electron density, see Chap. 6). More detailed analysis of the global propagation features is given in (Tsedilina 1974, Tushentsova et al. 1974, 1975; Fishchuck and Tsedilina 1975, Tushentsova et al. 1977a, b).

The Interlayer Ionospheric Duct. The interlayer duct *FE* is of great importance for very long-distance and RTW radio wave propagation. In this duct radio waves oscillate at the heights of $\sim 120-180$ km. The waves do not descend to the strongly absorbing layers *E* and *D*, and therefore, they usually undergo the absorption less than in the case of multihop propagation in the duct *F*. As far as energy is concerned, the propagation in the duct *FE* on the trajectories close to the sliding ones is the most advantageous. It is also important that in the night zone, the total volume of the ionospheric waveguide channels, *I*, usually becomes small, $I_m \to I_{min}$. Consequently, only the rays having the small values of the invariant, $I < I_m$, in the day side as well can pass through the night zone. It is just the volume of the duct *FE* that is small in the day zone (see the maps of Fig. 4.2b). Hence, the main role of the duct *FE* is to inject the energy at lower levels, i.e., the levels with the small value of *I*, allowing to pass through the night zone. But the radio waves which are not trapped in the duct *FE* usually have the great values of the invariant. That is why they are not confined by the ionosphere in the case of the narrowing of channels in the night zone.

Let us consider some general characteristics of the duct *FE*.

Capture into the Duct FE due to the regular variability of the ionosphere occurs at the frequencies $f \leqslant f_{\text{MUF-}E}$ (Sec. 3.4). That is why the curves $f = f_{\text{MUF-}E}$ restrict the regions where the direct capture on a given frequency is possible. For example, within the curve $f_{\text{MUF-}E} = 17$ MHz (see Fig. 4.5a) the capture into the duct *FE* at frequency $f = 17$ MHz and at frequencies $f < 17$ MHz is possible. It is seen that for the point $\theta = 60°$, the capture at the frequency $f = 17$ MHz is possible during 2 h, while for the point $\theta = 30°$, the capture is possible during $4-5$ h near the noon. The capture into the duct *FE* occurs at the emission angles $\alpha_0 = \alpha_{FE}$. Their values will be discussed below. Note that the capture angles α_{FE} decrease with increasing frequency.

Azimuthal Characteristics of the round-the-world channel existence (i.e., the channel *FE-F-FE* along the round-the-world path) are given in the diagrams (Fig. 4.7) for the latitudes $\theta = 0°$, $40°$, $60°$ at 9, 11, 12, 15 LT. The dark sectors in the diagram correspond to the presence of the round-the-world channel separated from the ground along the entire RTW path. The shaded sectors correspond to the conditions when through a part of the path radio waves pass on hop trajectories. It is seen from the diagram that on $f = 15$ MHz the round-the-world channel (with the conditions for entry and departure) is situated near the equator; at the middle latitudes the main directions are the latitudinal ones.

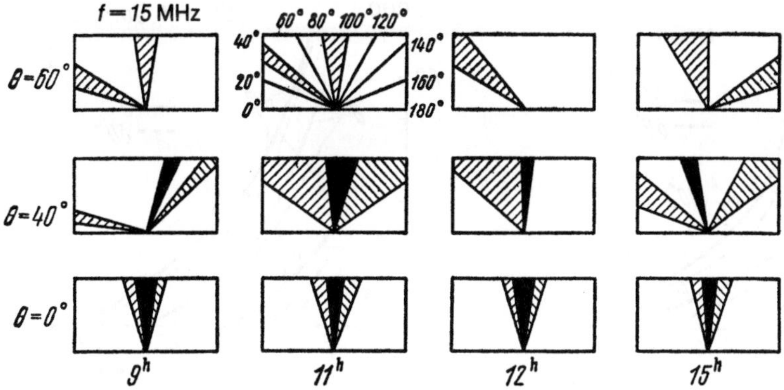

Fig. 4.7. The azimuthal characteristics of existence of the RTW duct on the frequency $f = 15$ MHz. The azimuth is read off clockwise from the northern direction

Angles of Radio Wave Capture into the IWD from the Earth and from the Sources Situated in the Ionosphere. First, let us consider the main features of capture angles into the IWD for radio waves, emitted from the Earth's surface (Fishchuk and Tsedilina 1975). When emitting in the angular range α_E from 0 up to α_E^m, a wave is trapped in the duct E (Fig. 4.8) and propagates in it by hop trajectories. When emitting in the angular range from α_E^m to α_F^m, radio waves are trapped in the duct F.

If a wave is emitted at the angle $\alpha \approx \alpha_E^m = \alpha_{FE}$, it can be trapped in the interlayer duct FE. In this case, the departure from the duct FE to the duct E and descent to the Earth occurs at the vertical arrival angles $\alpha_{FE} \approx \alpha_E^m$. From Fig. 4.8a it is seen that the capture angle α_{FE} at the latitude $\theta = 0°$ changes in the daytime from 0° on the frequency $f = 21$ MHz to $\alpha_{FE} = 31°$ on the frequency $f = 7$ MHz. Of course, in the morning and in the evening (Fig. 4.8b), the angles and frequency range of capture are much less than in the daytime. In this case, the morning capture angles are less than the evening ones. Thus, the direct capture into the duct FE and descent to the Earth are characterized by the vertical angles varying inversely with frequency, i.e., the lower frequencies correspond to the greater angles (Fig. 4.8a). This is confirmed by the results of experimental measurements of RTW arrival angles, shown in Fig. 4.8a.

Now let us consider the maximum capture angles, β^m, into ionospheric channels, when the transmitter is situated in the ionosphere at the altitude z_s. Here, z_s is the height of the sliding mode propagation in the ducts F and FE ($z_s = z_{sF}$) or in the duct E ($z_s = z_{sE}$). The maximum possible angles for radio wave capture, β^m, we shall denote, respectively: β_{FE} in the duct FE, β_F in the duct F, and β_E in the duct E. The capture angles β_i^m ($i = F, E, FE$), as well as the angles α, we shall read off from the horizontal. According to Snell's law, for each of the channels the angles are equal to (Carrara et al. 1970)

$$\beta^m = \arcsin \frac{\sqrt{\varepsilon_s' - \varepsilon_{\min}'}}{\sqrt{\varepsilon_s'}} . \quad (4.11)$$

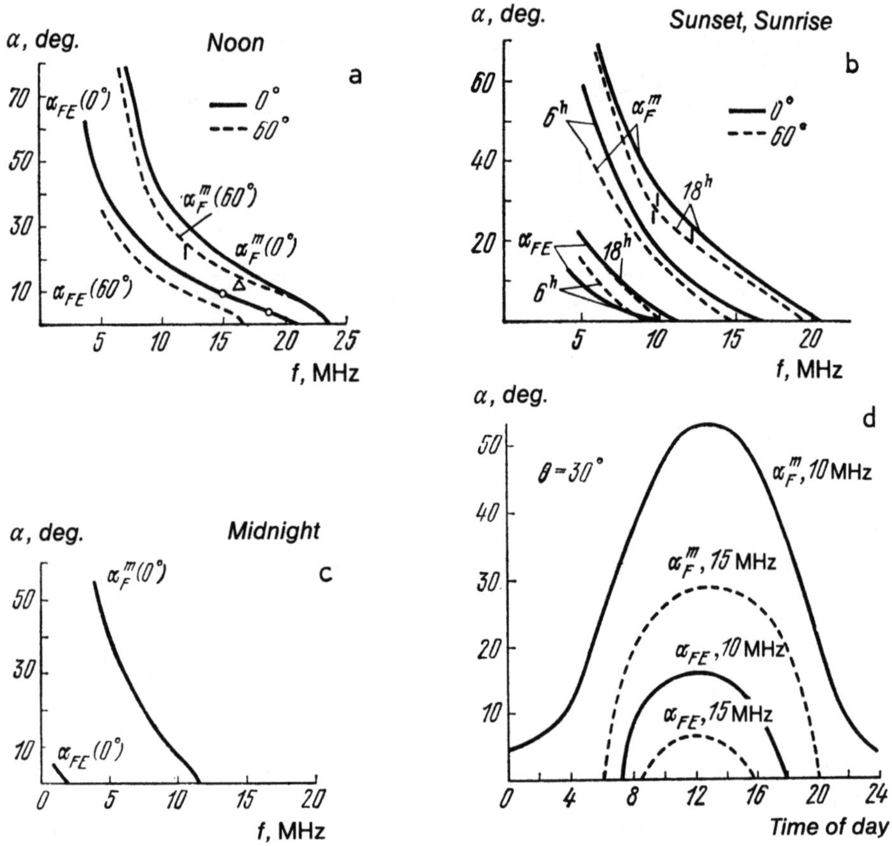

Fig. 4.8. a – c Capture angles α_{FE}, α_F^m vs wave frequency for the latitude 0 and 60°. *Vertical lines* indicate the arrival angle range measured at midlatitude short paths (the path length is about 1000 – 1500 km) (Grosskopf and Vogt 1941, Borowski et al. 1956): measurement results of RTW arrival angles are indicated by a *triangle* (Neyer and Rawer 1951) and by *circles* (Fenwick 1963). **d** Diurnal variation of the capture angles for $\theta = 30°$ and frequency $f = 10$ and 15 MHz

Here, ε_s' is the value of the dielectric constant at the height z_s, $\varepsilon_{\min}'$ is the minimum value of ε' for each duct (see Fig. 4.1a). Note that if the minimum value of ε' in the duct is less than 1 one should take $\varepsilon_{\min}' = 1$. As an example, Fig. 4.9 illustrates the latitudinal-longitudinal distribution of the capture angles β_{FE} and β_F. The dash – dotted line in Fig. 4.9a corresponds to the curve MUF_E. It appears to be the boundary at which the capture angles $\beta_{FE} = \beta_F$. Within the region restricted by this curve, radio waves can be trapped only in the duct *FE*. The broken curve restricts the region of the duct *FE* existence and, therefore, the capture angles β_{FE}. In the region between the dash – dotted curve and the broken one (Fig. 4.9a) and inside the broken curve (Fig. 4.9b), radio waves can be trapped both in the duct *FE* (at smaller emission angles) and in the duct *F* on the ricocheting trajectories (at greater emission angles). As it is seen from Fig. 4.9a, the capture angles β^m are minimum

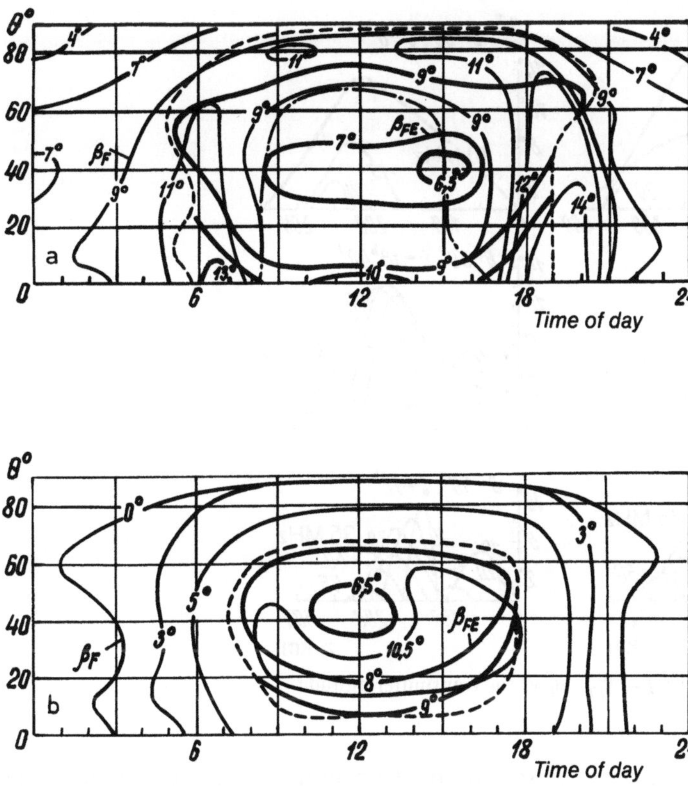

Fig. 4.9 a, b. Latitudinal – diurnal distribution of the capture angles β_{FE} and β_F: **a** for $f = 15$ MHz; **b** for $f = 25$ MHz

($\sim 6-7°$) at noon hours at the middle latitudes and at night at all the latitudes ($\sim 4-7°$), especially, at high latitudes ($\sim 3-5°$). In the transition region (twilight zone) the capture angles are much greater.

Figure 4.10 shows the capture angles $\beta^m(\beta_E, \beta_F, \beta_{FE})$ versus the height of the source. The wave emitted at an angle less than the capture angle β^m is trapped in the channel and does not descend to the Earth. Figure 4.10 shows how the transmitter height should vary on the day and night sides for radio waves to be trapped in a channel. For example, for a wave at the frequency 20 MHz to remain trapped in the channel on the night side, the transmitter should be situated at the heights of 300 – 350 km at the equator, and 220 – 280 km at the latitude $\theta = 50°$. In the daytime, the heights should be $\sim 100-250$ km for the latitude $\theta = 0°$ and 100 – 160 km for $\theta = 50°$. If the transmitter is situated at greater heights, a wave either descends to the Earth or leaves the ionosphere for the space.

Figure 4.11 shows the capture angles β_F and β_{FE} versus the frequency f. It is seen from the figure that at lower frequencies, $f < f_{\text{MUF-}E}$, the capture is determined by the angle β_{FE} decreasing with increase of frequency (Fig. 4.11a).

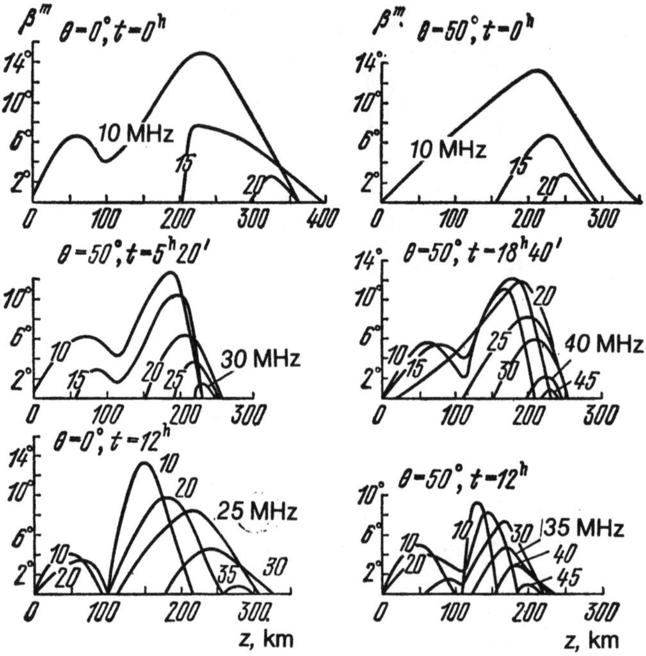

Fig. 4.10. Capture angles β_E, β_{FE}, β_F vs the transmitter altitude

When $f \geqslant f_{\text{MUF-}F}$, the wave emitted by the transmitter is trapped either in the duct *FE* or *F*, depending on the emission angle β^m. In this case, the value of β_F first increases with f and then decreases to 0° when the duct disappears. In the frequency range from MUF-*E* to the frequency at which the ducts *F* and *FE* disappear, radio waves are trapped in the ducts *FE* or *F*. It is seen from Fig. 4.11c that in the morning hours the maximum capture angles are observed at smaller frequencies than at the evening hours. At frequencies $f > 15$ MHz, the capture angles in the evening hours are greater than in the morning.

Ducts on Frequencies Higher than the MUF. The possibility for radio wave propagation in the waveguide channels on frequencies exceeding the MUF-*F* is of great interest.

Figure 4.12 shows the regions of duct existence on the frequency $f = 25$ MHz (EMI-81, $W = 10$). Hop propagation of radio waves on frequency $f = 25$ MHz is possible inside the curve MUF-*F*. However, the region is restricted by the day low latitude zone. But the duct F^r for $f = 25$ MHz exists above a greater part of the Earth's surface except the high latitude night zone where the channels with increasing frequency disappear rather quickly because of the effect of the "trough" in the *F*-layer. That is why the duct F^r can provide propagation in the night zone and even for RTW radio wave propagation on the trajectories close to the sliding ones. In this case, a greater part of the duct *F* is separated from the Earth, i.e., it is impossible to reach the

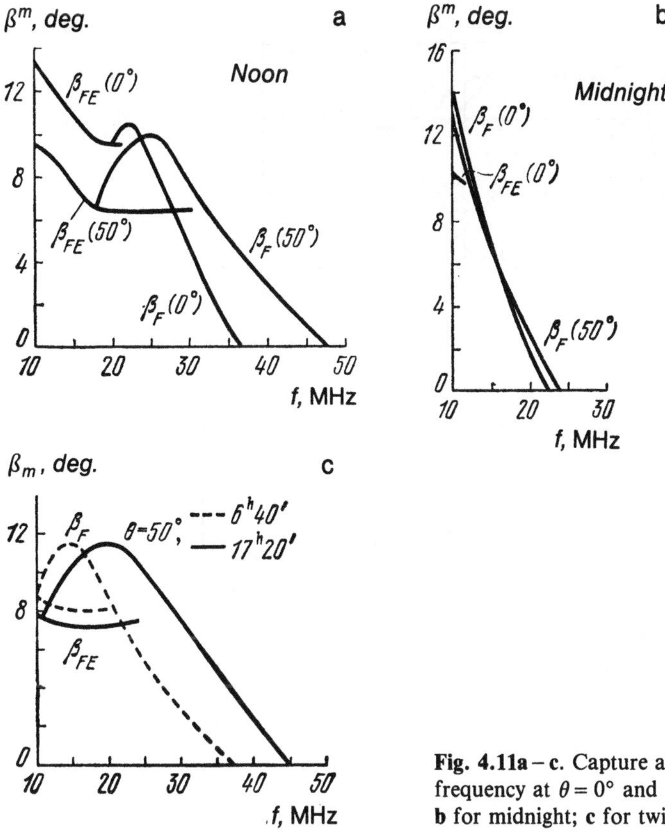

Fig. 4.11a – c. Capture angles β_{FE} and β_F vs wave frequency at $\theta = 0°$ and $\theta = 50°$: **a** for noon; **b** for midnight; **c** for twilight hours

duct from the Earth without an additional turn of a ray in the ionosphere. On the frequency $f = 25$ MHz, there is also the duct *E* which is separated everywhere from the Earth. There also exists the interlayer duct *FE* above a considerable area of low and middle latitudes in the daytime. The capture into the *FE* duct from the Earth due to the regular refraction is also impossible. The regions of existence of the ducts *E, F, FE* at high frequencies $f \geqslant 30$ MHz are shown in Fig. 4.12b. It is seen from the figure that high frequency communication channels (up to $f \approx 50$ MHz), exist above a considerable area of the Earth. The channels can be used both for the satellite and ground-based communications in the case of the special injection of energy from the Earth to a channel and output of the energy to the Earth (for example, by means of artificial heating of the ionosphere with powerful radio waves (see Sec. 7.1).

Variation of the channel volumes and the heights with frequency is seen from Fig. 4.13. In this figure, variation of the channel volumes *E, F, FE* and heights versus frequency *f* is shown for the latitude equal to 50° at noon conditions. The ducts *F, FE,* and *E* are shown by different types of shading. The broken curve shows the variation of the sliding mode height, z_{sF}, of the

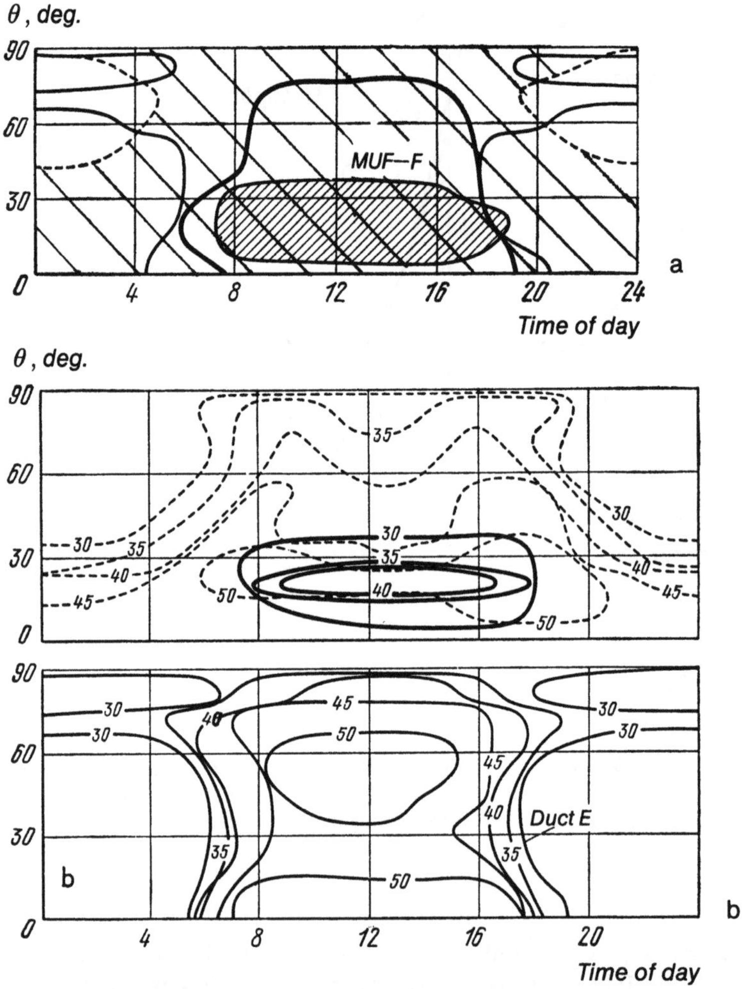

Fig. 4.12. **a** Regions of existence of the ducts *F* (*broken curve*); *FE* (*thin curve*); *E* (*solid curve*) at the frequency $f = 25$ MHz. The region of hop propagation is *shaded*. **b** Regions of existence of the ducts at the high frequencies indicated at the *curves* (in MHz)

duct *F* (or *FE*). It is seen from Fig. 4.13b that for the latitude $\theta = 50°$ and 12.00 LT, the MUF-*E* is equal to 17.2 MHz, while the MUF-*F* is equal to 27 MHz. The invariants of the ducts for these frequencies are $I_E \approx 16$, $I_F \approx 30$, respectively, i.e., the ducts *E* and *F* at the frequency $f > f_{\text{MUF}}$ have a significant volume. The ducts *E* and *F* exist up to the frequencies $f \approx 42.5$ MHz and $f \approx 47$ MHz, respectively. Note that for the duct *E* (at $\theta = 50°$ and 12^{h} LT) the frequency $f \approx 42.5$ MHz is by the factor of 2.5 greater than the MUF-*E*, and respectively, for the duct *F*, $f \approx 47$ MHz is by the factor of 1.74 greater than the MUF-*F*. The duct *FE* transforms to the duct *F* on the frequency $f \approx 33$ MHz.

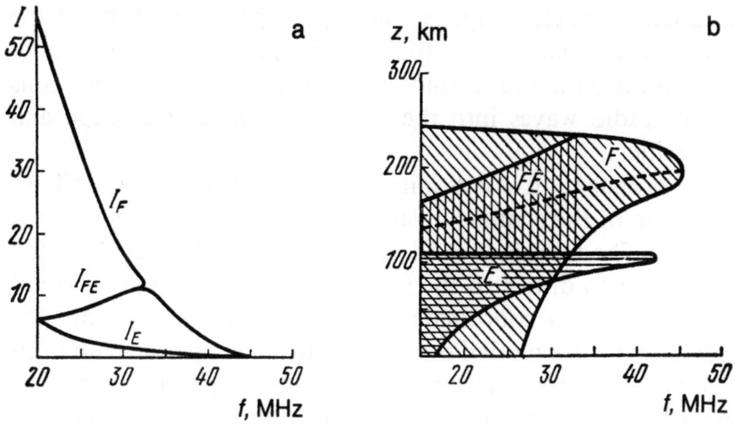

Fig. 4.13a, b. Invariants (**a**) and heights (**b**) of the ducts vs wave frequency for $\theta = 50°$, LT = 12^h

The fact of radio wave propagation at the frequencies exceeding the MUF was repeatedly pointed out in the scientific literature (Fenwick 1963, Golyan and Shlionskii 1971, Bubnov et al. 1975), and over the Okinawa – Guam path the experiment was carried out intentionally for evidence of the propagation over the night zone. The propagation occurs at the frequencies higher than the MUF in the Earth – detached channel. In the dual satellite experiment OV4-1R and OV4-1T (Barker and Grossi 1970), the RTW signals were observed on the frequency $f = 34.3$ MHz.

It is seen from Figs. 4.12, 4.12b, and 4.13 that the application of IWD at the frequencies, exceeding the MUF can significantly broaden the range of frequencies used for long-distance communications in the daytime when producing the artificial conditions for the injection of the energy into the IWD and output from it.

The Forecasting of Long-Distance HF Radio Communications. Maps of the total invariants of the ducts *F*; *FE*, *E* and also of the capture angles and duct heights (see Figs. 4.1 and 4.2) can be used for wave propagation forecasting over long and very long distances (including the RTW ones) at frequencies both lower and higher than the MUF. The prediction commonly used represents the global map of the MUF-*F* (Monthly prediction 1975). The maximum frequency on the path at a given instant or for the given time interval is determined by taking the values of MUF-*F* off the map. However, in such a prediction only the propagation on hop trajectories is taken into account. As the analysis carried out above has showed, other propagation modes are also possible over long-distance paths. Such modes involve the propagation in the ionospheric ducts at the frequencies significantly exceeding the hop MUF. The invariant maps enable one to determine these frequencies for the given paths. However, the maps of invariants and other IWD parameters allow to get much more information about radio wave propagation than the usual predictions. Besides, the determining of the maximum frequencies at which communication channels disappear on a given path, the study of the maps at various frequencies enables one to draw up a general picture of radio wave propagation on the path and to find out the most favorable conditions both for capture of radio waves into the IWD and for their passage and departure from the ducts.

To analyze the propagation conditions, it is necessary, first of all, to determine the variation of the channel invariants and other parameters along the given radio paths. The technique used is the same as in the usual predictions. The great circle map drawn to the same scale as that of the invariant maps is used. By means of a coordinate grid of the great circle map, the terminal points of the path are plotted on the tracing paper and the path itself is also plotted by means of the great circles. Next, bringing into coincidence the equator ($\theta = 0°$) on the tracing paper and on the maps, for example, Fig. 4.2, and moving the tracing paper along the equator in such a manner that a point chosen on the path with the known local time coincides with the corresponding meridian in the maps, one can take off the parameters of the ionospheric channels along the path for various operating frequencies. The analysis of the specific paths performed by means of the maps of the IWD parameters have been given earlier (Figs. 4.3 – 4.6) and will be considered in detail in Sec. 5.1.

Of course, to construct the maps for the prediction of radio wave propagation over long distances for different geophysical conditions is only possible on the basis of the corresponding ionospheric model. The model should reflect with sufficient reliability and completeness both the height variation of electron concentration and its latitudinal – longitudinal, diurnal, seasonal, sunspot cycle, and other variations. At the present time, there is a number of global ionospheric models (see Appendix 2).

The Group Time Delay of the RTW Signals is defined in the adiabatic approximation by Eq. (3.32)

$$t_{\mathrm{RTW}} = \frac{R_0}{c} \int_0^{2\pi} \frac{d\Psi}{\sqrt{\varepsilon'_{\min}}} \left(1 + \frac{2\bar{z}(\Psi)}{R_0}\right), \tag{4.12}$$

where

$$\bar{z}(\Psi) = \frac{2}{R_0 \Theta} \int_{z_{\min}}^{z_{\max}} \frac{z\,dz}{\sqrt{\varepsilon'(z) - \varepsilon'_{\min}}}. \tag{4.13}$$

$\varepsilon'_{\min}$ is the value of ε' at the points of ray reflection, $z = z_{\max}$, or $z = z_{\min}$. The time delay on a given path depends on the ray trajectory (for example, propagation in the interlayer duct, hop propagation) and on the value of the adiabatic invariant.

Figure 4.14a illustrates the result of the calculation of the group time delay of the RTW signal on various paths when the radio wave propagates in the interlayer duct (Borisov et al. 1977, 1978). In this case, the upper curves, $t_{\max}$,

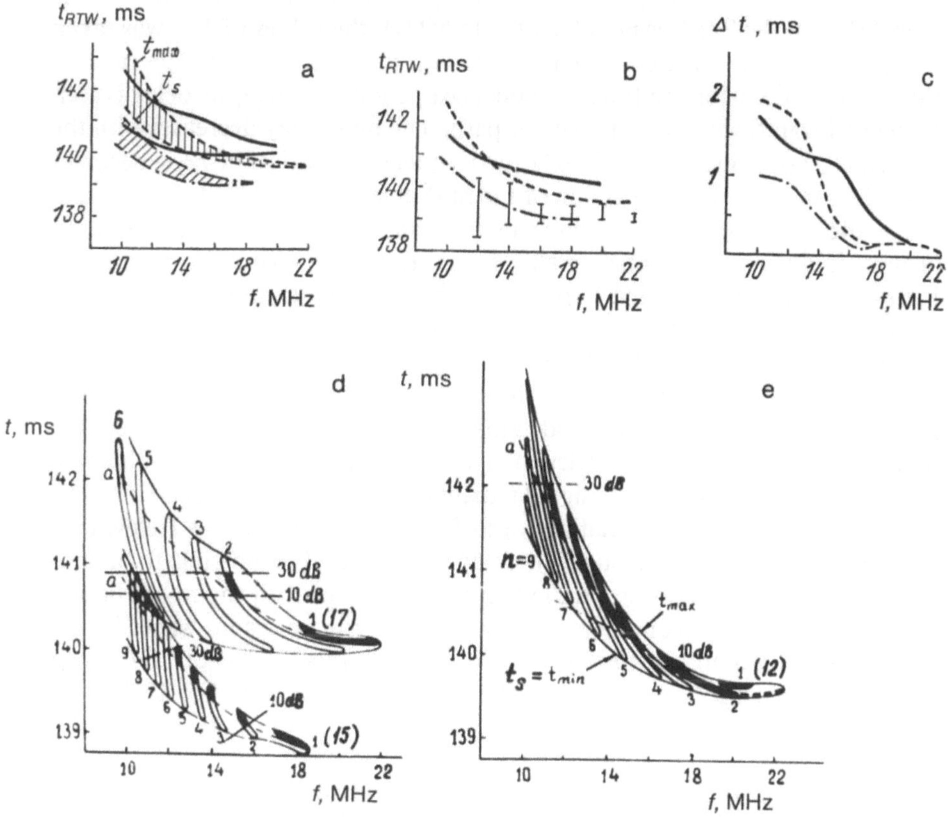

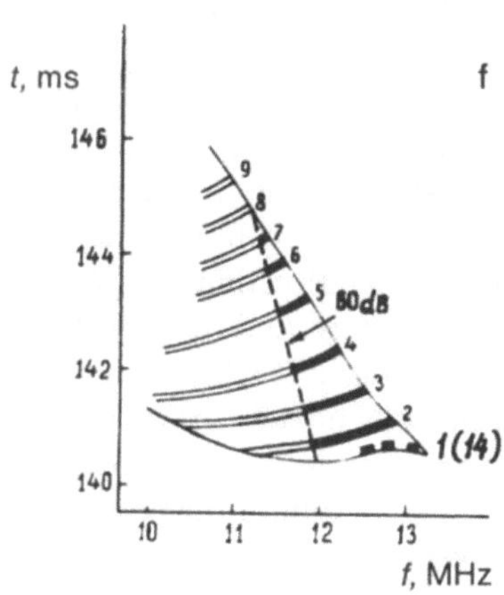

Fig. 4.14. **a** Maximum, t_{max}, and minimum, t_s, group time delay of RTW signals vs frequency for the noon–midnight meridional path (*dash-dotted line*), equatorial (*solid line*), and midlatitude (*dashed line*) paths.
b Quantity $t_{av} = (t_{max} + t_s)/2$ vs frequency. *Vertical lines* show the time delay of the RTW signal maximum τ_M, measured experimentally (Fenwick and Villard 1963c) (see also Fig. 1.3). **c** Dispersion $\Delta t = (t_{max} - t_s)$ for the above paths. **d–f** Ionograms of RTW signals. Mode numbers are indicated. *Numbers in parentheses* at the first mode number ($n = 1$) indicate the number of oscillations (hops) of the first mode. The *thick lines* show the modes or parts of the modes with smaller absorption

were determined for the maximum value of the invariant I_{max} on the path, the lower curves, t_s – for the minimum value, $I \sim 0$, i.e., for the sliding ray. To determine the maximum invariant, first, the total invariants of the ducts along the given paths were studied. It is the minimum value of total invariant of the duct F at the night side of a path, that is, the maximum invariant of the ray on the path for high frequencies. For low frequencies at which capture into the duct *FE* from the Earth is possible ($f \leqslant 15$ MHz), the values of I_{max} was taken to be equal to the value of the entry invariant into the duct *FE*. The calculation is carried out for the EMI-72 model (Appendix 2). The time delay is seen to depend significantly on the type of path. The time delay decreases with the frequency, however, at $f \geqslant 15$ MHz the frequency variation of t_{RTW} becomes slight (Fig. 4.14b). The dispersion of the time delay, $\Delta t = t_{max} - t_s$, is shown in Fig. 4.14c. The dispersion is seen to increase sharply for low frequencies, but at high frequencies it is very small. However, here one can also note small increase of dispersion at the frequencies close to the maximum frequency of the duct (analogous effect manifests itself more sharply in the spherically–symmetric ionosphere) (Chiu 1975).

Figure 4.14b illustrates the experimental results of the time delay measurements at Stanford (Fenwick and Villard 1963c). One can note a certain qualitative agreement of the calculated values [Eq. (4.12)] with the observed values of the time delay of the signal maximum and time delay dispersion. The observed time delay of the leading edge of the RTW signal at low frequencies is markedly less than the group time delay in the duct. It is possible to be connected with the propagation on the hop trajectories or with the ray deviation from the great circle plane. Apparently, the scattering of radio waves by the ionospheric inhomogeneities and also the Earth's magnetic field can noticeably influence the dispersion of the RTW signals (see Kim 1977, Kim et al. 1975).

The Structure of the RTW Signal Ionograms. Figure 4.14 d, e, and f show the RTW ionograms calculated for the model EMI-81, at $W = 10$ for various RTW paths with different propagation modes: d and e show the propagation by ricocheting trajectories over the entire RTW path in the ionospheric ducts *FE* and F^r; f shows the hop propagation in the duct F^h.

The ionogram calculation was carried out as follows. At fixed frequencies, in every duct, the parameters $\mathscr{P}$ [Eq. (3.27)] and t_{RTW} [Eq. (4.12)] were calculated as functions of I. This allowed to determine the propagation time $t^n_{RTW}(I_{0n})$ of each n–th mode [Eq. (3.29)] making $\mathscr{P}$ oscillations and having the invariant I_{0n}. Here, n is the mode number in the ionogram. The lowest mode in the ionogram corresponds to the least number of hops or oscillations $\mathscr{P}$ on the path. The low mode is usually called the first ($n = 1$), and the number of oscillations of every following mode is by unity more than that of the previous one. In the figures, the conventional width of a mode is accepted to be 0.1 ms. In Fig. 4.14 d and e the dash–dotted lines show regions of the ionograms (situated below the line, to the right or to the left) in which the mode absorption is less than 10 or 30 dB. In these figures, the dashed lines "a" cut off a part of the modes filling the upper levels in the ionospheric duct *FE*. Taking into account the conditions for capture into the duct (or for departure from the duct) it is most easy to output the energy to the Earth or inject it from the Earth to these levels. Shaded parts of the modes, restricted by the indicated curves, correspond to the signals arriving at the reception point with the minimum attenuation. In Fig. 4.14 f the modes with absorption less than 60 dB are shaded.

As it is seen from Fig. 4.14, the ionogram structure for the signals propagating in the ionospheric ducts F^{r}, FE on the trapped trajectories (Fig. 4.14 d, e) and in the hop duct F^{h} (Fig. 4.14b) is different. This is connected with the fact that in the ionospheric duct FE, the rays travelling on the bottom of the duct ($I_{\min} \approx 0$) with the minimum propagation time, $t_{\min} = t_{\mathrm{s}}$, have the maximum number of oscillations in the duct, $\mathscr{P} = \mathscr{P}_{\max}(I_{\min})$. On the other hand, in the hop duct F^{h} (or E^{h}) the low rays having the minimum propagation time $t_{\min}(I_{\min})$ have the minimum number of oscillations $\mathscr{P} = \mathscr{P}_{\min}(I_{\min})$, in contrast to the upper rays for which $t = t_{\max}(I_{\max})$, $\mathscr{P} = \mathscr{P}_{\max}(I_{\max})$. Note that the ionogram structure remains similar (Fig. 4.14f), if in the duct F the rays pass a part of the path on the ricocheting trajectories (see Sec. 5.1).

Comparing the experimental RTW ionograms (Fig. 1.4) with the calculated ones, one can conclude that the discrete structure of the experimental ionograms is accounted for by the mode structure of signals, similarly to the structure of the ionograms of oblique and oblique-back sounding. The general form of the ionogram shows that the ionograms in Fig. 1.4 a – f correspond to the hop propagation with the possible ricochet over a part of the path. But the structure of the ionogram 1.4 f is closer to the ionogram structure Fig. 1.4 d, e, i.e., here the propagation in the ionospheric interlayer duct seems to take place. Let us also note that the tracks with long time delay sometimes observed in the OS ionograms (see Fig. 1.10) are also like some separated first modes of the ionospheric duct or like the structure formed by all the modes of the duct. Due to little difference in time delay between the neighboring modes, all the modes of the duct appear to be merged in the ionograms (see also Sec. 5.1). Note that when comparing the ionograms, one should bear in mind that the experimental ionograms in contrast to the calculated ones are obtained for the conditions of high solar activity. The scattering seems to strongly influence the structure of the ionograms (see Sec. 6.2).

The Azimuthal Deviations of Trajectories. In the adiabatic approximation, the departure of trajectory from the great circle plane (lateral deviation) and also the deviation of bearing and initial azimuth are described by Eqs. (3.47) to (3.56), (3.60), and (3.62).

Figures 4.15 and 4.16 show the results of computations of the azimuthal deviations for the sliding ray propagating on the bottom of the duct FE and F over the meridional path (Manaenkova and Tsedilina 1981, 1982). The computations are performed by Eqs. (3.56), (3.60), and (3.62), on the basis of the model EMI-81, $W = 10$ (Appendix 2). The latitude of the emission point was accepted to be 55°, which corresponds to the geographical latitude of Moscow. Figure 4.15a illustrates the dependence of θ_{s} and β_{s} on the angle φ and latitude θ at $f = 15$ MHz and for the meridians $\varphi_1 = \pi/2$, $-\pi/2$ (the meridian evening – morning), $\varphi_2 = \pi/4$, $-3\pi/4$, $\varphi_3 = -\pi/2$, $\pi/2$ (the

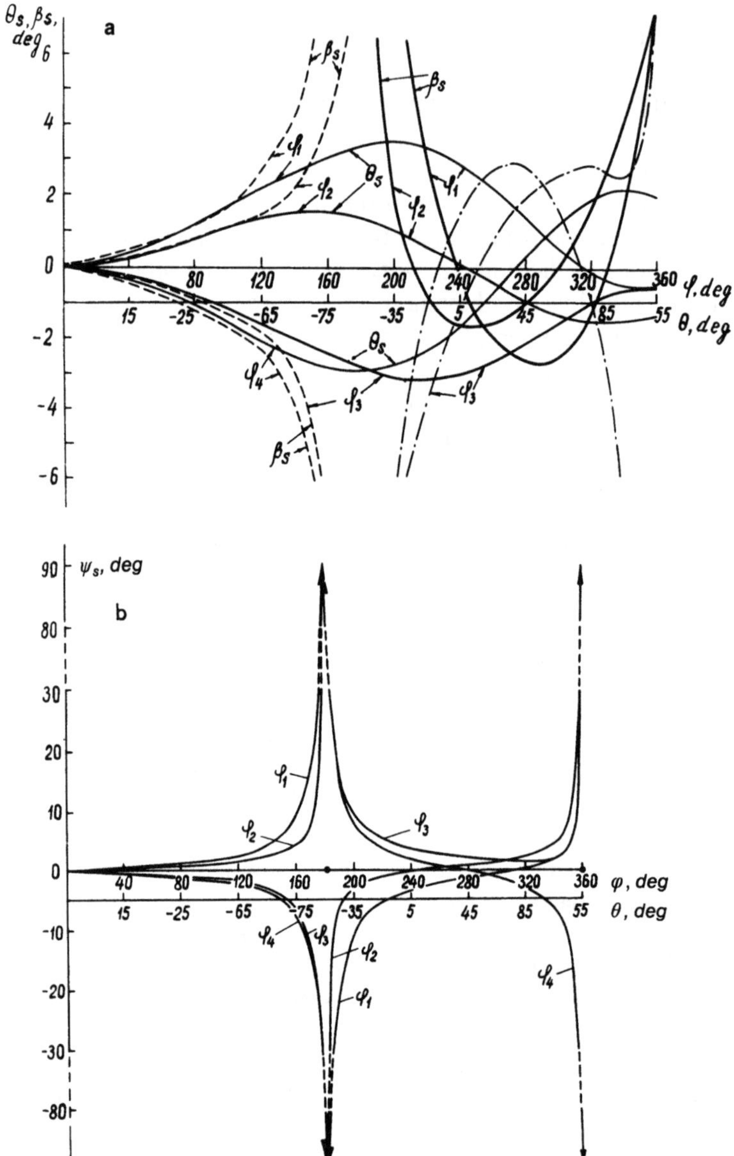

Fig. 4.15a, b. Variation of the sliding ray azimuthal deviations: **a** θ_s and β_s; **b** variation of ψ_s along the meridians φ_1, φ_2, φ_3 for $f = 15$ MHz

meridian morning – evening) and $\varphi_4 = -\pi/4, 3\pi/4$. Here, at the meridians φ_1 and φ_2, the values of θ_s and $\beta_s > 0$ correspond to the ray deviation towards the night zone, while at the meridians φ_3 and φ_4, they correspond to the ray deviation towards the day zone. The lateral deviations of the ray θ_s are seen to increase first with φ and then after passing of the pole, $\varphi = 145°$, $\theta = -90°$,

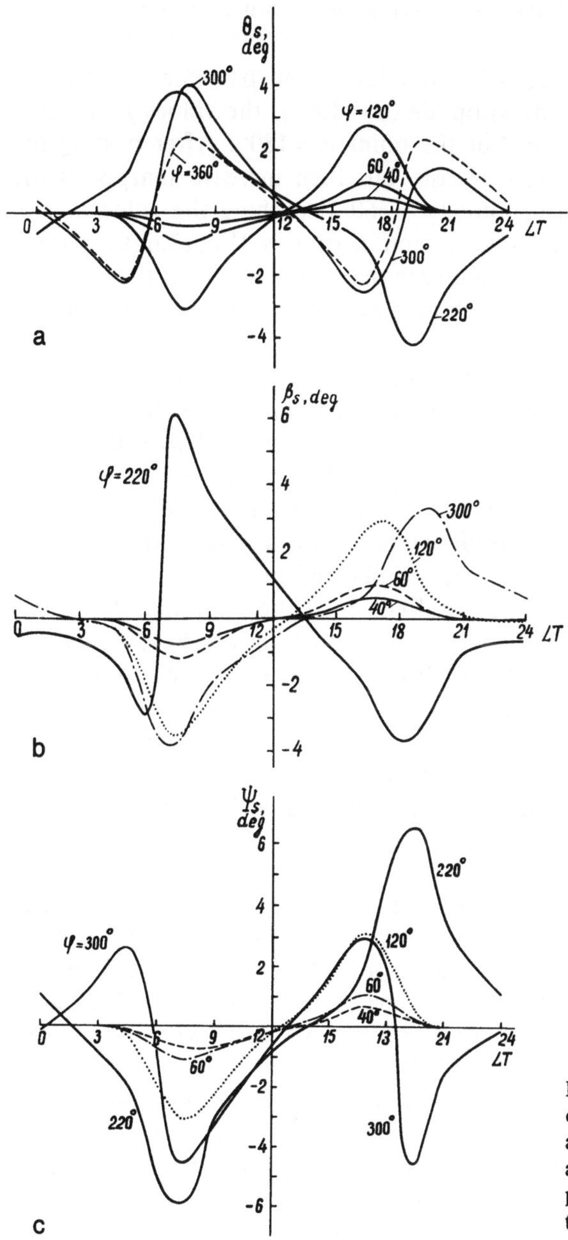

Fig. 4.16 a – c. Diurnal variations of the azimuthal deviation θ_s, β_s, and ψ_s on the meridional path. The angular distance from the emission point Ψ in degrees is indicated at the curves

the lateral deviations gradually decrease. The maximum values of $\theta_s \sim 3°$ are observed in the region shifted from the antipode ($\varphi = 180°$) by an angle of $\varphi \sim \pm 20°$. The maximum values of θ_s are observed as could be expected at the meridians morning – evening and evening – morning (the curves φ_1 and φ_3) or close to them, i.e., in the region of the terminator. When passing the entire round-the-world path ($\varphi = 360°$) the lateral deviations decrease in comparison with the region of antipode. However, there is no complete compensation in the lateral deviations.

The deviation of bearing, β_s, when $\varphi < 120°$ is of the same order as the value θ_s. When approaching the antipode ($\varphi = 180°$), the values β_s increase rapidly reaching the value of $\pi/2$ at the point $\varphi = 180°$. After passing the antipode, the values of β_s first decrease, but then increase sharply in the vicinity of the round-the-world point ($\varphi = 360°$) reaching the value $\sim\pi/2$. The dependence of deviation of the emission bearing ψ_s [Eq. (3.62)] on φ for $f = 15$ MHz is shown in Fig. 4.15b. It is seen that in the region of the antipode and round-the-world point, the discontinuity of ψ_s and change of the sign take place.

The diurnal distribution of θ_s, β_s, and ψ_s for $f = 15$ MHz is given in Fig. 4.16 a – c for different values of φ, indicated at the curves. It is seen that up to near the antipode ($\varphi \leqslant 180°$), the quantities θ_s, β_s, and ψ_s in the morning and evening time are opposite in sign. The change of the sign for $f = 15$ MHz, in contrast to $f = 10$ MHz, is shifted by 1 – 1.5 h towards the afternoon zone. Over the longer meridional paths, the azimuthal deviations of the sliding ray during 24 h have a more complicated structure.

Note that on the whole the calculated values of the azimuthal deviations correspond to the experimental ones (Fenwick and Villard 1963a, Agaryshev and Unuchkov 1977, Agaryshev 1981). However, the bearing deviation $\sim\pi/2$ were not observed in the experiments. It is likely to be accounted for by the fact that measurements were not carried out close enough to the antipodal or round-the-world point. Besides, one should bear in mind that for real ionospheric conditions the scattering influences greatly the bearing over the long-distance paths (Chap. 6).

Displacement of the Antipode Spot. Figure 4.17 a as shows the approximate energy density distribution of the radio wave field at the antipode, demonstrating the displacement of the antipodal "spot" (Manaenkova and Tsedilina 1981, 1982). For simplicity of the calculations, the transmitter was located at the pole ($\theta = \pm\pi/2$) of the spherical coordinate system in which the model $N_e(z, \theta, \varphi)$ has been constructed. In Fig. 4.17a the trajectories of radio waves for frequency $f = 15$ MHz, passing in the vicinity of the antipode ($\varphi = 180°$), are shown by the solid thick lines. The transmission was carried out along the meridians φ_1, the values of which are given at the corresponding trajectories. It is seen that due to the lateral deviation all the trajectories are intersected not at the antipode, but within a certain area displaced towards the night zone. The boundary of the region in which trajectories have the maximum number of intersections is shown by the curve 1. This region characterizes the spreading area of the antipode. The maximum transverse size of the area is equal to $\sim 3°$ or 300 km. However, some intensification of emission power can be observed in a greater region, the boundary of which is indicated by the number 2. The size of the region is much more. The calculated results for the antipodal spot sizes are in agreement with the experimental results (Pipp and

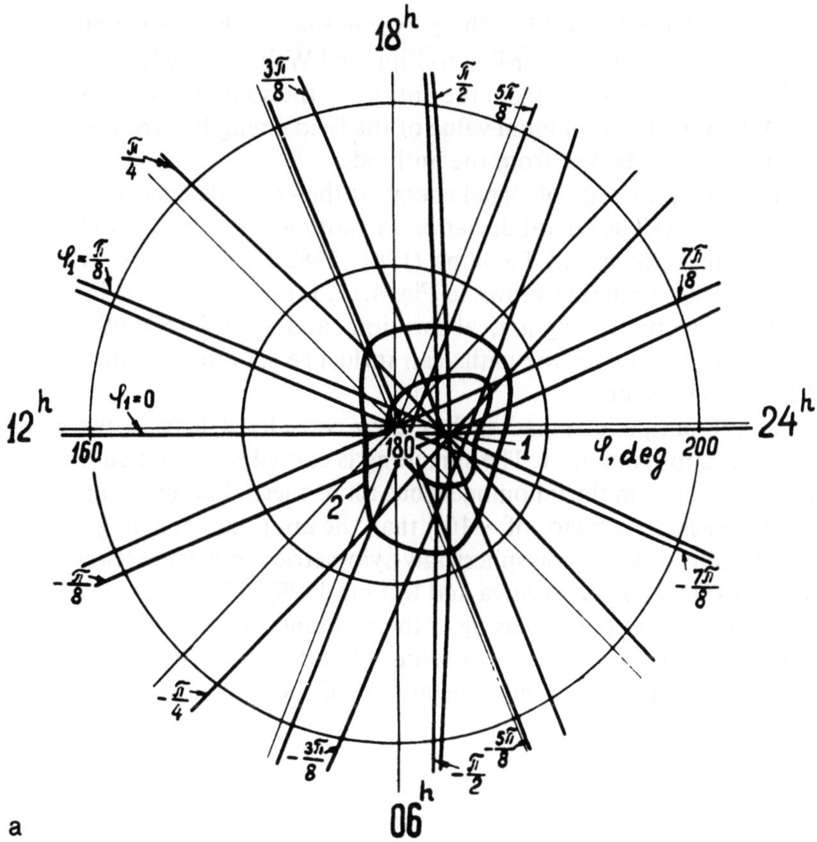

a

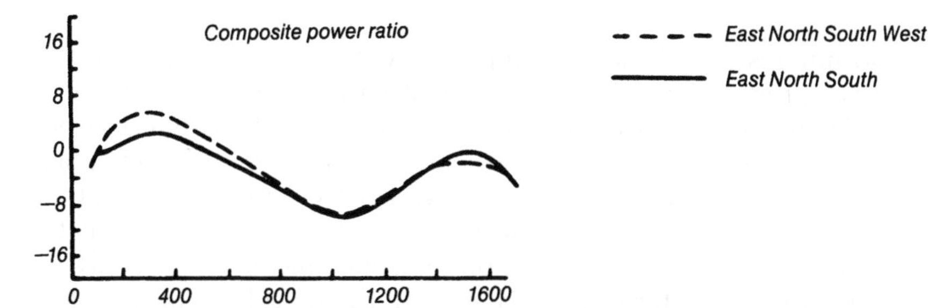

b

Fig. 4.17. a Trajectories of the sliding ray in the vicinity of the antipode for $f = 15$ MHz (the emission point is situated at the pole $\theta_0 = 0$). Spreading and displacement of the antipodal spot is illustrated. **b** Averaged ratio of the power measured with aircraft near the antipode to the power measured just at the antipode. The aircraft was moving away from the antipode to the north, south, east, and west

Webster 1964) shown in Fig. 4.17b. The curves in Fig. 4.17b have mainly two maxima. The first maximum according to Pipp and Webster (1964) is situated at a distance of ~200 – 400 km from the antipode, the second is at a distance of ~1400 – 1600 km. The minimum value of the field strength corresponds to a distance of ~800 – 1000 km from the antipode.

The spot size of spreading of signal energy at the antipode depends on the wave frequency, as well as lateral deviation of radio waves. For $f = 10$ MHz, according to Manaenkova and Tsedilina (1981, 1982), the spot size is by a factor of 2 – 3 greater than that shown in Fig. 4.17. It goes without saying that the transmitter directivity diagram, absorption, and specific conditions of radio wave propagation over the paths can influence significantly the structure of the antipodal spot.

Note that the antipode spreading has been considered here in the geometrical – optics approximation. The spreading is caused by the lateral deviations of the trajectories in the inhomogeneous ionosphere. This effect appears to be much stronger (by a factor of $\sim 10^3$) than the antipode spreading due to diffraction of radio waves in the spherically-symmetric ionosphere (Bremmer 1946, Furutsu 1951, Vsekhsvyatskaya and Kalinin 1979). Thus, the role of the wave effects for the antipodal focusing in the real ionosphere is insignificant. However, one can expect that the scattering of radio waves by ionospheric irregularities will influence markedly the antipodal focusing (see Sec. 6.1).

4.2 Radio Wave Absorption

The absorption is one of the most important characteristics of long-distance radio wave propagation. The absorption determined the intensity of the received signal. Among various possible radio paths, the minimum absorption path appears to be singled out. Consequently, the directions of minimum absorption give the optimum azimuth of RTW signals. Besides, the absorption determines to a great extent the frequency range of RTW as well. The cut-off at low frequencies is the consequence of absorption increase. The analysis of ionospheric absorption for equinoctial conditions on the basis of the calculations carried out for the analytical models of N_e and ν_e distribution (see Appendix 2) is given below.

The Technique for Calculation of Absorption. The absorption Γ which a radio wave undergoes over the half period of oscillations is equal to

$$\Gamma(\theta,\varphi) = \frac{2\pi e^2}{m c \omega^2} \int_{z_{\min}}^{z_{\max}} \frac{N_e(z,\theta,\varphi) \cdot \nu_e(z,\theta,\varphi)}{\sqrt{\varepsilon'(z,\theta,\varphi) - \varepsilon'_{\min}(\theta,\varphi)}}\, dz\,, \tag{4.14}$$

where the effective dielectric constant ε' is defined by Eq. (4.1).

If the absorption Γ is in dB, wave frequency is in Hz, N_e is in cm^{-3}, ν_e is in s^{-1}, and z is in km, then the factor before the integral in Eq. (4.14) is

$$\frac{2\pi e^2}{mc\omega^2}=0.582\cdot 10^3\frac{1}{f^2}\,.$$

When calculating the absorption of radio waves which propagate on multihop or trapped trajectories in the ionosphere with a great number of oscillations along the path, it is convenient to use the absorption K [Eq. (3.30)] averaged with respect to the oscillation period Θ

$$K=\frac{2\Gamma}{\Theta}\,,\qquad \Theta=\frac{2}{R_0}\int\limits_{z_{\min}}^{z_{\max}}\frac{dz}{\sqrt{\varepsilon'-\varepsilon'_{\min}}}\,. \tag{4.15}$$

Here, if Γ is in dB, Θ is in radians, then K is in dB/rad. For sliding waves, i.e., the waves travelling at the constant height $z=z_s$ with the invariant $I\approx 0$, the averaged absorption K_s is equal to

$$K_s=3.72\cdot 10^6\frac{N_e(z_s)\cdot \nu_e(z_s)}{f^2}\ \mathrm{dB/rad}\,. \tag{4.16}$$

The total absorption A of a radio wave on the path, the length of which is measured by the value of the central angle $\Psi>0$, can be represented in the form [Eq. (3.31)]

$$A\approx\int\limits_0^{\Psi}K\,d\Psi\approx\sum_i \Delta\Psi_i K_i\,, \tag{4.17}$$

where $K_i=K(\theta_i,\varphi_i)$ are the values of K calculated by Eqs. (4.15) and (4.16) at the fixed points θ_i, φ_i being spaced $\Delta\Psi_i$ apart along the path.

The calculation of the absorption global distribution in the IWD was carried out by Eqs. (4.14) – (4.16) on the basis of the analytical model EMI-72, EMI-81, $W=10$, constructed for equinoctial conditions (Appendix 2). The calculation techniques are as follows. First, the function $\varepsilon'(z)$ is analyzed at each given point with the coordinates θ, φ, the same as described in Sec. 4.1. The extreme points of the function $\varepsilon'(z)$ are determined. They are the maxima: $\varepsilon'_{\max E}=\varepsilon'_{sE}$, $\varepsilon'_{\max F}=\varepsilon'_{sF}$, and the minima: $\varepsilon'_{\min E}$, $\varepsilon'_{\min F}$, and the maximum, $z_{\max}$, and minimum, $z_{\min}$, values of the height z corresponding to these levels of ε'.

After performing such a preliminary analysis of the function $\varepsilon'(z)$, the values of absorption Γ and K, and the oscillation period Θ are calculated at each given point (θ,φ) for the levels corresponding to the values of ε' at the extreme points. Thus, for a given wave frequency at each fixed point, the following parameters are calculated: the values Γ_F, K_F, Θ_F, and Γ_{E+FE}, K_{E+FE}, Θ_{E+FE} for the "upper" and "lower" limiting rays, respectively, in the duct F, corresponding to the levels $\varepsilon'=\varepsilon'_{\min F}$ and $\varepsilon'=\varepsilon'_{\min E}$ (see Fig. 4.1a); Γ_E, K_E, Θ_E, Γ_{FE}, K_{FE}, Θ_{FE} for the upper limiting rays in the ducts E and FE with the level $\varepsilon'=\varepsilon'_{\min E}$; Γ_{SE}, K_{SE}, Θ_{SE}, and Γ_{SF}, K_{SF}, Θ_{SF} for the sliding rays travelling at the levels $\varepsilon' = \varepsilon'_{\max E} = \varepsilon'_{SE}$ in the duct E, and $\varepsilon' = \varepsilon'_{\max F} = \varepsilon'_{SF}$ in the duct F, Γ_{1E}, K_{1E}, Θ_{1E}, and Γ_{1F}, K_{1F}, Θ_{1F} are the absorption and the oscillation periods of the horizontal ray in the duct E or F, respectively, which correspond to the level $\varepsilon' = 1$.

Note that when calculating the absorption of the upper limiting rays in the ducts E, F, FE, in order to avoid the logarithmic divergence in Eqs. (4.14) and (4.15) at $\varepsilon' = \varepsilon'_{\min}$, the computations were performed not for the level $\varepsilon' = \varepsilon'_{\min}$, but for a certain level $\varepsilon'_{\Delta} = (1 + \Delta)\varepsilon'_{\min}$ which is slightly different from the $\varepsilon'_{\min}$, $\Delta \approx 0.002$ (Fishchuk and Tsedilina 1975, Fishchuk 1974, 1975).

After calculating the limit absorption at extreme points of the function $\varepsilon'(z)$, the values Γ, K, Θ in all the ducts for various intermediate values $\varepsilon'(\varepsilon' = \varepsilon'_i = \varepsilon_{\min})$, and appropriate values of the invariants I, are calculated. This allows to determine the parameters Γ, K, Θ, $z_{\min}$, $z_{\max}$ versus the invariant I, i.e., for various rays propagating in the ducts F, E, FE. Further, these parameters, in contrast to the limit ones, will be marked by the index^I, for example, Γ^I, K^I, Θ^I and so on.

Absorbing Properties of the IWD. The absorption distribution in the duct E, F, FE with respect to radio wave invariant I or height of the ray reflection, $z_{\max}$, was calculated for different frequencies. Figure 4.18 shows some results of the calculation for the ducts E, FE, F.

Let us consider variations of the parameters shown in Fig. 4.18. The value $z_{\max E} = z_{1E}$ in the given case is equal to 91 km and corresponds to the height of the horizontal ray reflection and gives the upper limit to the part of the duct E where radio waves propagate on the ricocheting trajectories (see Fig. 4.1a) without touching the Earth's surface. In this "ricochet" part of the duct E, the oscillation period Θ increases almost linearly with $z_{\max E}$ from $\Theta_s \sim 0.1$ rad to $\Theta_{1E} \sim 0.32$ rad. Such a variation of Θ corresponds to rapid widening of the duct E and to increasing of its volume. The variation of Γ^I_E and K^I_E in this part of the duct is more involved. The functions $\Gamma^I_E(z^I_{\max E})$ and $K^I_E(z^I_{\max E})$ have the peaks due to the maximum of the product $N_e(z)\,\nu_e(z)$ in the D region of the ionosphere (Fatkullin et al. 1981). The variation of K^I_E, Γ^I_E, Θ^I_E in the ground-based part of the duct E, where radio waves propagate by hop trajectories (i.e., within $z^I_{\max E}$ from 91 to 104 km) differs significantly from the previous one. So, the parameter Θ^I_E decreases from $\Theta_{1E} = 0.32$ to $\Theta = 0.11$, and then somewhat increases. The maximum value of Θ (or the maximum length of the hop) corresponds to the horizontal ray, while the minimum value of Θ corresponds to the ray, close to the limit one. When changing $z_{\max E}$ from $z_{1E} = 91$ km to $z^I_{\max E} = 104$ km, the parameter Γ^I_E first drops and then increases towards the duct boundary. The increase of the values of Θ and Γ at the boundary of all the ducts is connected with the logarithmic divergence in Eqs. (4.14) and (4.15). In addition, at the height close to the maximum height of the E-layer, the second maximum of the product $N_e(z)\,\nu_e(z)$ is situated. That is why, when reflecting at these heights, a wave undergoes the maximum attenuation. The quantity K^I_E involving the variation of the two parameters

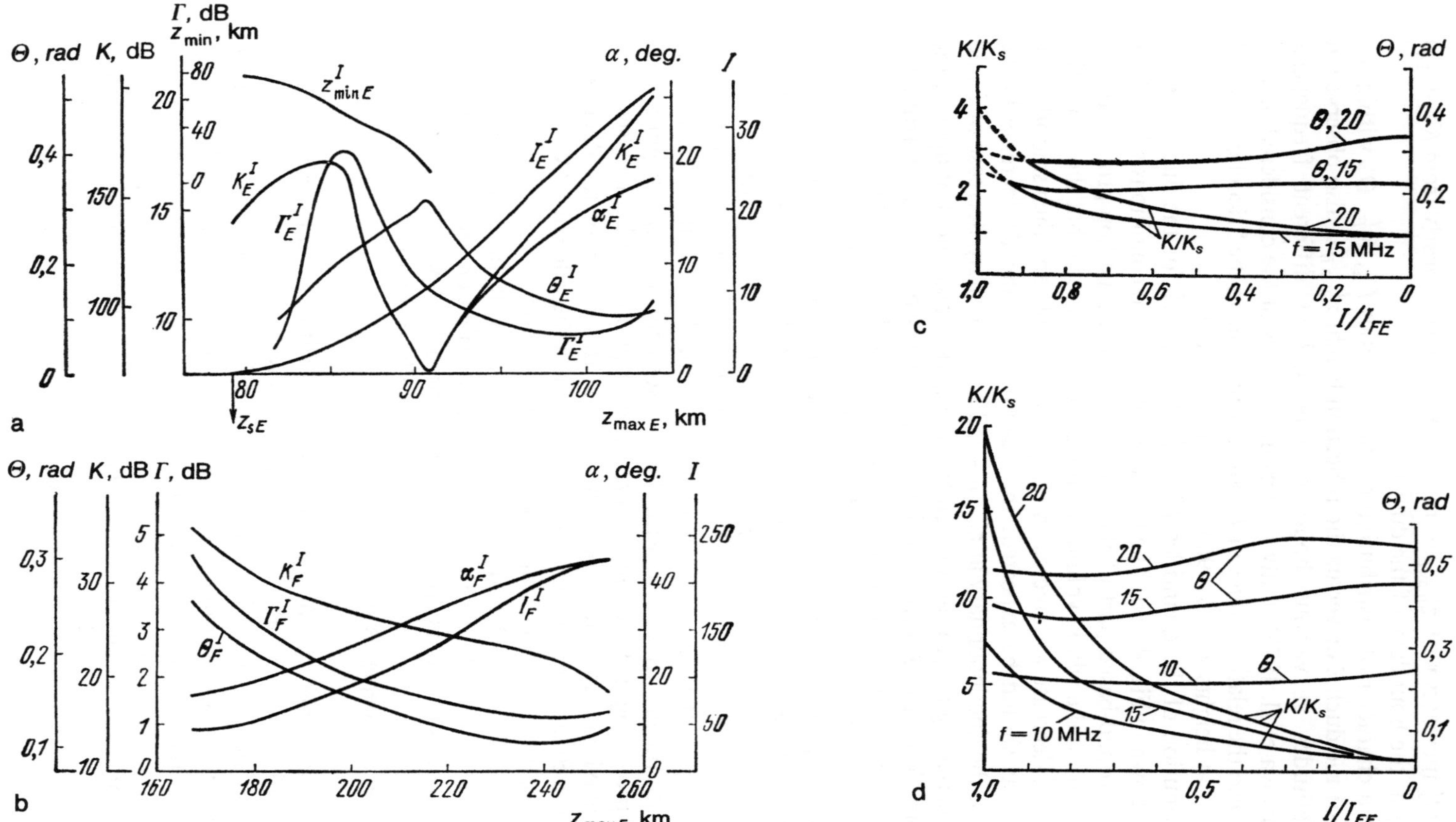

Fig. 4.18. a Quantities K, Γ, Θ, α, I in the duct E at the equator vs the maximum reflection height of a ray for $f = 10$ MHz, LT = 12^h; **b** Quantities K, Γ, Θ, α, I in the duct F at the latitude 40° vs the maximum reflection height of a ray for $f = 10$ MHz, LT = $10^h40'$; **c** Quantities K/K_s and Θ in the duct FE vs the ratio I/I_{FE} for the equator, LT = 12^h. **d** The same as in Fig. 4.18c for $\theta = 50°$

Γ_E^I and Θ_E^I increases with $z_{\max E}^I$ or the angle α_E almost linearly from $K_{1E}^I(z_{1E})$ to $K_{\max E}$ at the boundary of the duct E.

Such a character of distribution of the quantities Γ_E^I, K_E^I, Θ_E^I within the volume of the duct E or versus the reflection height $z_{\max E}^I$ is also observed at other latitudes at various hours of day and at different frequencies (Tushentsova et al. 1975). Only the absolute values of the considered quantities and also the relative values of Γ_E^I and K_E^I maxima in the "ricochet" part of the duct E at heights $z \sim 75-96$ km change markedly.

Distribution of the quantities Γ_F^I, K_F^I, Θ_F^I, I_F, α_F^I for the duct F versus $z_{\max F}^I$ is shown in Fig. 4.18b. The figure corresponds to conditions when there is only a ground-based part of the duct F, i.e., when the hop propagation is possible within the whole volume of the duct F. It is seen that the quantities Θ_F^I, Γ_F^I, K_F^I decrease with increasing $z_{\max F}^I$ first sharply and then smoothly. In this case, as should be expected, the quantities Θ_F^I and Γ_F^I increase in the vicinity of the duct boundary at $z_{\max F}^I \sim 252$ km. The sharp decrease of Γ and Θ with increasing $z_{\max F}^I$ is accounted for by rapid displacement of apex of the trajectory from the E-layer maximum (and, consequently, also from the region of the maximum product of $N_e(z)\,\nu_e(z)$ which defines the absorption).

Figure 4.18 c, d illustrates the distribution of the values of K/K_s and Θ characterizing the variation of the averaged absorption K and oscillation period Θ in the duct FE versus the ratio I/I_{FE} for the frequencies $f = 10$, 15, 20 MHz at noon hours. Here, I_{FE} is the total volume of the duct FE. Figure 4.18c is plotted for the equator, Fig. 4.18d is plotted for the latitude $\theta = 50°$. As is seen from the figures, the absorption in the duct FE first varies rather slowly with the growth of I or with the displacement from the height z_s, but when approaching the duct boundary, the absorption sharply increases.

The maximum rate of increase of K/K_s in this region of the duct takes place at the equator and at high latitudes for which the value of K/K_s increases by a factor of $7-20$ in comparison with the region $z \approx z_s$ for frequencies $10-20$ MHz (Fig. 4.18c) and by a factor of $3-4$ at the midlatitudes for frequencies $15-20$ MHz (Fig. 4.18b). The distribution of period Θ also differs somewhat from its distribution in the "ricochet" part of the duct E and F. This is connected with the form of the duct FE. Here, the distribution of Θ is smoother and depends slightly on I.

Global Regularities of the Absorption. According to the results of the previous subsection, the absorption, K, in the IWD lies between K_{SE} and K_{1E}, K_{1E} and K_E in the duct E, between K_{E+FE} and K_{1F}, K_{1F} and K_F in the duct F and between K_{SF} and K_{FE} in the duct FE. The same takes place for the absorption Γ. Latitudinal – longitude distribution of these parameters can give the possible lower and upper limits for the absorption variation, when radio waves propagate in the ducts E, F, FE along the specific paths. For investigation of latitude-longitude absorption variation, the values of absorption and corresponding oscillation period Θ at the fixed points θ, φ for a number of frequencies f were calculated. Maps of the global absorption

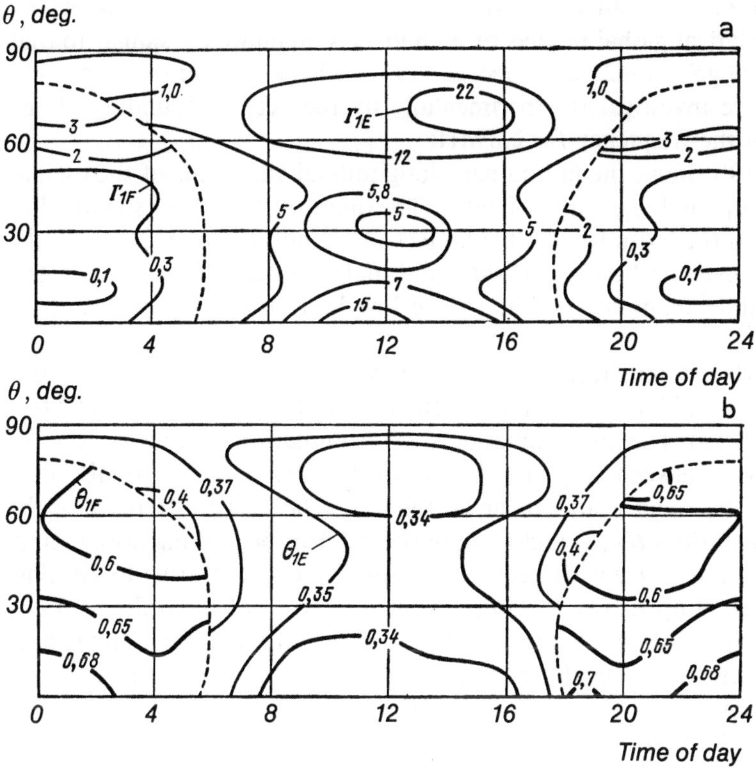

Fig. 4.19. a Curves of the latitudinal – diurnal distribution of Γ_{1E} and Γ_{1F} (in dB) for $f = 7$ MHz; **b** the same for Θ_{1E} and Θ_{1F} (in radians)

distribution are plotted according to this data (Tushentsova et al. 1975b, 1977c). Two such maps are shown in Fig. 4.19. They show the isopleths of Γ_{1E}, Γ_{1F} for $f = 7$ MHz and also the curves of corresponding values Θ_{1E} and Θ_{1F} in radians. The broken curves give the regions of wave reflection from the E and F-layers. As shown in Fig. 4.19, in the daytime at low and midlatitudes and also during 24 h at high latitudes, the horizontal ray is reflected from the E-layer, while at night at low and midlatitudes it is reflected from the F-layer. The main feature of the distribution of the absorption Γ (Fig. 4.19), as well as of all the other quantities Γ and K on all the frequencies is the noon minimum at the latitudes 30 – 40°. The maximum in absorption distribution is observed at high ($\theta \sim 60-70°$) and subequatorial latitudes ($\theta \approx 0°$). The night absorption in the middle and low latitudes makes up 1 – 3% from the day one, but it is considerable at high latitudes. The presence of minimum and maxima in latitude absorption variations is mainly accounted for by the fact that the maximum of the product $N_e(z)\,\nu_e(z)$ at the heights of the E-layer has the minimum value on the latitudes $\theta \sim 30-40°$, according to the ionospheric model used here. The values of the product $N_e(z)\,\nu_e(z)$ maximum at the latitudes $\theta \sim 30-40°$ differs from its maximum values at low and high latitudes by a factor of 5 – 10.

The picture of latitudinal – longitudinal absorption variations obtained by using the analytical global models of N_e and ν_e is qualitatively similar to that obtained in Fligel' (1962) and George (1971) where the global absorption variations were investigated experimentally by the vertical sounding of the ionosphere at the frequency $f = 2.2$ MHz.

Figure 4.19b shows the latitudinal – longitudinal variations of the oscillation periods Θ_{1E} and Θ_{1F} for $f = 7$ MHz. The quantity Θ_{1E} varies slightly, has a minimum on the latitudes $\theta \approx 0°$ and $\theta \approx 60°$ at noon, and increases to the morning and evening hours at middle and high latitudes. The value of Θ_{1F} increases mainly from the middle to the low latitudes (Tushentsova et al. 1975b, 1977c).

Figure 4.20 a, b illustrates the latitudinal variations of K and Θ in various ducts along the midday – midnight meridian for the frequency $f = 15$ MHz (EMI-81, $W = 10$). The night values of K at the midlatitudes are close to zero. The difference between K_{SF} and K_{FE} is of the order of 5 – 10 dB. It is maximum at the equator and at the latitudes 55 – 60°. The difference of the absorption quantities K_{E+FE} and K_F in the duct F for the limiting rays, mainly at all the latitudes except the equator, is of the order of 10 – 15 dB. The absorption of the upper ray K_{FE} in the duct FE can be of the order of K_F or somewhat higher than the absorption of the upper limiting ray, K_F, in the duct F. The difference between K_E and K_{SE} at all the latitudes, except the high ones, is no more than 5 – 10 dB, mainly, $K_{SE} > K_E$.

The diurnal variations of these quantities for the equator are seen from Fig. 4.20 c, d. Here, the difference between K_{SF} and K_{FE}, K_F and K_{E+FE} is of the order of 20 dB; K_F is somewhat less than K_{FE}; K_{SE} is significantly less than K_F at near noon hours, but at the evening and morning hours it is greater than K_E. The absorption is asymmetrical with respect to noon: in the evening hours (18^h) it is somewhat greater (by 5 – 7 dB) than in the morning hours (06^h).

The Absorption at the RTW Hop and Trapped Trajectories. The absorption of RTW signals propagating on *FE*, F^r trajectories trapped in IWD, and its comparison with the absorption of the signal propagating by hop modes, is given in Tables 4.1 and 4.2. In Table 4.1, for the frequencies 7, 10, 12, 15, 17, and 20 MHz for midday – midnight and twilight meridians at the equator, the values of the following parameters are given: Γ_E, A_{SE}, and A_{SF} which is the total absorption of the sliding ray on the RTW trajectory in the ducts F and *FE*, A_E and A_{FE} which is the total limiting absorption in the ducts E and *FE*, i.e., absorption of the rays which travel with the maximum invariants I_E and I_{FE}; the quantities A_{SF_0} and A_{FE_0} determining the absorption of the sliding mode and limiting ray in the duct *FE* with descent to the Earth, $A_{SF_0} = A_{SF} + 2\Gamma_E$, $A_{FE_0} = A_{FE} + 2\Gamma_E$. The quantities A_{SF_0} and A_{FE_0} are calculated for a comparison with the quantities $A_F(d)$ and $A_{1F}(d)$ which characterize the absorption of the upper and lower rays in the duct *FE* in the daytime (the absorption in the night part of the trajectory makes up 1 – 2% of the day one).

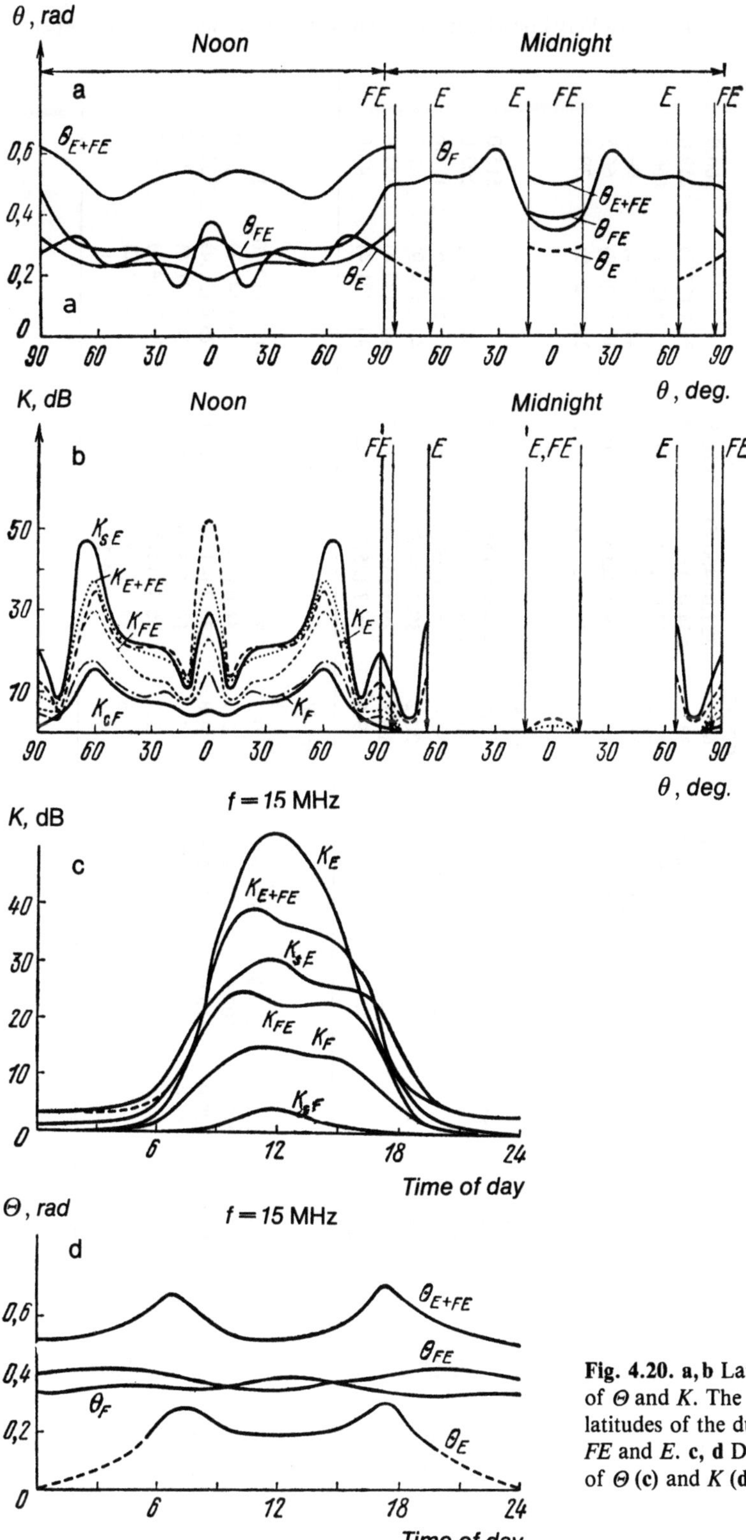

Fig. 4.20. a, b Latitude variations of Θ and K. The *arrows* indicate latitudes of the duct boundaries *FE* and *E*. **c, d** Diurnal variations of Θ (**c**) and K (**d**) at the equator

Table 4.1. Values of the total absorption A and quantity Γ_E in dB for various RTW paths

	f, MHz	Γ_E	A_{sE}	A_E	A_{sF}	A_{sF_0}	A_{FE}	A_{FE_0}	A_F (d)	A_{IF} (d)
		$(12^h$, $\theta=0°)$				12^h, $\theta=0°$		12^h, $\theta=0°$		
Equator	7	14.9 (10.1)	477.2		38.0 (144.8)	67.8	387.0 (460.5)	416.8		566.0
	10	10.7 (7.3)	286.0		11.8 (36.5)	33.2	154.0 (181.9)	175.4	85.7 (77.7)	259.0
	12	9.0 (5.8)	–	–	5.9 (15.4)	23.9	103.0 (112.4)	121.0	65.4 (54.7)	166.0
	15	7.2 (4.6)	–	–	2.5 (4.8)	16.9	52.7 (62.2)	67.1	44.2 (36.2)	94.5
	17	6.4 (4.0)	–	–	1.6 (2.6)	14.4	38.1 (44.8)	50.9	36.1 (28.8)	66.5
	20	5.5 (3.4)	–	–	0.9 (1.2)	11.9	24.5 (29.2)	35.5	29.5 (21.9)	39.2
		$(12^h$, $\theta=50°)$				12^h, $\theta=50°$		12^h, $\theta=50°$		
Day – night meridian	7	7.7 (6.4)	691.4	–	98.4 (295.4)	113.8	299.0 (418)	314.4	–	515.0
	10	5.6 (4.5)	–	–	35.5 (103.4)	46.7	119.0 (197.6)	130.2	94.8 (80.2)	207.0
	12	4.6 (3.7)	–	–	20.8 (56.0)	30.0	76.0 (100.2)	85.2	80.6 (53.6)	154.6
	15	3.8 (2.9)	–	–	12.6 (24.2)	20.2	39.2 (54.8)	46.8	48.2 (34.9)	81.2
	17	3.4 (2.5)	–	–	6.3 (14.8)	13.1	29.5 (38.2)	36.3	40.6 (27.4)	57.6
	20	3.0 (2.1)	–	–	3.5 (8.4)	9.5	16.7 (25.1)	22.7	30.2 (21.4)	43.0
		$(6^h 40'$ $\theta=50°)$				$6^h 40'$ $\theta=50°$		$6^h 40'$ $\theta=50°$		
Morning – evening meridian	7	3.0 (3.0)	406.0	299.0	3.3 (6.1)	9.3	86.2 (163.0)	92.2	117.0 (101.0)	181.0
	10	2.2 (1.9)	227.0	107.5	1.5 (2.4)	5.9	31.6 (66.3)	36.0	43.7 (43.3)	55.0
	12	1.9 (1.6)	208.0	115.0	1.0 (1.7)	4.8	20.2 (41.5)	24.0	26.4 (31.4)	32.2
	15	1.6 (1.2)	158.0	–	0.65 (0.98)	3.9	16.6 (23.6)	19.8	21.8 (21.6)	21.0
	17	1.6 (1.0)	–	–	0.52 (0.7)	3.7	6.4 (18.3)	12.8	14.7 (14.6)	20.2
	20	1.5 (0.9)	–	–	0.38 (0.57)	3.4	3.7 (7.3)	6.7	7.9 (11.0)	9.9
	22	– (0.8)	–	–	– (0.49)	–	– (4.1)	–	– (5.4)	–

Table 4.2. Values of the total absorption A and quantity Γ_E for midlatitude RTW paths

f, MHz	Γ_E, dB	A_{sF}, dB	A_{FE}, dB	A_{sF_0}, dB	A_{FE_0}, dB	A_F, dB	A_{1F}, dB
10	3.8	17.2	66.0	24.8	73.6	51.0	138.1
12	3.2	10.0	42.6	16.4	49.0	36.3	92.3
15	2.6	5.0	21.0	10.2	26.2	26.0	34.0
17	2.3	3.4	17.7	8.0	22.3	18.8	42.0
20	2.1	2.2	15.2	6.4	19.4	16.8	29.3

Note that in the horizontally – inhomogeneous ionosphere, the rays with the maximum invariants I_E, I_{FE} and I_F do not propagate over the RTW distances. The values for A_E, A_{FE}, and A_F are calculated here in order to obtain the limit numbers. The specific calculations of propagation for various rays (or invariants) over the long-distance paths are given in Sec. 5.2.

The data of Table 4.1 are calculated on the basis of the model EMI-72, $W = 10$. The second numbers indicated in parentheses for the quantities Γ_E, A_{sF}, A_{FE}, and A_F are calculated on the basis of the EMI-81 model. As is seen from the Table, the values of A_{sF} and A_{FE}, calculated on the basis of the EMI-81 model, are by a factor of 1.5 – 2 more than the values calculated on the basis of the EMI-72 model. This is connected with the fact that the height of the valley, $z_{\min FE}$, and the height of E-layer, $z_{\max E}$, is less for the EMI-81 model (for detail see Tushentsova et al. 1977c). The values of A_F and Γ_E are rather close.

In Table 4.2 the values of total absorption for the midlatitude RTW path (EMI-72) are given.

The dependence of RTW signal absorption on wave frequency plotted according to Tables 4.1 and 4.2 is seen from Figs. 4.21 – 4.23. Figure 4.21 illustrates the frequency dependence of the quantities A_{sF} and A_{FE} for the equatorial, midday – midnight, twilight, and midlatitude paths. These dependencies define the lower and possible upper limits of the total absorption variation in the RTW channel FE.

As is seen from Fig. 4.21, the ranges of the absorption variation in the ionospheric channel on a given RTW path and for a given frequency are rather wide. They significantly decrease with the growth of frequency and, consequently, with the decrease of channel volumes. Furthermore, in Fig. 4.21a the absorption, under conditions close to the discussed ones, on the midlatitude paths is given by the triangles. The data were obtained by Hess on the frequency 20 MHz and by Fenwick on the frequency 15.1 MHz as a result of comparison of the amplitude of RTW signals having circulated over the Earth once and twice. The experimental data are seen to be situated in the central part of the area restricted by the calculated limits in the absorption variation in the duct (the dash – dotted curves).

Figure 4.22a shows $A_{sF} = A_{\mathrm{RTW,min}}$ which are the minimum values of total absorption on various RTW paths (Fig. 4.21) and $A_{\mathrm{RTW,max}}$ which are the

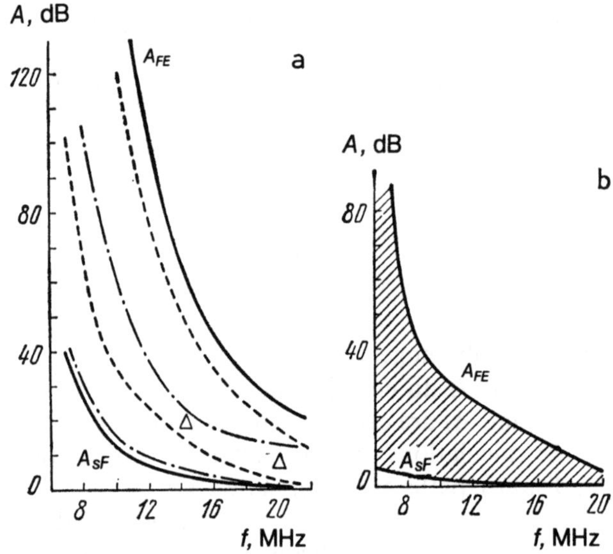

Fig. 4.21a, b. Frequency dependence of the total absorption A_{FE} and A_{sF} for various RTW paths: **a** equatorial path (*solid line*), noon – midnight meridional path (*dashed line*), midlatitudinal path (*dash-dotted line*); **b** meridional morning – evening path

maximum values of absorption in the case of propagation in IWD with the maximum invariant I_{max} (Tsedilina 1975, Tushentsova et al. 1977c). The values $A_{RTW,max}$ correspond to the ray which propagates with the invariant $I = I_{max}$ on the paths under consideration at the given frequencies (see Sec. 4.1; Fig. 4.6b). The values of $A_{RTW,max}$ are determined by means of calculation of the paths by the adiabatic invariant method (Tushentsova et al. 1977c, Fishchuk 1975). Figure 4.22b shows the frequency variation of the quantity $\bar{A}_{RTW} = (A_{max} + A_{sF})/2$ corresponding to Fig. 4.22a.

From comparison of Fig. 4.21 and the data of Tables 4.1 and 4.2 with Fig. 4.22, it is seen that the values of A_{FE} in the frequency range of 12 – 20 MHz mainly exceed by 10 – 20 dB the upper limit of absorption variation in the IWD on various paths.

As follows from Fig. 4.22, at the frequency $f < 15 - 16$ MHz, the RTW signal absorption is, on the average, higher (at the equator, by 20 – 30 dB and, at the day – night meridian, by 10 – 15 dB) than the absorption on the midlatitude paths. But at the frequencies $f > 16$ MHz, the absorption is small (less than 10 dB) and differs slightly on various paths. The value of A_{sF} has the minimum at the equator because here, according to the model of N_e, is the deepest valley.

Figure 4.23 illustrates the comparison of absorption, when radio waves propagate in the duct *FE* and *F* on trapped trajectories, and in the duct *F*, when the waves propagate by hop trajectories. It is seen that the upper limit of absorption on the trapped RTW trajectories [the value of $A_{FE_0}(12^h)$] is mainly higher than the lower limit when propagating by hops with reflections from the *F*-layer [the value of $A_F(d)$]. This means that, in principle, the absorption in the case of propagation by the two modes indicated are comparable. However, one should bear in mind that the passage by hop trajectories

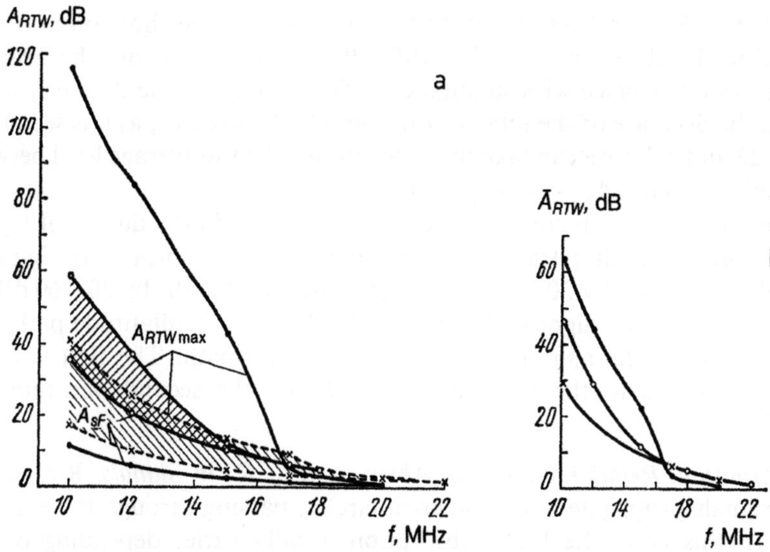

Fig. 4.22. **a** Minimum, A_{sF}, and maximum, $A_{RTW,max}$, absorption vs frequency for various RTW paths: equatorial path (*dots*), noon – midnight meridional path (*circles*), midlatitudinal path (*crosses*); **b** the same for averaged absorption $\bar{A}_{RTW}$

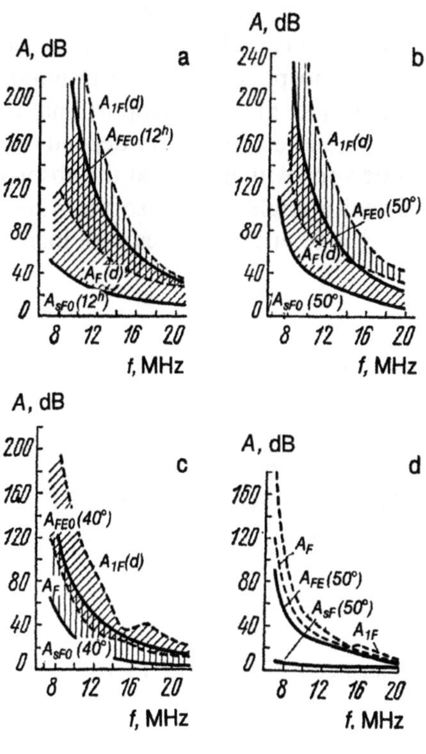

Fig. 4.23 a – d. Comparison of the total absorption on the hop trajectories with that on the trapped ones for various paths: **a** equatorial path; **b** noon – midnight meridional path; **c** midlatitudinal path; **d** meridional morning – evening path. The latitude of location of the ray descent to the Earth is indicated in *parentheses*

through the F-layer takes place at low angles α to the horizon, i.e., the absorption is close to $A_{1F}(d)$, while the propagation on the trapped trajectories takes place with simultaneous deepening into the duct FE, which leads to the decrease of the total absorption. On the average, as it is seen from Figs. 4.22 and 4.23, we can take the absorption to be the average level between the limiting values, $\bar{A}_{\mathrm{RTW}} = (A_{\max} + A_{sF})/2$.

From above we can conclude that the absorption in the duct F of signals propagating in multihop trajectories is higher than the absorption of signals in trapped trajectories by 20 – 100 dB for the equatorial path, by 20 – 60 dB for the meridional day – night path, by 15 – 50 dB for the midlatitude path, and by 2 – 30 dB for the meridional morning – evening path. The first value of absorption refers to the frequency $f = 20$ MHz, the second one refers to $f = 10$ MHz.

The Azimuthal Regularities of the Absorption of RTW Signals. Round-the-world signals propagate along the great circles, passing through the emission point. In this case, the RTW absorption usually varies depending on the emission azimuth. The direction which pertains to the minimum absorption appears to be optimum, i.e., when the conditions for the RTW propagation exist, the strongest signals must come from this direction.

The numerical calculations first carried out by Fenwick (Fenwick 1963, 1966) allowed to determine the optimum azimuth, $\beta^F_{\min}$. Here, this azimuth will be determined analytically on the basis of general assumptions about the character of radio wave absorption in the ionosphere. We bear in mind that the absorption primarily occurs on the day side, and it is proportional to the electron density in the lower ionosphere. The electron density depends on the position of the Sun, hence, we can assume approximately, that the absorption index, $\varkappa = \varkappa(\theta_s)$, where θ_s is the Sun's zenith angle. The absorption index $\varkappa$ has the maximum at the subsolar point, $\theta_s = 0$, and decreases with increasing θ_s. For example,

$$\varkappa \approx \varkappa_0 \cos\theta_s\,, \qquad 0 \leqslant \theta_s \leqslant \pi/2\,. \tag{4.18}$$

The total absorption on the RTW path is given by

$$A = \int \varkappa(\theta_s)\, ds\,. \tag{4.19}$$

To calculate the absorption A, it is convenient to draw the axis of the spherical coordinate system through the subsolar point $Q(\theta_s = 0)$ (Fig. 4.24); let us denote these coordinates θ_s, φ. We take into account that on the spherical surface of the radius R_0

$$ds = R_0 \sqrt{d\theta^2 + \sin^2\theta_s\, d\varphi^2}\,. \tag{4.20}$$

Besides, along the arc of any great circle, according to Eq. (3.10)

$$\left(\frac{d\varphi}{d\theta_s}\right)^2 = \frac{C^2}{\sin^2\theta_s(\sin^2\theta_s - C^2\cos^2\theta_s)}\,, \tag{4.21}$$

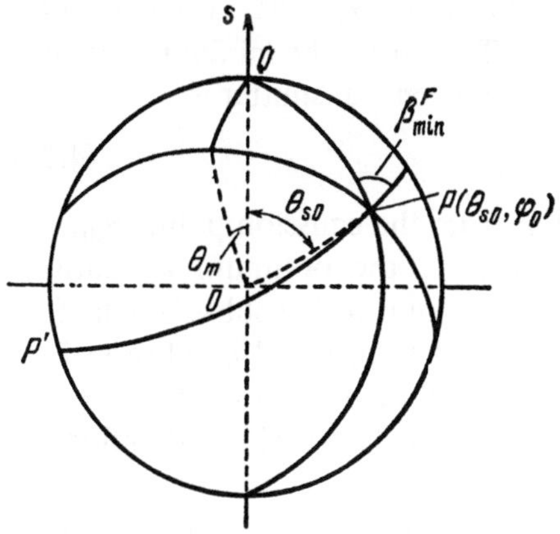

Fig. 4.24. The great circle arc passing through the point p with the coordinates θ_{s0}, φ_0 at the optimum azimuth $\beta = \beta^F_{\min}$; s is the direction of the subsolar point

where

$$\theta_s = \arctan\left[\frac{C}{\sin(\varphi-\beta)}\right]. \tag{4.22}$$

It is seen from Eq. (4.22) that the angle θ_s changes from $\pi-\theta_m$ to θ_m with changing angle φ, where θ_m is the minimum value of θ_s for a given great circle arc (Fig. 4.24). As follows from Eq. (4.22)

$$\theta_m = \arctan C. \tag{4.23}$$

Substituting Eq. (4.21) and Eq. (4.20) in Eq. (4.19) and taking into account Eq. (4.23), we obtain the following simple expression for the total absorption on the RTW path

$$A(\theta_m) = 2R_0 \int\limits_{\theta_m}^{\pi-\theta_m} \frac{\varkappa(\theta_s)\sin\theta_s\, d\theta_s}{\sqrt{\cos^2\theta_m - \cos^2\theta_s}}. \tag{4.24}$$

It is easy to make sure that if the function $\varkappa(\theta_s)$ decreases monotonically with increasing θ_s, then the total absorption A decreases monotonically with increasing θ_m. For example, in the case of Eq. (4.18), we have

$$A(\theta_m) = 2R_0\varkappa_0\cos\theta_m. \tag{4.25}$$

Thus, the RTW absorption is minimum for the ray travelling along the trajectory which corresponds to the maximum value of θ_m.

Now let us consider the transmitter situated at the point θ_{s0}, φ_0. All the great circles intersecting the emission point are described by Eq. (3.11)

$$\tan\theta_s = \frac{\tan\theta_{s0}\sin(\varphi_0-\beta)}{\sin(\varphi-\beta)}. \tag{4.26}$$

The minimum value of θ_m on a given trajectory, i.e., at fixed θ_{s0}, φ_0, and β, according to Eqs. (4.22), (4.23), and (4.26) is given by

$$\theta_m = \arctan[\tan\theta_{s0}\sin(\varphi_0-\beta)]. \tag{4.27}$$

Hence, as the azimuth β varies, the maximum value $\theta_m = \theta_m^{\max}$ is equal to θ_{s0} (or $\pi - \theta_{s0}$ if $\theta_{s0} > \frac{\pi}{2}$) at $\beta = \beta_{\min}^F = \varphi_0 \pm \pi/2$. Therefore, the minimum absorption azimuth or the optimum azimuth for the given transmitter is

$$\beta_{\min}^F = \varphi_0 \pm \pi/2 . \tag{4.28}$$

Let us bear in mind that φ_0 is the direction from the emission point, θ_{s0}, φ_0, towards the subsolar point, $\theta_s = 0$. Therefore, the optimum azimuth is orthogonal to the direction from the emission point to the subsolar one (see Fig. 4.24). Clearly, the minimum attenuation of signals will be observed on the paths intersecting the terminator at the small angles: $\theta_m \approx \pi/2$, $\beta = \beta_{\min}^F$ (Fenwick 1966, Golyan 1974, 1975; Ostrovskii 1974, Ostrovskii and Shlionskii 1975).

The simple relation, Eq. (4.28), is confirmed quite well by a number of experiments. However, more detailed investigation of features of RTW absorption shows possible marked deviations of the optimum azimuths from Fenwick's azimuth [Eq. (4.28)]. Below the calculated results of azimuthal dependence of the absorption for an equinoctial model of the ionosphere (EMI-72) are given (Tushentsova and Tsedilina 1976, 1977).

Figure 4.25 illustrates the azimuthal dependence of the total RTW absorption at various hours for the emission point at the latitude $\theta_0 = 40°$ on the operating frequency $f = 15$ MHz. Here, the quantity $\bar{A}_s = (A_{FE} + A_{sF})/2$ characterizes the average absorption of the signals propagating in the interlayer duct. The figure shows that at 11 and 15 o'clock, the curves of absorption have one deep minimum, while at 6 and 8 o'clock, they have two minima.

The presence of the two minima is also observed on other frequencies for the midlatitude emission point, while at high latitude emission points there is one minimum in azimuthal dependence of absorption. In Fig. 4.25 the direction of $\beta_{\min}^F$ for corresponding hours is shown by the arrow under the X-axis. In Fig. 4.25 one of the minima of A corresponds to the direction of $\beta_{\min}^F$. However, as is seen from Fig. 4.25, for midlatitude emission points, the azimuthal dependence of absorption $A(\beta)$ on the whole appears to be more complicated than in the simple model [Eqs. (4.25) and (4.28)].

This is also seen from Fig. 4.26 which shows the dependence of the minimum absorption azimuths for the sliding mode, $\beta_{s,\min}$, on local time for $f = 15$ MHz and $\theta_0 = 40°$. The dependence is plotted on the basis of numerical calculation. Here, the analogous dependence of the optimum azimuth, $\beta_{\min}^F$ [Eq. (4.28)], on time is shown (the thin line). Figure 4.27 illustrates the time dependence of absorption at the minima $A_{s,\min}(\beta_{s,\min})$. It is seen from Fig. 4.26 that there are two directions of the minimum absorption, namely, the latitudinal direction ($\beta_1 \approx 80 - 100°$) and the direction close to $\beta_{\min}^F (\beta_2 \approx \beta_{\min}^F)$. In Figs. 4.26 and 4.27, the direction of minimum absorption and corresponding values of $A_{s,\min}$ are shown by the solid thick curves. Thus, at noon hours the latitudinal direction is the main one, whereas at twilight hours the main direction is the meridional close to $\beta_{\min}^F$.

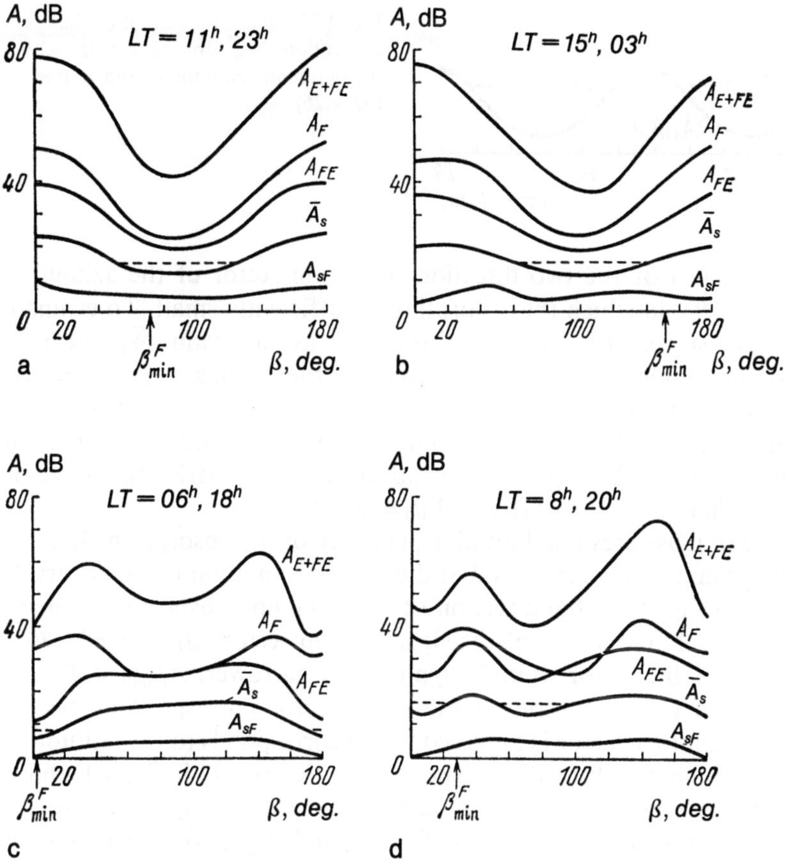

Fig. 4.25 a – d. Azimuthal dependence of the total absorption

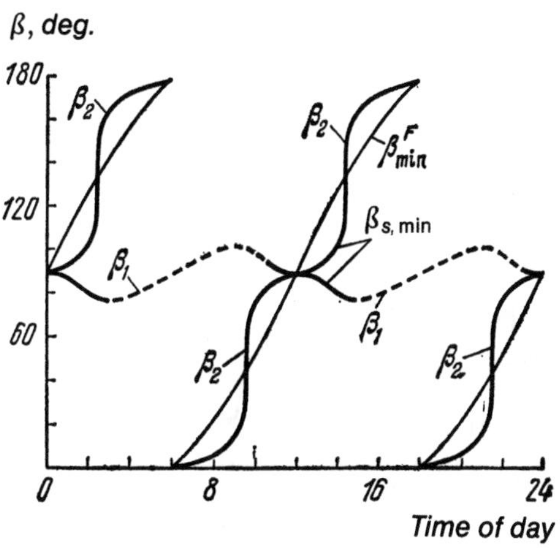

Fig. 4.26. Diurnal variations of the azimuths β^F_{min} and $\beta_{s,\,min}$

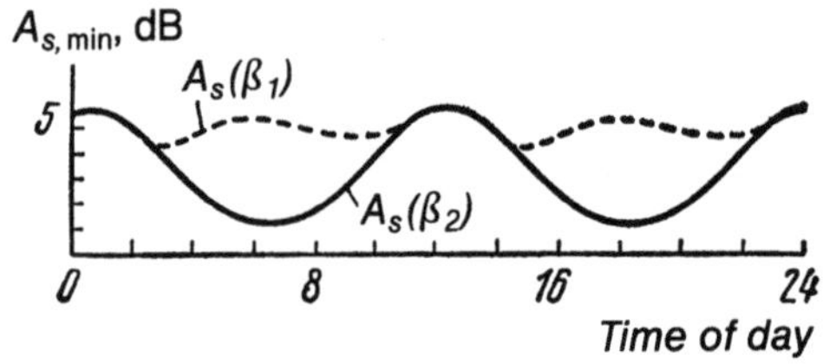

Fig. 4.27. Diurnal variation of the absorption $A_{s,\min}$ for RTW paths with the optimum azimuths β_1 and β_2 (see Fig. 4.26)

Clearly, in each of the two directions a certain sector of the azimuthal angles $\Delta\beta_{\min}$ exists, in which the absorption is sufficiently small. To estimate the value of the sector, it is convenient to use the average value $\bar{A}_s$. We have determined the azimuthal sector of arrival of the signals, $\Delta\beta_{\min}$, by the absorption at the level 1.3 $\bar{A}_{s,\min}$. This sector is shown in Fig. 4.25 by a broken line. Figure 4.28 shows the corresponding dependencies of minimum absorption directions, $\bar{\beta}_{s,\min}$ (solid line) and the azimuth restricting the sector $\Delta\bar{\beta}_{\min}$ (broken line) for $f = 15$ MHz and $\theta_0 = 40°$.

Figure 4.28b illustrates the diurnal dependence of the absorption $\bar{A}_{s,\min}$ at the main minimum (solid curve) and at the second minimum (broken curve). As is seen from Fig. 4.28, the direction of the minimum absorption at noon hours is latitudinal. The width of the signal arrival sector, $\Delta\bar{\beta}_{\min} \approx 60-80°$. At twilight hours, the sector width is significantly narrower: $\Delta\bar{\beta}_{\min} \approx 40°$. At intermediate hours, in addition to the main sector corresponding to the direction $\beta^F_{\min}$, another sector is observed which corresponds to the latitudinal direction of minimum absorption. The regularities of $\bar{A}_s$ indicated above correspond to the regularities of the absorption for the sliding signals, A_{sF}.

The ratios of the maximum values of $\bar{A}_{s,\min}$ and $A_{s,\min}$ to their minimum values during 24 h for $f = 15$ MHz and $\theta_0 = 40°$ are equal to 2 and 5, respectively (Figs. 4.27 and 4.28), i.e., the average distribution of the signal absorption in the interlayer duct *FE* during 24 h is quite smooth.

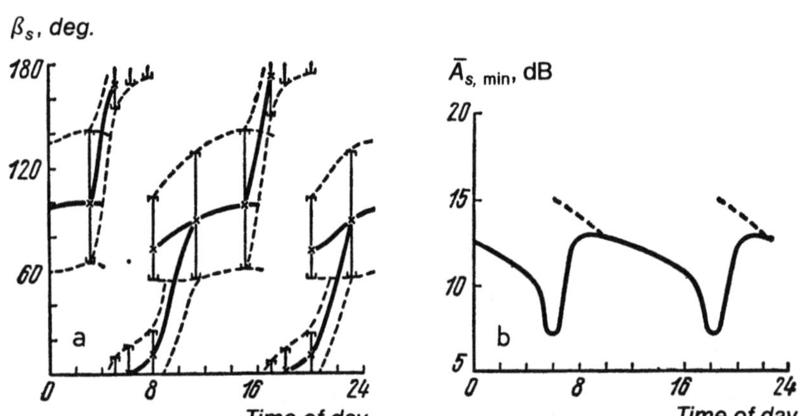

Fig. 4.28. **a** Diurnal variation of the optimum azimuths $\bar{\beta}_{s,\min}$ (*solid curves*) and their width (*dashed curves*); **b** diurnal variation of the absorption $\bar{A}_{s,\min}$

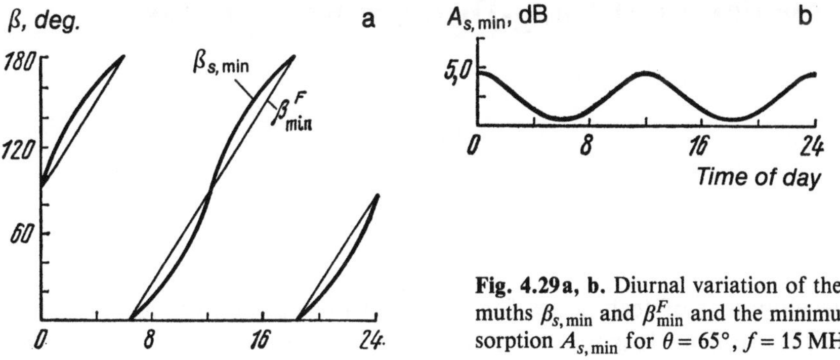

Fig. 4.29 a, b. Diurnal variation of the azimuths $\beta_{s,\min}$ and $\beta^F_{\min}$ and the minimum absorption $A_{s,\min}$ for $\theta = 65°$, $f = 15$ MHz

The indicated features of the azimuthal regularities of RTW absorption take place for midlatitudinal emission points and can be accounted for by the features of latitudinal – longitudinal absorption distribution (Fig. 4.19a). The direction of the first minimum β_1 corresponds to the directions close to latitudinal ones for the midlatitude RTW paths. These paths do not traverse the polar region with high absorption, consequently, the absorption on these paths varies insignificantly during 24 h (Figs. 4.27 and 4.28b). Note that the direction β_1 corresponds to the direction of existence of round-the-world channel for midlatitude emission points in the daytime. In addition to this direction, there is the second minimum absorption direction, which corresponds to the optimum azimuths $\beta^F_{\min}$. The second absorption minimum manifests itself most distinctly at twilight hours for the RTW paths passing at the minimum angle to the terminator and, consequently, having small absorption. It should be noted that the directions of minimum RTW absorption correspond to the absorption measurements (Fenwick 1963, Ostrovskii and Shlionskii 1975), carried out for the equinoctical conditions at midlatitude stations (Figs. 1.5 – 1.7). In these experiments, the amplitude curves of the RTW signals are obtained by means of azimuthal circular scanning. The curves can have two maxima, different or comparable, this is confirmed by the above calculations.

Analogous calculations performed for the transmitter situated at high latitudes ($\theta_0 = 65°$), showed that the first absorption minimum in the latitudinal direction does not manifest itself here (in contrast to the midlatitude stations). The minimum absorption directions, $\beta_{s,\min}$, for A_{sF} (the sliding signals) and $\bar{\beta}_s$ for $\bar{A}_s$ (the signals trapped into the duct *FE*) always correspond well enough to the optimum azimuth, $\beta^F_{\min}$ [Eq. (4.28); Fig. 4.29]. The width of the azimuthal sector of minimum absorption at the level 1.3 $\bar{A}_{s,\min}$ is almost the same during all 24 h, and is equal to 60 – 80°.

5 Calculation of Long-Distance Radio Paths

The development of techniques for radio path calculations is an important theoretical problem. It is seen from above that the horizontal inhomogeneity of the ionosphere greatly influences long-distance propagation. In the adiabatic approximation, the inhomogeneity is taken into account with sufficient completeness. That is why rather effective techniques for calculation of long-distance paths are developed on the basis of the adiabatic approximation.

In this chapter a description of the techniques is given. In Sec. 5.1, a parabolic model of the inhomogeneous ionosphere is considered. In this case, the calculation by the adiabatic method is most simple, and can be performed quickly and used successfully in practice. In Sec. 5.2 the method of radio path calculation for complex ionospheric models is given, taking into account the possibility of radio wave propagation in various waveguide channels. In Sec. 5.3 the results obtained by the adiabatic method and standard ray-tracing techniques are compared.

General Problems and Methods of Calculation. A radio path is considered, having the emission point, $\Psi_0 = 0$, $z = z_0$, and the reception point, $\Psi = \Psi_1$, $z = z_1$. The most interesting case is when the transmitter and receiver are situated on the ground: $z_0 = z_1 = 0$.

Which characteristics and parameters can be obtained from the radio path calculation? Let us formulate the aims which are of interest for experimental investigations and for radio communications.

1) Possibility for radio wave propagation over the path at a given frequency, i.e., possibility for the establishment of radio communication between the transmission and reception points.
2) The MUF of radio path, i.e., the maximum frequency propagating over the path under given ionospheric conditions.
3) The takeoff angles of waves reaching the reception point, angles of arrival.
4) Radio wave absorption on the path at a given frequency, and at various frequencies.
5) Height range in which waves propagate over the path.
6) The quantity and numbers of communication modes on the path for a given frequency and for various frequencies.
7) The field amplitude, signal energy at the reception point, energy distribution with respect to the modes. Knowledge of the calculated field

amplitude at a given transmitter power and sensibility of the receiver enables one to determine the lowest usable frequency (LUF) of the path.

8) The time delay for each mode. Knowledge of the mode time delay at various frequencies allows to construct the OS and LDOS ionograms (see Secs. 5.1; 5.2).
9) The azimuth of arrival, i.e., the horizontal bearing of a signal, lateral deviations of rays.
10) Change of the wave polarization when propagating over the path.
11) Fluctuations of the arrival angles in the horizontal and vertical directions.
12) Fluctuations of the pulse time delay, fluctuations of the pulse shape.
13) Field amplitude fluctuations at the reception point.
14) Fluctuations of wave polarization at the reception point.
15) Correlation of wave field amplitude and phase at different reception points close to each other, etc.

To reach these goals, it is necessary to have a specific ionospheric model. The model can be constructed both as the global averaged one, depending on a number of ionospheric conditions (time of day, season, solar activity cycle, magnetic activity, and so on), or as a specific model on a given path at a given time period. The latter model can be defined according to the results of vertical sounding at several points of the path. Of course, to answer the questions connected with field fluctuations, it is necessary to develop concepts about the structure and distribution character of the ionospheric irregularities (see Chap. 6).

For a spherically – symmetric (i.e., horizontally – homogeneous) ionosphere the systematic and perfectly developed methods of radio path calculation are well-known (for example, see Al'pert 1974, Shchukin 1940, Dolukhanov 1972, Chernii 1962, and Davies 1969). However, the methods are used successfully only for a one – two hop path, until the effects of horizontal inhomogeneity are not too strong. At long-distance paths the quasihomogeneous method can be used, allowing to approximately determine some parameters of the path. For instance, the maximum frequency on the path, f_{MUF}, according to this method, is defined by the equation

$$f_{\mathrm{MUF}} = (f_0 \sqrt{R_0/2z_m})_{\min} \,. \tag{5.1}$$

Here, $f_0 = (N_m e^2/\pi m)^{1/2}$ is the critical frequency of the F-layer, N_m is the electron density at the layer maximum, z_m is the height of the maximum. Equation (5.1) is obtained for the horizontally – homogeneous ionosphere (Al'pert 1974, Davies 1969 and others), in the quasihomogeneous case the minimum value of the function f_{MUF} [Eq. (5.1)] along the path is taken. Equation (5.1) appears to be approximately true for long-distance hop paths. When the capture on the ricocheting trajectories or into the interlayer duct exists, then Eq. (5.1) gives a considerable error in f_{MUF} calculation for long-distance paths (see Sec. 4.1).

Ray-tracing techniques based on the solution of geometrical – optics equations in the horizontally – inhomogeneous ionosphere are also well-developed

(Kerblay and Kovalevskaya 1974, Kiyanovskii 1971, Kelso 1968, Haselgrove 1955). However, these calculations are rather awkward and have much unnecessary information. They allow to determine reliably f_{MUF} on long-distance paths for simple hop and ricocheting trajectories. But the determination of complex trajectories, the identification of all the modes and their amplitudes at the reception point, the determination of takeoff and arrival angles, and, the more so, the mode time delay for various frequencies (i.e., the construction of ionograms) in different channels is, within the framework of this method, the complicated multiparametrical numerical problem. That is why the ray-tracing calculations are used to determine the propagation conditions over long-distance paths mainly by means of simple ionospheric models.

The adiabatic method seems to be most effective for the calculation of a long-distance radio path. The method takes into account quite accurately the horizontal inhomogeneity of the ionosphere. At the same time, the calculations in the adiabatic approximation are much simplified and involve no unnecessary information. This allows to quickly answer such questions as the mode identification, the determination of field amplitude, and the construction of ionograms.

5.1 Radio Path Calculation with a Parabolic Model of the Ionosphere

The present section deals in detail with the adiabatic method of long-distance path calculations using a parabolic model of the horizontally – inhomogeneous ionosphere (Gel'fond et al. 1982).

Ionospheric Model. A simple parabolic model of the ionosphere is defined in the form of one parabolic ionospheric layer

$$N_e(z, \theta, \varphi) = N_m(\theta, \varphi) \left\{ 1 - \frac{[z - z_m(\theta, \varphi)]^2}{y_m^2(\theta, \varphi)} \right\} . \tag{5.2}$$

Here, N_m is electron density at the F_2-layer maximum, z_m and y_m are its height and semithickness. Latitudinal – longitudinal distribution of the parabolic layer parameters, $N_m(\theta, \varphi)$, $z_m(\theta, \varphi)$, and $y_m(\theta, \varphi)$ in an averaged equinoctial model is given in Appendix 2. The model [Eq. (5.2)] can be used to describe the radio wave propagation in hop and ricocheting trajectories near the MUF of the F-layer, that is, for frequencies between the MUF-E and MUF-F_2 (in the absence of the F_1-layer), or between the MUF-F_1 and MUF-F_2 (in the presence of F_1-layer). An advantage of the parabolic model is that it is simple and defined by the minimum number of parameters. That is why the model is used successfully for radio path calculations when specific empirical data along the path are given.

Effective Potential. Let the parabolic distribution [Eq. (5.2)] of electron density along the path between the emission ($\Psi_0 = 0$) and reception ($\Psi = \Psi_1$) points be specified on the basis of the averaged model or specific empirical data

$$N_e(z, \Psi) = N_m(\Psi)\left\{1 - \frac{[z - z_m(\Psi)]^2}{y_m^2(\Psi)}\right\} \quad \text{if } |z - z_m| \leqslant y_m . \tag{5.2a}$$

Then the effective potential $U(z, \Psi)$ [Eq. (2.7)] for a radio wave of frequency f is given by

$$U(z, \Psi) = -1 - \frac{2z}{R_0} + \frac{f_0^2}{f^2}\left\{1 - \frac{[z - z_m(\Psi)]^2}{y_m^2(\Psi)}\right\} \quad \text{if} \quad |z - z_m| \leqslant y_m , \tag{5.3}$$

where f_0 is the critical frequency of the F_2-layer in MHz, $f_0 = 8.97 \cdot \sqrt{N_m} \cdot 10^{-3}$. Let us point out that Eqs. (5.2a) and (5.3) are true within the ionospheric layer, where $|z - z_m| \leqslant y_m$. Outside the layer, where $|z - z_m| > y_m$ we have

$$N_e = (0) , \qquad U = -1 - \frac{2z}{R_0} . \tag{5.4}$$

The potential shape, $U(z)$, is shown in Fig. 5.1. Its maximum value is given by

$$U_{\max} = -1 + \left(1 + \frac{\alpha^2}{4}\right)\frac{f_0^2}{f^2} - 2\frac{z_m}{R_0} , \qquad \alpha = 2\frac{y_m}{R_0}\frac{f^2}{f_0^2} \tag{5.5}$$

and the minimum value is given by

$$U_{\min} = -1 - \frac{2(z_m - y_m)}{R_0} . \tag{5.6}$$

According to Fig. 5.1, a wave trapped in the duct can propagate in hop trajectories, reflecting from the F-layer and from the Earth (channel F^{h}), and in ricocheting trajectories (channel F^{r}). The quantity E, characterizing the level of ray oscillations in the duct, can change in hop trajectories from $U_{\max}$ to -1

$$-1 \leqslant E \leqslant U_{\max} , \tag{5.7}$$

in ricocheting trajectories, E varies from -1 to $U_{\min}$

$$U_{\min} \leqslant E < -1 . \tag{5.8}$$

The potential $U(z)$ and other parameters, $U_{\min}$, $U_{\max}$, f_0, α, y_m, z_m, depend on Ψ, i.e., they vary along the path.

Adiabatic Invariant. [Eq. (3.17)]

$$I(E, \Psi) = \frac{4}{R_0}\int_{z_{\min}}^{z_{\max}} (E - U)^{1/2} dz . \tag{5.9}$$

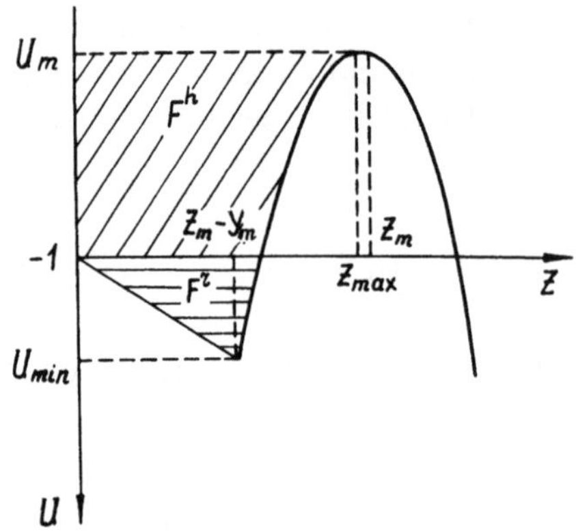

Fig. 5.1. Shape of the potential U for one – layer parabolic model of the ionosphere

Here, the integration is carried out from $z_{\min}(E, \Psi)$ to $z_{\max}(E, \Psi)$ defined by Eq. (3.24)

$$E = U(z, \Psi) . \tag{5.10}$$

Substituting here Eq. (5.3) for $U(z, \Psi)$ we find that

$$z_{\max} = z_m - \frac{y_m}{2}(\alpha + \sqrt{\alpha^2 + 4\beta}) , \tag{5.11}$$

where

$$\beta = 1 - \left(\frac{f}{f_0}\right)^2 \left(1 + E + \frac{2z_m}{R_0}\right) . \tag{5.12}$$

The parameter $z_{\min}$ for hop trajectories [Eq. (5.7)] is equal to zero

$$z_{\min} = 0 , \quad -1 \leqslant E \leqslant U_{\max} , \tag{5.13}$$

whereas for ricocheting trajectories [Eq. (5.8)]

$$z_{\min} = -\frac{E+1}{2} R_0 , \quad U_{\min} \leqslant E < -1 . \tag{5.14}$$

When integrating in Eq. (5.9) the integration range with respect to z is divided into two ones: from $z_{\min}$ to $z_m - y_m$, where $U(z, \Psi)$ is specified by the simple expression Eq. (5.4), and from $z_m - y_m$ to $z_{\max}$, where for $U(z, \Psi)$, Eq. (5.3) is used. After the integration we obtain

$$\begin{aligned} I &= \frac{4}{3}(g^{3/2} - b^{3/2}) + F(\alpha, \beta) \\ g &= E + 1 + 2(z_m - y_m)/R_0 \end{aligned} \tag{5.15}$$

$$\begin{aligned} b &= E+1 , \quad \text{if} \quad E \geqslant -1 \\ b &= 0 , \quad \text{if} \quad E < -1 \end{aligned} \tag{5.16}$$

$$F(\alpha,\beta)=\frac{y_m f_0}{R_0 f}\left[(2-\alpha)\sqrt{1-\alpha-\beta}-\frac{\alpha^2+4\beta}{2}\Lambda\right]$$

$$\Lambda=\ln\frac{2-\alpha+2(1-\alpha-\beta)^{1/2}}{(\alpha^2+4\beta)^{1/2}}. \tag{5.17}$$

Equation (5.15) gives in analytical form the dependence of the adiabatic invariant I on E and on the coordinate Ψ along the path. Taking into account the adiabatic invariant conservation [Eq. (3.22)]

$$I(E,\Psi)=I_0, \tag{5.18}$$

we can determine E as a function of Ψ from Eqs. (5.15) – (5.18) for every value of I_0. In principle, this solves the problem of the ray trajectory determination. Knowing $E(\Psi, I_0)$, we can find all the parameters in question.

The oscillation period $\Theta(E,\Psi)$ is defined according to Eq. (3.26). We have then

$$\Theta=2(g^{1/2}-b^{1/2})+2\frac{y_m f}{R_0 f_0}\Lambda, \tag{5.19}$$

where g, b, and Λ are defined by Eqs. (5.16) and (5.17). According to Eq. (2.8), the angles of takeoff, α_0, and arrival, α_1, for the transmitters and receivers situated on the Earth are equal to

$$\alpha_0=(E|_{\Psi=0}+1)^{1/2},\qquad \alpha_1=(E|_{\Psi=\Psi_1}+1)^{1/2}$$

$$E\geqslant -1. \tag{5.20}$$

A ray trajectory is situated in the height range

$$z_{\min}(\Psi)\leqslant z\leqslant z_{\max}(\Psi). \tag{5.21}$$

The minimum, $z_{\min}$, and maximum, $z_{\max}$, trajectory heights are defined by Eqs. (5.11), (5.13), and (5.14). The average height of trajectory [Eq. (3.33)] is

$$\bar{z}=\frac{2}{R_0\Theta}\int_{z_{\min}}^{z_{\max}}\frac{z\,dz}{\sqrt{E-U}}. \tag{5.22}$$

By integrating in Eq. (5.22), we find

$$\bar{z}=\frac{2R_0}{\Theta}\left\{\frac{1}{3}\left(\left[\left(\frac{z_m-y_m}{R_0}\right)-(1+E)\right]\sqrt{1+E+\frac{2(z_m-y_m)}{R_0}}+b^{3/2}\right)\right.$$

$$\left.+\frac{y_m^2 f}{R_0^2 f_0}\left[-\sqrt{1-\alpha-\beta}+\left(\frac{z_m}{y_m}-\frac{\alpha}{2}\right)\Lambda\right]\right\}. \tag{5.23}$$

Maximum Usable Frequency. The maximum adiabatic invariant of a duct plays an important role [Eq. (3.40)]. It is defined as the invariant $I(E,\Psi)$ [Eq. (5.15)], at $E=U_{\max}(\Psi)$ [Eq. (5.5)]

$$I_m(\Psi)=I(E,\Psi)|_{E=U_{\max}(\Psi)}. \tag{5.24}$$

The invariant I_m characterizes the wave confinement in the duct. A wave with $I > I_m$ is not confined in the duct, i.e., leaves the ionosphere. The quantity I_m depends on the coordinate Ψ along the path and also on the wave frequency f, $I_m = I_m(\Psi, f)$. Let us consider the variation of I_m vs Ψ at a given f along the whole path $0 \leqslant \Psi \leqslant \Psi_1$ and single out its minimum value $I_m^{\min}(f)$

$$I^m(f) = I_m^{\min}(f). \tag{5.25}$$

If the wave invariant, $I_0 > I^m(f)$, then the wave cannot be confined on a given path, $0 \leqslant \Psi \leqslant \Psi_1$. Hence, when propagating on a given path in the hop communication channel F^{h} or in the channel $F^{\mathrm{h}}F^{\mathrm{r}}F^{\mathrm{h}}$ with ricocheting trajectories on a portion of the path, the adiabatic invariant I_0 of wave frequency f does not exceed $I^m(f)$.

On the other hand, the invariant cannot be less than a certain value, $I_{\min}(f)$, which is defined as follows. Since we deal with the transmitter situated on the Earth, the invariant at the emission point cannot be less than the horizontal ray invariant $I_{1,\Psi=0}$ [Eq. (4.10)]. It is defined by Eq. (5.15) at $\Psi = 0$, $E = -1$ (because at $E = -1$ the takeoff angle $\alpha_0 = 0$ [Eq. (5.20)])

$$I_{1,\Psi=0} = I(E, \Psi)|_{E=-1, \Psi=0}. \tag{5.26}$$

The same refers to the reception point. A ray reaching the Earth at the point $\Psi = \Psi_1$ cannot have an invariant less than $I_{1,\Psi=\Psi_1} = I(E, \Psi)|_{E=-1, \Psi=\Psi_1}$. The desired minimum invariant of the path is the maximum of the two values $I_{1,\Psi=0}$, and $I_{1,\Psi=\Psi_1}$

$$I_{\min}(f) = \max(I_{1,\Psi=0}, I_{1,\Psi=\Psi_1}). \tag{5.27}$$

Thus, the range of the adiabatic invariant for waves which can propagate over the path is given by

$$I_{\min}(f) \leqslant I_0 \leqslant I^m(f), \tag{5.28}$$

where $I^m(f)$ and $I_{\min}(f)$ are defined according to Eqs. (5.24)–(5.27). As the frequency f increases, the range of the invariant I_0 converges (see Fig. 5.2). The frequency, at which the lower and upper limits in Eq. (5.28) join, is the maximum usable frequency, f_{MUF}, for a given path. Hence, it is defined by the relation

$$I_{\min}(f_{\mathrm{MUF}}) = I^m(f_{\mathrm{MUF}}). \tag{5.29}$$

In Fig. 5.2, $f_{\mathrm{MUF}} = 14.2$ MHz.

Modes of Radio Path. These modes are determined by the number of ray oscillations between the transmission ($\Psi = 0$) and reception ($\Psi = \Psi_1$) points [Eq. (3.27)]

$$\mathscr{P}(I_0) = \int_0^{\Psi_1} \frac{d\Psi}{\Theta(\Psi, I_0)}. \tag{5.30}$$

Here, $\Theta(\Psi, I_0) = \Theta(\Psi, E(\Psi, I_0))$ is the oscillation period of ray with the invariant I_0, defined according to Eqs. (5.19); (5.18), (5.17). For a wave

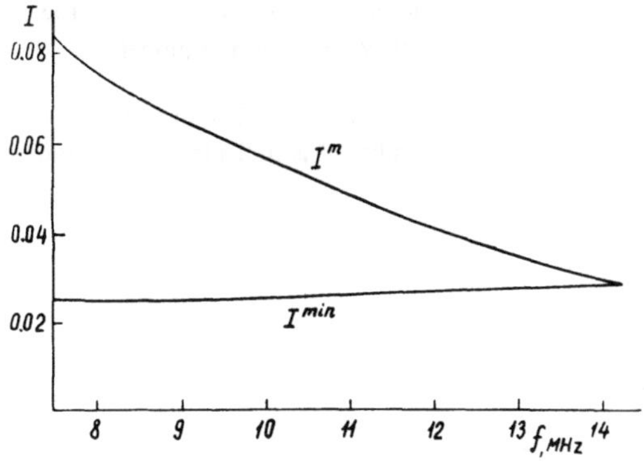

Fig. 5.2. Invariants I^m and $I_{\min}$ vs frequency on the path $\Psi_1 = 3$ for the parabolic model [Eqs. (5.3) and (5.38)]

emitted and received on the Earth, the number of oscillations must be integer [Eq. (3.29)]

$$n = \mathscr{P}(I_{0n}) = \int_0^{\Psi_1} \frac{d\Psi}{\Theta(\Psi, I_{0n})} . \tag{5.31}$$

Equation (5.31) defines such a value of the invariant $I_0 = I_{0n}$, which corresponds to the mode making n oscillations between the transmission ($\Psi = 0$) and reception ($\Psi = \Psi_1$) points. By changing the invariant I_0 within the range given in Eq. (5.28) at a fixed frequency f, we can find all the modes for this frequency on a given path, according to Eqs. (5.30) and (5.31).

The LDOS Ionograms. Time delay of a signal on a path is expressed as follows [Eq. (3.32)]

$$t = \frac{R_0}{c} \int_0^{\Psi_1} \frac{d\Psi}{\sqrt{-E(\Psi, I_0)}} \left(1 + \frac{2\bar{z}}{R_0}\right) , \tag{5.32}$$

where the quantity $\bar{z}$ is defined according to Eq. (5.23), $E(\Psi, I_0)$ is defined according to Eqs. (5.15)–(5.18). The time delay t depends on I_0, and consequently, it is different for various modes. Calculating the mode time delay at different frequencies, one can construct the LDOS ionograms, showing the time delay variation with frequency f for every mode. The ionograms are shown in Fig. 5.3.

The Field Amplitude. Average energy density for each mode, w_n, at the reception point Ψ_1 is defined by the expressions [Eq. (3.35) and (3.39a)]

$$\begin{gathered} w_n = \frac{W_0}{cR_0^2} \, \frac{D(\alpha_{0n}, \varphi_0) \cos \alpha_{0n} \cdot \exp\left[-2 \int_0^{\Psi_1} K_n \, d\Psi\right]}{\sin \Psi_1 \cdot \sin \alpha_{1n} \cdot Q_{1n}} \cdot p_n \\ Q_{1n} = 2\alpha_{0n} \, \Theta_{0n} \left| \int_0^{\Psi_1} \frac{1}{\Theta_n^2} \left(\frac{d\Theta}{dE}\right)_{E_n} d\Psi \right| . \end{gathered} \tag{5.33}$$

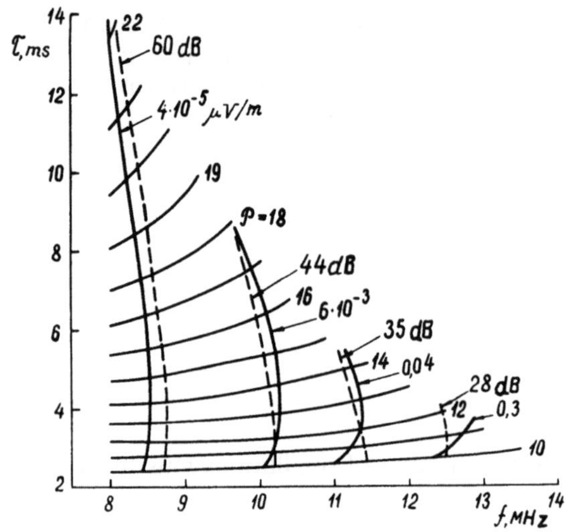

Fig. 5.3. LDOS ionogram for the path $\Psi_1 = 3$ and the model [Eqs. (5.3) and (5.38)]; $\tau = t - R_0 \Psi_1/c$. The oscillation number $\mathscr{P}$ is indicated at the modes

Here, $E_n = E(\Psi, I_{0n})$ and $\Theta_n = \Theta(\Psi, I_{0n})$ are the values of E and the period Θ for the $n-$th mode [Eqs. (5.18), (5.17), and (5.19)]; α_{0n}, α_{1n} are the angles of takeoff and arrival [Eq. (5.20)], Θ_{0n} is the initial oscillation period (at $\Psi = 0$) for the same mode, and $D(\alpha_{0n}, \varphi_0)$ is the directivity diagram of antenna [Eq. (2.45a)]. Next, the derivative $(d\Theta/dE)$ for the parabolic ionospheric model is equal to

$$\frac{d\Theta}{dE} = g^{-1/2} - b_1 + 2\frac{y_m f^3}{R_0 f_0^3}$$

$$\cdot \left\{\frac{2}{\alpha^2 + 4\beta} + \frac{1}{[2 - \alpha + 2(1 - \alpha - \beta)^{1/2}](1 - \alpha - \beta)^{1/2}}\right\} \tag{5.34}$$

$$b_1 = (E+1)^{-1/2} \quad \text{if} \quad E > -1\,; \qquad b_1 = 0 \quad \text{if} \quad E \leqslant -1\,.$$

According to Eq. (5.33), the differentiation is performed at $E = E_n$. The quantity K_n in Eq. (5.33) is the averaged absorption for the $n-$th mode [Eq. (3.30)]. Approximately, it can be represented as follows

$$K_n(E_n, \Psi) = \frac{K_0(\Psi)}{f^2 \Theta_n \sqrt{(E_n + 1) + 0.03}}$$

$$K_0(\Psi) = \frac{e^2}{\pi m c} \int N_e \nu_e dz\,. \tag{5.35}$$

Here, $\nu_e(z, \Psi)$ is the electron collision frequency, $N_e(z, \Psi)$ is electron density. It is taken into account that the integral in Eq. (5.35) is essentially "saturated" in the lower ionosphere within the range of 50 km $\lesssim z \lesssim$ 130 km. The quantity $K_0(\Psi)$ is not defined by the parabolic model [Eq. (5.2)], it should be specified additionally on the basis of calculated or empirical data. Finally, the factor p_n

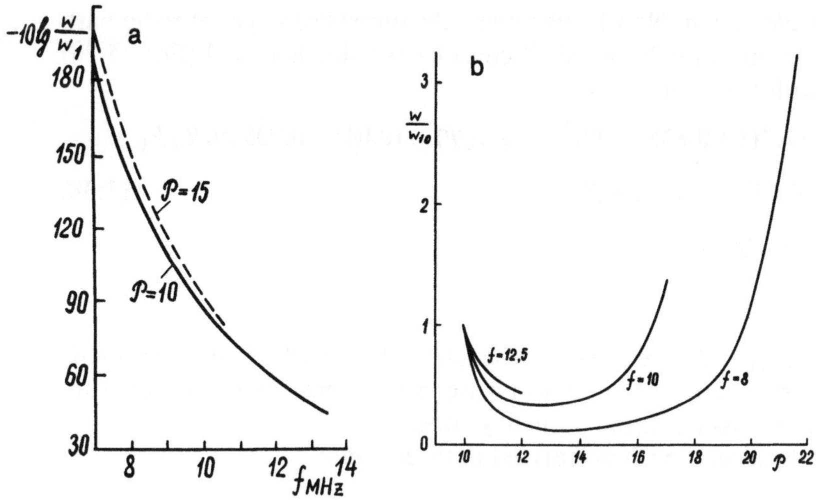

Fig. 5.4. Energy density w vs frequency (**a**) and vs oscillation number $\mathscr{P}$ (**b**): $w_1 = W_0/cR_0^2$; w_{10} is the power of the first mode ($\mathscr{P} = 10$) at the corresponding f

in Eq. (5.33) determines energy losses when the wave is reflected from the Earth.

The total energy density, w, of a wave and the average field amplitude $\mathscr{E}$ are given by Eq. (3.34)

$$w = \sum_n w_n\,, \qquad \mathscr{E}_n = \sqrt{8\pi w_n}\,, \qquad \mathscr{E} = \sqrt{8\pi w}\,. \tag{5.36}$$

Here, the summation is carried out over all the modes.

Figure 5.4 shows energy density for different modes at the reception point. The energy density is seen to be maximum near f_{MUF}, and rather rapidly decreases I with the wave frequency f (Fig. 5.4a). The dependence of energy density on the mode number is nonmonotonic (Fig. 5.4b).

Lateral Deviations and Bearing Variations at the reception point are defined by Eqs. (3.52), (3.56), and (3.60). To calculate them, it is necessary to know the derivative $\partial\varepsilon/\partial\chi(\Psi)$ along the direction of χ orthogonal to the path. In the case of the parabolic ionospheric model [Eq. (5.2)] we have

$$\frac{\partial\varepsilon}{\partial\chi} = -\frac{f_0^2}{f^2}\left\{\frac{\partial \ln N_m}{\partial\chi}\left[1-\frac{(z-z_m)^2}{y_m^2}\right]+\frac{2}{y_m^3}\,\frac{\partial y_m}{\partial\chi}(z-z_m)^2 + \frac{2}{y_m^2}\,\frac{\partial z_m}{\partial\chi}(z-z_m)\right\}. \tag{5.37}$$

Thus, it is necessary to define three functions along the whole path: $\partial \ln N_m(\Psi)/\partial\chi$, $\partial y_m(\Psi)/\partial\chi$, $\partial z_m(\Psi)/\partial\chi$. Substituting Eq. (5.37) in Eq. (3.48), and performing usual integration with respect to z, we find the quantity $\bar{F}(\Psi)$ in the analytical form and, consequently, the value of lateral deviation, $\Delta\chi$, and bearing variation, β_0, at the reception point, Ψ_1.

Radio Path Calculation. Now let us discuss the successive stages of radio path calculation. As an example, we shall consider the simple model [Eqs. (5.2a) and (5.35)] with the parameters

$$\begin{aligned} & f_0^2(\Psi) = 17.7(1+0.835\sin\Psi)^2\,, \qquad z_m(\Psi) = (0.045-0.005\sin\Psi)R_0 \\ & y_m(\Psi) = (0.010+0.002\sin\Psi)R_0 \\ & K_0(\Psi) = 10+20\sin\Psi \\ & 0 \leqslant \Psi \leqslant 3\,, \qquad \Psi_1 = 3\,. \end{aligned} \tag{5.38}$$

The frequencies f_0 and f are in MHz. The model [Eq. (5.38)] describes roughly the characteristic variation of ionospheric parameters along the hop sub-equatorial path, passing through the day zone.

Radio path calculation is performed in three stages.

1. The Determination of the Signal Propagation Conditions and the MUF of the Path. At this stage, the maximum adiabatic invariant of a channel, $I_m(\Psi)$, is determined by Eq. (5.24) at each point for a fixed frequency f. Changing Ψ from 0 to Ψ_1, we find the minimum value of the invariant $I_m^{\min}$, that is, the maximum adiabatic invariant of the path for the frequency f: $I^m(f)$ [Eq. (5.25)]. Next, the minimum invariant I_1 is determined at the points $\Psi=0$ and $\Psi=\Psi_1$ by Eqs. (4.10) and (5.26). The maximum of the two values of I_1 [Eq. (5.27)] is selected, it gives $I_{\min}(f)$. Thus, for a given frequency f, the range of the invariant I_0 [Eq. (5.28)] is established, at which the propagation of signals in the channel from the emission point ($\Psi_0=0$) to the reception point ($\Psi=\Psi_1$) is possible.

The same calculation is further performed for different frequencies f. With increasing f, the range of possible values of I_0 [Eq. (5.28)] is converged, and disappears altogether at a certain value of $f=f_{\mathrm{MUF}}$ [Eq. (5.29); see Fig. 5.2].

2. The Determination of the Path Parameters. At this stage, for a given frequency f, at every fixed point of the path, $\Psi=\Psi_i$, the following trajectory parameters are calculated: the oscillation period Θ [Eq. (5.19)], the takeoff angles α_0 and arrival angles α_1 [Eq. (5.20)], the trajectory heights $z_{\max}$ [Eq. (5.11)] and $z_{\min}$ [Eqs. (5.13) and (5.14)], and the height $\bar{z}$ [Eq. (5.23)]. To do this, any value of the invariant $I=I_0$ in the range (I^m, $I_{\min}$) is specified. Using Eq. (5.18) we find the variation of $E(\Psi, I_0)$ along the path and, consequently, all the other parameters, characterizing this trajectory.

3. The Mode Identification and Field Amplitude Determination. Construction of the Ionograms. First, at fixed frequency f, the value of $\mathscr{P}$ [Eq. (5.30)] is determined for different values of the invariant I_0 [Eq. (5.28)]. The integer values of $\mathscr{P}=n$ correspond to the modes of a given path. According to Eq. (5.31), we find the numerical values of $I=I_{0n}$ corresponding to various modes. Further, the energy density w_n and field amplitude, $\mathscr{E}_n$, of the modes, and also the total energy density, w, and the average field amplitude, $\mathscr{E}$, are calculated by Eqs. (5.35)–(5.36) (see Fig. 5.4). Besides, for each mode the pulse time delay $t_n(\Psi)$ is calculated by Eqs. (5.32) and (5.23). The same calculation is performed at various frequencies $f \leqslant f_{\mathrm{MUF}}$. This allows to construct the LDOS ionograms of the radio path (see Fig. 5.3). Analogously, one can determine the lateral deviation of rays [Eq. (5.37)].

Thus, in the approximation of the parabolic ionospheric model, it is easy to carry out the complete analysis of the path.

5.2 Calculation of Long-Distance Radio Paths, Using the Complex Ionospheric Model

Let us consider the calculation techniques of long-distance radio paths using the complex ionospheric models. These models describe real distribution of electron density N_e and electron collision frequency ν_e along the height z and in the horizontal direction Ψ along the path. For such calculations, the global ionospheric models are usually used (see Appendix 2).

We point out that in the case of real models of N_e and ν_e distributions, radio wave trajectories on long-distance paths are rather complicated and various in type. An advantage of the adiabatic invariant method is that it allows to find out and to investigate in detail all the possible propagation modes and trajectory types. However, such an analysis is rather time consuming.

Communication Channels of the Path (CCP). When using complex models, in addition to principal duct *F*, existing in the case of one-layer model (Sec. 5.1), the wave ducts *FE, E,* and others appear (see Chap. 4). Thus, one can identify the simple communication channels, CCP(smp) in which radio waves propagate along the whole path in one duct, and the complex communication channels of the path, CCP(cmp) in which radio waves propagate in two or more ducts. The simple CCP include the standard hop trajectories, reflecting from the *F* and *E*-layers (CCP $F^{\mathrm{h}}, E^{\mathrm{h}}$), and purely ricocheting trajectories, not descending to the Earth (F^{r}, E^{r}, *FE*). These CCP also include the trajectories with combined propagation in a duct ($F^{\mathrm{r}}F^{\mathrm{h}}$, $F^{\mathrm{h}}F^{\mathrm{r}}$, $E^{\mathrm{h}}E^{\mathrm{r}}$, and so on). In this case, radio waves pass a part of the path on hop trajectories and another part on ricocheting ones. In the designations of CCP used below the sequence of letters with the corresponding indices (h – hop, r – ricochet) will correspond to the sequence of the IWD passage on the path from the emission point. For example, the difference between CCP $F^{\mathrm{h}}F^{\mathrm{r}}$ and $F^{\mathrm{r}}F^{\mathrm{h}}$ is that in the first case ($F^{\mathrm{h}}F^{\mathrm{r}}$), from the emission point, the hop trajectory in the IWD changing then into the ricochet is observed. In the second case ($F^{\mathrm{r}}F^{\mathrm{h}}$), above the transmitter, the ground-detached channel F^{r} exists. In this channel radio waves propagate by ricocheting trajectories. When moving away from the emission point, the trajectories change into the hop ones. Since in the duct *FE* a wave always propagates by ricocheting trajectories, the index "r" for CCP *FE* will be omitted. If the capture of a wave from the Earth into the channel *FE* or the descent to the Earth takes place, then the designations $E^{\mathrm{h}}(FE)$, $(FE)E^{\mathrm{h}}$ are used for the corresponding CCP.

The use of complex ionospheric models for path calculation by the adiabatic method enables one, in addition to the standard hop CCP (F^{h}, E^{h}, $F^{\mathrm{h}}E^{\mathrm{h}}$), to identify channels in which radio waves pass a part of the path or the whole path on ricocheting trajectories (CCP $F^{\mathrm{h}}F^{\mathrm{r}}F^{\mathrm{h}}$, $E^{\mathrm{h}}(FE)E^{\mathrm{h}}$, $E^{\mathrm{h}}F^{\mathrm{r}}$, F^{r}, *FE*, and so on). Besides, this method allows to find out not only CCP F^{h}, E^{h}, $F^{\mathrm{h}}E^{\mathrm{h}}$, $F^{\mathrm{h}}F^{\mathrm{r}}F^{\mathrm{h}}$, $E^{\mathrm{h}}(FE)E^{\mathrm{h}}$, and others in which a wave passes directly from the emission point to the reception one (we shall call them CCP-I), but also the additional CCP: CCP-II, namely, as $E^{\mathrm{h}}(FE)$, $F^{\mathrm{h}}F^{\mathrm{r}}$, $E^{\mathrm{h}}F^{\mathrm{r}}$, $F^{\mathrm{r}}F^{\mathrm{h}}$, etc; and CCP-III, namely, $E^{\mathrm{r}}F^{\mathrm{r}}$, *FE*, $F^{\mathrm{r}}(FE)$, and so on. In spite of the regular refraction, the CCP-II have neither entry from the Earth, nor descent to the Earth from one side of the path, whereas the CCP-III have neither entry nor descent from both sides of the path. These channels are remarkable by the fact that in the presence of such additional factors as large-scale inhomogeneities, scattering, or nonlinear effects, a wave emitted from the Earth can be trapped in these channels when there are no conditions for the direct capture, or the wave can descend to the Earth when there are no conditions for the direct departure. When using the complex models, it is convenient to carry out the path calculations by several stages. At the first stage, the variation of total adiabatic invariants of IWD, I_m [Eq. (4.9)] along the path is investigated in order to analyze the possible propagation modes of the path (Tsedilina 1975, Tushentsova and Tsedilina 1983). The calculation of the total channel invariants I_m is enough to identify the presence of all the CCP along the path, and to determine the maximum, I^m, and the minimum, $I_{\min}$, invariants of every CCP on the path (see Sec. 5.1), and also to determine their volume $\Delta I = I^m - I_{\min}$. Such an analysis, carried out within the whole frequency range where the CCP exist, enables one (by the condition $\Delta I \to 0$; Sec. 5.1), to determine the limiting frequencies of all the CCP on the path, including the MUF for hop CCP.

The second calculation stage involves the determination of trajectory parameters in every identified CCP on the path, that is, the variations of the maximum and minimum trajectory heights, oscillation periods and numbers, entry and departure angles, and absorption and time delay. This stage enables one to determine all the above parameters for the lower and upper ray in every CCP, i.e., to determine the possible maximum and minimum parameters of every CCP. Besides, at this stage, by means of plotting the oscillation number $\mathscr{P}$ as a function of time delay t, the CCP modes are identified which are the integer values of $\mathscr{P} = n(I_{0n})$ (see Sec. 5.1) and the corresponding values $t(n)$. These data allow to construct the LDOS ionograms taking into account all the possible modes of the path. The dependence of absorption $A(I)$ in every CCP enables one to determine the absorption of every mode $A_n(I_{0n})$.

Note that the first stage is sufficient to estimate the general propagation conditions on the path and to determine the limiting propagation frequencies. Of course, even the performance of the total calculation makes it possible to study radio wave propagation on the path only for a certain time period. The total analysis of the path involves the calculations for different moments of time. This allows to completely investigate the propagation conditions on the path during 24 h. Such an analysis will be performed in the terminal subsection of the present section.

Now let us consider consecutively all the stages of path calculation.

Investigation of Radio Wave Trajectories on a Given Path and Identification of CCP. Let us consider in detail the variation along the around-the-world path of the following parameters: the total duct volumes I_E, I_{FE}, I_F, the horizontal ray invariant I_1, the minimum ($z_{\min E}$, $z_{\min FE}$, $z_{\min F}$) and maximum ($z_{\max E}$, $z_{\max FE}$, $z_{\max F}$) heights of the ducts E, FE, F; the sliding mode height z_s, the maximum capture angles into the duct E, $\alpha_{\max E} = \alpha_{\max FE}$, into the duct F, $\alpha_{\max F} = \alpha_F^m$. The calculation of these parameters is performed by means of the model EMI-81 at $W = 10$ and is described in Sec. 4.1.

Figures 5.5 and 5.6 show the invariants I_E, I_{FE}, I_{E+FE}, I_F, I_1 as a function of the angular distance Ψ (in degrees) along the RTW path. The coordinates of the terminator (T) and equator (E) are also shown in the figures.

Figure 5.5a is plotted for the transmitter located at the latitude $\theta_0 = 38°$ for LT $= 11^{\mathrm{h}}$, the emission azimuth is $\beta = 112°$, $f = 15$ MHz. This corresponds to the transmission from Stanford in March – May 1961. During this period, the optimum reception of the RTW signals with the azimuth $\beta = 112°$ on the frequency $f = 15$ MHz was observed at Stanford at 11 o'clock (Fenwick 1963).

It is seen from Fig. 5.5a that the maximum invariant I_E decreases with the increase of Ψ from the emission point $\Psi = 0$ and reaches zero at $\Psi = 135°$ near the terminator. The horizontal ray invariant I_1 is shown by the thick line. When I_E decreases to $I_E = I_{1E} = 12$ at $\Psi \approx 83°$, then the horizontal ray invariant changes by a jump up to the value $I_1 = I_{1F} = 28$, because at the point $\Psi = 90°$ the ray leaves duct E for duct F. Then the invariant I_1 increases up to the value $I_{1F} = I_F = 50$ at the point $\Psi \approx 145°$. The invariant I_{FE} first increases from $I_{FE} = 4$ at $\Psi = 0$ up to $I_{FE} = 19$ near the equator $\Psi = 60°$ and decreases again to $I_{FE} = 10$ near the terminator ($\Psi \approx 125°$). The duct FE vanishes when $I_E \to 0$, i.e., when duct E disappears. The maximum invariant I_F is not shown in Fig. 5.5a within $\Psi \approx 0 - 150°$ and $\Psi \approx 300 - 360°$, because here its values are much more than 50. Within the range of $145° \lesssim \Psi \lesssim 300°$, I_{1F} is absent, i.e., here the duct F is separated from the Earth. At the middle point of the path in the night zone, the volume of duct F decreases significantly, and I_F reaches the minimum value $I_{F\min} = 14$. In the terminal of the path the opposite variations of the invariants are observed between the terminator and reception point.

When analyzing the path, one should bear in mind that ricocheting trajectories with the invariant $0 < I < I_{1E}$ and hop trajectories $I_{1E} < I < I_E$ in

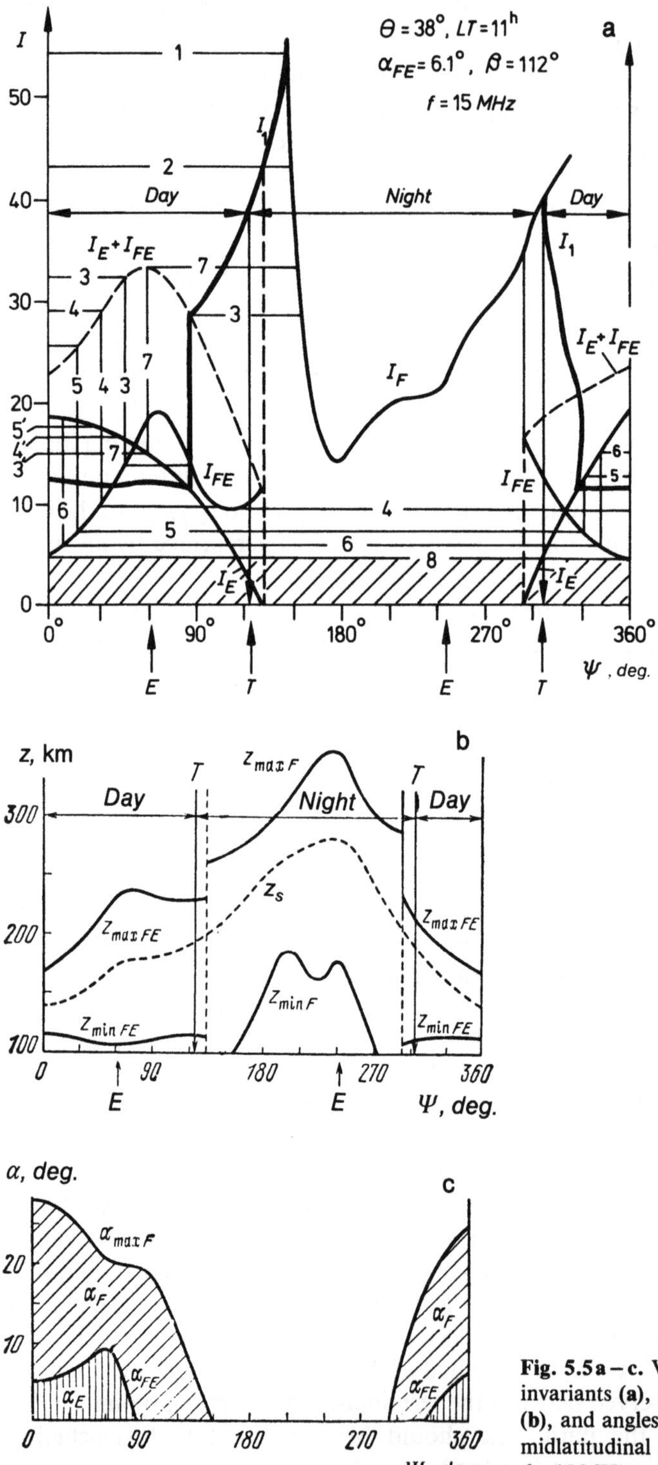

Fig. 5.5a–c. Variation of the total invariants (**a**), heights of the ducts (**b**), and angles of capture (**c**) over midlatitudinal RTW path for $f = 15$ MHz

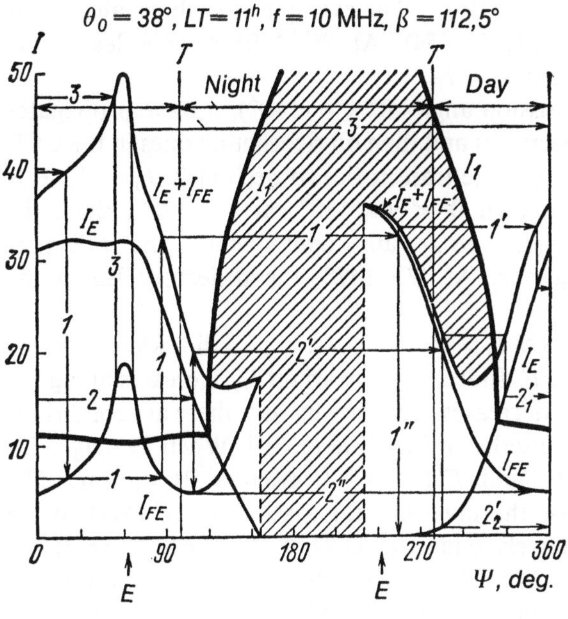

Fig. 5.6. Variation of the total invariants over midlatitudinal RTW path for $f = 10$ MHz

the duct E exist; in the duct FE there are ricocheting trajectories with the invariant $0 < I < I_{FE}$. In the presence of the E-layer there are the trajectories with $I_{FE} + I_E < I < I_F$ in the duct F. In this case, when $I > I_{FE} + I_E > I_{1E}$, or I_{1F}, the hop trajectories take place, and when $I_{FE} + I_E < I < I_{1F}$, then the ricocheting trajectories take place. When the E-layer is absent, the invariants in the channel F are within $0 < I < I_F$. The rules of change of invariants in transition from one IWD to another, when the channel volumes decrease, were discussed in Sec. 3.2. Note that according to the accepted classification (see Sec. 4.1), when the duct E and, consequently, the duct FE disappear, the trajectory invariants do not change in transition from the duct FE into the duct F (when $I < I_{FE}$, I_F; $I_{FE} = I_F$). The same takes place when the duct E together with the duct FE appears on the path. In the case of all the other trajectory transitions from duct to duct, the invariant changes by a jump (see Sec. 3.2).

Figure 5.5 b, c illustrates the variation along the path of the minimum ($z_{\min FE}$, $z_{\min F}$) and maximum ($z_{\max FE}$, $z_{\max F}$) duct heights (note that $z_{\min FE} \approx z_{\max E}$), the height of sliding z_s and the corresponding maximum and minimum capture angles. Figure 5.5 b, c allows to estimate approximately the channel heights and the ranges of the capture angles and descent angles for the CCP in question and also to determine the capture angle into the duct FE.

To identify the CCP over a given RTW path, let us first consider the possible trajectories when radio waves are emitted from the Earth. Such trajectories have the invariants $I_0 > I_{1E} = 12$ at the emission point $\Psi = 0$. Figure 5.5a shows that in this case the trajectory types are various. When $I_0 > 50$ (i.e., when the emission angle $\alpha_0 > 10°$), radio waves propagate in the duct F on hop trajectories in the CCP F^h. The example of such a trajectory is shown in Fig. 5.5a (trajectory 1, $I_0 = 54$). At $\Psi = 150°$ a wave leaves the ionosphere, because at this point $I_0 = I_F$.

When $33 < I_0 < 50$ (the emission angles $8° < \alpha < 10°$), a wave propagates in the duct F first on hop trajectories and then on ricocheting ones in the CCP F^hF^r. Figure 5.5a shows one of these trajectories (trajectory 2, $I_0 = 43$). It is of a hop type within the range of $0° < \Psi < 130°$, at $\Psi = 130°$, thus, the trajectory invariant becomes equal to I_{1F} and, consequently, the trajectory changes into the ricocheting one. At $\Psi = 147°$, the trajectory leaves the ionosphere.

When $22 < I_0 < 33$ and $19 > I_0 > 14$ ($5° < \alpha < 8°$), radio waves can be trapped in the interlayer duct FE. Indeed, the invariant I_{FE} increases along the path, whereas the invariant I_E decreases. That is why here there are conditions for the wave capture into the interlayer duct [Eq. (3.63)]. When $29 < I_0 < 33$, the CCP (cmp) $F^h(FE)F^hF^r$ and $F^h(FE)F^r$ (trajectory 3) appear. The maximum boundary values of the invariant for a sequence of IWD in the given CCP (cmp) are respectively equal to 33 (F^h), 18 (FE), 32 (F^r); the minimum values are equal to 29 (F^h), 10 (FE), and 20 (F^r). In these CCP, radio waves do not reach the end of the path. When $22 < I < 29$, the RTW interlayer CCP (cmp) $F^h(FE)F^r(FE)E^r$ or $F^h(FE)F^r(FE)E^h$ appears (trajectories 4, 5). When $19 > I_0 > 18$, then the emission angle $\alpha_{0E} - \alpha_{FE} \simeq 6°$ and the RTW CCP (cmp) $E^h(FE)F^r(FE)E^r$ and $E^h(FE)F^r(FE)E^h$ is formed (trajectories 4′, 4; 5′, 5; 6). The capture range width $\Delta\alpha$ [Eq. (3.72)] is about 1°. When $18 < I_0 < 14$, radio waves, after passing the equator, leave the duct FE for the duct F. They do not reach the terminal of the path (CCP (cmp) $E^h(FE)F^hF^r$ and $E^h(FE)F^r$, trajectory 3′, 3).

When $12 < I_0 < 14$, radio waves propagate in the CCP(cmp) $E^hF^hF^r$ (trajectory 7). At $\Psi \sim 150°$ they leave the ionosphere. These waves cannot be trapped in the channel FE because at their departure from the duct E the volume of the duct FE decreases.

Thus, on the RTW path two RTW channels appear: the CCP $E^h(FE)F^r(FE)E^h$ (trajectories 5′, 5; 6) and the CCP $F^h(FE)F^r(FE)E^h$ (trajectory 5), in which the radio waves emitted from the observation point (Stanford) cover the whole RTW path in the ionosphere at the heights $\sim 110 - 150$ km in the daytime and at the heights $180 - 300$ km at night (Fig. 5.5b). The radio waves descend to the Earth through the duct E at the reception point coinciding with the emission point. Of course, in this case they undergo weak absorption. Besides, at the emission and reception points the conditions of the ray capture and departure from the ionospheric duct are well satisfied. That is why the case in question is the most favorable for the RTW propagation. Note that on the path, besides the CCP considered above, there is the RTW CCP $(FE)F^r(FE)$ with the invariants $4 > I > 0$ (in Fig. 5.5a it is shown by shading). However, it is separated from the Earth at both terminals of the path and radio waves can be trapped in this channel or leave it only due to the additional factors such as the scattering, nonlinear effects, etc. (see Chaps. 6, 7).

Figure 5.6 illustrates variations of the invariants I_E, I_{FE}, I_E+I_{FE}, and I_1 for the same path as in Fig. 5.5, but for the lower frequency $f = 10$ MHz. The duct F volume is not shown in the figure. The volume is more than 50 and, consequently, in this duct the hop mode propagation is possible along the whole path at the emission angles $\alpha_0 \approx 15-20°$. In Fig. 5.6 the part of the duct F in which radio waves propagate on the ricocheting trajectories is shaded. Different trajectories on the path are shown by the numbers and arrows.

For example, let us consider the trajectory 2. On such a trajectory the ray emitted from the Earth at an angle of $\alpha_0 \approx 5°$ has the invariant $I = 15$. First, the ray propagates by hop modes in the duct E up to $\Psi = 100°$, then the ray leaves the duct E either for the channel F on the trajectory 2′ with the invariant $I = 20$ or for the duct FE and then F on the trajectory 2″ with the invariant $I = 5$. The ray on the trajectory 2′ at $\Psi = 285°$ leaves the duct FE either for the duct F with the departure to the Earth through the duct E (trajectory $2'_1$) or for the duct E on the ricocheting trajectory $2'_2$. The ray propagating on the trajectory 2″ comes out to the Earth and reaches the emission point at $\theta_0 = 38°$, LT = 11^{h} through the duct FE at an angle $\alpha_1 = 15°$. In this case on the trajectory 2, 2′, $2'_1$ the ray travels with the invariants 15, 20, 22, 15, while on the trajectory 2, 2″ it travels with the invariants 15, 5. In this instance, on the path for $f = 10$ MHz, besides the conventional hop channel F^{h} and channel FE (trajectory 2′, 2″), one can identify quite a number of complex channels ($E^{\text{h}}FE$, trajectories 2, 2″, $E^{\text{h}}F^{\text{h}}F^{\text{r}}(FE)F^{\text{r}}E^{\text{h}}$, trajectories 2, 2′, $2'_1$, etc.). Thus, Fig. 5.6 illustrates the presence of many rays, i.e., many CCP at low frequencies in the case of long-distance propagation. In this case the trajectory with the minimum losses can be singled out from all the possible trajectories. It should be expected that such a trajectory in Fig. 5.6 is the trajectory 1, 1′ for which the ray is trapped in the duct FE at an angle $\alpha_{FE} = 15°$. On this trajectory a wave covers the greater part of the day path in the duct FE. The wave deepens considerably in it, and, consequently, undergoes the minimum attenuation. Generally speaking, between the values $\Psi \sim 0°$ and $\Psi \sim 90°$, i.e., at a distance of about 2000 – 5000 km from the equator, reliable radio communications should be observed on the frequency $f = 10$ MHz. In this case, a ray passes this distance in the duct FE on the chordal trajectory. The angle of entry and descent from the duct FE is about $15-16°$.

Calculation of the Trajectory Parameters. Consider the calculation of the specific parameters of the trajectories: the maximum $z_{\max}$ and the minimum $z_{\min}$ heights, the angles of emission α_0 and descent to the Earth α_1, the oscillation period Θ, the number of ray oscillations $P_i = \Delta\Psi_i/\Theta_i$ over a given distance along the path $\Delta\Psi_i$ or over the whole path Ψ, and the total absorption on the path $A = \sum K_i \Delta\Psi_i + \Gamma_{\text{entry}} + \Gamma_{\text{departure}}$.

To calculate the ray trajectory parameters and wave field attenuation in the ionosphere, it is necessary to determine the invariants I_0 of ray entry into the IWD and subsequent changes of the invariant when transiting to other ducts. It is also necessary to calculate along the path the trajectory parameters and absorption corresponding to determined value of I_0 and to subsequent values of I, if transitions to other IWD are observed along the path.

One of the specific computational programs of trajectory parameters at a given distance is based on a simple principle which gives the complete information about possible modes of radio wave propagation through the IWD over the path in question (Fishchuk 1974). Besides, the total volume of IWD at the fixed points along the path, the invariants of intermediate levels $I(\varepsilon')$ are computed with a given step with respect to ε' in each duct from $I=0$ to I_{max}. Accordingly, the oscillation period Θ of these rays, the maximum and minimum heights, the angles of takeoff (or entry) α_0 and departure α_1, the total absorption Γ over half of the oscillation, and the averaged absorption K with respect to one oscillation at a given level of $\varepsilon'_{min}(I)$ are also computed. As a result, we obtain a set of values I, $z_{min}(I)$, $z_{max}(I)$, $\alpha(I)$, $K(I)$, $\Theta(I)$, $\Gamma(I)$ for every duct *E, FE, F* at each of the test points i. When calculating very long-distance paths and, in particular, the RTW ones it usually proves to be quite enough to determine the trajectory parameters at the test points, spaced from each other by a distance $\Delta\Psi = 30-15°$. Choice of the $\Delta\Psi$ value is defined by variation of parameters of the ionospheric model along the path.

To calculate the parameters of a specific trajectory on a given path it is necessary first to carry out the analysis of the total invariants of the channels *E, FE, F* (see above). After identification of the CCP and propagation modes of a given path the following parameters are tabulated: the ray invariants I for the trajectory in question and also the channels which they correspond to and the angular range $\Psi(I)$ at which the invariants are conserved. Next, from the tables of the trajectory parameters calculated at every fixed point, the additional table is compiled corresponding to the defined values of the invariant I (if it is conserved along the whole path), or to a given set of the values I, if a wave leaves the channel and goes into another one. In this case the intermediate values of the parameters are determined by means of interpolation.

The examples of variation of the trajectory parameters obtained by means of the above calculation for the midlatitude RTW path with the transmission point $\theta_0 = 38°$ (latitude), 11^h LT (longitude), $\beta = 112°$ are given in Tables 5.1 and 5.2. This path is shown in Figs. 5.5 and 5.6 and the transmission conditions correspond to the RTW observations at Stanford in spring 1962 (Fenwick 1963). The calculation was performed on the basis of the analytical models of electron density, EMI-72, and collision frequency ν_e, constructed for equinoctial conditions at the solar activity minimum (Appendix 2, $W = 10$).

Table 5.1 gives the trajectory parameters when the radio wave at the frequency 15 MHz propagates in the RTW ionospheric channel along the whole path ($I_0 = 5$, trajectory 8 in Fig. 5.5a). Here, as well as in the subsequent Tables, the coordinates are given of the points θ, φ on the path,

Table 5.1. Parameters of the trajectory 8 (Fig. 5.5) trapped into the interlayer duct; $\theta_0 = 38$, LT = 11^h, $f = 15$ MHz, $\beta = 112°$

θ, deg	φ, deg	Ψ, deg	z_{max}, km	z_{min}, km	Γ, dB	Θ, rad	K, dB/rad	$p_i = \Delta\Psi/\Theta$
38	−15	0	174	114	1.4	0.23	12.6	2.26
22	16	30	188	121	0.7	0.305	5.0	1.70
0	39	60	215	140	0.32	0.41	1.7	1.27
−19	60	90	210	135	0.3	0.38	1.6	1.37
−35	90	120	218	150	0.042	0.354	0.23	1.47
−42	126	160	250	173	0.07	0.39	0.035	1.34
−38	165	180	287	210	0.014	0.48	0.05	1.09
−22	−164	210	315	225	0.02	0.50	0.08	1.04
0	−141	240	325	237	0.01	0.496	0.044	1.05
19	−120	270	290	209	0.014	0.450	0.063	1.15
35	−90	300	232	160	0.015	0.379	0.08	1.37
42	−54	330	188	120	0.8	0.31	6.50	1.67
				$A_{RTW} = 14.6$ dB,		$\mathscr{P} = 16.8$,	$A_{RTW_0} = 16.8$ dB	

Table 5.2. Parameters of the trajectory 1.1′ (Fig. 5.6); $\theta_0 = 38°$, LT = 11^h, $f = 10$ MHz, $\beta = 112°$

Ψ, deg	z_{min}, km	z_{max}, km	Γ, dB	Θ, rad	K, dB/rad	z_s, km	p_i	α, deg	I	Channel
0	0	165	4.0	0.235	35.0	136	2.2	15.0	39	F^h
30	113	178	2.80	0.207	27.2	141	2.5	5.2	6.5	*FE*
60	130	195	1.0	0.25	7.0	159	2.08			
90	114	195	3.1	0.35	17.8	164	1.49			
120	0	220	2.0	0.40	7.5	176	1.3			
150	33	225	0.18	0.57	0.6	205	0.912		33	F^r, *FE*
180	70	293	0.13	0.606	0.45	229	0.858			
210	90	318	0.10	0.61	0.8	250	0.855			
240	112	327	0.05	0.55	0.15	260	0.945			
270	70	296	0.44	0.61	1.55	137	0.855		33.5	F^r
300	9	236	0.84	0.61	2.7	186	0.855			
330	0	180	3.4	0.32	25.0	160	1.63	18.5		F^h
								$A_0 = 65$ dB,		$\mathscr{P} = 16.5$

taken off from the map through $\Delta\Psi = 30°$ along the great circle for $\beta = 112°$. The value of the angle Ψ read off from the emission point is also given; as well as the absorption over half of oscillation Γ in dB; the oscillation period Θ in radian; the absorption $K = 2\Gamma/\Theta$ in dB/rad; the minimum, z_{min}, and maximum, z_{max}, trajectory heights and the number of oscillations within the range between the test points $p_i = \Delta\Psi_i/\Theta_i$.

As is seen from Fig. 5.5, in this case the ray emitted at an angle of $\alpha_0 = 6.1°$ with the invariant $I_0 = 5.5$ is trapped in the duct *FE* and covers the whole RTW path in the channels *FE*, F^r*FE*.

As is seen from Table 5.1, all the trajectory parameters change significantly along the path. The absorption is maximum on the day side when the capture into the duct *FE* occurs. The absorption decreases quickly when a wave deepens into the duct *FE*. The total absorption in the RTW channel, $A_{RTW} = \sum_i K_i \Delta\Psi_i = 14.6$ dB. With allowance for ray entrance into the channel and departure from it to the Earth, the absorption is equal to $A_{RTW_0} \approx A_{RTW} + 2\Gamma_E = 16.8$ dB. Note that for the similar day conditions at Stanford, the absorption of the RTW signal of the second circulation on the frequency $f = 15$ MHz (Fenwick 1966) was found to be within 15 – 20 dB, i.e., it is in good agreement with the calculated results.

As is seen from Table 5.1, the oscillation period Θ is maximum in the region of transition through the equator in the daytime ($\Psi \sim 60°$) and at night ($\Psi \sim 210-240°$), the trajectory rises considerably upwards due to the rise of the *F*-layer. The maximum value of Θ is equal to 0.42 and 0.5 rad in the daytime and at night, respectively, i.e., the "one hop" distance for the ray in the ducts *FE* and *F* is equal to 2680 and 3300 km, respectively. The minimum value of Θ takes place in the daytime near the reception point and is about 0.23 rad (or 1500 km). The oscillation number, p_i, over the range $\Delta\Psi$ is maximum by day (~ 2.26 at $\Delta\Psi = 30°$) and minimum (~ 1.04) on the night side. The whole number of oscillations over the given RTW trajectory is $\mathscr{P} = \sum_i p_i \approx 17$.

In Table 5.2 the results of calculation of the trajectory 1,1′ parameters for $f = 10$ MHz are given. Corresponding variation of the invariants along the path is shown in Fig. 5.6. On this trajectory the ray travels in the channels *F, FE, F, FE, F* with the invariants 39; 6.5; 33; 33.5. The total absorption is equal to 65 dB, the ray makes 16.5 oscillations, the entry angle is about 15°, the departure angle is about 11°.

It was found from the comparison of different trajectories on the RTW path for $f = 10$ MHz that the ray undergoes the minimum attenuation on the trajectory of 1,1′ type as it was expected previously. The propagation conditions on the trajectories close to the trajectory 1,1′ appear to be most favorable for radio wave propagation at $f = 10$ MHz on the path in question.

Note that as the calculation shows, the minimum absorption of a wave at $f = 10$ MHz is equal to 65 dB. It is by 48 dB greater than the absorption at $f = 15$ MHz on the RTW path in question (see also Lobachevskii et al. 1978).

A more complicated computer program (Tushentsova and Tsedilina 1983, Tushentsova et al. 1983) allows by means of a given value of I_0 in the simple CCP (smp) or by means of a given set of values of I in the complex CCP (cmp) to determine the following quantities: the angle of takeoff $\alpha_0(I)$ and descent $\alpha_1(I)$ [Eq. (4.3)], the total absorption $A(I)$ [Eqs. (3.30) and (3.31)], the oscillation number $\mathscr{P}(I)$ [Eq. (3.27)], and the group time delay $t(I)$ [Eqs. (3.32) and (3.33)]. The following subsection deals with the result of the complete analysis of the path, carried out by means of such a program by the technique described in the first and second part of the present section.

Complete Analysis of the Path. In this subsection an example of complete analysis of the Nikolaev – Havana path as long as $S = 9830$ km traversing the low and middle latitudes is given. Below, some results of the calculations performed for quiet equinoctial conditions at minimum solar activity are given (Appendix 2, EMI-81, $W = 10$) (Israetel et al. 1982, Lobachevskii et al. 1979a, b, c, Suares and Tushentsova 1980, Suares and Reborido 1979, Tushentsova and Tsedilina 1983, Lobachevskii et al. 1981, Tsedilina 1980, Vovk et al. 1981, Smirnov 1972, Golyan et al. 1981, Tushentsova et al. 1983). The complete analysis of the path allows to find out the features of diurnal variation of the main radio path characteristics.

First, let us consider diurnal variation of the main path parameters at fixed frequency. Figure 5.7 illustrates variations of the CCP volumes vs local time of the path middle point $LT_{m.p.}$ (52.3°N; 37.5°W) or vs the corresponding Moscow Legal Time (MLT) for $f = 10$ MHz, $\Delta I = I_{max} - I_{min}$, where $I_{max} = I^m$ and I_{min}, are the maximum and minimum invariants of the CCP – F^h, F^r, E^h, F^hF^r, FE, E^hF^r, etc. The volumes ΔI^r and $\Delta\bar{I}$ correspond to the CCP-III (when there is no direct capture and departure of waves from the channel), the volume $\overline{\Delta I}_{cmp}$ corresponds to the CCP-II and CCP-I. However, the fraction of $\overline{\Delta I}_{cmp}$ corresponding to the CCP-I (E^hF^h, $E^hF^hE^h$) is insignificant, consequently, these channels exist only at near noon hours. It is seen that at $f = 10$ MHz the hop propagation through the F-layer exists almost during 24 h except four night hours. The hop propagation through the E-layer takes place only at noon during 5 h. The complex CCP (two-three IWD) with the hop – ricochet propagation and the simple CCP (single IWD) with the ricochet propagation only exist during 24 h. However, the CCP-I exists

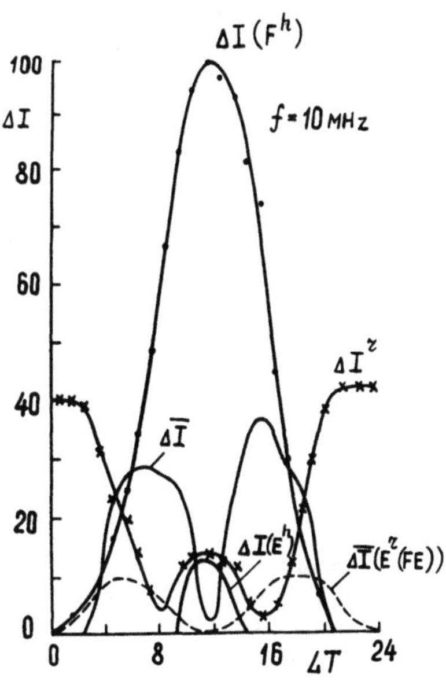

Fig. 5.7. Quantities $\Delta I(F^h)$, $\Delta I(E^h)$, ΔI^r, ΔI_{cmp}, $\Delta I(E^r, FE)$ vs local time of the middle point of the Nikolaev – Havana path for $f = 10$ MHz, where $\Delta I^r = \Delta I(F^r) + \Delta I(FE) + \Delta I(E^r)$, ΔI_{cmp} is the average (between the path terminals) value of the total volume of the CCP $F^h(FE)$, $E^h(FE)$, $E^hF^hF^r$, F^hF^r, E^hF^h, $F^hE^hF^h$, and others, $\overline{\Delta I/E^r(FE)/}$ is the smoothed value for the CCP $E^r(FE)$

mainly during intermediate hours, while the CCP-II exists at night. The conditions for direct propagation in the channels FE, F^r, i.e., in the CCP's such as $E^h(FE)E^h$ or $E^hF^hE^h$, $E^hF^rF^h$ were not observed on the path.

Figure 5.8 shows the calculated results of the maximum $\alpha_{max}(I = I_{max} = I^m)$ and minimum $\alpha_{min}(I = I_{min})$ takeoff angles α_0 and arrival angles α_1 for the CCP F^h and E^h (Lobachevskii et al. 1979a). It is seen that hop trajectories in the IWD F and E have the maximum takeoff angles α_0 and arrival angles α_1 and also the maximum angular ranges $\Delta\alpha_0$, $\Delta\alpha_1$ at noon hours of the path middle point, i.e., in the period of the maximum illuminance of the path.

Figure 5.9 a – f illustrates the calculated results of the minimum and maximum absorption A for various CCP at $f = 10$ MHz (Lobachevskii 1979b, c). The maximum absorption is seen to correspond to the noon, and is equal to 90 dB for the CCP F^h (Fig. 5.9a), and 70 dB for the CCP E^h. The minimum absorption in the CCP F^h is about 40 dB, in the CCP E^h it is ~25 dB. For the CCP F^h, the absorption of the lower modes ($I \approx I_{min}$, $\alpha_{0F} \approx \alpha_{min}$, $\mathscr{P} \geqslant \mathscr{P}_{min}$) is 50 – 30 dB higher than the absorption of the upper modes ($I \approx I_{max}$, $\alpha_{0F} = \alpha_{max}$, $\mathscr{P} \leqslant \mathscr{P}_{max}$). On the contrary, for the CCP E^h, the upper modes have greater absorption.

At noon hours for $f = 10$ MHz, there are 4 – 5 modes [Eq. (2.44)] in the CCP F^h (see Fig. 5.14), and there are 3 – 4 modes in the CCP E^h, hence (see Fig. 5.9a), at noon ($LT_{m.p.} = 12^h$) the absorption of every hop mode in the duct F on the average differs from the neighboring one by 10 – 8 dB, while in the duct E it differs by 10 – 12 dB. At night the hop mode absorption in the duct F (mainly one mode occurs) is negligible. The data show that in the daytime a number of hop modes in the ducts F and E cannot be observed experimentally due to rather high absorption.

It is seen from Fig. 5.9 that at noon the upper limits for absorption in all the possible CCP are equal to 70 – 100 dB, while the minimum lower limits are ~30 – 40 dB (Fig. 5.9 b, c, f). Hence, taking into account the additional losses at the capture and descent, we should assume that at $f = 10$ MHz for the path in question, the most probable are the hop modes in the IWD F and E (CCP F^h, E^h and F^hE^h, $F^hE^hF^h$).

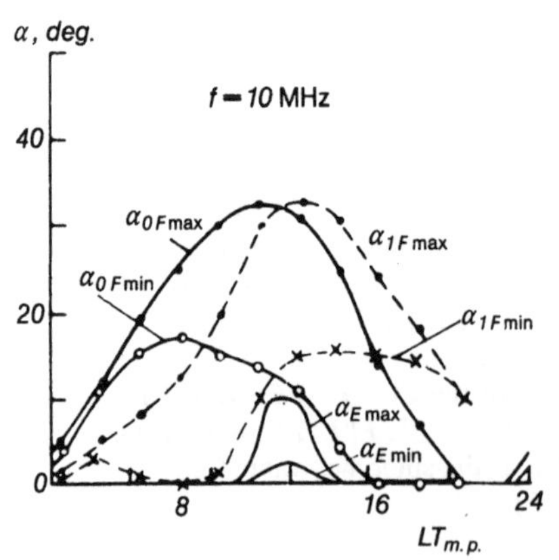

Fig. 5.8. Maximum and minimum angles of takeoff and arrival α vs local time of the middle point for the CCP F^h and E^h at $f = 10$ MHz on the Nikolaev – Havana path

Figure 5.10 shows diurnal variation of the calculated propagation time t of the modes in various CCP at $f = 12$ MHz (Tushentsova et al. 1982). Here, the thick solid lines illustrate the diurnal variation of the propagation time for separate modes in the CCP F^h and CCP E^h. The value of $\mathscr{P}$ is indicated by numbers at each mode. The maximum and minimum propagation time in the CCP F^h, E^h, and E^hF^h, F^hF^r, F^r is shown by a dash – dotted line. Of course, the main delay between the modes corresponds to the hop propagation in the duct $F(F^h)$. According to Fig. 5.10, in the daytime four modes (from four – hop mode, $\mathscr{P} = 4$, to seven – hop mode, $\mathscr{P} = 7$) and also the Pedersen ray (the mode $7P$ shown in Fig. 5.10 by a thick line at the boundary of CCP F^h) can be observed. As is seen from Fig. 5.10, the maximum widening of the signal (up to 4 ms) can be observed at noon. At night there is no channel F^h at $f = 12$ MHz. Possible additional modes in the CCP F^hF^r or F^rF^h with $\mathscr{P} = 3.4$ and 5 are shown by the thick dashed lines. These modes pass a portion of the path on the ricocheting trajectories and on one side of the path do not descend to the Earth. They can appear, for instance, in the case of strong scattering in the ionosphere. These are the lower modes in the duct F. In Fig. 5.10 the combined modes of CCP E^hF^h, $F^hE^hF^h$ and F^hE^h are also shown, which exist only at near noon hours.

In Fig. 5.10 the region I is shaded. Within this region additional modes can appear in the presence of the factors contributing to the capture or descent from the CCP to the Earth. They are the CCP-II modes: E^hFE, FE^rE^h, E^hF^r, F^rE^h, and others. These modes can provide an additional widening of the signal ($\sim$0.5 ms) at intermediate hours. In Fig. 5.10 purely ricocheting modes in the channel FE (CCP-III) are also shown. The probability of their occurrence is small, since the losses at the capture into the duct FE and at the descent from this duct must be rather considerable (more than 60 dB, see Sec. 6.1). It is seen from Fig. 5.10 that they give slight widening of the signal ($\sim$0.2 ms) at night.

Taking into account the energy losses at the entrance into the channels FE, F^r and at the departure from the channels, as well as the losses at the propagation in these channels, in addition to hop modes, it is most probable to expect the occurrence of weak low F-modes, shown by a broken line in Fig. 5.10 ($\mathscr{P}$– 4, $\mathscr{P}$– 5), and the modes in the region 1, while the modes of the FE-duct, which are not shown in Fig. 5.10, must be more weakened due to the additional losses both at the entrance and at the departure.

Figure 5.10 also shows the isopleths of absorption. It is seen that under conditions of the maximum illumination intensity of the path, the low modes in the CCP F^h ($\mathscr{P} = 4$, $\mathscr{P} = 5$) and the modes in the complex hop channels ($\mathscr{P} = 7$, CCP E^hF^h and others) are absorbed most strongly. Their absorption is more than 40 dB. The absorption of the mode $\mathscr{P} = 6$ in the CCP E^h is less than 40 dB. Hence, in this case at low frequencies when modes in the CCP-E^h, E^hF^h are present, the absorption does not influence the total broadening of the signal received, which is connected with the multipath propagation, because the absorption attenuates the modes in the middle of the total signal.

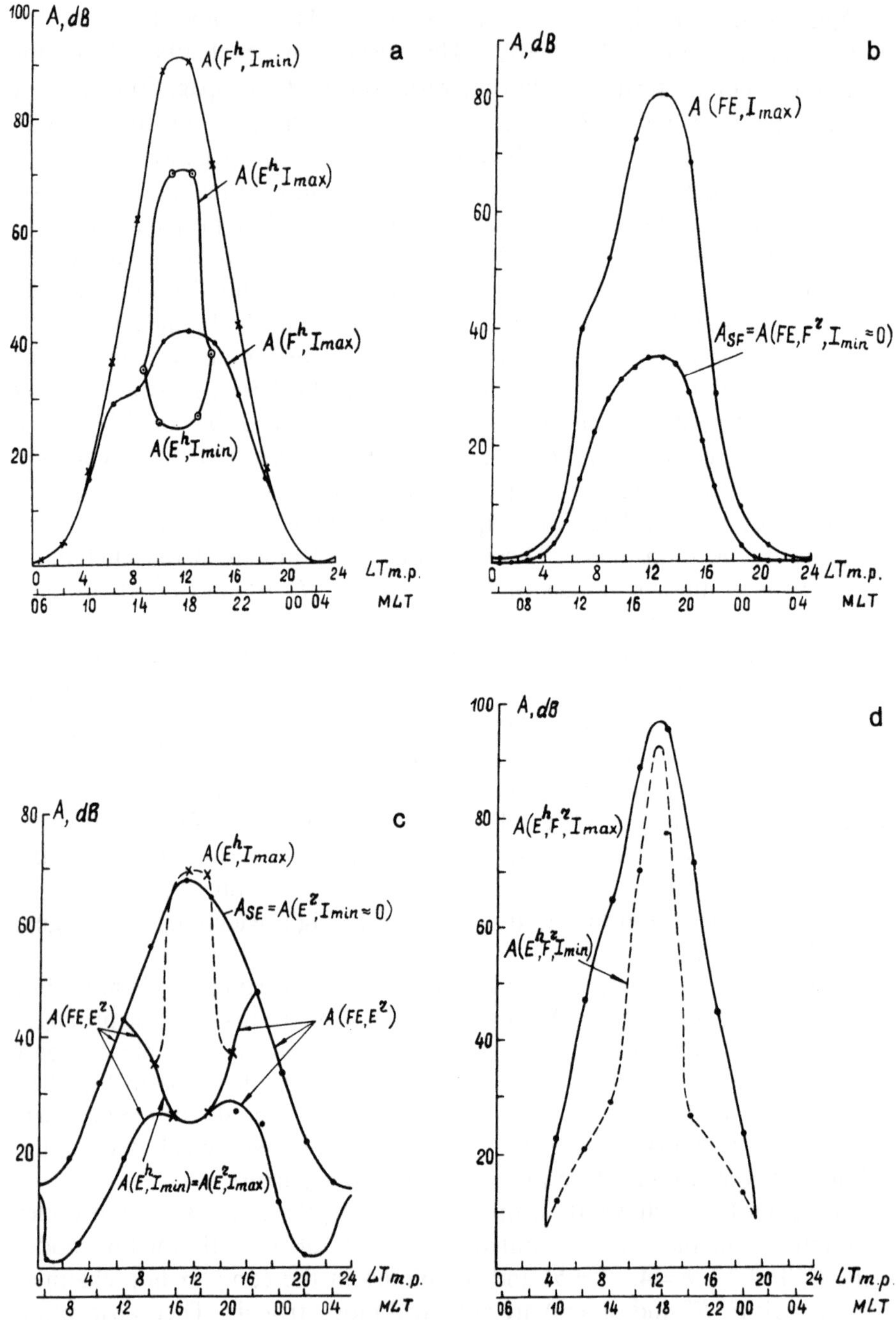

Fig. 5.9a–f. Diurnal variation of the total absorption for the limiting rays (I_{min} and I_{max}) at $f = 10$ MHz for the CCP **a** E^h, F^h; **b** FE; **c** E^h, $E^r(FE)E^r$; **d** E^hF^r, F^r, E^h; **e** $E^h(FE)$, $(FE)E^h$; **f** F^h, F^r on the Nikolaev – Havana path

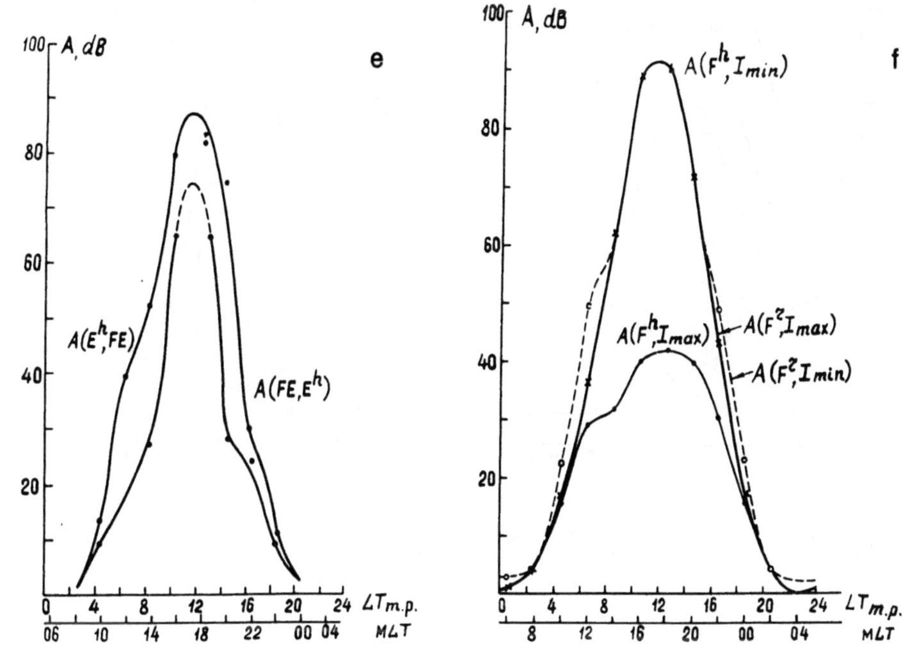

Fig. 5.9 e, f

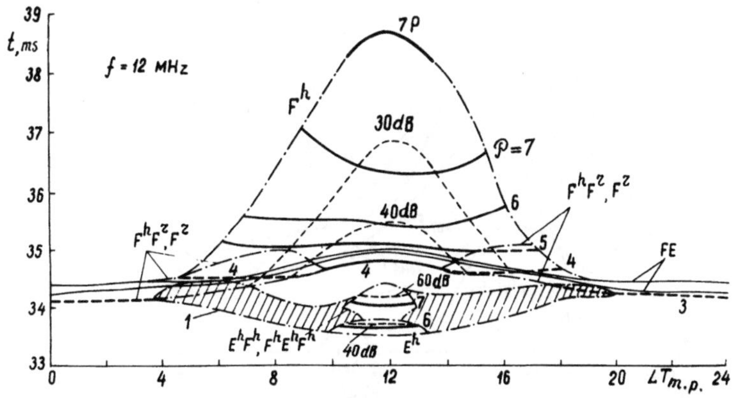

Fig. 5.10. Time delay of the modes in all the CCP vs midpoint local time or vs MLT on the Nikolaev – Havana path at $f = 12$ MHz

Let us consider the results of multifrequency analysis of the path.

Figure 5.11 a, b illustrates the diurnal variations of the boundary values of the maximum and minimum invariants of CCP F^h and E^h for different frequencies. It is seen from Fig. 5.11 a, b that the CCP E^h and F^h at all the frequencies have the maximum volumes at noon hours. According to the curves of Fig. 5.11 a, b and also according to the similar curves for the other CCP (according to the condition $\Delta I_{CCP} \rightarrow 0$), the hop channel MUF and limiting frequencies of the additional CCP were determined.

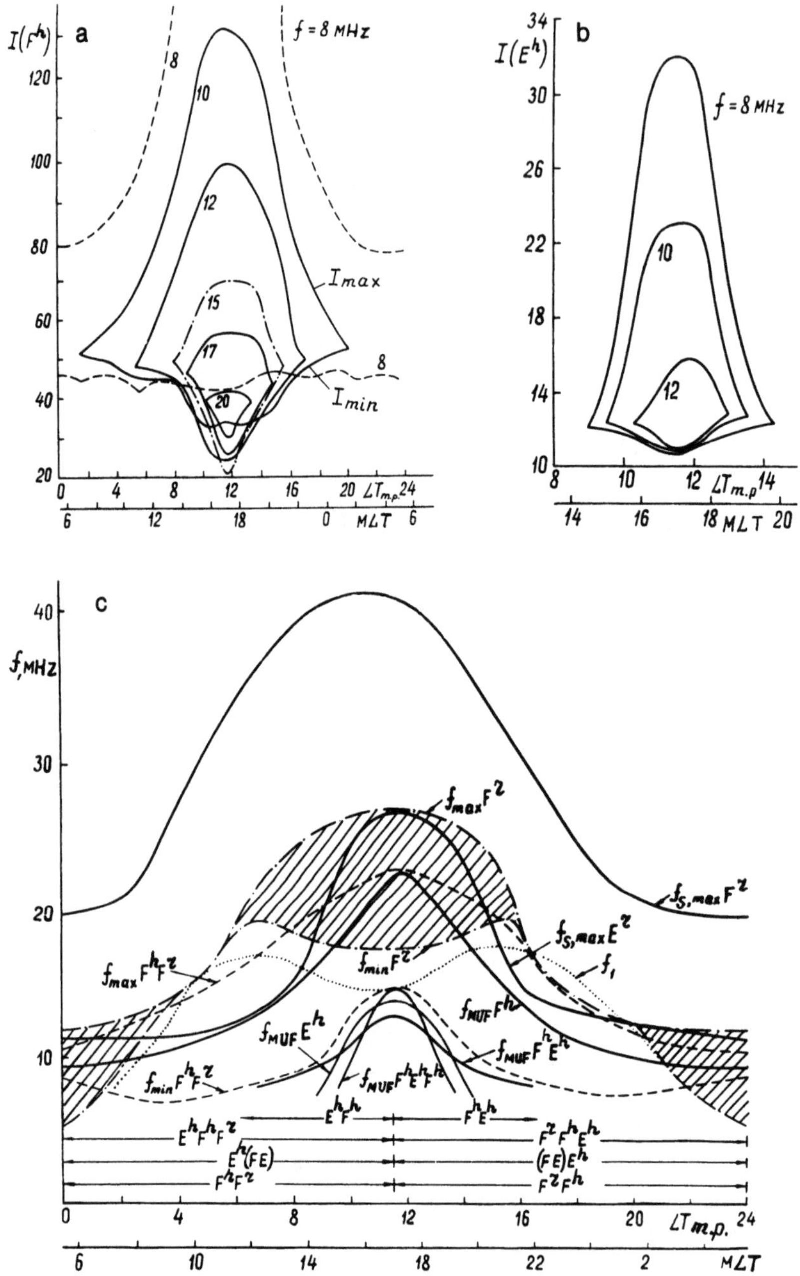

Fig. 5.11 a–c. Boundary values of the invariants vs midpoint local time or vs MLT on the Nikolaev–Havana path; **a** CCP F^h; **b** CCP E^h. Frequencies are indicated at the curves. **c** Diurnal variations of the maximum possible frequencies on the Nikolaev–Havana path: $f_{MUF}F^h$, $f_{MUF}E^h$, $f_{MUF}E^hF^h$, $f_{MUF}F^hE^hF^h$; $f_{s\,max}F^r$ is the maximum frequency of sliding mode in the duct F; f_1 is the maximum possible frequency of the CCP $F^hE^hF^r$, $E^h(FE)$, $(FE)E^h$, $F^rF^hE^h$; $f_{s\,max}E^r = f_{max}FE$ is the maximum frequency for existence of the channels FE and E^r or the maximum frequency of sliding in the duct E. The frequency range between $f_{max}F^r$ and $f_{min}F^r$ is *shaded* for the CCP F^r, $F^r(FE)$, $E^r(FE)E^r$; $f_{max}F^hF^r$ and $f_{min}F^hF^r$ is the maximum and minimum frequencies for CCP F^rF^h or F^hF^r. The time of CCP existence is indicated by the *horizontal lines*

The limiting frequencies of the radio wave propagation over the path during 24 h in all the possible communication channels are shown in Fig. 5.11 c. It is seen that the maximum frequency on the path can be increased in comparison with the MUF-F^h due to propagation in other additional channels. As noted above, the most probable increase of the frequency can be expected due to propagation in the CCP $F^r F^h$ or $E^h F^h F^r$ (CCP-II) when, for example, there is a considerable scattering in the ionosphere at night, in the morning, and in the evening. In this case the maximum increase of the MUF by 4 – 5 MHz, when a wave propagates in the CCP-II (see the curves f_1 and f_{max} of $F^h F^r$), can be expected during the periods when the terminator is situated near the middle of the path, i.e., at hours, close to $LT_{m.p.} = 6^h, 18^h$. The limiting propagation frequency on the path, $f_{s,max}$ of F^r (CCP-III) is 1.8 – 2.4 times more than the MUF.

The diurnal variation of absorption for the lower and upper rays in the hop CCP is shown in Fig. 5.12 a, b. It is seen that in the CCP F^h the lower rays are absorbed stronger, while in the CCP E^h the upper ones are absorbed stronger. The frequency variation of the total absorption A of lower and upper rays in the CCP F^h and FE for different moments of time is illustrated in Fig. 5.12 c, d. The figure shows that at the same frequencies, the absorption in the CCP FE is a bit smaller than the absorption in the CCP F^h. The path absorption in the morning hours is considerably greater than in the evening.

Figure 5.13 a, b illustrates the diurnal variation of the path absorption corresponding to the limiting frequencies shown in Fig. 5.11c. It is seen from Fig. 5.13a that the hop mode absorption is minimum at the frequency $f = f_{MUF\text{-}F^h}$. In the evening and in the morning, the absorption is greater than the noon one by 10 dB, because in the daytime at the middle point of the path, the MUF-F^h is 5 – 8 MHz higher than in the morning and evening time, when a considerable part of the path is still under the day conditions. These portions of the path define the absorption of radio waves at the frequencies $f \approx f_{MUF\text{-}F^h}$.

The absorption A at the frequency $f = 0.8\ f_{MUF\text{-}F^h}$ is more than the absorption at the MUF F^h by 5 – 15 dB. It increases significantly with decreasing frequency, i.e., as the frequency shifts from MUF F^h [see the curves of A $(0.5 \cdot f_{MUF\text{-}F^h})$].

In this case, as mentioned above, the upper ray absorption in the channel F^h (I_{max}) is less than the lower one (I_{min}) (see the corresponding curves for $0.8 \cdot f_{MUF\text{-}F^h}$ and $0.5\ f_{MUF\text{-}F^h}$).

Radio wave absorption in the CCP E^h and $E^h F^h$ with the maximum frequencies $f_{MUF\text{-}E^h}$ and $f_{MUF\text{-}E^h F^h}$ is considerably greater than the absorption in the CCP F^h at the frequency $f_{MUF\text{-}F^h}$. In contrast to the channel F^h, the absorption in the channel E^h is less for the lower rays (I_{min}). At the same time, on low frequencies at near noon hours the lower ray absorption in the CCP E^h proves to be less than the upper ray absorption (all the more so the lower ones) in the CCP F^h (compare the absorption at $f = 10$ MHz in various channels; Fig. 5.13).

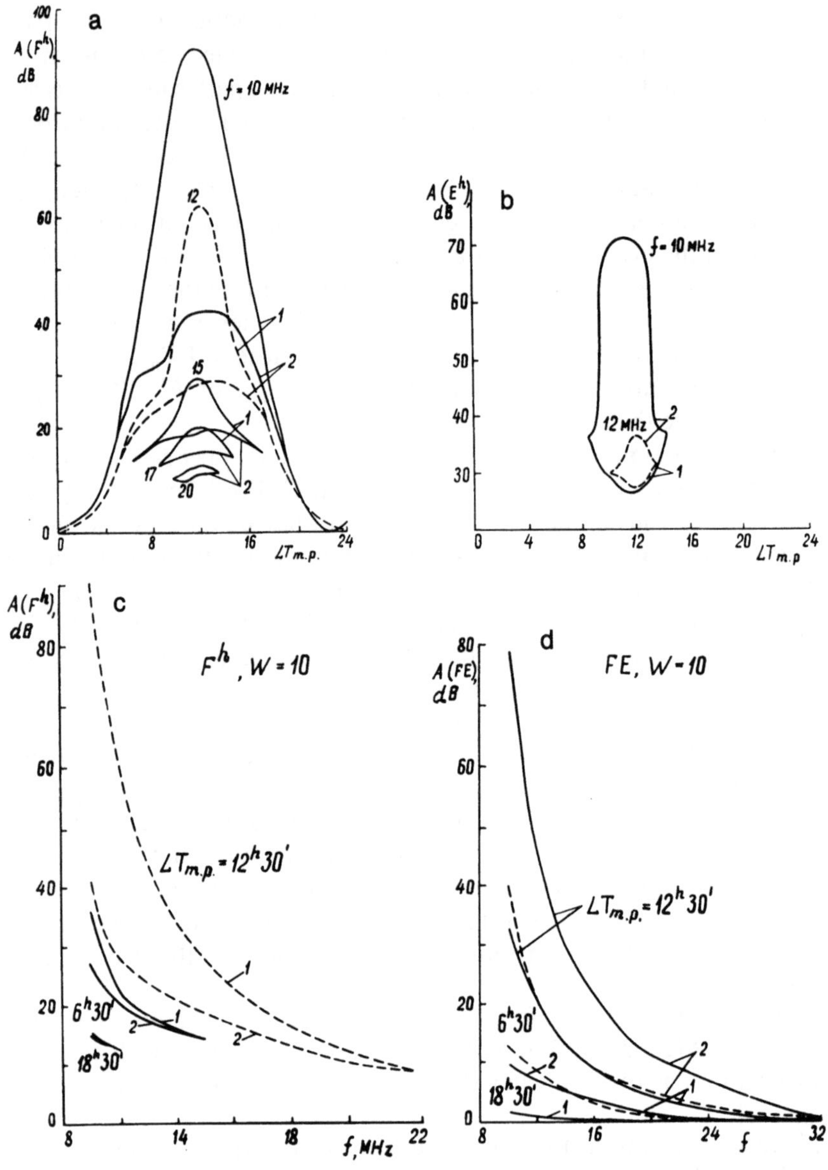

Fig. 5.12 a, b Diurnal variation of the total absorption A for the boundary values of invariants and frequencies shown in Fig. 5.11 a, b: **a** CCP F^h; **b** CCP E^h. The *number 1* indicates the lower ray absorption; the *number 2* the upper ray absorption. **c, d** Comparison of the frequency dependence of absorption for the lower rays (*number 1*) and the upper ones (*number 2*) in the CCP F^h (**c**) and (*FE*) (**d**). LT of the path midpoint is indicated at the curves

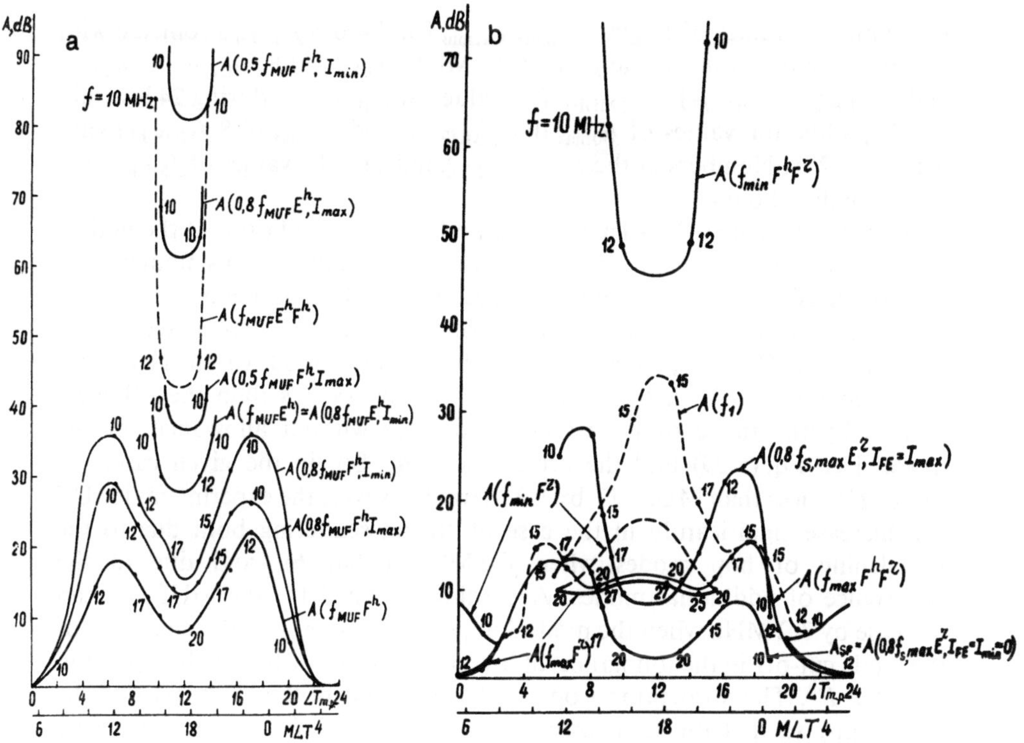

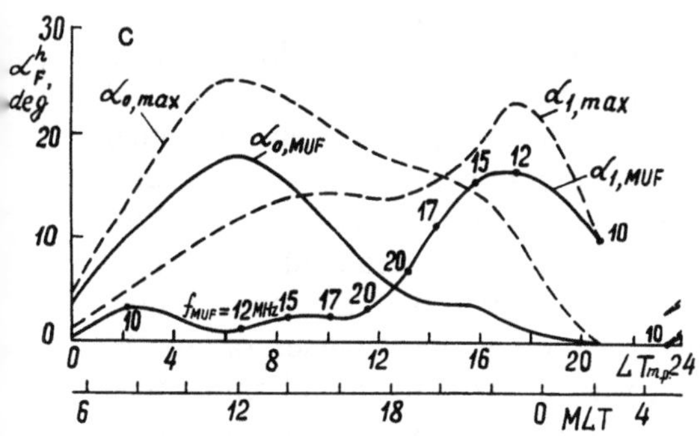

Fig. 5.13. a Diurnal variation of the absorption A on the Nikolaev – Havana path in the hop CCP at various frequencies for the upper (I_{max}) and lower (I_{min}) rays. The frequencies are indicated at the calculated points. **b** The same as in Fig. 5.13a for the hop – ricochet CCP and for CCP with ricocheting modes only. The symbols correspond to Fig. 5.11c. **c** Diurnal variations of the takeoff and arrival angles α_F^h for $f = f_{MUF}F^h$ and $f = 0.8\, f_{MUF}F^h$ on the Nikolaev – Havana path

The diurnal variation of absorption in the combined CCP (hops – ricochet) or in the ricochet CCP at the frequencies corresponding to Fig. 5.11c is shown in Fig. 5.13b. It is seen that the absorption is small in the CCP-*FE*, F^r, and at the frequencies $f \approx f_{\max F^h F^r}$ in the CCP-$F^h F^r$ [(curves A $(0.8 f_{s,\max E^r}$, $I_{FE} = I_{\max})$, A_{SF}, $A(f_{\min F^r})$, $A(f_{\max F^r})$, $(f_{\max F^h F^r})$].

It is of interest that because of the horizontal inhomogeneity of the ionosphere, the angles of takeoff, α_0, and descent, α_1, at the frequency $f = f_{MUF}$ of the CCP F^h are not equal to zero. For the same reason the vertical angles of arrival for one and the same ray ($I = I_0$) at the emission and reception point differ from each other (see also Fig. 5.8). Figure 5.13c illustrates the diurnal variation of α_0 and α_1 for $f = f_{MUF\text{-}F^h}$ (curves α_{0MUF}, α_{1MUF}). Here, the minimum values of angles $\alpha_{0,\min}$, $\alpha_{1,\min}$ for $f = 0.8\, f_{\text{MUF-}F^h}$ coincide with the corresponding values of angles for the MUF: $\alpha_{0,\text{MUF}}(0.8\, f_{\text{MUF-}F^h}) \approx \alpha_{0\text{MUF}}$, and $\alpha_{1\,\text{MUF}}(0.8\, f_{\text{MUF-}F^h}) \approx \alpha_{1\,\text{MUF}}$. The values of α_{MUF} vary during 24 h within $0-17°$, while the values of $\alpha_{0\max}(0.8\, f_{\text{MUF-}F^h})$ and $\alpha_{1,\max}(0.8\, f_{\text{MUF-}F^h})$ vary within $0-25°$. Numbers at the curve $\alpha_{1,\text{MUF}}$ indicate the values of $f_{\text{MUF-}F^h}$ for corresponding hours.

Figure 5.14 shows the LDOS ionogram for noon conditions at the middle point of the path (the maximum illuminance of the path). At this moment, the electron density distribution along the path is most homogeneous.

Let us consider the main features of the ionogram. As is seen from Fig. 5.14, the MUF$_1$-F^h, which is defined by the lower mode (in the given case by the mode $\mathscr{P} = 4$), in the hop CCP F^h is always less than the MUF-F^h defined by the range of the CCP-F^h existence without allowance for the modes [see Eq. (5.29) and the present section]. So, in the given case, the MUF$_1$-F^h is less than MUF$_2$-F^h by 1.3 MHz. However, the experimental MUF can increase significantly in the case of strong scattering both due to the broadening of hop modes when $f \gtrsim$ MUF (Chap. 6), and due to the occurrence of additional modes. According to Fig. 5.14, the MUF$_1$-F^h can increase by 3.2 MHz when the mode 3 appears. Of interest is the character of the mode cut-off in the ionograms at low frequencies due to the absorption in the CCP-F^h. The absorption isopleths have a bend in the lower part of the ionograms, so that the first lower mode in the CCP F^h can be cut-off by $1-2$ MHz more than the second one. All the subsequent modes are cut-off by a smaller value. This effect is often observed in LDOS ionograms for the long-distance path Moscow – Molodiozhnaya (Sec. 1.2; Fig. 1.9). Note also an irregularity of the intermode time delay. As is seen from Fig. 5.14, at $f = 10-12$ MHz the lower modes in the CCP F^h are spaced from each other by $\Delta t \sim 0.4-0.5$ ms. The delay between the upper modes at the same frequencies is considerably greater ($0.6-1$ ms; see also Sec. 1.2, Fig. 1.9). According to the calculation, under the conditions of weak inhomogeneity along the path (noon hours), there are always two rays with the same number of hops, but with different time delay, i.e., there is the Pedersen ray on the path (see Figs. 5.10 and 5.14). However, when there is strong inhomogeneity along the path (during the other hours), the Pedersen ray is not seen in the calculation (see Tushentsova et al. 1982, 1983 and Sec. 5.1, Fig. 5.3). The Pedersen ray is absent in the LDOS ionograms in 95% of the whole number of observations. This is the main feature of experimental ionograms (Lobachevskii et al. 1981, Golyan et al. 1981).

Thus, as shown above by means of the adiabatic approximation and the ionospheric model available, one can perform the complete analysis of propagation conditions over long-distance paths. On the basis of this analysis, one can also predict radio wave propagation over the path. Considerable advantage of this method over the standard ray-tracing techniques is the possibility of identification of all the communication channels, involving the Earth-detached ones. The latter has the propagation frequencies of about the

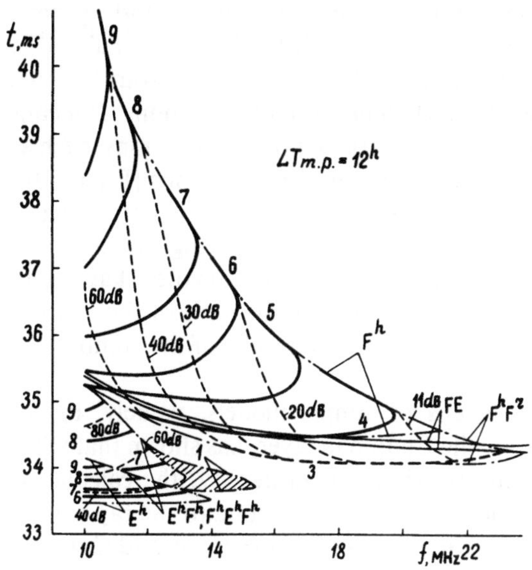

Fig. 5.14. A typical calculated ionogram of OS on the Nikolaev–Havana path. The symbols are the same as in Fig. 5.10

MUF or higher. These channels can be effective in the presence of significant ionospheric scattering (Chap. 6) and also at the artificial modification of the ionosphere by powerful radio waves (Chap. 7).

On the basis of the adiabatic approximation, the calculation of the ionograms involving additional modes have become possible. These modes can appear in ionograms under certain ionospheric conditions in addition to standard hop modes.

5.3 Comparison of the Adiabatic Approximation with the Ray-Tracing Techniques

Application of the adiabatic method to the long-distance radio path calculation has been considered above. In deriving the approximation, the fact is considered that the ionospheric inhomogeneity in the horizontal direction Ψ is small in comparison with inhomogeneity along the height z, i.e.,

$$\frac{1}{R_0 N_e}\frac{\partial N_e}{\partial \Psi} \ll \frac{1}{N_e}\frac{\partial N_e}{\partial z}. \tag{5.39}$$

Ionospheric properties in the horizontal direction (along Ψ) are assumed to vary quite slowly over the wave oscillation period Θ [Eq. (1.10)].

The well-developed standard ray-tracing techniques (Grossi and Langworthy 1966, Lukin et al. 1973, Carrara et al. 1970, Kerblay and Kovalevskaya 1974, Kiyanovskii 1971, Sazhin and Tinin 1975a, Danilin 1967, Kazantsev and Lukin 1966, Kazantsev et al. 1967, Woyk 1958, 1959, 1974;

Shlionskii 1965, 1974, Krasnushkin 1971, Danilin 1968, Egorov and Egorova 1976, Golyan et al. 1975, Gayvoronskaya et al. 1975, Boldovskaya et al. 1971, Kelso 1968, Haselgrove 1955) are applicable for any three-dimensional inhomogeneous medium described by sufficiently smooth functions. It means that in order to apply these calculations, the satisfaction of conditions, such as Eqs. (1.10) and (5.39), is not required. Although the conditions [Eq. (1.10)] are on the whole satisfied in the ionosphere, nevertheless, it is interesting to compare the adiabatic calculation results with the ray-tracing ones. Such a comparison has been carried out in a number of papers (Gayvoronskaya et al. 1975, Egorova et al. 1980a, b; Manaenkova 1980) on the basis of the analysis of both the numerical and analytical solutions. The results of the comparison are given below.

The analysis, carried out by means of simple models, namely, linear, parabolic, and quasiparabolic layer, the Chapman layer, has shown that when using average gradients of the ionospheric electron density, the calculation results obtained in the adiabatic approximation correspond to the ray-tracing calculations with high accuracy (Martinez Brunet 1980, Borisov and Gurevich 1980).

As shown in Sec. 5.2, in the case of complex ionospheric models, in the adiabatic approximation, the radio wave propagation is of much more various character. A whole number of trajectory types appears, which differ significantly from each other and characterize various CCP. Radio wave capture into the interlayer channels and on the ricocheting trajectories, radio wave departure from the channels, and transition among the channels occur. In the adiabatic approximation, the classification of the communication channels is quite natural. In the case of ray-tracing calculations, the situation is absolutely different. The communication channels, which are characterized by strongly different trajectory types, can have very close initial trajectory parameters. That is why it is difficult to identify the CCP by using the ray-tracing calculations. And so, before the adiabatic approximation method was developed, such an analysis, in fact, had not been carried out. Nevertheless, it is interesting to know to what extent the ray-tracing calculations confirm the adiabatic ones. It is also of interest to examine a possible occurrence of various communication channel types, and to refine the regions of radio wave capture and departure. The comparison of the adiabatic calculations with ray-tracing ones allows to examine the accuracy of the adiabatic calculations of trajectory parameters in the case of complex ionospheric models. Such a comparison based on the model EMI-81 at $W = 10$ will be carried out below for the around-the-world path and for the Nickolaev – Havana path.

Around-the-World Path. Let us consider the channel propagation of radio waves transmitted from the ground [the CCP (cmp): $E^{h}(FE)F^{r}(FE)E^{h}$] (Gayvoronskaya et al. 1975). Consider the CCP for the frequencies $f = 12$ and 15 MHz, the transmission point is situated at $\theta = 40°$, $\varphi = 10^{h}30'$, radio waves are emitted at an angle $\beta = 112°$, read off from the northern direction. The path is shown by the thick line in Fig. 5.15, in which the isopleths of the invariant I_{FE} and angle α_{FE} are given. In the figure a half of the great circle is shown, passing through the transmission point, indicated by a cross. The transmission direction is shown by an arrow. The propagation conditions along the path are close to these for Stanford, spring 1962 (see Fig. 5.5). The adiabatic analysis of propagation conditions along the path show that at $f = 12$ and 15 MHz, the waves trapped into the duct on the day side, remain trapped on the night side as well, and return to the Earth at a distance of about one hop from the transmission point (in the vicinity of the minimum of I_{FE} on the path). Thus, over the path, radio waves with $f = 12$ and 15 MHz propagate almost along the whole RTW trajectory in the duct *FE* and *F* without descending to the Earth.

The comparison of results of the trajectory calculations is given in Figs. 5.16 and 5.17. Here the trajectory height z in km is plotted as a function of distance S in km along the Earth's surface or as a function of the central angle Ψ in degrees. The figures also show the minimum, $z_{\min FE}$ and $z_{\min FE}$, and maximum, $z_{\max FE}$ and $z_{\max F}$, heights of the channels *FE* and *F*, the minimum, $z_{\min}$, and maximum, $z_{\max}$, trajectory heights (dashed lines) calculated by the adiabatic method. According to the calculation, for $f = 12$ MHz the capture into the interlayer duct *FE* must take place at the takeoff angle $\alpha_0 = 11.2°$; the trajectory invariant (or the entry invariant) I_0 is equal to 6 (see Fig. 5.15).

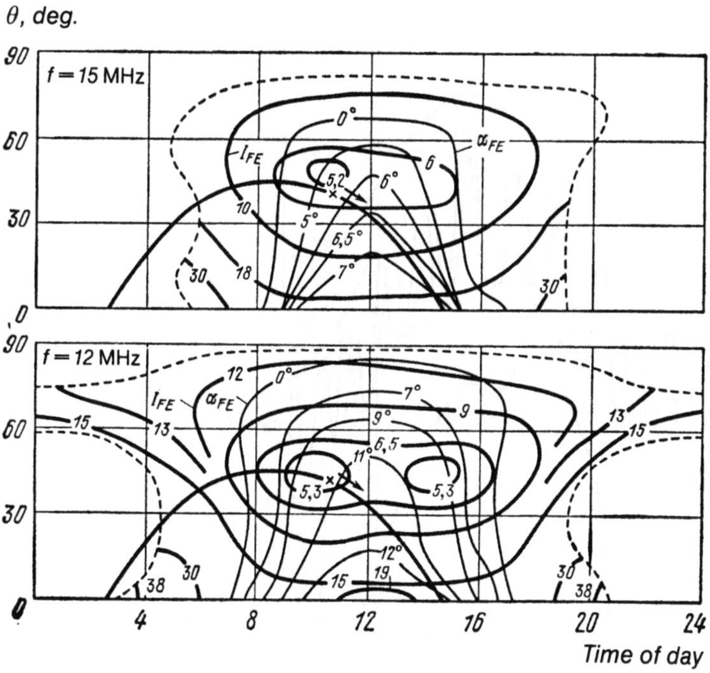

Fig. 5.15. Isopleths of the total adiabatic invariants I_{FE} and capture angles α_{FE} for $f = 12$ and 15 MHz

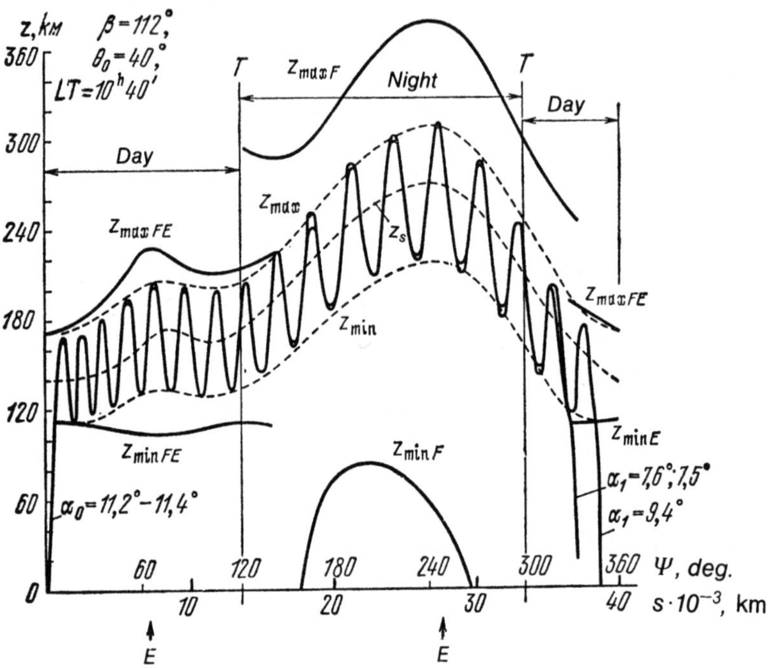

Fig. 5.16. Comparison of the trajectory parameters calculated by different methods for $f = 12$ MHz

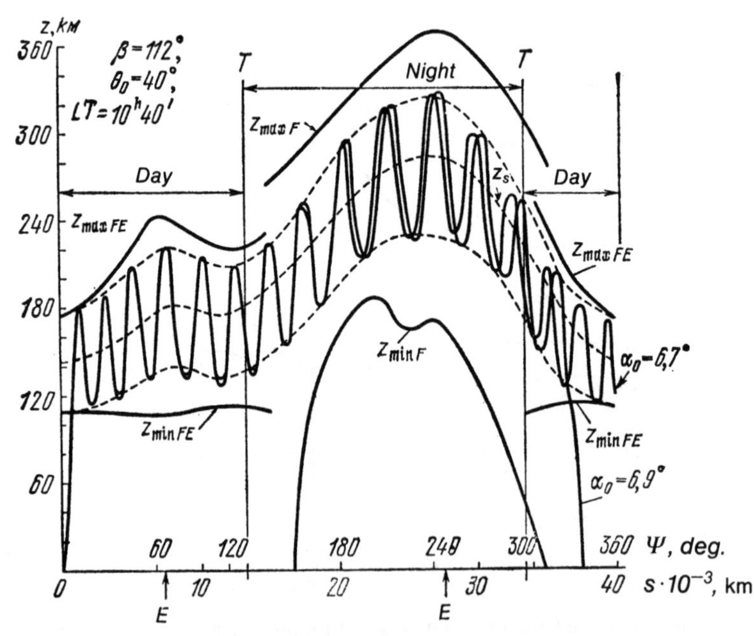

Fig. 5.17. The same as in Fig. 5.16 for $f = 15$ MHz

The parameters of such a trajectory for the path are presented in Table 5.3. They are: values of the minimum, z_{min}, and maximum, z_{max}, trajectory heights, the oscillation period Θ, the number of oscillations over the interval $\Delta\Psi = 30°$ between the test points, the test point coordinates Ψ_i taken off with the help of the great circle maps along the path, and also the sliding mode height, z_s. In Table 5.4 the same parameters are presented for $f = 15$ MHz (see Fig. 5.17). Here, the angle of capture, $\alpha_0 = 6.1°$; $I_0 = 5.9$.

Figures 5.16 and 5.17 also show the calculated results obtained by the standard ray-tracing techniques for the same ionospheric model. The trajectory calculation was performed within the range of $\alpha = \alpha_0 \pm 1°$ with the step $\Delta\alpha = 0.1°$, the values of $\alpha_0 = \alpha_{FE}$ were taken off from the maps for $f = 12$ and

Table 5.3. Parameters of trajectory shown in Fig. 5.16; $f = 12$ MHz, $\theta = 40°$, LT = $10^h40'$, $\alpha_0 = 11.2°$, $\beta = 112°$, $I_0 = 6$

Ψ, deg	z_{min}, km	z_{max}, km	z_s, km	Θ, rad	p_i	Channel
0	114	166	140	0.185	2.76	*FE*
30	113	182	144	0.24	2.13	
60	131	201	164	0.31	1.65	
90	130	204	170	0.35	1.46	
120	130	204	174	0.32	1.59	
150	154	231	204	0.38	1.34	F^r
180	185	268	234	0.44	1.16	
210	205	294	258	0.47	1.08	
240	219	308	271	0.47	1.08	
270	207	290	257	0.44	1.16	
300	164	242	213	0.39	1.31	
330	118	190	167	0.32	1.6	*FE*
						$\mathscr{P} = 18.3$

Table 5.4. Parameters of trajectory shown in Fig. 5.17; $f = 15$ MHz, $\theta = 40°$, LT = $10^h40'$, $\alpha_0 = 6.1°$, $I_0 = 5.9$, $\beta = 112°$

Ψ, deg	z_{min}, km	z_{max}, km	z_s, km	Θ, rad	p_i	Channel
0	114	174	144	0.24	2.16	*FE*
30	119	189	152	0.3	1.73	
60	135	216	176	0.32	1.63	
90	137	216	180	0.28	1.86	
120	138	212	182	0.36	1.45	
150	163	240	212	0.38	1.37	F^r
180	195	285	247	0.475	1.09	
210	227	313	274	0.5	1.04	
240	230	326	286	0.51	1.02	
270	218	306	270	0.46	1.13	
300	173	254	223	0.417	1.25	
330	126	190	170	0.33	1.58	*FE*
						$\mathscr{P} = 17.3$

15 MHz for the above mentioned coordinates of the transmitter. For $f = 12$ MHz, the capture turns out to occur within the range of $\alpha_0 = 11.2 - 11.4°$; for $f = 15$ MHz the capture occurs within the range of $\alpha_0 = 6.1 - 6.9°$, i.e., the angular width $\Delta\alpha$ of the capture for $f = 12$ MHz is equal to 0.2°, while for $f = 15$ MHz it is equal to 0.8°.

As is seen from Figs. 5.16 and 5.17 and from Tables 5.3 and 5.4, the capture conditions, determined by the ray-tracing technique, correspond completely to the ones, determined by the adiabatic method. The angular width of the capture also corresponds to variation of the values of $dI/d\Psi$ along the path. For instance, the ratio $(\Delta I/\Delta\Psi)_{15\,\mathrm{MHz}}/(\Delta I/\Delta\Psi)_{12\,\mathrm{MHz}} = 3$, which is close to the ratio $\Delta\alpha(15\,\mathrm{MHz})/\Delta\alpha(12\,\mathrm{MHz})$. The capture angular width calculated by Eq. (3.72) for $f = 12$ MHz is $\Delta\alpha(12\,\mathrm{MHz}) = 0.26°$, and for $f = 15$ MHz, $\Delta\alpha(15\,\mathrm{MHz}) = 0.77°$ which corresponds to the values obtained by the ray-tracing technique. For $f = 15$ MHz, it was the trajectory with $\alpha_0 = 6.0°$ that descended to the ground ($S = 37498$ km, the arrival angle, $\alpha_1 = 2.1°$). The other trajectories ($\alpha_0 = 6.0 - 6.7°$) circled the Earth twice without descending to the ground. At the second circulation, the arrival angles α_1 for $\alpha_0 = 6.1°$, 6.5°, 6.7° are equal to 3.9, 1.3, 4.8, respectively. In this case the point of descent to the Earth was distant: 1385, 186, and 1372 km from the transmission point. For $f = 12$ MHz, the trajectories with $\alpha_0 = 11.2$, 11.3, and 11.4° descended to the Earth. The corresponding values of S at the descent are equal to 38600, 37450 and 37374, the corresponding angles of descent α_1 are equal to 9.4, 7.8, 7.5°.

As is seen from the same Figures and Tables, the minimum, z_{min}, and maximum, z_{max}, trajectory heights calculated by both methods are close in magnitude. The oscillation number both on the whole RTW path and in the intervals $\Delta\Psi = 30°$ between the test points along the path also coincides. The conditions of departure and the angles of arrival are in a good agreement, as follows from the above mentioned Figures and Tables.

Nikolaev – Havana Path. A detailed comparison of the adiabatic calculations with the ray-tracing ones for the long-distance path has been given in Egorova et al. 1980a, b; Manaenkova 1980. All the possible CCP were identified by both methods, the detailed comparison was carried out for the angles of capture and arrival. The period and number of oscillations for various rays, their heights, etc. were also compared. Based on the calculations and comparison of calculated results obtained by both methods, the adiabatic approximation accuracy was estimated in the case of radio wave propagation in the simple channels and also in the combined ones, when transitions of a ray from channel to channel take place. A relation was obtained between the initial trajectory invariant and the takeoff angles. The main results are given below.

Figure 5.18 illustrates the variation of the adiabatic invariants for the IWD: *F, E,* and *FE* along the path, and also the variation of the maximum angles of capture α^m at $f = 10$ MHz, $LT_{m.p.} = 4^h30'$. By analyzing I as a function of Ψ, the main CCP were identified.

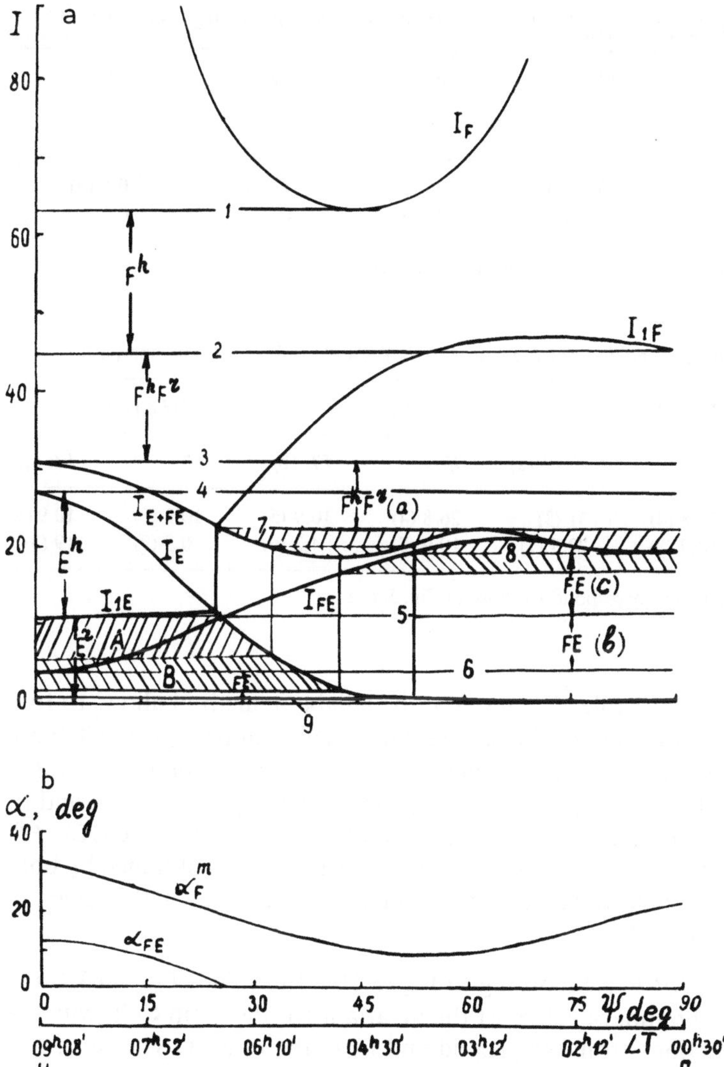

Fig. 5.18 a, b. Maximum invariants I and angles of capture α^m for the ducts F, E, FE vs angular distance Ψ over the Nikolaev – Havana path, or vs LT. The value $\Psi = 0$ (*point N* on the abscissa) corresponds to the emission point at Nikolaev, the value $\Psi = 90°$ (*point H*) corresponds to the reception point at Havana. *Numbers* in Fig. 5.18a indicate the CCP boundaries

The channel types and also the minimum, I_{min}, and maximum, $I^m = I_{max}$, invariants for every CCP are presented in Tables 5.5 and 5.6 and are shown in Fig. 5.18. Table 5.5 indicates the simple CCP with radio wave propagation in one of the IWD: *F, FE, E* over the whole path. Table 5.6 shows the main combined CCP corresponding to propagation in two IWD along the path. The volumes of separate parts of the complex CCP are marked by the letters a, b, c corresponding to a simple CCP (see Table 5.6 and Fig. 5.18). Besides the

Table 5.5. Values of I_{max} and I_{min} for the simple CCP on the Nikolaev – Havana path[a]

Type of CCP / I	F^h	F^hF^r	FE	E^r
I_{max}	63.1 (1)	45 (2)	4.2 (6)	0.5 (9)
I_{min}	45 (2)	31 (3)	0	0

[a] Numbers in parentheses correspond to those in Fig. 5.18a.

Table 5.6. Values of I_{max} and I_{min} for the complex CCP on the Nikolaev – Havana path[a]

Type of CCP	$E^hF^hF^r$ (a)		$E^h(FE)$ (b)		$E^r(FE)$ (c)	
I	E^h	F^hF^r (a)	E^h	FE (b)	E^r	FE (c)
I_{max}	26.8 (4)	31 (3)	26.8 (4)	10.9 (5)	10.9 (5)	10.9 (5)
I_{min}	10.9 (5)	22.5 (7)	10.9 (5)	4.2 (6)	0.5 (9)	19 (8)

[a] Numbers in parentheses correspond to those in Fig. 5.18a.

channels listed in the Tables 5.5 and 5.6, the adiabatic method allows to find out at least two more combined channels with small volumes. These CCP are CCP-$E^rF^r(FE)F^r$ and $E^rF^r(FE)$ which correspond to the volumes, indicated in Fig. 5.18a by different types of shading, and marked by the letters A and B. The analysis of calculated results in the case of propagation in the CCP F^h, E^hFE, and $E^hF^hF^r$, and also the comparison of the results obtained by both methods are given below. More detailed analysis of all the CCP is given in Egorova et al. (1980a, b).

Analysis of the CCP-F^h. According to Fig. 5.18a, the values $I_{max} = 63.1$ and $I_{min} = 45$ are the boundary values of the invariant for the main CCP with the radio wave reflections from the ground and from the F-layer. The value I_{max} defines the upper channel boundary. It corresponds to the minimum value of I_F on the path at $\Psi \approx 45°$. The value I_{min} defines the lower boundary of the channel. It corresponds to the value of I_{1F} at the reception point. The calculation of number of oscillations $\mathcal{P}_F$ on the path, i.e., at the point $\Psi = 90°$ as a function of the invariant I in the channel (Fig. 5.19) shows that in the hop part of the channel F over the path there are two modes, namely, the five – hop mode with $I_{0n} = 59$ and the four – hop mode with $I_{0n} = 45.6$.

Figure 5.20 illustrates typical trajectories obtained by the ray-tracing technique for the channel CCP-F^h. These trajectories define the four – hop and five – hop modes corresponding to the adiabatic analysis. The value of the emission angle is $\alpha_{0F} = 17°$ for the four – hop mode, while for the five – hop one, $\alpha_{0F} = 19.9°$. Figure 5.20 illustrates also the "upper" trajectory, i.e., the first trajectory which is not confined on the upper

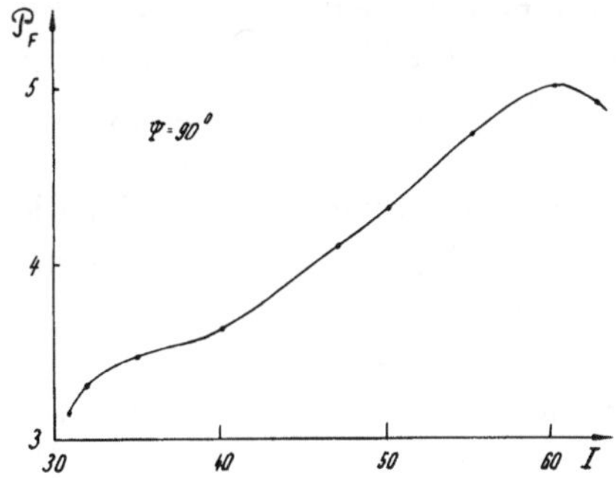

Fig. 5.19. Total number of oscillations in the CCP F^h and F^hF^r vs the invariant at the reception point $\Psi = 90°$

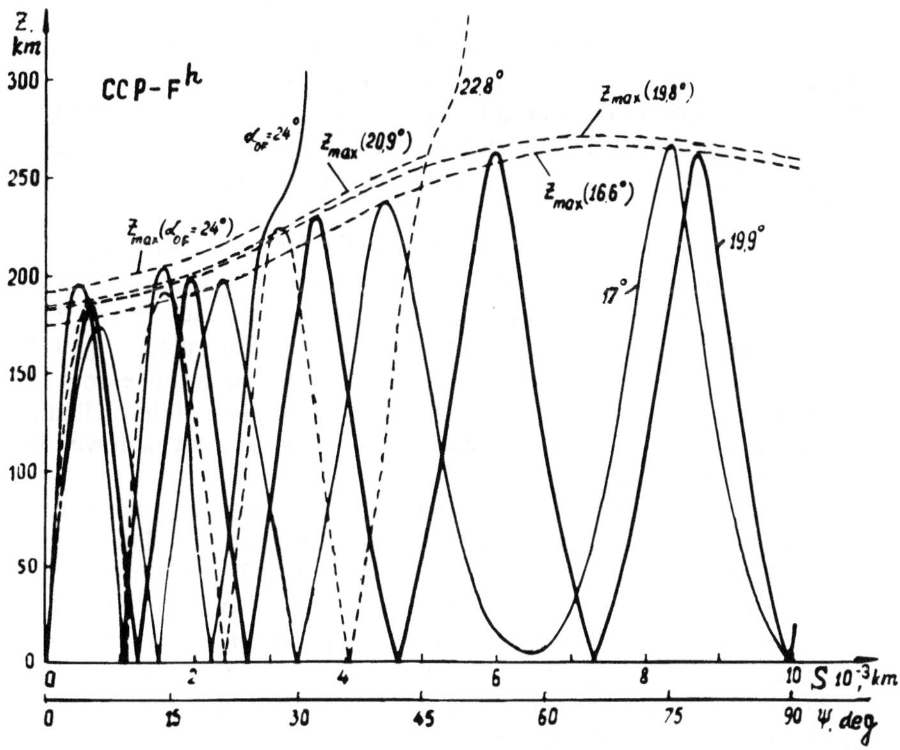

Fig. 5.20. Trajectory height z as a function of distance S or angular distance Ψ along the path for the CCP F^h. *Dashed lines* show the maximum trajectory heights, z_{max}, calculated by the adiabatic method for the takeoff angles indicated in *parentheses* and at the trajectories

boundary of the channel and has the maximum angle $\alpha_{0F} = 22.8°$ *calculated with an accuracy of* $\Delta\alpha_0 = 0.2°$. The region where the trajectory leaves both the channel and the ionosphere ($\Psi \approx 45°$) corresponds to the coordinate of the minimum I_F over the path (compare with Fig. 5.18a). In addition to this trajectory, Fig. 5.20 shows another one, leaving the ionosphere (and the channel F^h) at a shorter distance from the transmission point. It is obvious that this trajectory has a greater value of the takeoff angle, $\alpha_{0F} = 24°$. This also corresponds to the adiabatic calculation (see Fig. 5.18 a, b). The lower trajectory ($\alpha_{0F} = 16.5°$) in the channel is close to the four – hop mode, that is why the trajectory is not shown in Fig. 5.20.

Table 5.7 presents the values of the angle α_{0F} calculated by the adiabatic (α_{0F}, adiab.) and ray-tracing (α_{0F}, ray-tr.) techniques for the channel boundaries and for two channel modes. The results are seen to be in sufficiently good agreement. The maximum differences (about 8%) between the ray-tracing and adiabatic values are obtained for the angle α_{0F} defining the upper boundary of the channel. It is connected with the fact that at smaller values of the angle α_{0F}, the trajectory in the vicinity of the I_F minimum approaches the ground, i.e., is outside the ionosphere. Consequently, the trajectory departure from the duct F takes place, in fact, at the angle α_{0F}, which is greater than the adiabatic value of α_{0F}. The greater angle corresponds to the trajectory approach to the upper boundary of the channel at $I_F = I_{F\min}$.

Figure 5.20 shows also the heights z_{max} for the invariants corresponding to the values of α_{0F} indicated in parentheses. It is seen that for close values of α_{0F}, the adiabatic values of z_{max} differ from the ray-tracing ones only by 2 – 8%.

Note that departure from the Earth for the trajectory with $\alpha_{0F} = 17°$ between the third and fourth oscillations at $\Psi = 60 - 70°$ also corresponds to the adiabatic approximation (see Fig. 5.18). Within this interval of Ψ the parameter $I_{1F}(\Psi)$ has the maximum and the corresponding value of I_0 of the trajectory is less than the I_{1F}. That is why the four – hop mode does not descend to the Earth after the third oscillation. The variation of the oscillation period of the trajectories along the path is also in a good agreement with the adiabatic approximation.

According to the adiabatic calculation, the descent angles α_{1F} are equal to 2° for the four – hop mode, and 7.7° for the five – hop one. For the ray-tracing calculations the angles are equal to 1.6° and 7.19°, respectively.

As is seen from Fig. 5.19, the curve $\mathscr{P}_F(I)$ has the maximum $\mathscr{P} = 5.04$ at the invariant close to the boundary value (the upper boundary of the channel F). Such a behavior of the curve $\mathscr{P}_F(I)$ corresponds to the presence of the five – hop Pedersen ray on the path. This is also confirmed by the ray-tracing calculations (Egorova et al. 1980a).

Table 5.7. Comparison of the takeoff angles calculated by the adiabatic and ray-tracing techniques for the CCP F^h

Mode type	$\alpha_{0F(\text{adiab})}$, in degrees	$\alpha_{0F(\text{ray-tr})}$, in degrees	I_{0n}
Upper boundary of channel	20.7	22.6	63.1
Five – hop mode	19.8	19.9	59
Four – hop mode	16.5	17	45.4
Lower boundary of channel	16.4	16.5	45

Analysis of the CCP $E^hF^hF^r$ *and the CCP* E^hFE, E^hE^rFE. Let us consider the combined CCP (see Table 5.6) involving propagation in the duct E and further the capture into the ducts FE or F.

According to Fig. 5.18, the ray emitted within the range of $\alpha_{0E} = 0-13.1°$ after 0.5 – 2.5 oscillations in the duct E goes into the duct F (the CCP $E^hF^hF^r$ is formed) or into the duct FE (the CCP E^hFE or E^hE^rFE are formed). The analysis of trajectories with the takeoff angles, $\alpha_0 = \alpha_{0E}$ (a part of the trajectories is shown in Fig. 5.21 a, b) shows that for the CCP in question ($\alpha_{0E} = 0-13.1°$) the angular range of capture determined by the adiabatic method for the transmission point $\Psi = 0$ (Fig. 5.18) is in a good agreement with the angular range calculated by the ray tracing technique.

By means of the ray-tracing calculations, the angular ranges α_{0E} were determined, for which the trajectories are trapped in the duct F or FE. The results of these calculations are presented in table 5.8. The table gives the angular ranges of α_{0E}, the description of trajectory character within the angular ranges, and the CCP types. As is seen from Table 5.8, when calculating the trajectories with the takeoff angles $\alpha_0 < \alpha_{0E}^m$, there are three regions of radio wave capture into the duct FE at the takeoff angles $\alpha_{0E} = 0-3°$, $9-11.6°$, and $12.6-13°$. The value of the capture range, $\Delta\alpha_{0E}$ (ray-tr.) for the values of $\alpha_{0E} \approx 0°$; 10.6°; 12.8° is equal to 3; 2.6; 0.4°, respectively. We bear in mind that as α_{0E} (or invariant I_0) changes, both the oscillation period and the distance Ψ_{max} also change (the latter is the distance covered by the wave within the duct E, it is restricted by the curve $I_0 = I_E$ within the range where $I_E > I_{1E}$, see Fig. 5.18a). Thus, one can understand why in the angular range from $\alpha_{0E}^{min} = 0$ to $\alpha_{0E}^{max} = 13°$ there are several regions of capture into the duct FE. The distance Ψ_{max} and corresponding oscillation number of the ray in the duct E vs α_{0E} are shown in Fig. 5.22. It is seen from Fig. 5.22 that the presence of three capture regions is explained by the variation character of the curve $\mathscr{P}_E(\alpha_{0E})$. It is clear that the capture into the duct FE can take place as the trajectory approaches z_{min} of the duct FE, i.e., at the oscillation number $\mathscr{P}_E \approx 0.5$ and 1.5. Since, within the range of $0 < \alpha_{0E} < 13°$, the curve $\mathscr{P}_E(\alpha_{0E})$ has the values $\mathscr{P}(\alpha_{0E} = 0) = 1.3$ and $\mathscr{P}_E(\alpha_{0E} \lesssim 13°) = 0.5$ and the maximum $\mathscr{P}_E^{max} > 1.5$ the curve traverses the level $\mathscr{P}_E = 1.5$ twice and the level $\mathscr{P}_E = 0.5$ once. Thus, three capture regions (numbers 1, 2, 3) appear. The thickenings of the curve $\mathscr{P}_E$ in Fig. 5.22 show the ranges $\Delta\alpha_{0E}$ (ray-tracing) given in Table 5.8 in the case of the capture into the duct FE. According to the adiabatic approximation [Eq. (3.36)], we obtain: $\Delta\alpha_{0E}(\alpha_{0E} \approx 0°) = 7.5°$, $\Delta\alpha_{0E}(\alpha_{0E} = 10.6°) = 1.6°$ and $\Delta\alpha_{0E}(\alpha_{0E} = 12.8°) = 0.7°$. It is seen that the variation character of $\Delta\alpha_{0E}$ (adiabat.) corresponds, in general, to the variations of $\Delta\alpha_{0E}$ (ray-tr.) obtained by the ray-tracing techniques.

As an example, Fig. 5.21 shows z_{min} and z_{max} for various values of the invariant I_{FE} and the takeoff angles α_{0E}, calculated by the adiabatic method and also by the ray-tracing techniques. The comparison showed the difference in heights to be no more than 20%. In this case, the heights z_{min} differ more

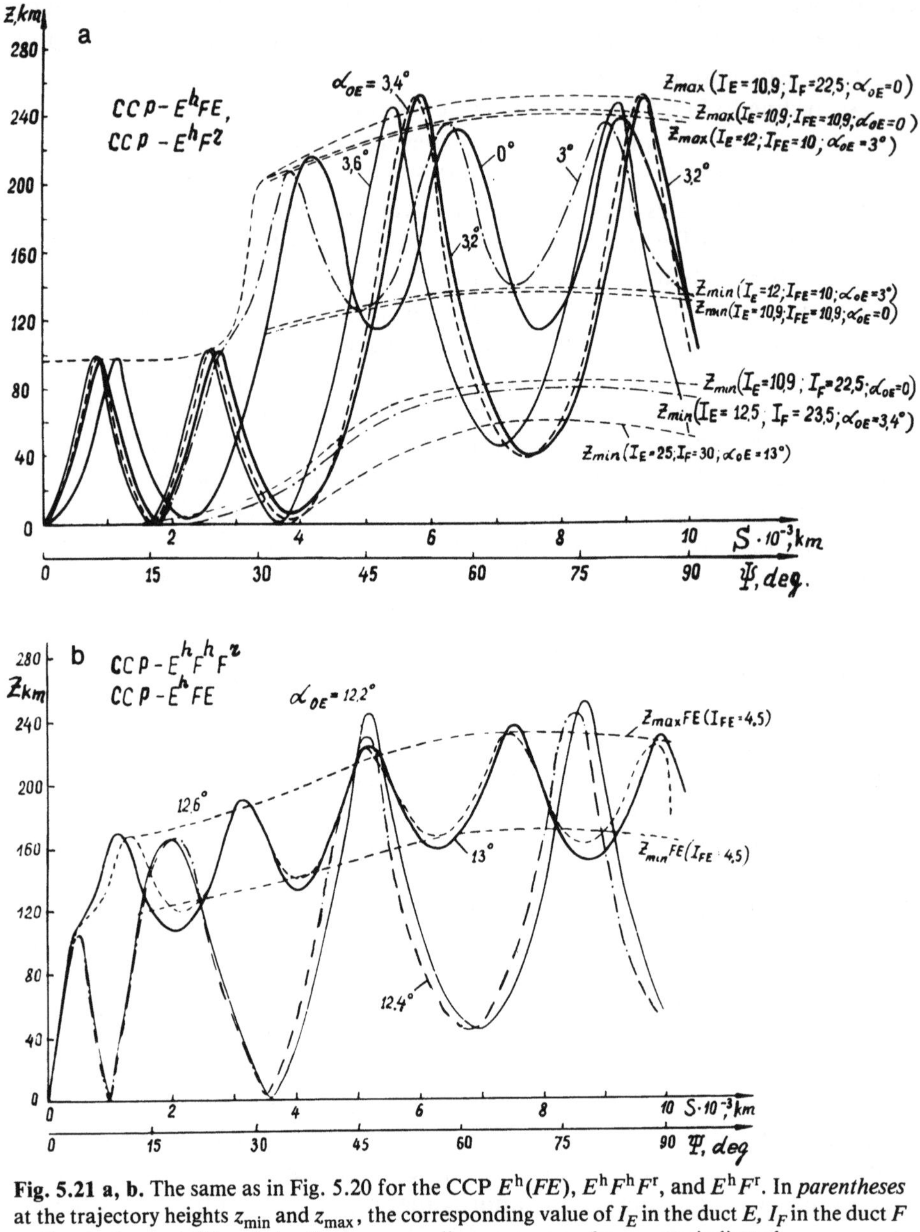

Fig. 5.21 a, b. The same as in Fig. 5.20 for the CCP $E^h(FE)$, $E^hF^hF^r$, and E^hF^r. In *parentheses* at the trajectory heights z_{min} and z_{max}, the corresponding value of I_E in the duct E, I_F in the duct F or I_{FE}, in the duct FE and also the corresponding capture angles α_{0E} are indicated

than z_{max}. This corresponds to the character of dependence of the heights on α_{0E}. In the case of capture into the duct F, the difference in z_{min} reaches 40 km.

The Adiabatic Invariant and the Adiabaticity Parameter. If directly calculating the parameter I by the ray-tracing technique, one can verify the adiabatic

Table 5.8. Range of the takeoff angles α_{0E} calculated by the ray-tracing techniques for various CCP

Degrees	Trajectory type	Type of CCP
0–1.4	1.5 oscillation in duct E, capture into duct FE	$E^{h}E^{r}(FE)$
1.6–3.0	1.5 hop in duct E, capture into duct FE	$E^{h}(FE)$
3.2–3.4	1 hop and 1 oscillation in duct E, ricochet in duct F	$E^{h}E^{r}F^{r}$
3.6–8.8	2 hops in duct E, ricochet in duct F	$E^{h}F^{r}$
9.0–11.6	1.5 hop in duct E, capture into duct FE	$E^{h}(FE)$
11.8–12.0	1 hop in duct E, ricochet in duct F	$E^{h}F^{r}$
12.2–12.4	1 hop in duct E, 1 hop in duct F, ricochet in duct F	$E^{h}F^{h}F^{r}$
12.6–13.0	0.5 hop in duct E, capture in duct FE	$E^{h}(FE)$

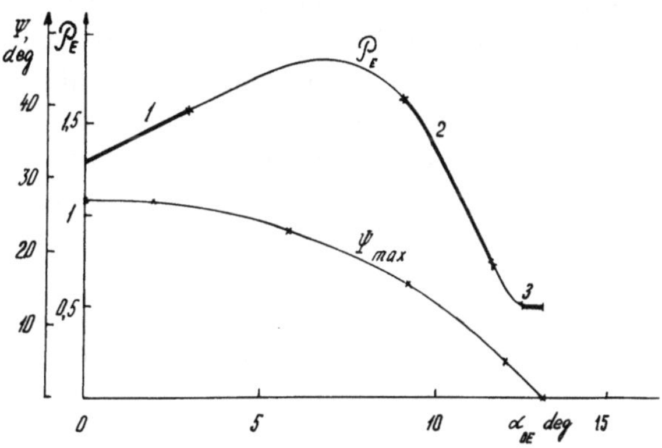

Fig. 5.22. Total number of oscillations $\mathcal{P}$ and angular distance Ψ_{max} covered by a wave in the duct E from the emission point to the point of departure from the duct vs the takeoff angle α_{0E}. The numbers 1, 2, 3 indicate the regions of capture

invariant conservation [Eq. (3.17)] which forms the basis of the adiabatic approximation. The calculations show that in simple channels the adiabatic invariant is conserved with high accuracy, $\Delta I/I \lesssim 5\%$.

The adiabatic invariant changes sharply within the regions of capture into the ducts or during transition among the channels. Within these regions, the adiabatic invariant is not conserved. Its "jumplike" change when passing through the separatrix, in the adiabatic approximation is defined by Eq. (3.44a). The ray-tracing calculations show marked deviations from this equation. When a wave goes into the duct FE from the duct E, these deviations essentially do not exceed 10%. However, sometimes they can be as large as 30%. The similar deviations in the case of transition from the duct E into the duct F can be 20–30%.

The adiabatic invariant can also suffer a marked perturbation in the case of the trajectory transition through the terminator where the ionospheric parameters change sharply and, thus, the condition of adiabaticity [Eq. (1.10)] can be violated.

The results of calculation of the adiabaticity parameter [Eq. (1.10)]

$$\mu_{1,2} = \frac{1}{2\pi} \frac{\Theta}{z_{1,2}} \frac{dz_{1,2}}{d\theta}$$

by the ray-tracing technique are shown in Fig. 5.23 for the CCP F^hF^r ($\alpha_{0F} = 9.2°$), $E^hF^h(FE)$ ($\alpha_{0E} = 0°$; 5.8°), $E^hE^r(FE)$ ($\alpha_{0E} = 0°$, broken line), and $E^h(FE)$ ($\alpha_{0E} = 8.8°$). Here, $z_{1,2}$ are the values of the averaged maximum (z_1) and minimum (z_2) trajectory heights. It is seen that in the simple channel F^hF^r ($\alpha_{0F} = 9.2°$) the parameters μ_1 and μ_2 are small even when passing through the terminator. Their maximum values are: $\mu_1 = 0.05$, $\mu_2 = 0.4$. In the complex CCP the parameters μ_1 and μ_2 are of the same order, if in the terminator region a wave remains in one and the same IWD. But, sometimes, if the path traverses the terminator, the wave goes from one duct into another one. It is mainly connected with the disappearance or appearance of the duct E. Figure 5.23 illustrates these very examples, where for $f = 15$ MHz (solid curves) the terminator traverses the path at $\Psi = 45°$ and for $f = 10$ MHz (broken line) at $\Psi = 30°$; and on the path in the CCP $E^hF^h(FE)$ the wave transition from the channel E^h takes place first into the channel F^h and then into the duct FE; in the CCP $E^r(FE)$ the transition takes place from the channel E^r into the channel FE; in the CCP $E^h(FE)$ the transition takes place from E^h into FE. In this case the adiabaticity parameter is seen to reach the great values $\mu \gtrsim 1$. In such cases the variation of the wave invariant F undergoes the greatest deviations from Eq. (3.44a). Consequently, in the above mentioned cases, the trajectory parameters in the second channel can also differ significantly from the values obtained by the adiabatic approximation.

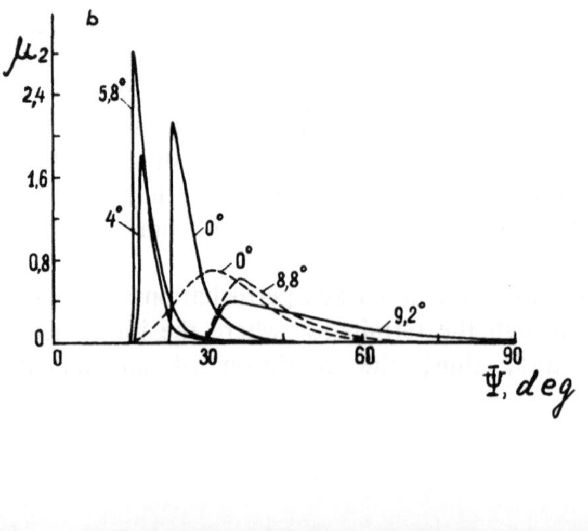

Fig. 5.23. Variation of the adiabaticity parameters μ_1 and μ_2 along the path for the middle point LT = $0.6^h30'$, $f = 15$ MHz ($\alpha_0 = \alpha_{0E} = 0°$, 5.8°; $\alpha_{0F} = 9.2°$; *solid curve*) and for the middle point LT = $0.4^h30'$, $f = 10$ MHz ($\alpha_{0E} = 0°$, $\alpha_{0F} = 8.8°$, *dashed line*)

General Conclusions. The comparison of calculated results obtained by the ray-tracing technique and the adiabatic method allows to make the following conclusions:

1. The detailed ray-tracing calculations for the path confirm completely the general picture for radio wave propagation obtained in the adiabatic approximation. In particular, all the CCP found out naturally in the adiabatic method are confirmed by the ray-tracing calculations.
2. The agreement between the main propagation parameters obtained by both methods in the main simple CCP is rather good. The parameters include oscillation number over the path, range of the takeoff angles, range of the trajectory heights, etc. The differences are within 5%.
3. In the complex channels, when the transitions from duct to duct take place in the region of sharp horizontal gradient in the ionosphere (the terminator zone), a significant difference in value of the adiabatic invariant and the trajectory parameters is possible after the transition. In this case, the calculated results by the ray-tracing technique can differ from the adiabatic approximation by 10–30%.

It should be emphasized that on the whole the observed differencies are small. They are within the range of trajectory variation, which takes place at rather small changes of initial takeoff angles. The directivity diagram width for actual transmitters is usually much greater.

In conclusion of this chapter, let us point out again that the real problem of long-distance radio path calculation is to describe properties of the signal which propagates in the medium, where not exact electron density distribution is known, but the effective one (model of the ionosphere). In such a medium, travelling large-scale inhomogeneities (so-called smooth ionospheric disturbances, Afraimovich 1982) and a complex spectrum of the irregularities of various scales exist. Under these conditions, the exact calculation of all the possible ray trajectories for a complex ionospheric model is not adequate to the real problem. We believe the path calculation for the simple ionospheric model (Sec. 5.1) with the subsequent identification of the main CCP for the complex model to be a first approximation to reality. The CCP are characterized by the significantly varying trajectory types (the first part of Sec. 5.2). The detailed comparison of such calculated results with the experimental data would allow to improve the ionospheric model and to make further steps in refining the calculation.

6 Radio Wave Scattering

Ionospheric inhomogeneity has been assumed previously to be regular. Actually, random irregularities occur as well. Scattering of radio waves by ionospheric irregularities, as was indicated in Sec. 1.2, influences essentially long-distance propagation. The present chapter is devoted to the investigation of the phenomena arising in these situations.

6.1 Effect of Scattering on the Capture of Radio Waves into the Interlayer Duct

Capture Coefficient. We consider first the capture at single scattering (Gurevich et al. 1975). The wave frequency will be assumed to exceed the maximum usable frequency $f_{\text{MUF-}F}$. The potential functions $U(z)$ in these cases are shown in Fig. 6.1. There are the ducts *FE* (Fig. 6.1a) and *F* (Fig. 6.1b) between the altitudes z_1 and z_2. If neglecting the ionospheric regular horizontal inhomogeneity, in order that radio waves emitted from the Earth be trapped in these ducts under conditions $f > f_{\text{MUF-}E}$ or $f > f_{\text{MUF-}F}$, an additional ray turn through a finite angle is needed.

The trajectory of a ray, emitted from the Earth at an angle α_0 with respect to the horizon, is shown for $f > f_{\text{MUF-}F}$ in Fig. 6.2. The ray undergoes no reflections and, in general, leaves the Earth's ionosphere. However, during its ionospheric propagation in the trajectory part (a_1, b_1) a scattering by irregularities can cause a capture of a portion of wave energy into the duct. Thus, the region (a_1, b_1) can be treated as a set of reradiators which inject scattered rays onto different trajectories within the duct.

The ray trajectory in the duct will be specified by the angle β_s^α between the ray direction and horizontal at the sliding wave level z_s. The angle β^α at any height z for a given trajectory is connected with β_s^α by Snell's law

$$\cos\beta^\alpha(z) = \cos\beta_s^\alpha \sqrt{U(z_s)/U(z)}\,. \tag{6.1}$$

The trapped ray trajectory is shown in Fig. 6.2 by the dashed curve. The ray has two turning points, z' and z'', defined by

$$U(z) - U(z_s)\cos^2\beta_s^\alpha = 0\,. \tag{6.2}$$

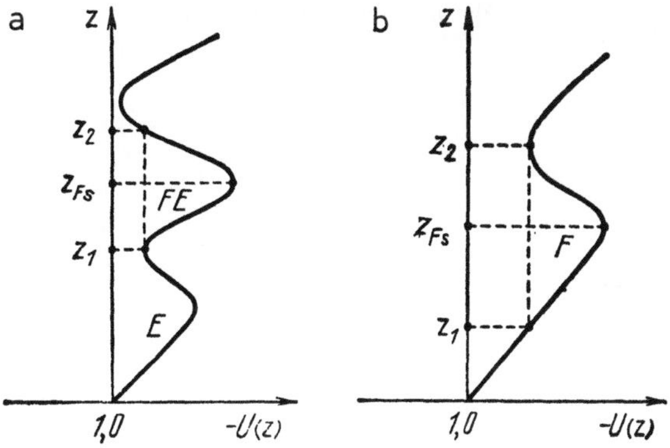

Fig. 6.1a, b. Various potential types

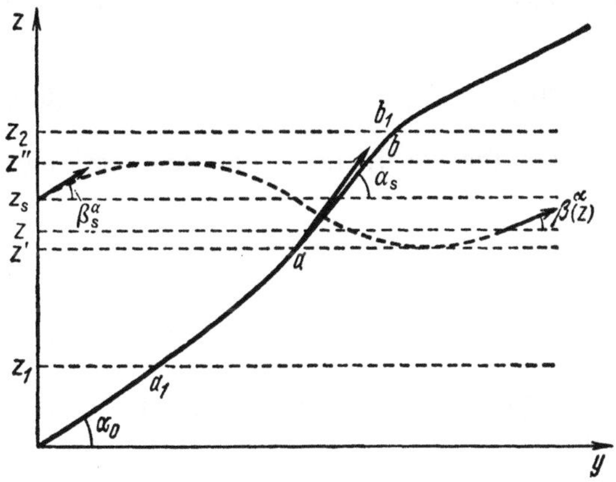

Fig. 6.2. Trajectory of a trapped ray in an ionospheric duct (*dashed line*) and the trajectory of a ray emitted from the Earth at $f > f_{\text{MUF}}$ (*solid line*)

The levels z' and z'' cut off the interval (a, b) on the original ray trajectory where the scattering contributes essentially to the intensity of the scattered ray β_s^α. Thus, to evaluate the energy of the ray β_s^α, the contribution of each element of the original ray trajectory within the interval (a, b) should be taken into account, the interval length being dependent on β_s^α. Note that the angle β_s^α for the trapped ray is less than a critical angle which is defined by the condition for a capture into the duct F

$$U(z_2) - U(z_s)\cos^2\beta_{s0}^\alpha = 0\,, \quad \beta_s^\alpha \leqslant \beta_{s0}^\alpha\,. \tag{6.3}$$

For FE-duct $U(z_2)$ should be replaced by $U(z_1)$.

Let us consider now the process of scattering. Let a wave with the wave vector $\boldsymbol{k}_0$ (Fig. 6.3) be incident on the scattering volume dv. We choose the coordinate system in such a way that the axis $0y$ be placed in the plane $\boldsymbol{k}_0, z(|z|$ is the altitude), i.e.,

$$\boldsymbol{k}_0 = k_0\{0; \cos\alpha; \sin\alpha\}\,. \tag{6.4}$$

The wave vector of a scattered wave, $\boldsymbol{k}_{sc}$, is determined by angles φ, $\beta^\alpha(z)$, i.e., it has the components $(k_{sc} = k_0)$

$$k_{sx} = k_0\cos\beta^\alpha\sin\varphi\,; \quad k_{sy} = k_0\cos\beta^\alpha\cos\varphi\,; \quad k_{sz} = k_0\sin\beta^\alpha\,. \tag{6.5}$$

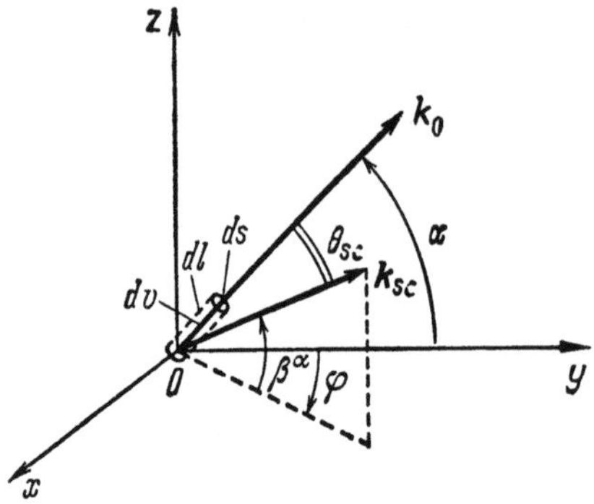

Fig. 6.3. The scattering process

Consequently, the scattering angle θ_{sc} (the angle between $\boldsymbol{k}_{sc}$ and $\boldsymbol{k}_0$) is given by

$$\cos\theta_{sc} = \cos\alpha\cos\beta^\alpha\cos\varphi + \sin\alpha\sin\beta^\alpha\,. \tag{6.6}$$

Here, φ is read off in the plane xy from the axis $0y$, i.e., from the direction of the original ray

$$\varphi = \varphi_0 - \varphi'\,, \tag{6.7}$$

where φ_0 and φ' are the azimuths of the original and scattered rays, respectively.

The scattering cross-section in the Born approximation is given by Landau and Lifshitz (1984) and Rytov (1966)

$$\sigma = \frac{\pi}{2}\,k_0^4\rho_\varepsilon(\mathscr{K})\sin^2\Psi\,, \tag{6.8}$$

$$\mathscr{K} = \boldsymbol{k}_{sc} - \boldsymbol{k}_0\,, \quad |\mathscr{K}| = 2k_0\sin(\theta_s/2)\,; \quad \mathscr{K} = \mathscr{K}(\boldsymbol{r})\,, \quad \Psi = \Psi(\boldsymbol{r})\,. \tag{6.9}$$

Here, Ψ is the angle between the vector $\boldsymbol{k}_{sc}$ and dipole moment of a reradiator; $\rho_\varepsilon(\mathscr{K})$ is the spectral density of ε-fluctuations

$$\rho_\varepsilon = \frac{\overline{(\Delta\varepsilon)^2}}{8\pi^3}\int R_\varepsilon(\xi)e^{i\mathscr{K}\xi}d\xi\,; \tag{6.10}$$

$\overline{(\Delta\varepsilon)^2}$ are the dielectric constant fluctuations, $R_\varepsilon(\xi,\boldsymbol{r})$ is the correlation function of ionospheric irregularities. First, we shall consider the case of isotropically distributed irregularities, $R_\varepsilon = R_\varepsilon(\xi)$, when

$$\rho_\varepsilon = \frac{\overline{(\Delta\varepsilon)^2}}{2\pi^2}\int_0^\infty R_\varepsilon(\xi)\frac{\sin\mathscr{K}\xi}{\mathscr{K}}\cdot\xi d\xi\,,$$

$$\sigma = \frac{k_0^4}{4\pi}\overline{(\Delta\varepsilon)^2}I(z)\,, \quad I(z) = \sin^2\Psi\int_0^\infty\frac{\sin\mathscr{K}\xi}{\mathscr{K}}R_\varepsilon(\xi)\xi d\xi\,. \tag{6.11}$$

The flux of a power scattered by the volume dv at an angle β^α into a solid angle $d\Omega_1$ is given by

$$\Phi(\beta^\alpha, z) = \Pi_0 \sigma(z) \frac{d\Omega_0}{ds} dv\, d\Omega_1 , \qquad \Pi_0 = \frac{c}{2} \mathscr{E}_0^2 . \tag{6.12}$$

Here, Π_0 is the energy flux density of the original wave, $d\Omega_0/ds$ is the factor taking into account the spherical divergence in the incident beam. Equation (6.12) is a well-known relation for the scattering energy in the ray approximation (Al'pert 1974). Dividing Eq. (6.12) by $\Pi_0 d\Omega_0$ and taking into account that

$$dv = ds\, dl = ds \frac{dz}{\sin\alpha(z)} \tag{6.13}$$

(see Fig. 6.3), we obtain

$$d^3 G_0(\beta^\alpha, z) = \frac{k_0^4}{4\pi} \overline{(\Delta\varepsilon)^2} \cdot I(z) \frac{dz\, d\Omega_1}{\sin\alpha(z)} . \tag{6.14}$$

The quantity $d^3 G_0$ describes the power fraction of the wave incident in a solid angle $d\Omega_0$ which is scattered by the element dz near the point z into the unit of solid angle $d\Omega_1$

$$d\Omega_1 = \sin\theta_{sc} d\varphi_1 d\theta_{sc} = \sin\left(\frac{\pi}{2} - \beta^\alpha\right) d\beta^\alpha d\varphi = \cos\beta^\alpha d\beta^\alpha d\varphi .$$

For convenience, here a transition has been made from scattering angles θ_{sc}, φ_1 to angles β^α, φ. Integrating now Eq. (6.14) over z and taking into account the dependence of β^α on β_s^α, z for the given trajectory, we obtain

$$G = \frac{\partial^2 G_0}{\partial\beta_s^\alpha \partial\varphi} = \frac{1}{4\pi} \int_{z'(\beta_s^\alpha)}^{z''(\beta_s^\alpha)} \frac{k_0^4 \overline{(\Delta\varepsilon)^2} \cos\beta^\alpha \cdot I(z)}{\sin\alpha(z)} \left(\frac{d\beta^\alpha}{d\beta_s^\alpha}\right)_z dz . \tag{6.15}$$

Here, $d\beta^\alpha/d\beta_s^\alpha$ is defined by Eq. (6.1), $G = G(\varphi, \beta_s^\alpha; \varphi_0, \alpha_0)$ is the coefficient of the energy capture into the trajectory β_s^α, φ' ($\varphi' = \varphi_0 - \varphi$) for the ray emitted from the Earth at the angles α_0, φ_0. This quantity is naturally to be termed the differential capture coefficient.

The total energy fraction of the wave α_0, φ_0 trapped in IWD, i.e., the total capture coefficient G_0 is clearly

$$G_0 = \frac{1}{4\pi} \int_{-\pi}^{\pi} d\varphi \int^{\beta_{s0}^\alpha} d\beta_s^\alpha \int_{z'(\beta_s^\alpha)}^{z''(\beta_s^\alpha)} \frac{k_0^4 \overline{(\Delta\varepsilon)^2} \cos\beta^\alpha \cdot I(z)}{\sin\alpha(z)} \left(\frac{d\beta^\alpha}{d\beta_s^\alpha}\right)_z dz . \tag{6.16}$$

Here, β_{s0}^α is the critical angle [Eq. (6.3)]. Finally, if the antenna directivity diagram $D(\alpha_0, \varphi_0)$ is specified, then the total power trapped in the duct is given by

$$W = W_0 \int D(\alpha_0, \varphi_0) G_0(\alpha_0, \varphi_0) d\Omega_0 , \quad (6.17)$$

where W_0 is the transmitter power.

Approximation of Small Scattering Angles. Equations (6.15)–(6.17) determine, in the general case, the radio wave energy capture into IWD for $f > f_{\text{MUF}}$. Under real ionospheric conditions both the characteristic angles (α, β_s^α, φ) and effective scattering angles θ_{sc} are usually small (see Fig. 6.2). This enables Eqs. (6.15)–(6.17) to be simplified. Indeed, assuming

$$\alpha \ll 1 , \quad \beta^\alpha \leqslant \beta_s^\alpha \ll 1 , \quad \varphi \ll 1 , \quad (6.18)$$

we obtain from Eqs. (6.6) and (6.11)

$$\begin{aligned} &\cos\theta_{sc} \approx 1 - \frac{(\alpha-\beta^\alpha)^2}{2} - \frac{\varphi^2}{2} , \qquad \cos\beta^\alpha \approx 1 , \\ &\sin^2\Psi \approx 1 - \sin^2\beta^\alpha \sin^2\varphi \approx 1 , \\ &\sin\frac{\theta_{sc}}{2} \approx \frac{1}{2}[(\alpha-\beta^\alpha)^2 + \varphi^2]^{1/2} , \\ &\mathcal{K}(z) \approx k_0[(\alpha-\beta^\alpha)^2 + \varphi^2]^{1/2} . \end{aligned} \quad (6.19)$$

Finally, from Eqs. (2.29) and (6.1) we have

$$\begin{aligned} &\alpha(z) = \sqrt{\alpha_0^2 - U(z) - 1} = \sqrt{\alpha_0^2 - \frac{\omega_0^2}{\omega^2} + \frac{2z}{R_0}} , \\ &\alpha_s = \sqrt{\alpha_0^2 - U(z_s) - 1} , \qquad \beta^\alpha(z) = \sqrt{(\beta_s^\alpha)^2 - U(z) + U(z_s)} , \\ &\left(\frac{d\beta^\alpha}{d\beta_s^\alpha}\right)_z = \frac{\beta_s^\alpha}{\beta^\alpha} . \end{aligned} \quad (6.20)$$

Equations (6.19) and (6.20) can be applied for the calculation of G_0 and G.

Let us bear in mind now that the scattering cross-section [Eq. (6.11)] for HF radio waves in the ionosphere usually drops sharply as the scattering angle θ_{sc} increases. In other words, the scattering through small angles is most effective and, thus, the main contribution in the integrals in Eqs. (6.15) and (6.16) is made by the range of z where, in order that the original ray become trapped, it should be turned through the minimum angle $\theta_{sc_{\min}} = \gamma$. One can easily make sure that this range of z is $\Delta z \sim z_s$. Indeed, according to Eq. (6.19)

$$\theta_{sc} = [(\alpha-\beta^\alpha)^2 + \varphi^2]^{1/2}$$

and, thus, the angle θ_{sc} has its minimum at $\varphi = 0$ and when the angle difference $(\alpha - \beta^\alpha)$ has its minimum. The former condition is true in all cases when the scattering occurs in the vertical plane alone [see Eq. (6.7)]. The latter is valid at $z = z_s$ since

$$\alpha - \beta^\alpha = \sqrt{\alpha_s^2 - U(z) + U(z_s)} - \sqrt{(\beta_s^\alpha)^2 + U(z_s) - U(z)} \geqslant \alpha_s - \beta_s^\alpha .$$

Here, we have taken into account that always $\alpha_s > \beta_s^\alpha$. Thus, the minimum scattering angle for the capture onto the trajectory β_s^α, φ is given by

$$\gamma = \alpha_s - \beta_s^\alpha . \tag{6.21}$$

It is reached at $z = z_s$ and $\varphi = 0$.

In the neighborhood of $z = z_s$ the function $U(z)$ can be represented in the form

$$U(z) = U(z_s) + \frac{1}{2}\left(\frac{\partial^2 U}{\partial z^2}\right)_{z_s} \cdot (z - z_s)^2 = U(z_s) + \frac{1}{2}\,\frac{\omega_{0s}^2}{\omega^2}\,\varkappa_s (z - z_s)^2 ,$$

$$\omega_{0s} = \omega_0(z_s) , \qquad \varkappa_s = \left(\frac{1}{N_e(z)}\,\frac{d^2 N_e(z)}{dz^2}\right)_{z_s} . \tag{6.22}$$

Substituting Eq. (6.22) into Eqs. (6.19) and (6.20), we obtain

$$\beta^\alpha = \beta_s^\alpha - \frac{\omega_{0s}^2 \varkappa_s}{4\beta_s^\alpha \omega^2}(z - z_s)^2 , \qquad \alpha = \alpha_s - \frac{\omega_{0s}^2 \varkappa_s}{4\alpha_s\,\omega^2}(z - z_s)^2 ;$$

$$\mathscr{K}(z, \varphi) = k_0 \left\{ \alpha_s - \beta_s^\alpha + \frac{\varphi^2}{2(\alpha_s - \beta_s^\alpha)} + \frac{\omega_{0s}^2 \varkappa_s}{4\omega^2}\,\frac{(\alpha_s - \beta_s^\alpha)}{\alpha_s \beta_s^\alpha}(z - z_s)^2 \right\} , \tag{6.23}$$

$$\left(\frac{d\beta^\alpha}{d\beta_s^\alpha}\right)_{z_s} = 1 .$$

The function $\mathscr{K}(z, \varphi)$ is seen from Eq. (6.23) to reach its minimum value at $\varphi = 0$, $z = z_s$. Substituting Eq. (6.23) into Eqs. (6.15) and (6.11) and bearing in mind that $\sin(\mathscr{K}\xi)$ in Eq. (6.11) is a rapidly oscillating function of z, the integration over z can be performed by means of the method of stationary phase. This yields

$$G = \frac{\overline{(\Delta\varepsilon)^2_{zs}}\, k_0^{5/2}\, \omega \sqrt{\beta_s^\alpha}}{2\sqrt{2\pi\alpha_s \varkappa_s (\alpha_s - \beta_s^\alpha)}\,\omega_{0s} A} \int_0^\infty \sqrt{r}\, R_\varepsilon(r)\,[\sin(k_0 r A) + \cos(k_0 r A)]\, dr ,$$

$$A = [(\alpha_s - \beta_s^\alpha)^2 + \varphi^2]^{1/2} \approx \alpha_s - \beta_s^\alpha + \frac{\varphi^2}{2(\alpha_s - \beta_s^\alpha)} . \tag{6.24}$$

Here, we have taken into account that

$$\sin(\mathscr{K} r) = \sin(k_0 r A)\cos B + \cos(k_0 r A)\sin B ,$$

$$B = k_0 r \frac{\omega_{0s}^2 \varkappa_s}{4\omega^2}\,\frac{(\alpha_s - \beta_s^\alpha)}{\alpha_s \beta_s^\alpha}(z - z_s)^2 ,$$

$$\int_{-\infty}^{\infty} \sin(p x^2)\, dx = \int_{-\infty}^{\infty} \cos(p x^2)\, dx = \sqrt{\frac{\pi}{2p}} .$$

Equation (6.24) is valid under conditions

$$\sqrt{k_0 r}(z'-z_s)\frac{\omega_{0s}}{2\omega}\sqrt{\varkappa_s}\sqrt{\frac{\alpha_s-\beta_s^\alpha}{\alpha_s\beta_s^\alpha}} \gg 1\,, \qquad \frac{1}{k_0 r(\alpha_s-\beta_s^\alpha)} \ll 1\,. \tag{6.25}$$

To obtain the total capture coefficient G_0, Eq. (6.24) must be integrated over φ and β_s^α. The integration with respect to φ is carried out in the same manner as with respect to z. We obtain, finally

$$G_0 = \frac{k_0^2\overline{(\Delta\varepsilon)^2_{z_s}}\,\omega}{\sqrt{2}\,\omega_{0s}\sqrt{\varkappa_s}\,\alpha_s}\int\limits_{-\beta_{s0}^\alpha}^{\beta_{s0}^\alpha}\frac{\sqrt{|\beta_s^\alpha|}}{\alpha_s-\beta_s^\alpha}\,d\beta_s^\alpha\int\limits_0^\infty R_\varepsilon(r)\cos[k_0 r(\alpha_s-\beta_s^\alpha)]\,dr\,. \tag{6.26}$$

Let us consider by way of example

$$R_\varepsilon(r) = \exp(-r^2/a^2)\,. \tag{6.27}$$

In this case we have

$$\begin{aligned} G_0 &= \frac{k_0^2\overline{(\Delta\varepsilon)^2_{z_s}}\,\omega a}{\sqrt{2}\,\omega_{0s}\sqrt{\varkappa_s}\,\alpha_s}\,\frac{\sqrt{\pi}}{2}\int\limits_{-\beta_{s0}^\alpha}^{\beta_{s0}^\alpha}\frac{\sqrt{|\beta_s^\alpha|}\exp\left\{-\dfrac{a^2k_0^2}{4}(\alpha_s-\beta_s^\alpha)^2\right\}}{(\alpha_s-\beta_s^\alpha)}\,d\beta_s^\alpha \\ &\approx \frac{\sqrt{\pi}}{\sqrt{2}}\,\frac{\overline{(\Delta\varepsilon)^2_{z_s}}\,\omega\sqrt{\beta_{s0}^\alpha}}{\omega_{0s}\sqrt{\varkappa_s}\,\alpha_s a(\alpha_s-\beta_{s0}^\alpha)^2}\cdot\exp\left\{-\frac{a^2k_0^2}{4}(\alpha_s-\beta_{s0}^\alpha)^2\right\} \\ &\quad\cdot[1-\exp\{-a^2k_0^2\beta_{s0}^\alpha\alpha_s\}]\,. \end{aligned} \tag{6.28}$$

This formula is convenient for the estimation of the wave capture in the ionosphere [the conditions in Eq. (6.25) are usually satisfied, if $k_0^2a^2(\alpha_s-\beta_{s0}^\alpha)^2 \gg 1$]. The capture coefficient is seen to be determined essentially by the turning angle $\gamma = \alpha_s-\beta_{s0}^\alpha$.

Above, in the derivation of Eqs. (6.26) and (6.28), in addition to Eq. (6.25), an assumption has been used essentially that $\overline{(\Delta\varepsilon)^2_z}$ and $R_\varepsilon(r,z)$ are sufficiently smooth and gradual functions of z within the duct. Under real conditions, sometimes it can be that this is not satisfied (for instance, when trapped trajectories descend into the E-region where the scattering increases strongly). In this case, the general Eqs. (6.15) – (6.17) should be applied. These equations are simplified considerably if the function $R_\varepsilon(r,z)$ is of the Gaussian type [Eq. (6.27)]. Then, making an integration in Eq. (6.11) with respect to ξ, we obtain

$$\begin{aligned} I(z) &= \frac{\sqrt{\pi}}{4}a^3\exp\left\{-\mathscr{K}^2(z)\frac{a^2}{4}\right\}, \\ G &= \frac{1}{16\pi^{1/2}}\int\limits_{z'(\beta_s^\alpha)}^{z''(\beta_s^\alpha)}\frac{k_0^4\overline{(\Delta\varepsilon)^2_z}a^3}{\alpha(z)}\left(\frac{d\beta^\alpha}{d\beta_s^\alpha}\right)_z\cdot\exp\left\{-\mathscr{K}^2(z)\frac{a^2}{4}\right\}dz\,. \end{aligned} \tag{6.29}$$

Here, $\alpha(z)$, $\mathscr{K}(z,\varphi)$, $d\beta^\alpha/d\beta_s^\alpha$ are defined by Eqs. (6.19) – (6.21). When calculating G_0 [Eq. (6.16)], first, the integration over $d\varphi\,d\beta_s^\alpha$ may be carried out which yields

$$G_0 = \frac{1}{8}\int_{z_1}^{z_2} \frac{k_0^3 a^2 \overline{(\Delta\varepsilon)_z^2}}{\alpha(z)}\,dz\left[\int_{-\beta_s^\alpha}^{\beta_s^\alpha} \cdot \exp\left\{-\left(\frac{k_0 a}{2}\right)^2(\alpha-\beta^\alpha)^2\right\}d\beta^\alpha\right]$$
$$= \frac{k_0^2 a\sqrt{\pi}}{8}\int_{z_1}^{z_2}\frac{\overline{(\Delta\varepsilon)_z^2}}{\alpha(z)}\,dz\left\{\Phi\left[\frac{k_0 a}{2}(\alpha+\beta_s^\alpha)\right]-\Phi\left[\frac{k_0 a}{2}(\alpha-\beta_s^\alpha)\right]\right\},$$
$$\Phi(x) = \frac{2}{\sqrt{\pi}}\int_0^x e^{-t^2}dt. \tag{6.30}$$

Here from the integration with respect to β_s^α we have gone over to that with respect to β^α; $\beta_s^\alpha(z)$ is the value of $\beta^\alpha(z)$ for the critical ray β_{s0}^α, i.e.,

$$\beta_s^\alpha(z) = \sqrt{(\beta_{s0}^\alpha)^2 - U(z) + U(z_s)}\,.$$

We have taken into account also the fact that the small angle approximation [Eqs. (6.19) and (6.20)] is valid provided

$$(k_0 a)^2 \gg 1\,. \tag{6.31}$$

If a stronger condition, $k_0 a(\alpha_s - \beta_{s0}^2) \gg 1$, is satisfied and the main contribution is made by the neighborhood of $z = z_s$, then Eq. (6.30) reduces to Eq. (6.28).

It was the scattering by isotropic random irregularities that was considered above. Actually, in the upper ionosphere anisotropic irregularities can also exist. They are elongated along the magnetic field (field-aligned irregularities). The general Eqs. (6.15) – (6.17) for the coefficient of a scattering-induced radio wave capture remain unchanged in the presence of an anisotropy. It only must be borne in mind that the function $I(z)$ is defined, in general, by

$$I(z) = \frac{\sin^2\Psi}{4\pi}\int R_\varepsilon(\xi)e^{i\mathscr{K}\xi}d\xi\,, \tag{6.32}$$

where $R_\varepsilon(\xi)$ is the irregularity correlation function (with allowance for the anisotropy). In the isotropic case $R_\varepsilon = R_\varepsilon(\xi)$, and Eq. (6.32) transforms to Eq. (6.11) (Filipp 1974). Specific calculations of the capture coefficient for an anisotropic Gaussian correlation function are presented in Gurevich and Kim (1976). The effect of the anisotropy becomes important at q as large as 0.1 – 0.5; here, $q = a_\perp/a_\parallel$ is the anisotropy coefficient, i.e., the ratio between the typical transverse and longitudinal (with respect to the Earth's magnetic field) irregularity scales. Unlike the isotropic case when it was the scattering in the neighborhood of the sliding point z_s [Eq. (6.21)] that made the principal contribution to the energy capture, now two principal regions appear near two singled out points, one of them being higher and the other lower than the sliding point. This is caused by a high aspect – ratio sensibility of scattering by field-aligned irregularities, when the capture direction in question gets the scattering cone only at definite heights within the duct. Note, in addition, that now the azimuth of the maximum of captured energy also does not coincide with the incident ray direction. This is seen from Fig. 6.4, where at $q = 0.1$

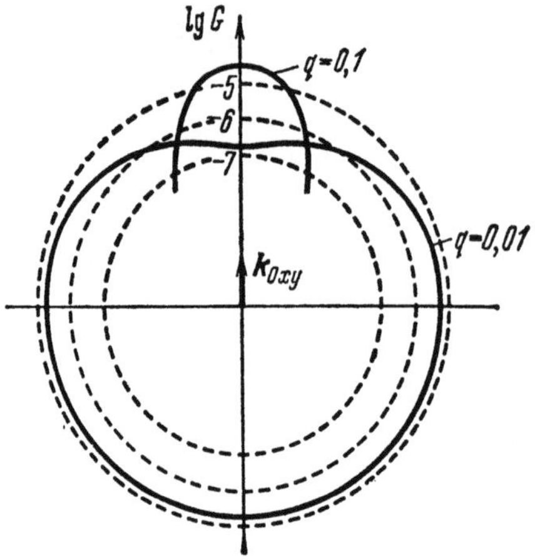

Fig. 6.4. Azimuthal distribution of the capture coefficient in the horizontal plane xy, $\boldsymbol{k}_{Oxy}$ is the incident wave vector component in this plane, $f = 17$ MHz, $a_{\|} = 0.3$ km, γ_H (the angle between the magnetic field $\boldsymbol{H}$ and the vertical) is equal to 27°

both the directions are coincident, but at $q = 0.01$, the maximum of the energy capture coefficients is strongly smeared and has another direction.

Especially significant deviations from the isotropic case appear for strongly field-aligned irregularities $q \sim 10^{-2} - 10^{-3}$. Such irregularities occur in the polar regions, as well as in ionospheric regions artificially disturbed by intense radio waves. They will be considered specially in Sec. 7.2.

We considered above scattering into a duct only from a wave passing through the ionosphere (see Fig. 6.1). If a wave is reflected from the ionosphere, $f < f_{\text{MUF-}F}$, then the scattering is also of importance from the reflected wave. This results in the doubling of the capture coefficient. Note that in the presence of the reflected wave, a caustic is formed, in the neighborhood of which the scattering intensifies (Brekhovskikh 1957, Vas'kov 1969). The study shows, however, that the total contribution into the capture coefficient due to this intensification is small enough, so that to the accuracy ~10% the presence of the reflected ray can be thought only to produce the doubling of the capture coefficient (Kim 1975). Note, in addition, that we deal here only with a single scattering, the effect of multiple scattering will be included in Sec. 6.2 (see also Sec. 7.2).

Turning Angles and Capture Coefficient in the Ionosphere. Let us proceed now to the evaluation of the role of scattering-induced capture in the ionospheric conditions. Figure 6.5 shows the loss coefficient $\Gamma = -10 \log G_0$, calculated by Eq. (6.30) for the morning and nighttime equinoctial ionosphere at midlatitudes, as a function of a typical Gaussian irregularity scale. The calculation of the capture into IWD is made for the ray emitted from the Earth at an angle $\alpha_0 = 0$. Relative electron density fluctuations, $\delta N = \sqrt{\langle(\Delta N)^2\rangle}/N_e$, were assumed to be equal to 10^{-3}. The structure of the regular duct was found on the basis of the ionospheric model EMI-72 (Appendix 2), and the values of z_1, z_2 in Eq. (6.30) were put to be equal to the heights of the upper and lower duct boundaries, respectively.

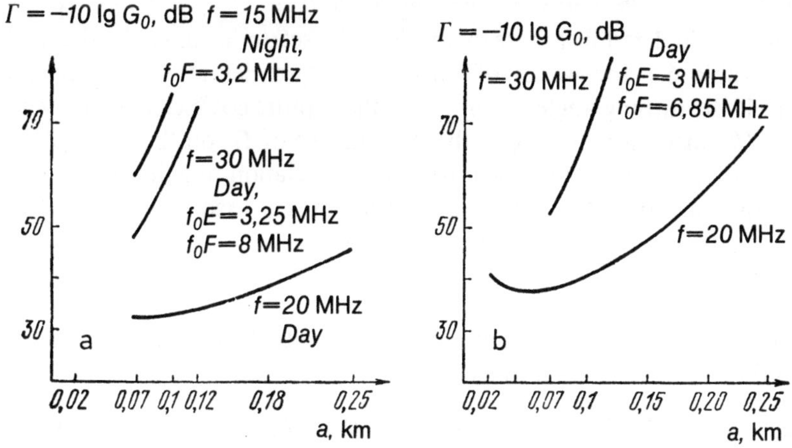

Fig. 6.5 a, b. Loss coefficient Γ vs the irregularity scale; f_0E and f_0F are the critical frequencies of the E and F layers, respectively: **a** $\theta = 30°$; **b** $\theta = 50°$

The figures demonstrate that the optimal radio wave capture takes place when irregularities with a scale $a \sim 50 - 100$ m are present at the duct altitudes. In the ionosphere, such irregularities seem to be present by night in the neighborhood of the geomagnetic equator (Wernik and Liu 1974), and at any time at the auroral latitudes during the periods of strong ionospheric disturbances (Rufenach 1974). The latter indicates that at these latitudes scattering-induced radio wave capture into IWD must be especially efficient.

At midlatitudes, a small-scale ionospheric structure seems to be pronounced much weaker. The observations of the ionospheric radio wave sounding from the natural cosmic sources and man-made satellites show the presence in the midlatitude F-region of irregularities $\delta N \sim 10^{-3} - 10^{-2}$ with scales $a \sim 0.7 - 1.5$ km, and occasionally (particularly during disturbed periods) also of weak irregularities with scales of the order of $200 - 400$ m and even less (Getmantsev and Erukhimov 1969). The presence of irregularities with $a \sim 200$ m in the midlatitude F-region is evidenced also by the data on the vertical sounding (Al'pert 1974) as well as on the oblique one.

Unfortunately, due to the difficulties in the interpretation of the data it is not possible to evaluate δN for 100 m scale irregularities. We note, however, that for $a > 150 - 200$ m, the major variations in G_0 occur due to the exponential factor in Eq. (6.28), which is defined by the ratio between the

Table 6.1. Scattering angle in the ionosphere, θ_{sc} (in degrees)

f, MHz	10	15	20	30	40	50
$a = 200$ m	2.7	1.8	1.35	0.9	0.675	0.54
$a = 1000$ m	0.54	0.36	0.27	0.18	0.135	0.11

turning angle $\gamma = \alpha_s - \beta_{0s}^{\alpha}$ and the scattering angle $\theta_{sc} = 2/k_0 a = \lambda_0/\pi a$. Below we shall compare in detail just these quantities. The scattering angle (in degrees) is listed in Table 6.1 for waves of different frequencies at $a = 200$ m and $a = 1000$ m. If the turning angle $\gamma > \theta_{sc}$, then the capture coefficient drops very sharply as γ/θ_{sc} increases. Note that the dependence of G_0 on the ratio γ/θ_{sc}, when $\gamma/\theta_{sc} > 1$, is determined by the type of the correlation function $R_\varepsilon(r)$. For the Gaussian function $R_\varepsilon(r)$, the coefficient is $G_0 \sim \exp[-(\gamma/\theta_{sc})^2]$ [see Eq. (6.28)]. Thus, the capture is efficient provided

$$\gamma \lesssim n \cdot \theta_{sc}, \quad n \sim 1-2 .$$

The turning angle γ has its minimum for rays emitted from the Earth in the horizontal direction, $\alpha_0 = 0$. In this case

$$\alpha_s = \alpha_{s0} = \arccos \sqrt{1/\varepsilon'(z_s)} , \quad \gamma = \gamma_{\min} = \alpha_{s0} - \beta_{s0}^{\alpha} . \tag{6.33}$$

The minimum turning angles γ for wave capture into the ducts *FE* and *F*, calculated by Eq. (6.33) in the model of equinoctial ionosphere EMI-73 (Appendix 2), are shown in Fig. 6.6 (Fishchuk and Tsedilina 1975). Figure 6.6a shows the latitudinal – longitudinal distribution of the angles γ_{FE} and γ_F for $f = 15$ MHz. The curve $\gamma_{FE} = 0$ corresponds to the curve $f_{\text{MUF-}E} = 15$ MHz. Interior to this curve the capture into the *FE*-duct is most effective (Tushentsova et al. 1974, 1975) (Sec. 3.4 and 4.1). The dashed curve limits the area of the *FE*-duct existence.

In the area between the curve $\gamma_{FE} = 0$ and the dashed curve, the scattering-induced capture is possible: in order that ray emitted horizontally ($\alpha_0 = 0$) from the Earth be trapped in the duct *FE*, it must undergo an extra turn in the ionosphere through an angle $\gamma \geqslant \gamma_{FE}$. The minimum turning angles, γ_{FE}, are seen to be small, particularly at high latitudes $\theta \sim (65° - 85°)$. At $a \sim 200 - 300$ m, the angles γ_{FE} are close to θ_{sc} (see Table 6.1). In other words, the conditions for an efficient scattering-induced capture [Eq. (6.32)] to the duct *FE* are satisfied in a broad range.

Figure 6.6b shows the angles γ_F and γ_{FE} for $f = 25$ MHz. Within the region restricted by the dashed curve, the cut-off *FE*-duct exists. The figure demonstrates that for the wave capture into this duct, a wave turn through $3-4°$ is needed. Between the curve $\gamma_F = 0$ and dash – dot curve, the Earth – separated *F*-duct exists. For the capture into it, an extra turn through $2-8°$ at day, and through $8-10°$ at night are necessary.

Figure 6.6 demonstrates that the scattering by ionospheric irregularities leads to a time expansion for the capture into this duct by 1 – 2 h on both sides of the direct capture region $\gamma_{FE} = 0$. In addition, the conditions for the capture at high latitudes are satisfied much easier. Therefore, due to a scattering, both the frequency ranges for the reception of RTW signals and the MUF for paths crossing over the region of small γ_{FE} must be expanded. This is seen from Fig. 6.7 where calculated curves are plotted for the conditions of Stanford (Fenwick 1963). The curve MUF-*E* (solid line) refers to the region of efficient direct capture into the duct *FE*, this region is especially favorable for

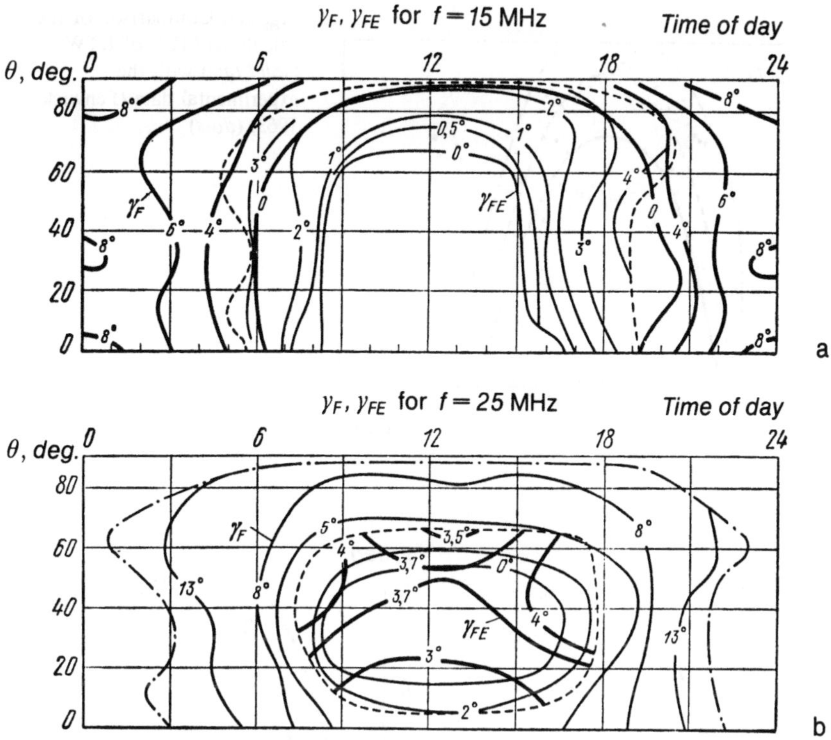

Fig. 6.6a, b. Curves of the turning angles: **a** $f = 15$ MHz; **b** $f = 25$ MHz

long-distance and RTW propagation. The curves $\gamma_{FE} = 2-3°$ (dashed and dash – dot lines) give the upper limit of the frequency range for the efficient scattering-induced capture into the duct *FE* [Eq. (6.32); Table 6.1]. The points show the experimental data by Fenwick in Stanford, the light points show the minimum RTW frequencies, whereas the dark points show the maximum ones (Fenwick 1963). The agreement between the calculation and experiment gives evidence that the RTW frequency range can be governed to a great extent by scattering-induced effects.

The turning angle γ_F for capture into the *F*-duct, also shown in Fig. 6.6, is greater than γ_{FE}. At night, the angle γ_F is much greater than the scattering angle θ_{sc}. In addition, the angle γ_F increases from $\gamma_F = 0$ to $\gamma_F \approx \theta_{sc}$ much more rapidly than the angle γ_{FE}, i.e., the region, where scattering is essential, is smaller here. Therefore, for the capture of a radio wave into the duct *F* at the frequency under discussion, i.e., $f = 15$ MHz, a far greater ionospheric disturbance is required.

Figure 6.8 shows the turning angles γ_F, γ_{FE} for latitudes $\theta = 0°$, $\theta = 50°$ for the midday (a) and midnight (b). It is seen that at daytime for $f \lesssim 20-30$ MHz, an efficient scattering-induced capture into the interlayer duct *FE* is possible. At night, as well as in the daytime, for high frequencies, $f \sim 40-50$ MHz, in principle, a capture into the *F*-duct also becomes possible.

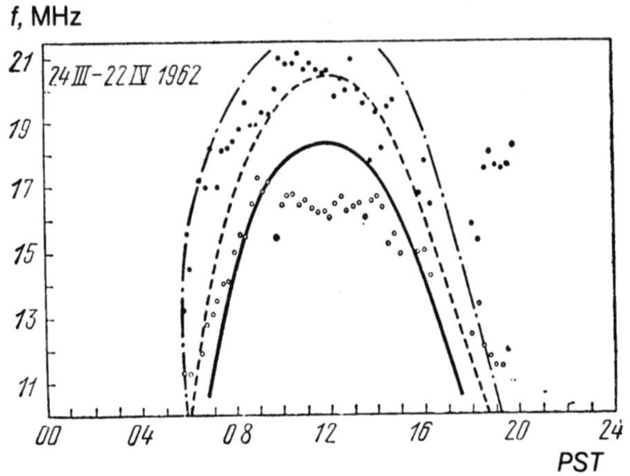

Fig. 6.7. Comparison of the calculated MUF of RTW (*solid line*) with the experimental data (Fenwick 1963) (*dots*)

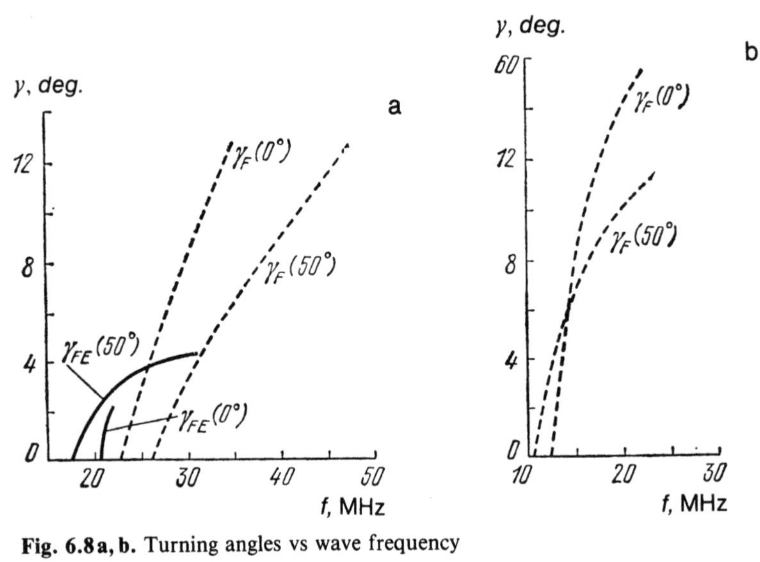

Fig. 6.8 a, b. Turning angles vs wave frequency

However, the turning angles in the latter case are great, so that this process is essential only when the ionosphere is strongly disturbed, that is, more precisely, when irregularities with the scale $a \sim 10-20$ m are efficiently generated. It is of interest to note that the characteristic scale ~ 10 m in the ionosphere is equal to the ion cyclotron radius. Thus, when ion cyclotron waves are generated in the ionosphere (in particular, due to artificial disturbance), an intensification of scattering is possible which results in efficient capture of radio waves into the duct F with their ensuing long-distance propagation along the Earth's surface at frequencies exceeding the MUF-F. Note that small-scale irregularities are generated in the ionosphere when the latter is

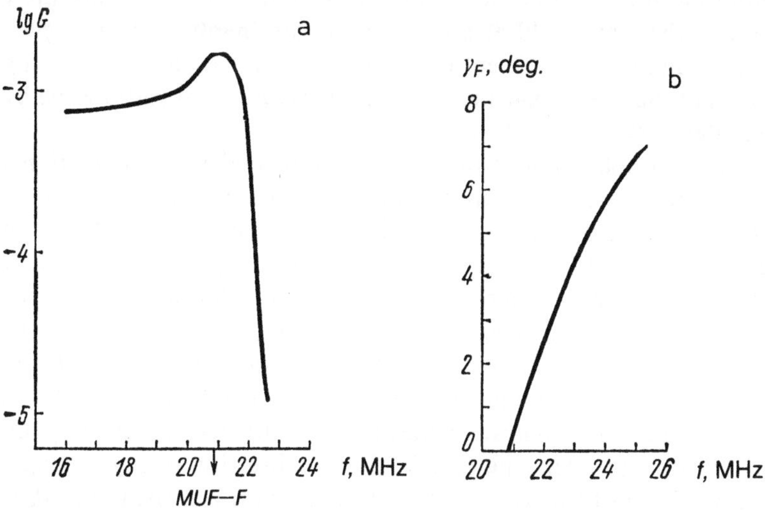

Fig. 6.9. **a** Frequency dependence of the capture coefficient; the *arrow* indicates $f_{\text{MUF-}F}$; **b** frequency dependence of the turning angle γ_F

disturbed by intense radio waves (see Sec. 7.2). Figure 6.8 also gives a possible limit for an extension of MUF, caused by a small scattering-induced turn of a ray emitted horizontally from the Earth. For example, when the ray turns through 2°, the MUF increases in the daytime by 2 – 2.5 MHz, at night by 0.5 – 1 MHz, and at twilight by 2 – 3 MHz (Fishchuk and Tsedilina 1975).

The capture coefficient G as a function of the incident wave frequency, f, for the electron density model given in Herbert (1967) is shown in Fig. 6.9a, which indicates that G somewhat increases in the neighborhood of MUF-F (Kim 1974).

Note, in addition, that radio wave capture into the ducts and escape from the ducts can be influenced by the effects of multiple scattering considered in the next section.

6.2 Effect of Multiple Scattering on Radio Wave Propagation in a Duct

Radiation Transfer Equation. We considered above the process of capture induced by a single radio wave scattering by ionospheric irregularities. At the same time, when propagating in the ionosphere, which includes random irregularities of different types and scales, the wave also undergoes a multiple scattering. As a result, the wave field acquires in the ionosphere an essentially stochastic character. For instance, if the propagation medium is regular, then the wave vector $\boldsymbol{k}$ at a given point r, θ, φ has a fixed direction, but after multiple scattering in a random medium the direction of $\boldsymbol{k}$ is not absolutely fixed. Moreover, the wave field at a given point can involve a few or even many waves propagating in different directions (such waves come into being as a result of multiple scattering, i.e., multiple reradiation of radio wave by different irregularities).

Let us consider propagation of radio wave beams under such conditions. Note that waves trapped in IWD propagate essentially in a horizontal direction along the Earth's surface. Let the axis $\boldsymbol{e}_\theta$ be directed along the beam intensity peak direction, $\boldsymbol{k}_0$. Ray deviations on vertical z are small: $k_z \ll k_0$ [Eqs. (1.4) and (3.8)]. We assume that the horizontal deviations from the beam axis are also small, $k_\varphi \ll k_0$. Then it is convenient to describe the intensity distribution in the beam in terms of a distribution function $f(\theta, z, \varphi, \alpha, \beta)$. Here, φ is the azimuthal angle in the direction orthogonal to the beam axis, together with the coordinates θ, z the angle φ characterizes a given point of the space (ionosphere). The angles α, β give the direction of ray propagation, i.e., the direction of the wave vector of the given ray at the point (θ, z, φ): α is the angle with respect to the horizontal $z=$ const, β is the angle with the axis $\varphi=$ const in the horizontal plane. The distribution function f gives the intensity of a ray which at a given point (z, θ, φ) has a given direction α, β. The function is determined by the radiation transfer equation. In our case, by virtue of $k_z \ll k_0$, $k_\varphi \ll k_0$, the equation can be written in the small angle approximation (Tatarskii 1967, Klyatskin 1975). In addition, let us take into consideration the spherical geometry of the problem as well as the fact that for the radio waves in question, which have a sufficiently high frequency, the ionospheric dielectric constant, $\varepsilon \approx 1$. The radiation transfer equation takes the form (Gurevich and Tsedilina 1976)

$$\begin{aligned}
&\frac{1}{R_0}\frac{\partial f}{\partial \theta}+\alpha\frac{\partial f}{\partial z}+\frac{\beta_1}{R_0 \sin^2\theta}\frac{\partial f}{\partial \varphi}-\frac{1}{2}\frac{\partial U}{\partial z}\frac{\gamma f}{\partial \alpha}+\frac{1}{2R_0}\frac{\partial \varepsilon}{\partial \varphi}\frac{\partial f}{\partial \beta_1}=S\,,\\
&S=-\alpha^* f+\iint G(\alpha-\alpha';\ \beta-\beta')\,f(\theta, z, \varphi;\ \alpha', \beta')\,d\alpha'\,d\beta'\ ,\\
&U=-\varepsilon-\frac{2z}{R_0}\,. \qquad\qquad (6.34)\\
&\beta_1=\frac{\beta}{\cot\theta_0}\,.
\end{aligned}$$

Here, θ_0 is the initial point where originally the distribution function is assumed to be specified. The left-hand side of Eq. (6.34) describes the translation of f along the geometrical-optics ray trajectories. Indeed, the characteristic equations for Eq. (6.34) are given by

$$\begin{aligned}
&\frac{1}{R_0}\frac{dz}{d\theta}=\alpha\,, \qquad \frac{d\alpha}{d\theta}=-\frac{R_0}{2}\frac{\partial U}{\partial z}\,, \qquad \frac{d^2 z}{d\theta^2}=-\frac{R_0^2}{2}\frac{\partial U}{\partial z}\,, \qquad (6.34\text{a})\\
&\frac{d\varphi}{d\theta}=\frac{\beta_1}{\sin^2\theta}\,, \qquad \frac{d\beta_1}{d\theta}=\frac{1}{2}\frac{\partial \varepsilon}{\partial \varphi}\,, \qquad \frac{d}{d\theta}\left(\sin^2\theta\frac{d\varphi}{d\theta}\right)=\frac{1}{2}\frac{\partial \varepsilon}{\partial \varphi}\,.
\end{aligned}$$

They are seen to coincide identically with the ray trajectory equations in the geometrical optics [Eqs. (3.12) and (3.14a)]. In the right-hand side of Eq. (6.34), α^* is the coefficient of the total wave attenuation (caused by the absorption and scattering), $G(\alpha, \beta)$ is the scattering function. They can be expressed in terms of spectral fluctuation density of the dielectric constant for a random medium ρ_ε [Eq. (6.10)] as follows

$$\alpha^* = 2\Gamma_0 + G_0 \qquad \Gamma_0 = \frac{d\Gamma}{d\theta} = \frac{2\pi e^2 N_e \nu_e}{m c \omega^2}, \tag{6.35}$$

$$G(\gamma) = \frac{\pi k_0^4}{2} \rho_\varepsilon(k_0 \gamma, 0), \qquad \gamma = (\alpha, \beta) = \frac{\mathcal{K}_\perp}{k_0} = \frac{(\boldsymbol{k}_{sc} - \boldsymbol{k}_0)_\perp}{k_0} \tag{6.36}$$

$$G_0 = \int G(\gamma)\, d\gamma\,. \tag{6.37}$$

Here, the sign $\perp$ denotes a direction orthogonal to the beam axis.

Typical ionospheric irregularities making the main contribution to multiple scattering are great in comparison with the wave length

$$ka \gg 1\,. \tag{6.38}$$

Due to this, the main contribution to the integral in Eq. (6.34) is made by scattering through small angles α, β. The integral can be then represented in a differential form

$$S = -2\Gamma_0 f + \frac{\partial}{\partial \alpha}\left(D_{\alpha\alpha}\frac{\partial f}{\partial \alpha}\right) + \frac{\partial}{\partial \alpha}\left(D_{\alpha\beta}\frac{\partial f}{\partial \beta}\right)$$
$$+ \frac{\partial}{\partial \beta}\left(D_{\alpha\beta}\frac{\partial f}{\partial \alpha}\right) + \frac{\partial}{\partial \beta}\left(D_{\beta\beta}\frac{\partial f}{\partial \beta}\right) \tag{6.39}$$

$$D_{\alpha\alpha} = -\frac{\overline{(\Delta\varepsilon)^2}}{8}\left[\frac{\partial^2}{\partial \xi_z^2}\int R_\varepsilon(\xi)\, d\xi_\theta\right]_{\xi=0}$$

$$D_{\alpha\beta} = -\frac{\overline{(\Delta\varepsilon)^2}}{8}\left[\frac{\partial^2}{\partial \xi_z \partial \xi_\varphi}\int R_\varepsilon(\xi)\, d\xi_\theta\right]_{\xi=0}$$

$$D_{\beta\beta} = -\frac{\overline{(\Delta\varepsilon)^2}}{8}\left[\frac{\partial^2}{\partial \xi_\varphi^2}\int R_\varepsilon(\xi)\, d\xi_\theta\right]_{\xi=0}.$$

Here, $\overline{(\Delta\varepsilon)^2} = \overline{(\Delta N_e/N_0)^2} \cdot \omega_0^4/\omega^4$ are dielectric constant fluctuations, $R_\varepsilon(\xi)$ is the correlation function of ionospheric irregularities [see Eq. (6.10)], ξ_θ, ξ_φ, ξ_z are the components of the vector ξ referred to the axes θ, φ, z, respectively. The symbol $\overline{}$ or $\langle\ \rangle$ denotes an ensemble averaging with respect to continuous random quantity $\tilde{\varepsilon}$. It should be emphasized that the expressions for diffusive spreading depend, in general, not only on the correlation function, but on the form of ray trajectories as well. Here, they are given for parallel trajectories. If the trajectories diverge, the coefficients decrease, otherwise, they increase (Rytov et al. 1978). When Eq. (6.38) is satisfied, scattering results in a diffusive spreading of the distribution function over the angles α, β. For isotropic irregularities, as well as for anisotropic irregularities reduced to the principal axes, the coefficient is $D_{\alpha\beta}=0$.

The linear density of wave energy at a given θ is defined through the distribution function f by the relation

$$w_0=\int f d\alpha d\beta dz d\varphi . \tag{6.40}$$

If the transmitter is located at the point $\theta_0=0$, $z=z_0$, then the distribution function is here

$$f=f_0(\theta_0,z_0,\varphi_0;\ \alpha_0,\beta_0)=\frac{W_0}{c}D(\alpha_0,\varphi_0)\,\delta(z_0)\,\delta(\beta_0)\cos\alpha_0 . \tag{6.41}$$

Here, W_0 is the total radio wave power, $D(\alpha_0,\varphi_0)$ is the directivity diagram with respect to the vertical, α_0, and azimuthal, φ_0, angles, $\int D(\alpha_0,\varphi_0)\cdot\cos\alpha_0 d\alpha_0 d\varphi_0=1$ [see Eq. (3.65)]. The distribution function in Eq. (6.41) serves as the initial (boundary) condition for the radiation transfer in Eq. (6.34). The random medium (ionosphere) is assumed to exist from $\theta=\theta_0$. No scattering is at $\theta<\theta_0$, and the distribution function [Eq. (6.41)] translates here along the characteristics

$$f(\theta_0,z,\varphi,\alpha,\beta)=f_0(0,z_0,\varphi_0,\alpha_0,\beta_0) . \tag{6.41a}$$

Here, z_0, φ_0, α_0, β_0 are unambiguous functions of z, φ, α, β, θ_0 which are determined by the characteristics in Eq. (6.34a). So, if the distribution function is given at $\theta=0$ in accordance with Eqs. (6.41) and (6.41a), then its variation caused by refraction and multiple scattering of radio waves can be determined through Eq. (6.34).

In the absence of scattering, at any given point the wave vector has a fixed direction (one or several, if reflections occur). In this case the distribution function on the initial condition [Eq. (6.41)] can be sought in the form

$$f=w(\theta,z,\varphi)\,\delta[\alpha-F_1(\theta,z,\varphi)]\,\delta[\beta-F_2(z,\theta,\varphi)] , \tag{6.41b}$$

where w is the wave energy density. This corresponds to the standard geometrical – optics approximation, which, thus, represents a particular case of the radiation transfer equation. The field of application of the latter is much broader than of the geometrical – optics one. In a special formulation of boundary conditions, the energy transfer equation can describe roughly the diffraction effects as well (Klyatskin and Tatarski 1973, Kravtsov and Rytov 1975, Klyatskin 1975).

In the geometrical – optics approximation [Eq. (6.41b)], on the boundaries of ray intersection volumes, caustic surfaces are formed, where the energy density $w(\theta,z,\varphi)$ goes to infinity. With allowance for scattering, the caustics disperse, so the energy density remains finite. Note that caustics also spread due to wave effects (Ginzburg 1970, Landau and Lifshitz 1977).

The radiation transfer equation is substantiated within the framework of the general theory of radio wave propagation in random media (Tatarski 1967, Klyatskin 1975). Let us note here only the fact that by Fourier transformation with respect to α, β, Eq. (6.34) can be reduced to an equation for the function Γ (Kravtsov and Rytov 1975, Dolin 1968, Vorobiov 1971, Tatarski 1971)

$$\Gamma(\theta,\boldsymbol{R},\boldsymbol{\rho}) = \iint f(\theta,\boldsymbol{R},\gamma)\, e^{i\rho\gamma k} d^2\gamma\,, \qquad \gamma = \alpha, \beta$$

where Γ is the coherence function for the radio wave random field $\mathscr{E}(\theta,\boldsymbol{\rho})$

$$\Gamma(\theta,\boldsymbol{R},\boldsymbol{\rho}) = \langle \mathscr{E}(\theta,\boldsymbol{\rho}_1)\, \mathscr{E}^*(\theta,\boldsymbol{\rho}_2)\rangle\,.$$

Here, $\boldsymbol{R} = (z, R_0\varphi) = (\boldsymbol{\rho}_1+\boldsymbol{\rho}_2)/2$; $\boldsymbol{\rho} = k(\boldsymbol{\rho}_1-\boldsymbol{\rho}_2)$. This equation for Γ can also be derived directly, when considering the wave electric field $\mathscr{E}(\theta,\rho)$ in a medium with continuously distributed random irregularities, $\mathscr{E}(\theta,\boldsymbol{\rho})$ being described in the parabolic equation approximation. This fact enables α^* and $G(\gamma)$ to be expressed in terms of the medium correlation function $R_\varepsilon(\xi)$ [Eqs. (6.35) – (6.37) and (6.39)].

Note, in addition, that the radiation transfer equation in the small-angle approximation [Eq. (6.34)] is similar to the two-dimensional kinetic equation where the coordinate θ along the beam axis plays the role of time, angles α, β are used instead of the particle velocities, and z, φ play the role of coordinates. The scattering operator is equivalent to the collision integral.

Diffusion Coefficients in the Ionosphere. A specific form of the diffusive coefficients $D_{\alpha\alpha}$, $D_{\alpha\beta}$, and $D_{\beta\beta}$ is determined through the correlation function of ionospheric irregularities $R(\xi)$. For example, assuming the function $R(\xi)$ to be isotropic Gaussian

$$R(\xi) = \exp\left(-\frac{\xi_\theta^2+\xi_\varphi^2+\xi_z^2}{a^2}\right)$$

we obtain from Eq. (6.39)

$$\begin{aligned} &D_{\alpha\alpha} = D_{\beta\beta} = \frac{\sqrt{\pi}\,\overline{(\Delta\varepsilon)^2}}{4a}\,, \qquad D_{\alpha\beta} = 0 \\ &\overline{(\Delta\varepsilon)^2} = \overline{\left(\frac{\Delta N_e}{N_0}\right)^2}\,\frac{\omega_0^4}{\omega^4}\,. \end{aligned} \tag{6.42}$$

Ionospheric irregularities are usually anisotropic, namely, they are elongated along the magnetic field direction $\boldsymbol{H}$, i.e., in the general case

$$R(\xi) = R(\xi_\parallel, \xi_\perp) \tag{6.42a}$$

where $\xi_{\|}$ is the coordinate along $\boldsymbol{H}$, and $\xi_{\perp}$ is a coordinate transverse to $\boldsymbol{H}$. The components of the magnetic field in the coordinate system (θ, φ, z) have the form

$$\boldsymbol{H} = (H\sin\gamma_H\cos\varphi_0\,,\ H\sin\gamma_H\sin\varphi_0\,,\ H\cos\gamma_H)\,.$$

Here, $\gamma_H = \pi/2 - I_H$ is the inclination angle of $\boldsymbol{H}$ with respect to the vertical z, I_H is the dip angle, φ_0 is the azimuth of the axis $\boldsymbol{e}_\theta$, i.e., the axis of the beam, with respect to the magnetic meridian plane. Hence,

$$\begin{aligned}\xi_{\|} &= \frac{\boldsymbol{\xi H}}{H} = \xi_\theta\sin\gamma_H\cos\varphi_0 + \xi_\varphi\sin\gamma_H\sin\varphi_0 + \xi_z\cos\gamma_H\\ \xi_{\perp} &= \sqrt{\xi^2-\xi_{\|}^2} = (\xi_\theta^2+\xi_\varphi^2+\xi_z^2-\xi_{\|}^2)^{1/2}\,.\end{aligned} \tag{6.42b}$$

If substituting Eqs. (6.42b) and (6.42a) into Eq. (6.39), one can easily obtain the diffusion coefficients. In particular, for the Gaussian anisotropic correlation function

$$R(\xi) = \exp\left[-\left(\frac{\xi_{\|}^2}{a_{\|}^2}+\frac{\xi_{\perp}^2}{a_{\perp}^2}\right)\right] \tag{6.42c}$$

we obtain

$$\begin{aligned}D_{\alpha\alpha} &= \frac{\sqrt{\pi}}{4}\overline{(\Delta\varepsilon)^2}\,\frac{a_{\|}\,[a_{\|}^2\sin^2\gamma_H + a_{\perp}^2\cos^2\gamma_H + (a_{\perp}^2-a_{\|}^2)\sin^2\gamma_H\cos^2\varphi_0]}{a_{\perp}\,[a_{\|}^2(1-\sin^2\gamma_H\cos^2\varphi_0)+a_{\perp}^2\sin^2\gamma_H\cos^2\varphi_0]^{3/2}}\,,\\ D_{\beta\beta} &= \frac{\sqrt{\pi}}{4}\overline{(\Delta\varepsilon)^2}\,\frac{a_{\|}\,[a_{\|}^2\cos^2\gamma_H + a_{\perp}^2\sin^2\gamma_H + (a_{\perp}^2-a_{\|}^2)\cos^2\gamma_H\cos^2\varphi_0]}{a_{\perp}\,[a_{\|}^2(1-\sin^2\gamma_H\cos^2\varphi_0)+a_{\perp}^2\sin^2\gamma_H\cos^2\varphi_0]^{3/2}}\,,\\ D_{\alpha\beta} &= -\frac{\sqrt{\pi}}{4}\overline{(\Delta\varepsilon)^2}\,\frac{a_{\|}\,(a_{\|}^2-a_{\perp}^2)\sin\gamma_H\cos\gamma_H\sin\varphi_0}{a_{\perp}\,[a_{\|}^2(1-\sin^2\gamma_H\cos^2\varphi_0)+a_{\perp}^2\sin^2\gamma_H\cos^2\varphi_0]^{3/2}}\,.\end{aligned} \tag{6.42d}$$

It is seen from Eq. (6.42d) that in this case the diffusion coefficients essentially depend on the angle γ_H and on the direction of beam propagation, φ_0. For isotropic irregularities, $a_{\|} = a_{\perp}$, Eq. (6.42d) coincides with Eq. (6.42). For strongly elongated irregularities, $a_{\|}^2 \gg a_{\perp}^2\tan^2\gamma_H$, $\cot^2\gamma_H$, the main role is played by the transverse scale $a_{\perp}$

$$\begin{aligned}D_{\alpha\alpha} &= \frac{\sqrt{\pi}}{4a_{\perp}}\,\frac{\overline{(\Delta\varepsilon)^2}\sin^2\gamma_H\sin^2\varphi_0}{(1-\sin^2\gamma_H\cos^2\varphi_0)^{3/2}}\\ D_{\beta\beta} &= \frac{\sqrt{\pi}}{4a_{\perp}}\,\frac{\overline{(\Delta\varepsilon)^2}\cos^2\gamma_H\sin^2\varphi_0}{(1-\sin^2\gamma_H\cos^2\varphi_0)^{3/2}}\\ D_{\alpha\beta} &= \frac{\sqrt{\pi}}{4a_{\perp}}\overline{(\Delta\varepsilon)^2}\,\frac{\sin\gamma_H\cos\gamma_H\sin\varphi_0}{(1-\sin^2\gamma_H\cos^2\varphi_0)^{3/2}}\,.\end{aligned} \tag{6.42e}$$

In this case, at the polar, $\gamma_H \to 0$, and equatorial, $\gamma_H \to \pi/2$, latitudes the scattering coefficients become strongly anisotropic.

Under real ionospheric conditions, the principal contribution to scattering is made by F-region irregularities. Generally speaking, the correlation function of these irregularities is not of a Gaussian type, and in the large-scale region its scale spectrum has an intensifying factor of a power type. The correlation function is commonly represented in the form

$$R(\xi_\parallel, \xi_\perp) = C_0 \int (\varkappa_\parallel^2 + \varkappa_\perp^2)^{-p/2} \exp\left(-\frac{\varkappa_\parallel^2 a_\parallel^2}{4} - \frac{\varkappa_\perp^2 a_\perp^2}{4} + i\varkappa\xi\right) d\varkappa . \qquad (6.42\text{f})$$

Here, C_0 is a normalizing constant defined by the condition $R(0,0) = 1$, p is the exponent of the power part of the irregularity spectrum. At $p = 0$, the correlation functions [Eqs. (6.42c) and (6.42f)] coincide. Actually observed spectra have the exponent $1 \lesssim p \lesssim 3$ (Erukhimov et al. 1977, Gdalevich et al. 1980).

As p increases, the role of large-scale irregularities becomes more important. This leads to the change in value of the diffusion coefficients. For example, at $p = 2$, we obtain from Eqs. (6.42f) and (6.39) that at $\varphi_0 = \pi/2$ (i.e., for beam propagation in the magnetic meridian plane)

$$D_{\alpha\alpha} = \frac{\sqrt{\pi}\,\overline{(\Delta\varepsilon)^2}}{4a_\perp} Q \frac{a_\parallel \sin^2\gamma_H + a_\perp \cos^2\gamma_H}{a_\parallel + a_\perp}$$

$$D_{\beta\beta} = \frac{\sqrt{\pi}\,\overline{(\Delta\varepsilon)^2}}{4a_\perp} Q \frac{a_\parallel \cos^2\gamma_H + a_\perp \sin^2\gamma_H}{a_\parallel + a_\perp}$$

$$D_{\alpha\beta} = -\frac{\sqrt{\pi}\,\overline{(\Delta\varepsilon)^2}}{4a_\perp} Q \frac{a_\perp (a_\parallel - a_\perp)}{a_\parallel (a_\parallel + a_\perp)} \sin\gamma_H \cos\gamma_H \qquad (6.42\text{g})$$

$$Q = \left(1 - \frac{a_\perp^2}{a_\parallel^2}\right)^{1/2} \Bigg/ \sqrt{2}\ln\left[\frac{1 + (1 - a_\perp^2/a_\parallel^2)^{1/2}}{a_\perp/a_\parallel}\right].$$

Comparing these equations with Eq. (6.42d), we see that the dependence of the coefficients on the angle γ_H has remained unchanged, but their values are somewhat changed. For instance, for isotropic irregularities ($a_\parallel = a_\perp = a$), the diffusion coefficients are, by Eq. (6.42g), by a factor of $2\sqrt{2}$ less than those in the Gaussian case [Eq. (6.42)]. For strongly elongated irregularities, $a_\parallel^2 \gg a_\perp^2$, the diffusion coefficients are weaken by a factor of $\sqrt{2}\ln(2a_\parallel/a_\perp)$.

To estimate the diffusion coefficients in the ionosphere, we assume for typical relative perturbations of electron density in the ionosphere, the magnitude $\delta N = \Delta N_e/N_0 \approx (1-3) \cdot 10^{-3}$, and for irregularity scale, the value $a \sim 0.7 - 1.5$ km (see Sec. 6.1). Then the characteristic value of the diffusion coefficients is, according to Eq. (6.42)

$$D_0 \approx (2 \cdot 10^{-11} - 10^{-14})\ \text{cm}^{-1} . \qquad (6.42\text{h})$$

This value will be used below for numerical evaluations.

Diffusion over Adiabatic Invariants. For consideration of wave propagation in IWD, it is convenient to make a transition from z, α to new variables: adiabatic invariant I and phase Φ. The variables I, Φ are canonically conjugated: I is the action variable, and Φ is the angle variable (Landau and Lifshitz 1976, Sec. 50) both are in common use for kinetic equations (Belyaev 1958). For simplicity, we assume that the duct is spherically symmetric, $U = U(z)$. We shall determine the ray oscillation phase, making a transition in Eq. (3.28) from the integration over θ to the integration over z

$$\Phi(z) = \begin{cases} \dfrac{2\pi}{\Theta R_0} \displaystyle\int_{z_{\min}}^{z} \frac{dz}{\sqrt{E-U}}, & \text{at} \quad \Phi \leqslant \pi, \\ \pi - \dfrac{2\pi}{\Theta R_0} \displaystyle\int_{z_{\max}}^{z} \frac{dz}{\sqrt{E-U}} & \text{at} \quad \pi \leqslant \Phi < 2\pi \end{cases} . \tag{6.43}$$

The phase Φ is seen to be double-valued function of the height z. This corresponds to the two signs of $dz/d\theta = \pm R_0\sqrt{E-U}$. The sign before the square root alters in the trajectory turning points $z_{\min}$ and $z_{\max}$; at $z = z_{\min}$, the phase $\Phi = 0$, at $z = z_{\max}$, $\Phi = \pi$. The constant E, by Eq. (2.29), is given by

$$E = \alpha^2 + U(z) . \tag{6.44}$$

Assuming $f = f(\theta, I, \Phi; \beta, \varphi)$, and noting that $I = I(E)$ is defined by Eq. (3.17) and the phase $\Phi = \Phi(z, E)$ is defined by Eq. (6.43), we obtain

$$\alpha\frac{\partial f}{\partial z} - \frac{1}{2}\frac{\partial U}{\partial z}\frac{\partial f}{\partial \alpha} = \left(\alpha\frac{\partial E}{\partial z} - \frac{1}{2}\frac{\partial U}{\partial z}\frac{\partial E}{\partial \alpha}\right)\left(\frac{dI}{dE}\frac{\partial f}{\partial I} + \frac{\partial \Phi}{\partial E}\frac{\partial f}{\partial \Phi}\right)$$
$$+ \alpha\frac{\partial \Phi}{\partial z}\frac{\partial f}{\partial \Phi} = \alpha\frac{\partial \Phi}{\partial z}\frac{\partial f}{\partial \Phi} . \tag{6.45}$$

Here, we have taken into account Eq. (6.44). Substituting Eq. (6.45) into Eq. (6.34), we can go over to new variables. To this end, multiplying Eq. (6.34) by $\delta(I - I(E))\,\delta(E - \alpha^2 - U(z))\,\delta(\Phi - \Phi(z))$, we integrate it over the whole phase volume $d\alpha\,dz\,dE$. In so doing, we take into account that

$$A = \int \delta(I - I(E))\,\delta(E - \alpha^2 - U(z))\,\delta(\Phi - \Phi(z))\,d\alpha\,dz\,dE$$
$$= \int \delta(I - I(E))\,dE \int \frac{dz}{2\sqrt{E-U}}\,\delta(\Phi - \Phi(z))$$
$$= \frac{R_0}{8\pi}\int \delta(I - I(E))\,\Theta(E)\,dE = \frac{R_0}{8\pi} . \tag{6.46}$$

Here, we have taken into account that Φ is a double-valued function of z, so that

$$\int_{z_{\min}}^{z_{\max}} \frac{dz}{\sqrt{E-U}}\,\delta(\Phi - \Phi(z)) = \frac{R_0\Theta}{4\pi} .$$

To make sure that the latter is true, one should integrate it over Φ from 0 to 2π. Next, we obtain

$$\frac{\partial f}{\partial \theta}+\frac{2\pi}{\Theta}\frac{\partial f}{\partial \Phi}+\frac{\beta_1}{\sin^2\theta}\frac{\partial f}{\partial \varphi}+\frac{1}{2}\frac{\partial \varepsilon}{\partial \varphi}\frac{\partial f}{\partial \beta_1}=8\pi\bar{S}. \tag{6.47}$$

The coefficient at $\partial f/\partial I$ in the left-hand side of Eq. (6.47) has become a zero, which should be anticipated, since I is the integral of motion for the characteristic Eq. (6.34a). The dash above S means the transition to new variables I, Φ through the integration given in Eq. (6.46). The normalization condition [Eq. (6.40)] in the new variables is

$$w_0=\int f\, d\beta d\varphi dI d\Phi .$$

The initial condition [Eq. (6.41)] in this case takes the form

$$f=f_0(0,I,\Phi_0,\varphi_0,\beta_0)$$
$$=\frac{W_0}{2c\Theta\sqrt{E(I)+1}}D(\sqrt{E(I)+1},\varphi_0)\cos(\sqrt{E(I)+1})\,\delta(\Phi_0)\,\delta(\beta_0)\,, \tag{6.48}$$

where $D(\alpha_0,\varphi_0)$ is, as before, the directivity diagram of the transmitter antenna. When going to the variables I, Φ, it has been taken into account that, by Eq. (6.44), $\alpha_0=\sqrt{E-U(z_0)}$, $d\alpha_0=\dfrac{dE}{2\sqrt{E-U(z_0)}}=\dfrac{dI}{2\Theta\sqrt{E-U(z_0)}}$ and, for the transmitter located on the Earth, $U(z_0)=-1$. The initial distribution function [Eq. (6.48)] satisfies the normalization condition near the transmitter, $w_0=W_0/c$.

Further simplifications are connected with a supposition about the character of scattering in the ionosphere. Let us assume that the initial distribution of the beam with respect to the horizontal angle φ_0 (the distribution is determined by the transmitter properties) is smooth, i.e., the beam is broad. If now ionospheric scattering is weak enough, then the scattering in β does not essentially change the initial distribution with respect to φ and, hence, affects slightly the distribution of intensity even at a long-distance path. In this case

$$f(\theta,I,\Phi;\beta,\varphi)\approx f(\theta,I,\Phi,\beta_0,\varphi(\theta))\,, \tag{6.49}$$

where the function $\varphi(\theta)$ is determined by the characteristic Eq. (6.34a)

$$\frac{d}{d\theta}\left(\sin^2\theta\frac{d\varphi}{d\theta}\right)=\frac{1}{2}\frac{\partial\varepsilon}{\partial\varphi}, \qquad \varphi(\theta_0)=\varphi_0\,, \qquad \frac{d\varphi}{d\theta}(\theta_0)=\beta_{10}=0\,. \tag{6.50}$$

Here, $\partial\varepsilon/\partial\varphi$ is taken at given $\theta, I, \Phi(\theta)$. We note that $\partial\varepsilon/\partial\varphi$ at fixed I may be averaged over the phase Φ [see Eq. (3.48)], i.e., may be replaced by $\overline{(\partial\varepsilon/\partial\varphi)}_I$. In the spherically-symmetric case $\partial\varepsilon/\partial\varphi=0$ and

$$\varphi=\varphi_0\,, \qquad \beta=\beta_0=0\,. \tag{6.51}$$

It is to be emphasized, of course, that the scattering in β can be ignored in not at all cases. First of all, it is important in focusing regions: in the antipode $\theta=\pi$ and round-the-world $\theta=2\pi$ points. Both the scattering in β and refraction, $\partial\varepsilon/\partial\varphi$, can influence also the RTW time delay. The scattering over β also determines the variation of the bearing and broadening of narrow beams. These questions will be treated in the following section. Here we concentrate our attention only on the effects of vertical scattering, which are predominant. Therefore, for the distributions in β and φ, the weak scattering approximation will be used. Of most importance in this case are the distributions in I and Φ.

Next, let us pay attention to a significant difference between the effects of scattering on the adiabatic invariant distribution and the phase one. Whereas the adiabatic invariant distribution in the absence of scattering remains altogether unchanged, the phase manages, independently of scattering, to run its whole range between 0 and 2π at a relatively small interval of θ equal to one ray oscillation period Θ. Assuming that the scattering operator varies slightly over the range $\sim\Theta$ (in the spherically-symmetric case it does not change with θ at all) one can average Eq. (6.47) over the interval $\Delta\theta\sim\Theta$, or, in other words, over the oscillation phase Φ. Equation (6.47) with allowance for Eqs. (6.49) and (6.51) then takes the form

$$\frac{\partial\bar{f}}{\partial\theta}=4\int_0^{2\pi}\bar{S}(\bar{f})\,d\Phi;\quad \bar{f}=\bar{f}(\theta,I)=\frac{1}{\Theta}\int_{\theta-\Theta/2}^{\theta+\Theta/2}f\,d\theta'=\frac{1}{2\pi}\int_0^{2\pi}f\,d\Phi. \quad (6.52)$$

Here, the operator $\bar{S}$, defined by Eqs. (6.34), (6.46), and (6.47), describes variations of the distribution function caused by absorption and scattering. Recognizing that in the ionosphere the major contribution is made by small-angle scattering [Eq. (6.39)] and taking into account that

$$\frac{\partial}{\partial\alpha}=2\alpha\Theta\frac{\partial}{\partial I}$$

$$\frac{\partial}{\partial\alpha}\left(D_{\alpha\alpha}\frac{\partial\bar{f}}{\partial\alpha}\right)=4\int\alpha\Theta\frac{\partial}{\partial I}\left[D_{\alpha\alpha}\alpha\Theta\frac{\partial\bar{f}}{\partial I}\right]\delta(I-I(E))\,\delta(E-\alpha^2-U)$$

$$\cdot\,\delta(\Phi-\Phi(z))\,dz\,d\alpha\,dE\,d\Phi$$

$$=2\int\Theta\frac{\partial}{\partial I}\left[D_{\alpha\alpha}\sqrt{E-U}\,\Theta\frac{\partial\bar{f}}{\partial I}\right]\delta(I-I(E))\,dz\,dE$$

$$=2\frac{\partial}{\partial I}\left[\left(\int_{z_{\min}}^{z_{\max}}D_{\alpha\alpha}\sqrt{E-U}\,dz\right)\Theta\frac{\partial\bar{f}}{\partial I}\right]$$

we obtain

$$4\cdot\int_0^{2\pi}\bar{S}(\bar{f})\,d\Phi=4\int\left[-2\Gamma_0\bar{f}+\frac{\partial}{\partial\alpha}\left(D_{\alpha\alpha}\frac{\partial\bar{f}}{\partial\alpha}\right)\right]\delta(I-I(E))\,\delta(E-\alpha^2-U)$$

$$\cdot\,\delta(\Phi-\Phi(z))\,dz\,d\alpha\,dE\,d\Phi=\frac{\partial}{\partial I}\left(\bar{D}\Theta\frac{\partial\bar{f}}{\partial I}\right)-2\bar{K}(I)\bar{f},\tag{6.53}$$

$$\bar{D}=8\int_{z_{min}}^{z_{max}}D_{\alpha\alpha}\sqrt{E(I)-U(z)}\,dz,\tag{6.54}$$

$$\bar{K}(I)=\frac{4\pi e^2}{mc\omega^2\Theta}\int_{z_{min}}^{z_{max}}\frac{N_e(z)\,\nu_e(z)\,dz}{\sqrt{E(I)-U(z)}}.\tag{6.55}$$

The terminal equation describing the change of the adiabatic invariant distribution of rays induced by multiple scattering takes the form (Gurevich and Tsedilina 1976)

$$\frac{\partial f}{\partial\theta}=\frac{\partial}{\partial I}\left[D\Theta\frac{\partial f}{\partial I}\right]-2K(I)f.\tag{6.56}$$

Henceforth the averaging symbol is omitted. The coefficients D, K are defined by Eqs. (6.54) and (6.55).

Note that the opportunity for the use of the scattering equation under ionospheric conditions in the differential form [Eq. (6.56)] is provided by a great number of modes, $\sim 10^2-10^3$, in ionospheric ducts. Scattering dispersion over the distance of the order of oscillation period, virtually, always significantly exceeds an intermode interval. Therefore, the mode distribution of intensity is bound to be smoothed, enabling the diffusion approximation to be used. The substantiation of the equation of adiabatic invariant diffusion in the wave theory is given in Borisov (1978).

The initial condition [Eq. (6.48)], after the averaging [Eq. (6.52)], takes the form

$$f=f_0(0,I)=\frac{W_0}{2c\Theta\sqrt{E(I)+1}}D_{\varphi_0}(\sqrt{E(I)+1})\cos(\sqrt{E(I)+1}).\tag{6.57}$$

Here, $D_{\varphi_0}(\alpha)=D(\alpha_0,\varphi_0)$ is the directivity diagram with respect to vertical angles α_0 at a fixed azimuthal direction φ_0. The normalization condition has a simple form

$$w(\varphi_0)=\int f\,dI,\qquad w_0=\int w(\varphi_0)\,d\varphi,\tag{6.58}$$

where w_0 is the linear density of radio wave energy.

Boundary conditions for Eq. (6.56) are specified on the duct bottom, $I\to 0$, and at its boundary, $I\to I_m$. On the bottom, there is no energy flux, i.e., at $I\to 0$

$$D\Theta\frac{\partial f}{\partial I}\to 0.\tag{6.59}$$

Insofar as in the neighborhood of its bottom the well has a parabolic form, i.e., $D\Theta \to CI$ (see below), then the condition [Eq. (6.59)] is rewritten in the form

$$\left(I\frac{\partial f}{\partial I}\right)_{I\to 0} \to 0\,. \tag{6.60}$$

The condition at the duct boundary, $I = I_m$, depends on the situation in the adjacent ducts, i.e., whether there are radio waves and what conditions are present for their propagation. We shall consider two kinds of boundary conditions, distinguishing between the propagation and capture regions. In the propagation region, we neglect a radiation outside the duct (for instance, by reason of enhanced radio wave absorption in the lower ionosphere, or its release beyond the upper limit of the ionosphere). Here,

$$f|_{I=I_m} = 0\,, \tag{6.61}$$

since beyond the duct $f \to 0$, and the diffusion equation permits no discontinuity in f. In the capture region, the form of the boundary condition depends on the initial condition [Eq. (6.57)] (see below).

Oscillation Phase Diffusion. The averaged equation for the adiabatic invariant diffusion [Eq. (6.56)] does not describe an important process of spreading of the initial distribution over ray oscillation phases. Therefore, let us derive a more general equation involving an oscillation phase Φ diffusion. To this end, we return to Eq. (6.47) which with allowance for Eqs. (6.39) and (6.49) – (6.51), can be written down in the form

$$\frac{\partial f}{\partial \theta} + \frac{2\pi}{\Theta}\,\frac{\partial f}{\partial \Phi} = 8\pi\,\overline{\left[\frac{\partial}{\partial \alpha}\left(D_{\alpha\alpha}\frac{\partial f}{\partial \alpha}\right) - 2\Gamma_0 f\right]}\,. \tag{6.61a}$$

Here, $f = f(\theta, I, \Phi)$, and the averaging sign (dash line), as before, means the transition to the new variables I, Φ through Eq. (6.46).

Note that

$$\frac{\partial}{\partial \alpha} = \frac{\partial I}{\partial E}\,\frac{\partial E}{\partial \alpha}\,\frac{\partial}{\partial I} + \frac{\partial \Phi}{\partial E}\,\frac{\partial E}{\partial \alpha}\,\frac{\partial}{\partial \Phi} = 2\alpha\left(\Theta\frac{\partial}{\partial I} + \frac{\partial \Phi}{\partial E}\,\frac{\partial}{\partial \Phi}\right) \tag{6.61b}$$

and, consequently,

$$S = \frac{\partial}{\partial \alpha}\left(D_{\alpha\alpha}\frac{\partial f}{\partial \alpha}\right) = 4\alpha\Theta\frac{\partial}{\partial I}\left[\alpha D_{\alpha\alpha}\left(\Theta\frac{\partial f}{\partial I} + \frac{\partial \Phi}{\partial E}\,\frac{\partial f}{\partial \Phi}\right)\right]$$
$$+ 4\alpha\frac{\partial \Phi}{\partial E}\,\frac{\partial}{\partial \Phi}\left[\alpha D_{\alpha\alpha}\left(\Theta\frac{\partial f}{\partial I} + \frac{\partial \Phi}{\partial E}\,\frac{\partial f}{\partial \Phi}\right)\right].$$

Averaging the latter equation in accordance with Eq. (6.46), we obtain the terminal equation

$$\frac{\partial f}{\partial \theta}+\frac{2\pi}{\Theta}\frac{\partial f}{\partial \Phi}=\frac{\partial j_1}{\partial I}+\frac{\partial j_2}{\partial \Phi}-2\Gamma_0 f,$$

$$j_1=D_{11}\frac{\partial f}{\partial I}+D_{12}\frac{\partial f}{\partial \Phi},$$

$$j_2=D_{12}\frac{\partial f}{\partial I}+D_{22}\frac{\partial f}{\partial \Phi}, \tag{6.61c}$$

$$D_{11}=4R_0\Theta(E-U)D_{\alpha\alpha}$$

$$D_{12}=4R_0\frac{\partial \Phi}{\partial E}(E-U)D_{\alpha\alpha}, \qquad D_{22}=4R_0(E-U)\left(\frac{\partial \Phi}{\partial E}\right)^2 D_{\alpha\alpha}.$$

Here, $E=E(I)$, $U=U(z(I,\Phi))$, $D_{\alpha\alpha}=D_{\alpha\alpha}(z(I,\Phi))$. Being averaged over phase, Eq. (6.61c), naturally, transforms to Eq. (6.56).

Parabolic Duct. Near the duct bottom, i.e., in the vicinity of the "sliding" point z_s, the potential well is always parabolic [Eq. (2.9)]

$$U(z)=U(z_s)+\frac{1}{2}U''(z_s)(z-z_s)^2. \tag{6.62}$$

In this case, according to Eqs. (3.19), (3.24), and (6.43)

$$I=\frac{4}{R_0}\int_{z_{min}}^{z_{max}}\sqrt{E-U(z_s)-\frac{1}{2}U''(z-z_s)^2}\,dz=(E-U(z_s))\,\Theta,$$

$$\Theta=\frac{dI}{dE}=\frac{2\sqrt{2}\,\pi}{R_0\sqrt{U''(z_s)}}, \qquad z_{min}=z_s-a(I), \qquad z_{max}=z_s+a(I), \tag{6.63}$$

$$a(I)=\frac{R_0}{2\pi}(\Theta I)^{1/2}, \qquad z(\Phi,I)=z_s-a(I)\cos\Phi.$$

The ray oscillation period in a parabolic duct is independent of I, so Eq. (6.56) takes a simple form

$$\frac{\partial f}{\partial \tau}=\frac{\partial}{\partial I}\left(I\frac{\partial f}{\partial I}\right). \tag{6.64}$$

For simplicity, we have neglected here an absorption and assumed the distribution of irregularities to be uniform, $D_{\alpha\alpha}=\text{const}=D_0$. The dimensionless variable τ is

$$\tau=2D_0R_0\Theta(\theta-\theta_0). \tag{6.65}$$

The parabolic duct Eq. (6.64) is precise near the well bottom. However, it can also be applied as a model equation for the whole duct. A model equation enables to study the properties of ray diffusion and to recognize the basic parameters of the problem. Therefore, let us assume henceforth the duct to be parabolic.

The plot of the potential $U(z)$ for the parabolic model of F-duct is shown in Fig. 6.10a. Below the barrier peak z_m, $U(z)$ has a parabolic form, above z_m its form is arbitrary. The transmitter is located within the duct, so the initial condition for f at $\tau=0$ is written in the form

$$f|_{\tau=0}=\begin{cases} f_0(I), & \text{at} \quad E(I)+1\geqslant 0 \\ 0, & \text{at} \quad E(I)+1<0 \end{cases} \tag{6.66}$$

where the function $f_0(I)$ is defined by Eq. (6.57). The boundary condition at $I=I_m$ has the form of Eq. (6.61). In this case, as a wave travels in the duct, the initial adiabatic invariant distribution of the wave [Eq. (6.66)] varies and a portion of its energy leaves the ionosphere, both processes being induced by multiple scattering.

The potential $U(z)$ for the parabolic model of FE-duct is shown in Fig. 6.10b. Below the barrier $z=z_m$, the function $U(z)$ is arbitrary, and when $z>z_m$, the duct is parabolic. The initial condition [Eq. (6.57)] defines now the type of the boundary condition at $I=I_m$. Indeed, waves emitted from a point $\theta=0$, $z=0$ at an angle $\alpha\approx\alpha_m=\sqrt{U_m+1}$, gain the barrier peak, z_m, at some point $\theta=\theta_m$. Thus, in Eq. (6.65) $\theta_0=\theta_m$. The boundary condition [Eq. (6.57)] at $\theta=\theta_m$, i.e., at $\tau=0$, is rewritten in the form

$$f_{\tau=0}=\begin{cases} 0 & \text{at} \quad I<I_m, \\ f_0 & \text{at} \quad I>I_m, \end{cases} \qquad f_0=f_0(I)\exp(-2\Gamma_m)\,. \tag{6.66a}$$

Here, the function $f_0(I)$ is defined by Eq. (6.57), Γ_m is the total wave absorption on the path in the lower ionosphere up to the duct boundary

$$\Gamma_m=\frac{2\pi e^2}{mc\omega^2}\int\limits_0^{z_m}\frac{N_e(z)\,\nu_e(z)\,dz}{\sqrt{E-U(z)}}\,, \qquad E\approx U_m\,.$$

The angular width of the capture region is typically much less than the width of the transmitter directivity diagram [see Eq. (6.77a)]. Therefore, Eq. (6.57) for $f_0(I)$ can be rewritten in the form

$$f_0(I)=\frac{W_0 D_{\varphi_0}(\alpha_m)\cos\alpha_m}{2c\Theta_m\sqrt{\alpha_m^2+E(I)-U_m}}\,, \qquad \alpha_m=\sqrt{U_m+1}\,. \tag{6.67}$$

Hence, when the critical angle α_m is sufficiently great, one can think that $f_0(I)=\text{const}=f_0$. At rather small α_m, when the wave frequency becomes close in value to the MUF of the layer, the dependence of f_0 on I becomes important.

The solution of Eq. (6.64) with boundary conditions [Eqs. (6.66a) and (6.67)] describes the wave energy capture into the duct, caused by multiple scattering. This solution is valid, however, only during the first oscillation period from $\theta=\theta_m$ to $\theta=\theta_m+\Theta$, i.e., when

$$0\leqslant\tau\leqslant\tau_0=2D_0R_0\Theta^2\,. \tag{6.68}$$

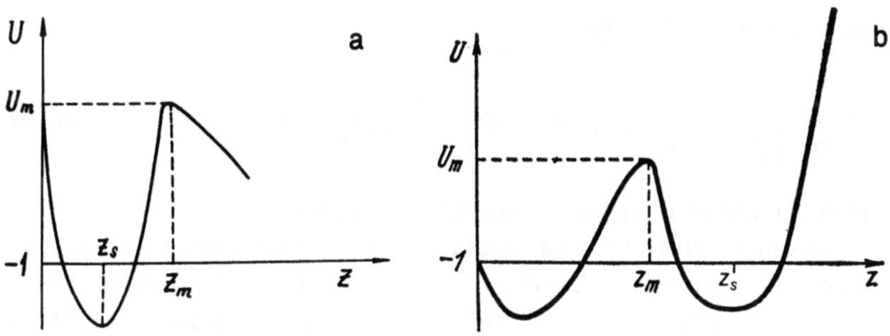

Fig. 6.10. **a** Form of the potential $U(z)$ for the parabolic duct F. **b** Form of the potential $U(z)$ for the parabolic duct FE

It is the capture region; when $\theta > \theta_m + \Theta$, radio waves which are not trapped in the duct propagate in the absorbing region below the barrier peak, $z = z_m$. If the waves are supposed to be strongly absorbed or scattered, so that their further capture can be neglected, then when $\theta > \theta_m + \Theta$, i.e., when $\tau > \tau_0$, the boundary condition at $I = I_m$ takes the form of Eq. (6.61); it is the propagation region. In this region the problem for the parabolic FE-duct is the same as for the F-duct.

The solution of Eq. (6.64) in both regions will be considered below. It describes both radio wave capture into a duct and an adiabatic invariant redistribution of the waves with their multiple – scattering – induced release from the duct. To estimate the effects in the ionosphere, the following parametric values can be used [see Eqs. (6.42h) and (6.65)]

$$D_0 \approx (2 \cdot 10^{-11} - 10^{-14})\ \mathrm{cm}^{-1}\,, \qquad R_0 = 6.37 \cdot 10^8\ \mathrm{cm}\,, \qquad \Theta_0 \approx 0.2 - 0.6\,,$$
$$I_m \lesssim 0.1\,, \qquad \tau_0 \approx 3 \cdot 10^{-2} - 3 \cdot 10^{-4}\,. \tag{6.69}$$

Capture into the Interlayer Duct. Let us obtain, first, the solution of Eq. (6.64) in the capture region [Eq. (6.68)]. To this end, we go over to a new variable, $x = I - I_m$

$$\frac{\partial f}{\partial \tau} = (I_m + x)\frac{\partial^2 f}{\partial x^2} + \frac{\partial f}{\partial x}\,. \tag{6.70}$$

Assuming that during the small "time" τ_0, mainly the upper duct levels are filled up, the quantity x in Eq. (6.70) can be neglected in comparison with I_m, and, in addition, the boundary condition [Eq. (6.54)] at $I \to 0$ can be replaced by a condition for a solution regularity at $x \to -\infty$

$$f \to 0 \quad \text{at} \quad x \to -\infty\,. \tag{6.71}$$

Simultaneously, the last term in Eq. (6.70) should be ignored, since $\partial f/\partial x$ is of the same order of magnitude as $x\,\partial^2 f/\partial x^2$.

The diffusion Eq. (6.70) takes then a canonical form

$$\frac{\partial f}{\partial \tau} = I_m \frac{\partial^2 f}{\partial x^2}\,. \tag{6.72}$$

Its solution is well-known to be

$$f(x,\tau)=\frac{1}{2\sqrt{\pi I_m\tau}}\int_{-\infty}^{\infty} f_0(I_m+x_0)\exp\left[-\frac{(x-x_0)^2}{4I_m\tau}\right]dx_0\,, \tag{6.73}$$

where $f_0(I)$ is the initial distribution function [Eq. (6.67)].

First, we assume that the critical duct angle α_m is great enough, so that in Eq. (6.67) the expression $(U_m-E(I))$ can be neglected in comparison with α_m^2, i.e., at $x_0>0$, $f_0=\text{const}$ can be put, and, at $x_0<0$, $f_0=0$. In this case, from Eq. (6.73) it follows that

$$f=\frac{f_0}{2}\{1+\Phi(p)\}\,,\qquad \Phi(p)=\frac{2}{\sqrt{\pi}}\int_0^p e^{-t^2}dt\,,\qquad p=x/2\sqrt{I_m\tau}\,. \tag{6.74}$$

Equation (6.74) describes the intensity variation within the duct and out of it when $0\leqslant\tau\lesssim\tau_0$.

In the terminal of the capture region, $\tau=\tau_0$, by Eq. (6.74), we have for the intensity distribution in the duct

$$f=\frac{f_0}{2}\left\{1-\Phi\left(\frac{I_m-I}{2\sqrt{I_m\tau_0}}\right)\right\}. \tag{6.75}$$

This equation is true provided the argument of the function Φ is great enough, $p>1$, i.e., if

$$\tau_0<I_m/4\,. \tag{6.76}$$

Otherwise, the condition for applicability of Eq. (6.71) is not satisfied.

The total power trapped in the duct is given by

$$\Delta W=c\int_0^{I_m} f\,dI=\frac{f_0c}{2}\left\{I_m\left(1-\Phi\left(\sqrt{\frac{I_m}{4\tau_0}}\right)\right)+\frac{2\sqrt{I_m\tau_0}}{\sqrt{\pi}}\left(1-e^{-I_m/4\tau_0}\right)\right\}$$

$$\approx\frac{cf_0\sqrt{I_m\tau_0}}{\sqrt{\pi}}\,. \tag{6.77}$$

If the condition inverse to Eq. (6.76) is satisfied, then for any I in the duct we have

$$f\approx f_0\,,\qquad \Delta W\approx cf_0I_m\,.$$

In this case, the duct is filled up at most.

The value of τ_0 under ionospheric conditions is, by Eq. (6.69), $\tau_0\approx3\cdot10^{-2}-3\cdot10^{-4}$. Hence, Eq. (6.76) is violated for very narrow ducts only. We note that near the boundary I_m, a real duct profile differs strongly from the model parabolic one. Therefore, to investigate an actual situation in the capture region, a supplementary analysis of the precise Eq. (6.56) is needed; it is carried out below.

The capture coefficient is given by

$$G=\frac{\Delta W}{E_0}\approx\frac{\Delta\alpha}{\bar{\alpha}}\,,\qquad \Delta\alpha=\frac{be^{-2\Gamma_m}}{2\Theta_m\alpha_m}\cos\alpha_m\,,$$

$$b=\begin{cases}\sqrt{\dfrac{I_m\tau_0}{\pi}} & \text{at}\quad \tau_0<\dfrac{I_m}{4}\,,\\[2ex] I_m & \text{at}\quad \tau_0>\dfrac{I_m}{4}\,.\end{cases} \tag{6.77a}$$

Here, $\bar{\alpha}=1/D_{\varphi_0}(\alpha_m)$ is the angular width of the antenna directivity diagram, $\Delta\alpha$ is the effective range of capture angles ($\Delta\alpha\ll\bar{\alpha}$ is assumed). Under ionospheric conditions, by Eq. (6.69), $\Delta\alpha\sim(10^{-2}-10^{-4})/\alpha_m$, so the capture coefficient is rather great. Thus, multiple scattering leads to an effective duct excitation.

The capture coefficient increases with decreasing α_m, i.e., as the wave frequency approaches the maximum usable frequency of the layer, e.g., MUF-*E* or MUF-*F*. However, near this frequency, the dependence of f_0 on I in the boundary condition [Eq. (6.67)] is of importance. The analysis of the solution [Eq. (6.73)], with allowance for the dependence of f_0 on I at the wave frequency close to the MUF, i.e., at $\alpha_m\approx 0$, demonstrates that when Eq. (6.76) is satisfied, the trapped energy increases roughly by a factor of $\sqrt{\Theta_m}\alpha_m/(I_m\tau_0)^{1/4}$ compared to the total energy [Eq. (6.77)]. Hence, Eqs. (6.77) and (6.77a) are true only at $\alpha_m\gtrsim\alpha_{mk}=(I_m\tau_0)^{1/4}/\sqrt{\Theta_m}$. At $\alpha_m\lesssim\alpha_{mk}$, the capture coefficient, $G\approx\Delta\alpha/\bar{\alpha}$, $\Delta\alpha\approx(I_m\tau_0)^{1/4}\Theta_m^{-1/2}\exp(-2\Gamma_m)$ (Gurevich and Tsedilina 1980).

Duct Wave Mode Distribution of Intensity. Let us consider the propagation region, i.e., suppose that there is no radiation beyond the duct, and thus, Eq. (6.61) is valid at its boundary. Let us examine the character of the adiabatic invariant distribution of the wave field. The solution of Eq. (6.64) can be represented in the form of a series

$$f(\theta,I)=\sum_k C_k e^{-\lambda_k\tau}f_k(I)\,. \tag{6.78}$$

For the functions $f_k(I)$, from Eq. (6.64) the equation follows

$$\frac{d^2f_k}{dI^2}+\frac{1}{I}\frac{df_k}{dI}+\frac{\lambda_k}{I}f_k=0\,. \tag{6.79}$$

Its solution satisfying Eq. (6.60) is given by

$$f_k=\mathscr{J}_0(2\sqrt{\lambda_k I})\,, \tag{6.80}$$

where $\mathscr{J}_0$ is the Bessel function of the zero order, λ_k are constants defined by the boundary condition [Eq. (6.61)]

$$2\sqrt{\lambda_k I_m}=\beta_k\,,\qquad \lambda_k=\frac{\beta_k^2}{4I_m}\approx\frac{\pi^2}{4I_m}\left(k-\frac{1}{4}\right)^2. \tag{6.81}$$

Here, β_k are the zeroes of $\mathcal{J}_0(z)$: $\beta_1 = 2.4048$; $\beta_2 = 5.5201$, and so on. The value of λ_k sharply increases with k. This means that as the wave propagates in the duct, its high harmonics ($k \gg 1$) rapidly dissipate, so the first harmonics: $f_1, f_2, \ldots$, predominate.

The principal harmonic

$$f_1 = C_1 \mathcal{J}_0(\beta_1 \sqrt{I/I_m}) \tag{6.82}$$

gives the most stable, "quasistationary" adiabatic invariant distribution of energy, or characterizes the relative extent of mode generation in the ionospheric waveguide. The quasistationary distribution [Eq. (6.82)] is shown in Fig. 6.11a.

The energy of a wave travelling in the duct dissipates due to the ray absorption and ray release from the duct. The released energy flux is given by

$$S = -\left(I \frac{\partial f}{\partial I}\right)_{I_m} \cdot 2 D_0 \Theta_0 c R_0 . \tag{6.83}$$

Here, c is, as usual, the velocity of light. For the principal harmonic [Eq. (6.82)] we have

$$S = 1.25\, C_1 D_0 \Theta_0 c R_0 . \tag{6.84}$$

The decrease of energy due to its release is described by the factors $\exp(-\lambda_k \tau)$. The decrease is important for the k-th harmonic, if

$$\lambda_k \tau = \frac{\beta_k^2}{2} \, \frac{D_0 \Theta_0 R_0}{I_m} \, \theta \gtrsim 1 . \tag{6.85}$$

Taking Eq. (6.69) into account, we see that under ionospheric conditions all over the RTW path, $\theta = 2\pi$, a noticeable attenuation is also possible of the principal harmonic.

Variation in Intensity Distribution during Wave Propagation in the Duct. As the wave propagates in the parabolic duct F, its initial adiabatic invariant distribution [Eq. (6.66)], due to multiple scattering, transforms gradually into the quasistationary distribution [Eq. (6.82)]. The same takes place for a wave in the FE-duct. The transformation process is described by the solution of Eq. (6.64) with boundary conditions [Eqs. (6.60) and (6.61)], the solution has the form [Eqs. (6.78) – (6.81)]

$$f(\tau, I) = \sum_{k=1}^{\infty} C_k \mathcal{J}_0\left(\beta_k \sqrt{\frac{I}{I_m}}\right) \exp\left[-\frac{\beta_k^2}{4 I_m} \tau^*\right] . \tag{6.86}$$

Here, $\tau^* = \tau$ for the F-duct, and $\tau^* = \tau - \tau_0$ for the FE-duct, τ_0 is the instant of the capture completion [Eq. (6.68)]. Thus, we only need to find the coefficients C_k. Using the orthogonality property of Bessel functions, we obtain

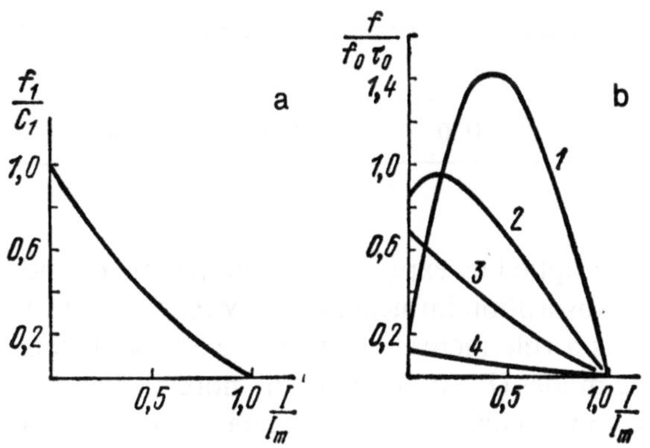

Fig. 6.11. a Quasistationary distribution of the wave energy in adiabatic invariants. **b** Distribution of energy at various values of the parameter $b = (\tau - \tau_0)/4I_m$. The curves *1, 2, 3, 4* refer to the values of $b = 0.05$; 0.1; 0.2; 0.5, respectively

$$C_k = \frac{1}{q}\int_0^{I_m} f_0(I)\,\mathscr{I}_0\left(\beta_k\sqrt{\frac{I}{I_m}}\right)dI = \frac{2I_m}{q}\int_0^1 f_0(x^2)\,\mathscr{I}_0(\beta_k x)\,x\,dx\,,$$
$$q = 2I_m\int_0^1 \mathscr{I}_0^2(\beta_k x)\,x\,dx = I_m\mathscr{I}_1^2(\beta_k)\,. \tag{6.87}$$

Here, $f_0(I)$ is the initial distribution [Eq. (6.66)] for the *F*-duct, and $f_0(I) = f_0(\tau_0, I)$ for the *FE*-duct.

Let us consider, for example, the transformation process for the initial distribution in the duct *FE*, when scattering-induced capture occurs and the function $f_0(I)$ is defined by Eq. (6.75), i.e.,

$$f_0(x^2) \approx \frac{f_0}{2}\left\{1 - \Phi\left(\frac{1-x^2}{a}\right)\right\}, \quad a = 2\sqrt{\frac{\tau_0}{I_m}}\,.$$

When Eq. (6.76) is valid, $a \ll 1$. Taking this into account, we can carry out the integration in Eq. (6.87). We have

$$C_k = \frac{C_{k0}}{2I_m}\tau_0 f_0\,, \quad C_{k0} = \frac{\beta_k}{2\mathscr{I}_1(\beta_k)}\,. \tag{6.88}$$

A few first coefficients C_{k0} are listed in Table 6.2. The distribution $f(I, \tau)/f_0\tau_0$ at various values of the parameter $(\tau - \tau_0)/4I_m$, defined by Eqs. (6.86) and (6.88), is shown in Fig. 6.11b. The figure shows that the initial distribution $f_0(I) = f(I)\,|_{\tau=\tau_0}$ dissipates and degenerates rather rapidly, tending to the "quasistationary" distribution [Eq. (6.82)]. The amplitude of the latter is $C_1 = 2.23\,f_0\tau_0$. A greater part of the trapped energy [Eq. (6.77)] in a horizontally – homogeneous duct decreases rapidly due to its release from the duct.

Above the effect of wave absorption on the wave energy distribution in the duct was ignored. This is correct if the condition is satisfied

$$I_m|K(I_m) - K(0)| \lesssim 1\,, \tag{6.89}$$

where $K = K(I)$ is the averaged wave absorption coefficient [Eq. (6.55)]. In particular, if $K = \text{const}$, then the distribution function $f(I)$ in all the expres-

Table 6.2. Coefficients C_{k0}

k	1	2	3	4	5
C_{k0}	2.23	−8.14	16.0	−25.4	36.2

sions obtained above is simply multiplied by $\exp(-2K\tau)$. On the other hand, if Eq. (6.89) is violated, then the absorption influences greatly the form of the distribution function. Since, K as a rule increases with I (see Sec. 4.1), then due to absorption, nearly sliding modes ($I \to 0$) become still more distinct: the function $f(I)$ drops faster when approaching the duct boundary. This leads to weakening of the energy release effect (for detail, see Gurevich and Tsedilina 1976).

Region near the Duct Boundary. The parabolic duct profile approximation is unapplicable in the vicinity of the duct boundary, $I \approx I_m$, where the period $\Theta(I)$ has a logarithmic singularity

$$\Theta(I)\,|_{I\to I_m} \sim \ln\left(\frac{I_m}{I_m-I}\right). \tag{6.90}$$

Therefore, let us discuss some features of $f(I)$ near the boundary, $I = I_m$, with allowance for Eq. (6.90). Here, the solution of Eq. (6.64) is given by

$$D\Theta\frac{\partial f}{\partial I} = S_0\,,$$

where S_0 is a constant. Hence, taking Eq. (6.90) into account, we obtain

$$f \approx C_0\frac{I_m-I}{\ln[(I_m/I)-1]^{-1}}\,. \tag{6.91}$$

Thus, the intensity distribution near the boundary increases slower than in the case of the parabolic duct, when $f \sim (I_m - I)$.

Eqation (6.72) describing the capture into the duct now takes the form

$$\frac{\partial f}{\partial \tau} = I_m\frac{\partial}{\partial x}\left[\ln\frac{C_0}{|x|}\,\frac{\partial f}{\partial x}\right]. \tag{6.92}$$

Its solution coincides to a logarithmic accuracy with Eq. (6.74), it is only the argument of the function $\Phi(p)$ that changes,

$$p = \frac{x}{2\sqrt{\tau I_m}\sqrt{\ln[C_0/(I_m-I)]}} = \frac{I_m-I}{2\sqrt{\tau I_m}\sqrt{\ln[C_0/|I_m-I|]}}\,. \tag{6.93}$$

The capture is seen to be somewhat intensified owing to the logarithmic divergence [Eq. (6.93)].

Phase Oscillation and Spreading. It is only the adiabatic invariant distribution of the wave field that was considered above. But at the same time, of great interest is also the field distribution in the oscillation phase Φ (in particular, for the establishment of the communication channel, see Sec. 3.4). Indeed, the transmitter produces the initial distribution of a field which has a δ-function type with respect to Φ [Eq. (6.48)]. How does this distribution diffuse during wave propagation in the duct? First of all, multiple scattering leads to the phase diffusion which results in a dispersal of the initial δ-function. This process is described by Eq. (6.61c). For the parabolic duct it takes the form (Gurevich and Tsedilina 1979)

$$\frac{\partial f}{\partial \tau} = \frac{\partial}{\partial I}\left(I\frac{\partial f}{\partial I}\right) + \frac{1}{2I}\frac{\partial^2 f}{\partial \Phi^{*2}}, \qquad f = f(\tau, I, \Phi^*), \quad \Phi^* = \Phi - 2\pi\tau/\tau_0. \tag{6.94}$$

Its solution is naturally to be sought by the Fourier expansion with respect to Φ^*

$$f = f_0(\tau, I) + \sum_{p=1}^{\infty} [f_p^{\mathrm{s}}(\tau, I)\sin p\Phi^* + f_p^{\mathrm{s}}(\tau, I)\cos p\Phi^*]. \tag{6.95}$$

Then for the function $f_0(\tau, I)$ describing the adiabatic invariant distribution of the wave energy, the averaged Eq. (6.64) results, as would be expected. For harmonics $f_p^{\mathrm{s},c}$ describing the oscillation phase distribution, we have

$$\frac{\partial f_p^{\mathrm{s},c}}{\partial \tau} = \frac{\partial}{\partial I}\left(I\frac{\partial f_p^{\mathrm{s},c}}{\partial I}\right) - \frac{p^2}{2I} f_p^{\mathrm{s},c}. \tag{6.96}$$

The solution of this equation in the propagation region [Eqs. (6.60) and (6.61)] can be easily found by separation of variables [Eq. (6.78)]. In this case, the harmonics prove to be damped rather rapidly. The most slow damped harmonic is the first one, for which we have

$$f_1^{\mathrm{s},c} \sim e^{-\lambda_{11}\tau}, \quad \lambda_{11} = \frac{\beta_{11}^2}{4I_m}, \quad \beta_{11} = 3.83. \tag{6.97}$$

Comparing Eq. (6.97) with Eqs. (6.81) and (6.85), we see that the damping rate for this first harmonic exceeds the corresponding rate for the principal harmonic [Eq. (6.82)] by a factor of 2.5. Consequently, under conditions of Eq. (6.69), all over the RTW path, the initial δ-function distribution with respect to the oscillation phase Φ is to be strongly dispersed. This is seen from Fig. 6.12 where the radiation flux from the duct is shown $S_1 = SS_2$, $S_2 = 4R_0I_m/D_0\Theta c\tau_0 f_0$. Here S is defined by Eq. (6.83). Note that the adiabatic invariant distribution of harmonics, $f_p(I)$, in the propagation region is given by Bessel functions of $p-$th order. On the duct axis, when $I \to 0$, all of them become zero. The only nonvanishing harmonic here is the zero one, $f_0(I)$ (see Fig. 6.11a). This means that close to the sliding mode, the distribution with respect to Φ rapidly flattens, which is also seen from Eq. (6.94).

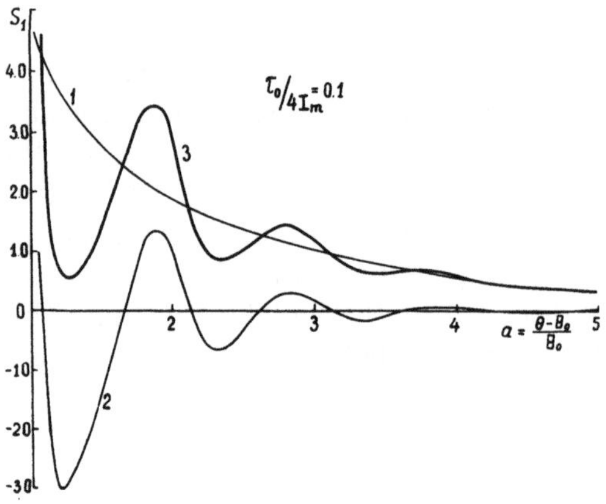

Fig. 6.12. Emission flux S_1 from the duct vs distance: *1* with no phase dispersal; *2* variation of the flux due to the phase dispersal; *3* the total flux

We note also another process of flattening of the phase distribution which is connected with the diffusion over adiabatic invariants. Let us bear in mind the fact that the ray oscillation period in an actual duct depends on the invariant I, the dependence is especially strong near the duct boundary [Eq. (6.90)]. This means that the oscillation phase given initially to be equal for different I [Eq. (6.48)], after a few ray oscillations, becomes strongly different at different I: $\Delta\Phi(I) \sim p\Delta\Theta(I)$, where p is the number of oscillations, $\Delta\Theta = \Theta(I_1) - \Theta(I_2)$ is the difference between periods. However, scattering induces a noticeable redistribution of wave field in adiabatic invariants. Therefore, it leads at the same time to a spreading in the phase distribution as well. The spreading is especially effective in the vicinity of the duct boundary, i.e., for rays entering the duct or leaving it. This circumstance can be also important for the establishment of the communication channel (see Sec. 3.4).

Large-Scale Irregularities. In the present subsection we consider an influence on the distribution of wave field over adiabatic invariants and phases of large-scale irregularities (i.e., irregularities whose scales can be compared with the oscillation period Θ or a typical scale of the regular ionospheric height inhomogeneity) (Borisov 1978, 1980; Gurevich and Borisov 1982). Actually such irregularities always occur in the neighborhood of the terminator, travelling ionospheric disturbances (TID) are included in this group, etc.

Assuming Eqs. (6.49) and (6.50) to be satisfied, we shall consider here only the scattering with respect to vertical angles α. Equation (6.34) for the distribution function $f(\theta, z, \alpha)$ is written then in the form

$$\frac{1}{R_0}\frac{\partial f}{\partial \theta} + \alpha\frac{\partial f}{\partial z} - \frac{1}{2}\frac{\partial U}{\partial z}\frac{\partial f}{\partial \alpha} = 0\,. \tag{6.98}$$

Next, let an irregularity $\Delta N_e(z, \theta)$ be present in a horizontally – homogeneous ionosphere $N_e(z)$. The effective potential $U(z, \theta)$ is composed then of the regular part $U_0(z)$ and a small perturbation $\tilde{U}(z, \theta)$. The potential $U_0(z)$ refers to the horizontally – homogeneous ionosphere, and $\tilde{U}(z, \theta)$ refers to the large-scale irregularity. Let us present the distribution function $f(z, \theta, \alpha)$ in the same manner as a sum of the undisturbed part f_0 and a small additive $\tilde{f}$. The equation for $\tilde{f}(z, \theta, \alpha)$ can be easily obtained from Eq. (6.98)

$$\frac{1}{R_0}\frac{\partial \tilde{f}}{\partial \theta}+\alpha\frac{\partial \tilde{f}}{\partial z}-\frac{1}{2}\frac{\partial U_0}{\partial z}\frac{\partial \tilde{f}}{\partial \alpha}=\frac{1}{2}\frac{\partial \tilde{U}}{\partial z}\frac{\partial f}{\partial \alpha}. \tag{6.99}$$

In a zero-order approximation with respect to the small perturbation, the distribution function f in the right-hand side of Eq. (6.99) should be replaced by its undisturbed part, f_0. In certain cases, however, the terms of higher orders must be also involved.

Let us go over in Eq. (6.99), as well as in Eq. (6.47), to new variables: the adiabatic invariant I and oscillation phase Φ. Then Eq. (6.99) can be rewritten in the form

$$\frac{\partial \tilde{f}}{\partial \theta}+\frac{2\pi}{\Theta}\frac{\partial \tilde{f}}{\partial \Phi}=\alpha R_0\frac{\partial \tilde{U}}{\partial z}\left(\Theta\frac{\partial f}{\partial I}+\frac{\partial \Phi}{\partial E}\frac{\partial f}{\partial \Phi}\right). \tag{6.100}$$

To obtain simple analytical results, we consider now the parabolic duct [Eq. (6.62)]. For it, by Eq. (6.63), we have

$$\alpha=\left(\frac{I}{\Theta}\right)^{1/2}\sin\Phi\,, \qquad \frac{\partial \Phi}{\partial E}=\frac{\Theta}{2I}\cot\Phi\,. \tag{6.101}$$

We assume that the irregularity ΔN_e results from a variation in the sliding height, i.e., it is connected with the duct bending

$$U=U(z_s)+\frac{U''(z_s)}{2}(z-z_s)^2+\tilde{U}(z,\theta)\,,$$
$$\tilde{U}(z,\theta)=\frac{b(\theta)}{R_0}(z-z_s)\,. \tag{6.102}$$

Next, let the wave field in the duct be described by the distribution function $f_0(\theta, I, \Phi)$. When passing through the irregularity, the latter changes by the quantity $\tilde{f}$, which, by Eqs. (6.100) – (6.102), is given by

$$\tilde{f}=\int_{\theta_0}^{\theta} d\theta'\, b(\theta')\left\{\sqrt{I\Theta}\sin\left[\Phi-\frac{2\pi(\theta-\theta')}{\Theta}\right]\frac{\partial f_0}{\partial I}\right.$$
$$\left.+\frac{1}{2}\sqrt{\frac{\Theta}{I}}\cos\left[\Phi-\frac{2\pi(\theta-\theta')}{\Theta}\right]\frac{\partial f_0}{\partial \Phi}\right\}. \tag{6.103}$$

It is easy to see that being independent of θ', the function f_0 in Eq. (6.103) can be taken out of the integral resulting in a simple form for Eq. (6.103)

$$\tilde{f}=\frac{\partial f_0}{\partial I}\Delta I+\frac{\partial f_0}{\partial \Phi}\Delta\Phi\,. \tag{6.104}$$

Here, ΔI and $\Delta\Phi$ represent variations in the adiabatic invariant and oscillation phase caused by the irregularity. In particular, for an irregularity of the Gaussian type

$$b(\theta) = b_0 \exp\left[-\left(\frac{\theta-\theta^*}{\Delta\theta}\right)^2\right], \tag{6.105}$$

we obtain

$$\begin{aligned} \Delta I &= 2\sqrt{\pi I\Theta}\, b_0 \Delta\theta \exp\left[-\left(\frac{\pi\Delta\theta}{\Theta}\right)^2\right] \sin\left[\Phi - \frac{2\pi(\theta-\theta^*)}{\Theta}\right] \\ \Delta\Phi &= \sqrt{\frac{\pi\Theta}{I}}\, b_0 \Delta\theta \exp\left[-\left(\frac{\pi\Delta\theta}{\Theta}\right)^2\right] \cos\left[\Phi - \frac{2\pi(\theta-\theta^*)}{\Theta}\right]. \end{aligned} \tag{6.106}$$

The expressions obtained [Eq. (6.106)] show that the perturbations ΔI and $\Delta\Phi$, at the beginning, increase with the dimensionless irregularity scale, $\Delta\theta$, up to the magnitudes $\Delta\theta_k \approx \Theta/\sqrt{2}\pi$, i.e., in real lengths, up to the scales $l_k \approx R_0\Theta/\sqrt{2}\pi \sim 500-1000$ km; after this they decrease. If the irregularity scale exceeds the quantity l_k (the latter being of the order of the oscillation scale), the perturbations ΔI and $\Delta\Phi$ drop exponentially rapidly. This is completely consistent with the adiabatic approximation [Eqs. (3.17), (3.18) and (3.20)]. Of interest is also the fact that the perturbation magnitude alters periodically depending on the irregularity location, $\theta = \theta^*$, with respect to the wave trajectory. This effect of periodical "sway" of rays was indicated in Kinber et al. (1979).

Let us draw our attention to the situation, when random irregularities, $\tilde{U}(z,\theta)$, are distributed along the path; we assume that their mean value is zero: $\langle \tilde{U}(z,\theta)\rangle = 0$. Then in the equation for $\langle \tilde{f}\rangle$, the first-order small terms in the right-hand side of Eq. (6.100) become zero. To take into account the next-order effects, we reduce the equation for the distribution function to an integral equation and find the quantity

$$\begin{aligned} R_0\left\langle \frac{\partial\tilde{U}}{\partial z}f\right\rangle &= \sqrt{I\Theta}\int_{\theta_0}^{\theta} \rho(\theta-\theta')\sin\left[\Phi - \frac{2\pi(\theta-\theta')}{\Theta}\right]\left\langle\frac{\partial f}{\partial T}\right\rangle d\theta' \\ &+ \frac{1}{2}\sqrt{\frac{\Theta}{I}}\int_{\theta_0}^{\theta}\rho(\theta-\theta')\cos\left[\Phi - \frac{2\pi(\theta-\theta')}{\Theta}\right] \\ &\cdot \left\langle\frac{\partial f}{\partial\Phi}\right\rangle d\theta'\,. \end{aligned} \tag{6.107}$$

Here, $\rho = R_0\left\langle \frac{\partial\tilde{U}}{\partial z}(\theta,z)\cdot\frac{\partial\tilde{U}}{\partial z}(\theta',z')\right\rangle$ is the correlation function. For simplicity, it has been assumed to depend on the longitudinal coordinate θ

only. Equation (6.107) corresponds to a so-called ladder approximation (Rytov et al. 1978). Substituting Eq. (6.107) into the equation for the distribution function, we obtain

$$\frac{\partial\langle f\rangle}{\partial\theta}+\frac{2\pi}{\Theta}\frac{\partial\langle f\rangle}{\partial\Phi}=\sqrt{I\Theta}R_0\sin\Phi\cdot\frac{\partial}{\partial I}\sqrt{I\Theta}\int_{\theta_0}^{\theta}\rho(\theta-\theta')$$
$$\cdot\sin\left[\Phi-\frac{2\pi(\theta-\theta')}{\Theta}\right]\cdot d\theta'\left\langle\frac{\partial f}{\partial I}\right\rangle+\frac{R_0}{2}\Theta$$
$$\cdot\cos\Phi\frac{\partial}{\partial\Phi}\int_{\theta_0}^{\theta}\rho(\theta-\theta')\sin\left[\Phi-\frac{2\pi(\theta-\theta')}{\Theta}\right]$$
$$\cdot\left\langle\frac{\partial f}{\partial I}\right\rangle d\theta'+\frac{R_0}{2}\sqrt{I\Theta}\sin\Phi\frac{\partial}{\partial I}\sqrt{\frac{\Theta}{I}}$$
$$\cdot\int_{\theta_0}^{\theta}\rho(\theta-\theta')\cdot\cos\left[\Phi-\frac{2\pi(\theta-\theta')}{\Theta}\right]$$
$$\cdot\left\langle\frac{\partial f}{\partial\Phi}\right\rangle d\theta'+\frac{R_0}{4}\frac{\Theta}{I}\cos\Phi\frac{\partial}{\partial\Phi}$$
$$\cdot\int_{\theta_0}^{\theta}\rho(\theta-\theta')\cos\left[\Phi-\frac{2\pi(\theta-\theta')}{\Theta}\right]$$
$$\cdot\left\langle\frac{\partial f}{\partial\Phi}\right\rangle d\theta'\ . \tag{6.108}$$

Equation (6.108) is of a diffusion type. It describes variations in adiabatic invariant and phase distributions, which are induced by random large-scale irregularities. The solution of Eq. (6.108) is convenient to be sought in the form of a series

$$f=f_0(\theta,I)+\sum_{p=1}^{\infty}\left[f_p^c(I,\theta)\cos(p\Phi)+f_p^s(I,\theta)\sin(p\Phi)\right]\ . \tag{6.109}$$

The substitution of Eq. (6.109) in Eq. (6.108) leads to a chain of equations for the coefficients $f_p(I,\theta)$. It is easy to show that as well as in the preceding subsection, all the coefficients at $p\neq 0$ rapidly damp along the path. For the zero harmonic (after the damping of all the higher ones) we obtain a simple equation

$$\frac{\partial f_0}{\partial\theta}=\frac{\partial}{\partial I}\left(D\Theta\frac{\partial f_0}{\partial I}\right)$$
$$D=\frac{R_0 I}{4}\int_0^{\infty}\rho(\xi')\cos\frac{2\pi\xi'}{\Theta}d\xi'\ . \tag{6.110}$$

Here, we have taken into account that in the parabolic duct the oscillation period Θ is independent of I. Besides, the function $f_0(I, \Theta)$ has been assumed to change slightly over the correlation length $a_\parallel$ for the function $\rho(\xi')$. This allowed the function $f_0(I, \theta)$ to be taken out of the integral.

For irregularities with scales $a_\parallel$ which are small compared to the oscillation period $R_0\Theta$, in the expression for D [Eq. (6.110)] the factor $\cos(2\pi\xi'/\Theta)$ can be omitted. Then Eq. (6.110) becomes identical to Eq. (6.64), considered above. We emphasize that here also the diffusion coefficient

$$D^* = I\frac{R_0}{8}\frac{\partial^2}{\partial r_\perp^2}\int_{-\infty}^{\infty}\langle\tilde{U}(r_1)\tilde{U}(r_2)\rangle d(\theta_1-\theta_2)\,, \qquad r_\perp = (r_1-r_2)_\perp$$

identically coincides with that in Eqs. (6.54) and (6.64).

For very large-scale irregularities, however, we must take into account in Eq. (6.110) the factor $\cos(2\pi\xi/\Theta)$ which plays here a role of a cut-off factor. This factor, in particular, results in a sharp diminishing of $D(I)$ on gradually varying perturbations, when $a_\parallel > R_0\Theta$. This fact is a consequence of the adiabatic invariant conservation. Thereby, the ray diffusion equation [Eqs. (6.54) and (6.110)] proves to be valid for random irregularities of various scales.

In conclusion, we make specific evaluations for deviations from the adiabatic approximation, which can originate as the wave travels through real large-scale ionospheric irregularities. First, we consider the terminator. As before, we shall use the parabolic duct model. The change of the duct height, Δz_s, connected with the termintor is approximated by

$$\Delta z_s = a_0\tanh\frac{\theta-\theta^*}{\Delta\theta} \tag{6.111}$$

where the parameters a_0 and $\Delta\theta$ characterize the magnitude of bending and the width of the transient region, respectively. In this model, we have for the oscillation amplitudes of ΔI, $\Delta\Phi$

$$\Delta I_{\max} = \Delta\Phi_{\max} = \frac{\sqrt{I\Phi}}{2}\frac{a_0R_0}{\rho_s^4k_0^2}\frac{\Delta\theta}{\sinh\dfrac{\pi^2\Delta\theta}{\Theta}}$$

$$\rho_s = [k_0^2U''/2]^{-1/4}\,. \tag{6.112}$$

The variation in adiabatic invariant, ΔI, can be readily associated with the number of duct eigenmodes, Δn, excited in the transient region [Eq. (6.111)]

$$\Delta I = \frac{4\pi}{k_0R_0}\Delta n\,.$$

Hence, [Borisov 1978, 1980]

$$\Delta n \approx \frac{a_0}{\rho_s}\sqrt{2n}\frac{\pi^2}{2}\frac{\Delta\theta}{\Theta}\frac{1}{\sinh\dfrac{\pi^2\Delta\theta}{\Theta}}\,. \tag{6.113}$$

For large-scale travelling disturbances, one can make use of the Gaussian irregularity model [Eq. (6.105)]. In this case, using Eq. (6.106), we obtain for the number of excited modes

$$\Delta n = b_0 \Delta\theta k_0 \rho_s \sqrt{2n} \exp\left[-\left(\frac{\pi \Delta\theta}{\Theta}\right)^2\right]. \tag{6.114}$$

Next, using the maps of global distribution of the parameters in IWD (Tushentsova et al. 1977c), we obtain typical values for the parameters involving in Eq. (6.112)

$$a_0 \approx 30\,\text{km}\,, \qquad \Delta\theta \cdot R_0 \approx 1.5 \cdot 10^3\,\text{km}\,,$$

$$\rho_s \approx 1\,\text{km}\,, \qquad \Theta R_0 \approx (2-2.5) \cdot 10^3\,\text{km}\,.$$

Inserting these values into Eq. (6.113), we estimate the number of excited modes in the terminator case as $\Delta n \approx 10-40$. Similar estimations using data of Kulikova et al. 1980 for large-scale travelling disturbances give a close result, $\Delta n \approx 10-30$.

Both these estimations demonstrate that in deep ducts, at smooth mode distribution of energy, the errors resulting from neglect of variations of adiabatic invariants on large-scale irregularities are relatively small, $\Delta I/I \approx \Delta n/n \sim 1/\sqrt{n}$. At the same time, for small duct depths or a small number of excited duct levels, the allowance for a mode redistribution of energy associated with large-scale irregularities becomes important. This is completely in agreement with the numerical results given in Sec. 5.3.

6.3 Signal Fluctuations in the Reception Point

Radio wave scattering induces intense wave field fluctuations in the reception point. Fluctuations of bearing, angles of arrival, field amplitude, and pulse time delay, as well as the correlation of the fluctuations in separated reception points, etc. form a broad class of effects actually observed. A theory of fluctuations for long-distance radio wave propagation is at present in development. It is based on the general theory of radio wave propagation in random media (Tatarskii 1967, Rytov 1966, Chernov 1958, Rytov et al. 1978). However, the ionospheric propagation is unique enough, and in order that the observational data be adequately interpreted, a further development of the theory is necessary.

In the present Section we restrict ourselves to the consideration of fluctuations of bearing alone.

Note that above (Sec. 6.2) the effects connected with the scattering-induced diffusion of rays with respect to adiabatic invariant and oscillation phase were considered. A scattering in the horizontal plane has been ignored [Eqs. (6.49) – (6.51)]. Now, on the contrary, we consider the ray diffusion over horizontal angles, which leads to a beam broadening, distortion of azimuthal angles of arrival, and intensification of fluctuations of bearing (Gurevich and Tsedilina 1983).

Let, for simplicity, the distributions in the beam over adiabatic invariants I and phases Φ be sufficiently smooth, so that the diffusion in I and Φ is unimportant [as in Eqs. (6.49) and (6.50)]. Then Eq. (6.47) describes a beam distribution in horizontal angles φ and azimuths β. With allowance for Eq. (6.39), this equation is rewritten in the form

$$\frac{\partial f}{\partial \theta}+\frac{\beta_1}{\sin^2\theta}\frac{\partial f}{\partial \varphi}+\frac{1}{2}\frac{\partial \varepsilon}{\partial \varphi}\frac{\partial f}{\partial \beta_1}=\frac{\partial}{\partial \beta}\left(D\frac{\partial f}{\partial \beta}\right)$$

$$\beta_1=\frac{\beta}{\cot\theta_0}\,. \tag{6.115}$$

Here, D is the diffusion coefficient with respect to β

$$D=R_0D_{\beta\beta}=R_0D_0\,. \tag{6.116}$$

The boundary condition [Eq. (6.41)] takes the form

$$f=f_0(\theta_0,\varphi_0,\beta_0)=\frac{W_0}{c}D(\varphi_0)\,\delta(\beta_0)\,. \tag{6.117}$$

Here, $D(\varphi_0)$ is the antenna directivity diagram with respect to φ_0. The angle β_0 is read off from the direction of φ_0.

Let us consider, for simplicity, a spherically-symmetric ionosphere ($\partial\varepsilon/\partial\varphi=0$) and assume, as in Eqs. (6.64) and (6.65), that $D_{\beta\beta}=D_0=\text{const}$. In this case, Eq. (6.98) is rewritten in the form

$$\frac{\partial f}{\partial \theta}+\frac{\beta_1}{\sin^2\theta}\frac{\partial f}{\partial \varphi}=D_{10}\frac{\partial^2 f}{\partial \beta_1^2}\,, \tag{6.118}$$

$$D_{10}=\frac{D}{\cot^2\theta_0}\,. \tag{6.119}$$

In Eq. (6.118) we have gone over to the angle β_1. Then the boundary condition [Eq. (6.117)] takes the form

$$f_0(\theta_0,\varphi_0,\beta_0)=\frac{W_0}{c}D(\varphi_0)\frac{\delta(\beta_{10})}{\cot\theta_0}\,. \tag{6.120}$$

The solution of Eq. (6.118) is naturally to be sought in the form of the Fourier integral

$$f=\frac{1}{(2\pi)^2}\int e^{i(k_\varphi\varphi+k_{\beta_1}\beta_1)}f_k(\theta,k_\varphi,k_{\beta_1})\,dk_\varphi dk_{\beta_1}\,. \tag{6.121}$$

Taking Eq. (6.121) into account, we obtain from Eq. (6.118)

$$\frac{\partial f_k}{\partial \theta}-\frac{k_\varphi}{\sin^2\theta}\frac{\partial f_k}{\partial k_{\beta_1}}=-D_{10}k_{\beta_1}^2 f_k\,. \tag{6.122}$$

The characteristic equations for Eq. (6.122) are given by

$$\frac{dk_{\beta_1}}{d\theta} = -\frac{k_\varphi}{\sin^2\theta}, \quad \frac{df_k}{d\theta} = -D_{10}k_{\beta_1}^2 f_k .$$

Hence,

$$\begin{aligned} k_{\beta_1} &= k_{\beta_{10}} + k_\varphi(\cot\theta - \cot\theta_0) \\ f_k &= f_{k_0}, \exp\left[-D_{10}\int_{\theta_0}^{\theta}(k_{\beta_1})^2 d\theta\right] . \end{aligned} \tag{6.123}$$

Here, f_{k_0} is a function of $k_{\beta_{10}}$ and k_{φ_0}, specified in the initial point $\theta = \theta_0$. To determine f_{k_0}, we use the boundary condition [Eq. (6.120)] and obtain

$$\begin{aligned} f_{k_0} &= \int e^{-i(k_{\beta_{10}}\beta_{10} + k_\varphi\varphi_0)} f_0(\varphi_0, \beta_{10})\, d\varphi_0 d\beta_{10} \\ &= \frac{W_0}{c\cdot\cot\theta_0}\int D_0(\varphi_0)e^{-ik_\varphi\varphi_0}d\varphi_0 . \end{aligned} \tag{6.124}$$

Let us assume, for definiteness, that the original beam has the angle distribution of a Gaussian form

$$D_0(\varphi_0) = \frac{1}{\sqrt{\pi}\, g_0}\, \mathrm{e}^{-\varphi_0^2/g_0^2} . \tag{6.125}$$

The parameter g_0 characterizes here the angle width of the beam. Then, we have

$$f_{k_0} = \frac{W_0}{c\cdot\cot\theta_0}\exp\left(-\frac{k_\varphi^2 g_0^2}{4}\right) . \tag{6.126}$$

Substituting Eqs. (6.126) and (6.123) into Eq. (6.121) and carrying out all the necessary transformations, we obtain finally

$$f(\theta, \beta, \varphi, \theta_0) = \frac{W_0}{4\pi c\sqrt{ab - d^2}}\exp\left\{-\left[\frac{\varphi^2}{4a} + \frac{a\left(\beta + \frac{d}{a}\varphi\right)^2}{4(ab - d^2)}\right]\right\}, \tag{6.127}$$

where

$$\begin{aligned} a &= \frac{g_0^2}{4} + D\left[(\theta - \theta_0)\left(1 - \frac{1}{\cot^2\theta_0}\right) + \left(\frac{1}{\cot\theta_0} - \frac{\cot\theta}{\cot^2\theta_0}\right)\right. \\ &\qquad \left. - \frac{2}{\cot\theta_0}\cdot\ln\frac{\sin\theta}{\sin\theta_0}\right], \\ b &= D(\theta - \theta_0) \end{aligned} \tag{6.128}$$

$$d = -D\left[(\theta-\theta_0) - \frac{1}{\cot\theta_0}\ln\frac{\sin\theta}{\sin\theta_0}\right]. \tag{6.128}$$

In Eqs. (6.127) and (6.128) we have gone over to β and D.

When $\theta_0 \to 0$, Eqs. (6.127) and (6.128) take a simple form

$$f = \frac{W}{2\pi c g_0 \sqrt{D(\theta-\theta_0)}} \exp\left\{ -\frac{\varphi^2}{g_0^2+4D(\theta-\theta_0)} - \frac{\left(\beta+\frac{d}{a}\varphi\right)^2}{\left[\frac{g_0^2 D(\theta-\theta_0)}{g_0^2/4+D(\theta-\theta_0)}\right]} \right\} \tag{6.129}$$

$$a = \frac{g_0^2}{4} + D(\theta-\theta_0), \qquad b = -d = D(\theta-\theta_0).$$

Note that above we have used the Gaussian angle distribution in the original beam [Eq. (6.125)]. The general solution at arbitrary $D_0(\varphi_0)$ can be obtained as well. In particular, when $\theta_0 \to 0$, it has the form

$$f(\theta,\varphi,\beta) = \frac{W_0}{2c\sqrt{\pi D(\theta-\theta_0)}} e^{-\beta^2/4D(\theta-\theta_0)} D_0(\varphi-\beta). \tag{6.130}$$

For Gaussian $D(\varphi_0)$, Eq. (6.130) coincides with Eq. (6.129).

The form of the distribution function [Eqs. (6.128) and (6.129)] shows, first, that due to scattering with respect to horizontal angles, the original beam gradually expands in φ. So, the mean square value throughout the whole beam cross-section, $\overline{\varphi^2}$, is given by

$$\overline{\varphi^2} = \frac{\int \varphi^2 f\, d\varphi d\beta}{\int f\, d\varphi d\beta} = \frac{1}{2}[g_0^2 + 4D(\theta-\theta_0)]. \tag{6.131}$$

Hence, it is seen that the expansion is especially important for narrow original beams, when $g_0^2 < 4D(\theta-\theta_0)$.

A beam expansion arises also with respect to azimuthal angles of arrival (broadening of fluctuations of bearing, β, in the reception point θ, φ). Evaluating the dispersion of bearing, according to Eq. (6.129),

$$D_\beta^2 = \overline{\beta^2} - \bar{\beta}^2,$$

$$\bar{\beta} = \int \beta f\, d\beta / \int f\, d\beta, \qquad \overline{\beta^2} = \int \beta^2 f\, d\beta / \int f\, d\beta, \tag{6.132}$$

we obtain

$$D_\beta = \sqrt{\overline{\beta^2} - \bar{\beta}^2} = \sqrt{\frac{2 g_0^2 D(\theta-\theta_0)}{g_0^2 + 4D(\theta-\theta_0)}}. \tag{6.133}$$

Hence, it is seen that in a broad beam, $g_0^2 \gg 4D(\theta-\theta_0)$

$$D_\beta = \sqrt{2D(\theta - \theta_0)} \tag{6.134}$$

i.e., the dispersion of bearing increases with distance, being proportional to the diffusion coefficient. In a narrow beam, $g_0^2 \ll 4D(\theta - \theta_0)$, the dispersion is given by

$$D_\beta = g_0/\sqrt{2}$$

i.e., the scatter in bearing is established as constant, being determined by the directivity diagram width. Finally, Eq. (6.129) shows that in a scattering medium a shift of the average bearing direction appears, which depends on the distance and diffusion coefficient. Indeed, the peak of bearing and its mean value, $\bar{\beta}$, as is seen from Eqs. (6.129) and (6.132), corresponds not to $\beta = 0$ (i.e., radiation bearing), but to

$$\beta = \bar{\beta} = \frac{d}{a}\varphi\,. \tag{6.135}$$

At greater θ, under conditions $4D(\theta - \theta_0) \geqslant g_0^2$, we have $d/a \to -1$, and mean bearing shift, $\bar{\beta} \to \varphi$. Along the beam axis, $\varphi = 0$, no bearing shift occurs, but in lateral directions, $\varphi \neq 0$, the shift can become significant.

7 Nonlinear Phenomena

It was shown previously that the principal properties of long-distance radio wave propagation are connected with either wave refraction on regular ionospheric horizontal inhomogeneity or wave scattering by ionospheric irregularities. At the same time, both regular and irregular inhomogeneities can be produced in the ionosphere by means of artificial disturbance of the ionosphere with high-power radio waves. In this case, the propagation properties mentioned above are governed by the nonlinear process which depends on the power of the modifying wave. We emphasize that the problem of nonlinear ionospheric disturbances is in itself very involved and far from being completely studied. The goal of the present chapter is to outline the main nonlinear processes which can lead to radio wave capture into or escape from IWD, to wave refraction or scattering on long-range radio paths.

7.1 Regular Nonlinear Refraction

Character of Regular Nonlinear Ionospheric Perturbations. Nonlinear refraction is caused by a perturbation of ionospheric plasma dielectric constant ε. The perturbation of ε is caused mainly by modification in the electron density, N_e.

Regular variations in the electron density, as the electrons are heated by intense radio waves, develop as a result of modification of the ionization balance or expulsion of plasma from the heated region. The former process predominates at rather low heights z, i.e., below 200 km, the latter one is dominant in the F-region and above, i.e., at $z \gtrsim 250$ km.

A modification of the ionization balance leads to an increase in the electron density, $\Delta N_e > 0$. In this case (Gurevich 1978)

$$\frac{\Delta N_e}{N_{e0}} = \gamma_1 \left(\frac{\mathscr{E}_0}{\mathscr{E}_p}\right)^2 \varphi \,. \tag{7.1}$$

Here, N_{e0} is the undisturbed value of the electron density, ΔN_e is perturbation of the electron density, $\mathscr{E}_0$ is the wave electric field amplitude, φ is a combined polarization – temperature factor ($\varphi \approx 1$), $\mathscr{E}_p$ is a characteristic plasma field

$$\mathscr{E}_p = \sqrt{3\, T_{e0}\, m\, \delta_0}\, \frac{\omega}{e} \,, \tag{7.2}$$

where e, m are the electron charge and mass, T_{e0} is the undisturbed electron temperature, δ_0 is the mean fraction of the electron energy which is lost by an electron during one collision (Ginzburg and Gurevich 1960, Gurevich 1978). The factor $\gamma_1(z)$ characterizes the change of the ionization balance. As shown for the lower ionosphere in Fig. 7.1, the factor $\gamma_1(z)$ is close to 0.5 up to $z \sim 160-170$ km and sharply decreases with increasing z to nearly zero when $z \gtrsim 200$ km. Hence, it can be approximated as $\gamma_1 = 0.5$ for $z \leqslant 180$ km and $\gamma_1 = 0$ for $z > 180$ km. Equation (7.1) can be applied for energies up to the value as great as $\mathscr{E} \approx \mathscr{E}_p$. Note that the electron density perturbation [Eq. (7.1)] is established during the electron lifetime, τ_N; in the ionosphere, $\tau_N \sim 10^2$ s.

An important feature of density perturbation at small altitudes is their local dependence on the wave field amplitude: Eq. (7.1) shows that ΔN_e is proportional to $\mathscr{E}^2$, given in the same point z. At greater altitudes, $z \gtrsim 150-200$ km, transport processes, namely, the plasma heat conductivity and diffusion along the Earth's magnetic field, become important. In this case, a variation in the density is no longer determined by the ionization balance, but by hydrodynamical equilibrium. This results in a decrease of the plasma density in the heated volume. In addition, ΔN_e is here nonlocally related to the field amplitude. This is seen from Fig. 7.2, where the density perturbation is shown vs the heights on the axes of beams of various width. ΔN_e is shown to be negative at great altitudes, being strongly dependent on beam dimensions.

Note that under certain conditions, at altitudes $z \sim 150-300$ km, the increase in density is also possible which is induced by an enhancement of the ionizing effectiveness of photoelectrons due to their additional acceleration in the intense radio wave reflection region. This is a stimulated ionization (Vas'kov et al. 1982).

Note that in the upper ionosphere, where $\Delta N_e < 0$, the plasma readily becomes unstable, resulting in irregular density perturbations. Nonlinear effects in such cases will be considered in the final subsection of this section, as well as in the next section.

Thus, when analyzing nonlinear effects connected with regular refraction, we can restrict ourselves to the lower ionosphere with $z \lesssim 200-250$ km, where radio waves produce the increase in the electron density [Eq. (7.1)]. In this region, radio wave beams expand, due to nonlinearity, i.e., they are defocused.

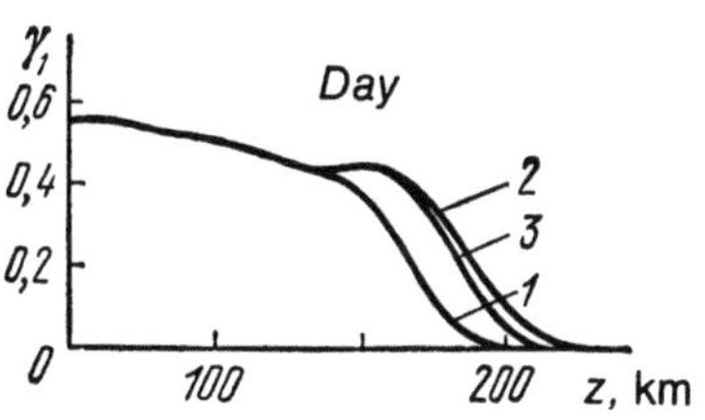

Fig. 7.1. Electron density perturbations on the beam axis in the lower ionosphere. Curves *1, 2, 3* correspond to effective beam witdth $a = 10$, 30, 100 km, respectively

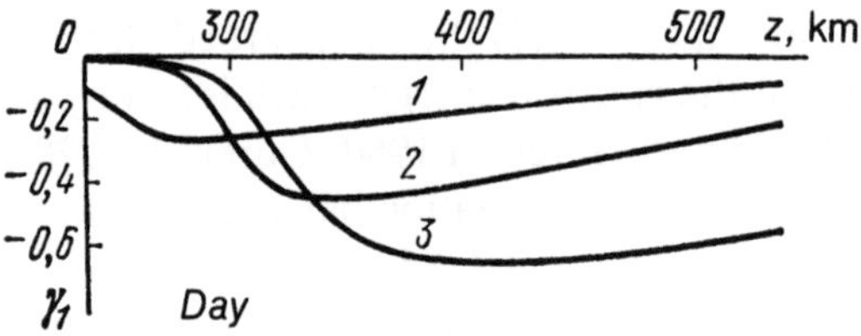

Fig. 7.2. The same as in Fig. 7.1, but for the upper ionosphere

Defocusing of Narrow Radio Wave Beams. Geometrical – optics Eq. (3.4), when taking into account nonlinear perturbations of the dielectric constant, $\Delta\varepsilon_n$, is written in the form

$$(\boldsymbol{k}\,\nabla)\boldsymbol{k} = \frac{k_0^2}{2}\,\nabla(\varepsilon + \Delta\varepsilon_n)\,. \tag{7.3}$$

Insofar as $\Delta\varepsilon_n$ depends on wave intensity w, Eq. (7.3) should be supplemented with an energy transport equation

$$\nabla(\boldsymbol{k}\,w) = 0\,. \tag{7.4}$$

This equation follows from Eq. (6.34), if the distribution function f has the form of Eq. (6.42)

$$f(\theta, z, \varphi; \alpha, \beta) = w(\theta, z, \varphi)\,\delta(\alpha - \alpha_0)\,\delta(\beta - \beta_0)\,,$$

$$\alpha_0 = \frac{k_z}{k_0}\,, \qquad \beta_0 = \frac{k_\varphi}{k_0}\,,$$

i.e., if there is only one ray of a given direction (α_0, β_0) at any point θ, z, φ. Equation (7.4) is true when radio wave dissipation and scattering are unimportant.

Now let us take into account the particular form of the dielectric constant perturbation $\Delta\varepsilon \sim \Delta N_e$ [Eq. (7.1)]. Extracting, as usual, the direction along the beam axis, taking into account that $\varepsilon \approx 1$, and assuming the beam to be narrow, $k_\perp \ll k_0$, we rewrite Eqs. (7.3) and (7.4) in the form

$$\left(\frac{1}{R_0}\,\frac{\partial}{\partial\theta} + \boldsymbol{u}\,\nabla_\perp\right)\boldsymbol{u} + \gamma_2 \nabla_\perp w = -\frac{1}{2}\nabla_\perp U\,, \tag{7.5}$$

$$\frac{1}{R_0}\,\frac{\partial}{\partial\theta}\,w + \nabla_\perp(\boldsymbol{u}\,w) = 0\,,$$

$$U = -\varepsilon(z, \theta, \varphi) - \frac{2z}{R_0}\,, \qquad \boldsymbol{u} = \frac{\boldsymbol{k}_\perp}{k_0}\,, \qquad \gamma_2 = \frac{4\pi\,\omega_0^2\gamma_1}{\omega^2\,\mathscr{E}_p^2}\,,$$

$$w = \frac{\mathscr{E}_0^2}{8\pi}\,, \qquad \varepsilon \approx 1\,. \tag{7.6}$$

Here, $\boldsymbol{k}_\perp$ is the vector in the perpendicular plane, $\boldsymbol{k}_\perp = (k_z, k_\varphi)$, the operator $\nabla_\perp = (\partial/\partial z, \partial/\partial y)$, $dy = R_0 \sin\theta d\varphi$. The term in Eq. (7.5) proportional to γ_2 describes the regular nonlinear refraction effects, just those we are interested in. To extract them, we neglect, first, a linear refraction ($\nabla_\perp U = 0$) and assume, for simplicity, that $\gamma_2 = \text{const}$. Then Eqs. (7.5) and (7.6) are equivalent to equations for the dynamics of two-dimensional perfect gas with the specific heat ratio equal to 2. Here, the energy density w is used instead of the gas density; the flux velocity $\boldsymbol{V}$ is replaced by the vector of ray direction, $\boldsymbol{u} = \boldsymbol{k}_\perp / k_0$, the coordinate along the beam axis, θ, plays a role of the time. The vector $\boldsymbol{u}$ is positioned in the plane orthogonal to the direction of ray propagation and gives the ray direction projection on this plane. If all the vectors $\boldsymbol{u}$ are directed towards the beam axis, then the rays converge, i.e., the beam is focused. If, on the contrary, the vectors $\boldsymbol{u}$ are directed from the axis, the rays diverge and the beam is defocused.

Let us give some simple examples of the solution of Eqs. (7.5) and (7.6) in a homogeneous medium, $\nabla_\perp U = 0$, which enable one to conceive a character of nonlinear refraction. The problem is formulated as follows. Let an initial distribution of energies and ray directions be given at the boundary of the nonlinear region, $\theta = \theta_0$

$$w|_{\theta=\theta_0} = w_0(\rho)\,, \qquad \boldsymbol{u}|_{\theta=\theta_0} = \boldsymbol{u}_0(\rho)\,, \qquad \rho = \boldsymbol{r}_\perp\,. \tag{7.7}$$

One needs to find w and $\boldsymbol{u}$ at any cross-section of the beam by the plane $\theta = \text{const}$.

The Parabolic Beam (Talanov 1965, Gurevich 1972). Let us consider an axisymmetric beam and put $\theta = \theta_0$

$$w_0(\rho) = w_0 \begin{cases} 1 - \rho^2/a^2\,, & \rho < a\,, \\ 0\,, & \rho \geqslant a\,, \end{cases} \qquad u_0(\rho) = \frac{\rho}{R}\,. \tag{7.8}$$

These initial conditions correspond to a spherical wave, diverged from the point $\theta = \theta_0 - R/R_0$ and possessed by a parabolic distribution of initial intensity (Talanov 1965). If the wave is supposed to be emitted from the point $\theta = 0$, then

$$\theta_0 = R/R_0\,. \tag{7.9}$$

We shall seek the solution of Eqs. (7.5) and (7.6) at any θ in the form

$$w(\theta,\rho) = \frac{w_0}{f_1^2(\theta)} \begin{cases} 1 - \dfrac{\rho^2}{a^2 f_1^2}\,, & \rho \leqslant a f_1\,, \\ 0\,, & \rho \geqslant a f_1\,, \end{cases} \qquad u(\theta,\rho) = \varphi(\theta) \cdot \rho\,. \tag{7.10}$$

Substituting Eq. (7.10) in Eqs. (7.5) and (7.6), we obtain the following equations for $f_1(\theta)$, $\varphi(\theta)$

$$\frac{1}{R_0^2}\,\frac{d^2 f_1}{d\theta^2} = 2\gamma_2 \frac{w_0}{f_1^3}\,, \tag{7.11}$$

$$\varphi(\theta) = \frac{1}{R_0}\,\frac{\partial}{\partial\theta}\ln f_1\,. \tag{7.12}$$

Boundary conditions [Eq. (7.8)] with allowance for Eqs. (7.10) – (7.12) are rewritten in the form

$$\varphi|_{\theta=\theta_0} = \frac{1}{R_0}\,, \qquad f_1|_{\theta=\theta_0} = 1\,, \qquad \left.\frac{df_1}{d\theta}\right|_{\theta=\theta_0} = \frac{R_0}{R}\,. \tag{7.13}$$

The solution of Eq. (7.11) with boundary conditions [Eq. (7.13)] is given by

$$f_1 = \left[\left(1+\frac{R_1}{R}\right)^2 + \frac{2\gamma_2 w_0 R_1^2}{a^2}\right]^{1/2}\,. \tag{7.14}$$

Here, $R_0(\theta-\theta_0) = R_1$ is the distance covered by the ray in the nonlinear medium. In the absence of nonlinearity

$$f_1 = 1 + \frac{R_0}{R}(\theta-\theta_0) = \frac{R_0}{R}\,\theta\,. \tag{7.15}$$

In the latter equation, Eq. (7.9) for θ_0 has been taken into account. Equation (7.14) shows that the beam is broadened (defocused) due to nonlinearity.

Let us introduce a beam divergence angle

$$\chi = \frac{a}{R_0}\,\frac{df_1}{d\theta} = \frac{a}{R}\,\frac{1+\dfrac{R_1}{R}+\dfrac{R_1}{R}\,\dfrac{2\gamma_2 w_0 R^2}{a^2}}{\left[\left(1+\dfrac{R_1}{R}\right)^2+\dfrac{2\gamma_2 w_0 R_1^2}{a^2}\right]^{1/2}}\,, \qquad R_1 = R_0(\theta-\theta_0)\,, \tag{7.16}$$

i.e., the angle between the boundary ray direction and beam axis. In the absence of nonlinearity, the divergence angle is constant, i.e., is independent of θ

$$\chi = \chi_0 = \frac{a}{R}\,.$$

Equation (7.16) shows that owing to nonlinearity, the divergence angle increases with θ, tending at great θ, $\theta \gg \theta_0$, to

$$\chi_n = \sqrt{\chi_0^2 + 2\gamma_2 w_0}\,. \tag{7.17}$$

Accordingly, the beam radius increases due to nonlinearity by a factor of $(1+2\gamma_2 w_0/\chi_0^2)^{1/2}$. At the same time, the intensity on the beam axis decreases by a factor of $(1+2\gamma_2 w_0/\chi_0^2)$. If, initially, the beam was parallel ($\chi_0 = 0$), then, due to nonlinearity, it becomes divergent with

$$\chi = \chi_n = \sqrt{2\gamma_2 w_0}\,. \tag{7.18}$$

If the nonlinearity is weak, $\gamma_2 w_0 \ll \chi_0^2$, then at any θ, as follows from Eq. (7.16), we have

$$\chi = \chi_0 \left[1 + \frac{\gamma_2 w_0}{\chi_0^2} \left(1 - \frac{\theta_0^2}{\theta^2} \right) \right]. \tag{7.19}$$

Hence, the divergence angle χ_n [Eq. (7.17)] reaches its maximum at θ as small as $\theta = (2-3)\,\theta_0$ (see Fig. 7.3).

Breakdown of the Intensity Discontinuity. Equation (7.5) shows that the influence of nonlinearity is intensified with increasing gradient of the beam intensity, $|\nabla w|$. Therefore, let us consider the case of abrupt beam boundary, i.e., the case of intensity discontinuity (Gurevich 1972). So, let at $\theta = \theta_0$

$$w|_{\theta=\theta_0} = w_0 \begin{cases} 1 & \text{at} \quad z < 0\,, \\ 0 & \text{at} \quad z > 0\,. \end{cases} \tag{7.20}$$

Here, the coordinate z is a distinguished direction in the plane orthogonal to the beam axis: this direction is orthogonal to the line of intensity discontinuity. Without restriction of generality, we obtain

$$u_z|_{\theta=\theta_0} = 0\,. \tag{7.21}$$

Equations (7.5) and (7.6) with neglect of linear refraction take the form

$$\begin{aligned} &\left(\frac{1}{R_0} \frac{\partial}{\partial \theta} + u_z \frac{\partial}{\partial z} \right) u_z + \gamma_2 \frac{\partial w}{\partial z} = 0\,, \\ &\left(\frac{1}{R_0} \frac{\partial}{\partial \theta} + u_z \frac{\partial}{\partial z} \right) w + w \frac{\partial u_z}{\partial z} = 0\,. \end{aligned} \tag{7.22}$$

Here, we have taken into account the fact that when Eqs. (7.20) and (7.21) are satisfied, then the coordinate z transversal to the beam axis is of importance only.

Neither Eq. (7.22), nor boundary conditions [Eqs. (7.20) and (7.21)] involve a characteristic beam size. Therefore, the solution should be sought in a self-similar form

$$u_z = u(\tau)\,, \qquad w = w(\tau)\,, \qquad \tau = \frac{z}{R_0(\theta - \theta_0)}\,. \tag{7.23}$$

Substituting Eq. (7.23) into Eq. (7.23), we obtain the equations

$$(u-\tau)\frac{dw}{d\tau} + w\frac{du}{d\tau} = 0\,, \qquad (u-\tau)\frac{du}{d\tau} + \gamma_2 \frac{dw}{d\tau} = 0\,. \tag{7.24}$$

Their solution is given by

$$\begin{aligned} &w = \frac{1}{9\gamma_2}(C-\tau)^2\,, \qquad u = \frac{2\tau + C}{3}\,; \\ &w = w_0\,, \quad u = 0\,; \quad w = 0\,, \quad u = \tau\,. \end{aligned} \tag{7.25}$$

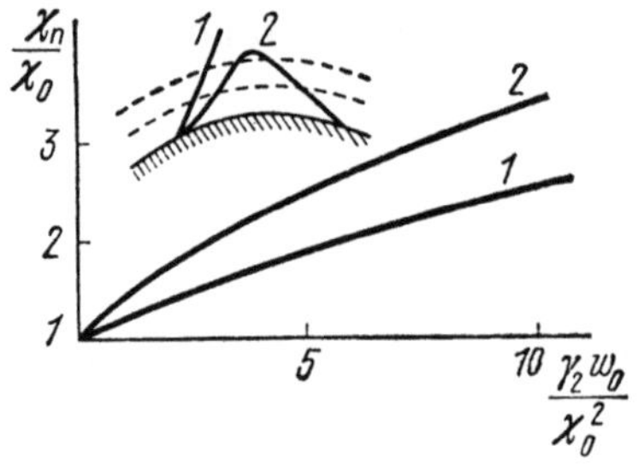

Fig. 7.3. Increase of the beam divergence angle for two different trajectories (shown at the *top*)

Boundary conditions [Eqs. (7.20) and (7.21)] in terms of self-similar variables [Eq. (7.23)] are written in the form

$$\begin{aligned} &w \to 0 \quad \text{when} \quad \tau \to +\infty \\ &\left.\begin{aligned} w &\to w_0 \\ u &\to 0 \end{aligned}\right\} \quad \text{when} \quad \tau \to -\infty . \end{aligned} \tag{7.26}$$

Taking Eq. (7.26) into account, we obtain desired distributions of intensity and ray directions

$$w = w_0 \begin{cases} 1 \quad \text{at} \quad \dfrac{z}{R_0(\theta-\theta_0)} < -\sqrt{\gamma_2 w_0}, \\ \left(\dfrac{2}{3} - \dfrac{z}{3\sqrt{\gamma_2 w_0}\,R_0(\theta-\theta_0)}\right)^2 \quad \text{at} \quad -\sqrt{\gamma_2 w_0} \leqslant \dfrac{z}{R_0(\theta-\theta_0)} \leqslant 2\sqrt{\gamma_2 w_0}, \\ 0 \quad \text{at} \quad \dfrac{z}{R_0(\theta-\theta_0)} > 2\sqrt{\gamma_2 w_0}, \end{cases} \tag{7.27}$$

$$u = \begin{cases} 0 \quad \text{at} \quad \dfrac{z}{R_0(\theta-\theta_0)} \leqslant -\sqrt{\gamma_2 w_0}, \\ \dfrac{2}{3}\left(\dfrac{z}{R_0(\theta-\theta_0)} + \sqrt{\gamma_2 w_0}\right) \quad \text{at} \quad \dfrac{z}{R_0(\theta-\theta_0)} > -\sqrt{\gamma_2 w_0}. \end{cases}$$

It is seen that the initial intensity discontinuity (at $\theta = \theta_0$) breaks down, and the beam in the vicinity of the discontinuity line rapidly diverges. The divergence angle reaches its maximum at the beam boundary ($\tau = 2\sqrt{\gamma_2 w_0}$)

$$\chi_n = 2\sqrt{\gamma_2 w_0}. \tag{7.28}$$

It is seen that the angle χ_n in the case of breakdown of abrupt boundary is by a factor of $\sqrt{2}$ greater than in the case of axisymmetric parabolic beam [Eq. (7.18)].

The examples presented above show that under conditions of lower ionosphere [Eq. (7.1)], radio wave beams are defocused (broadened) due to nonlinear refraction.

Mutual Defocusing. Above we considered the auto-action of a radio wave beam. However, of no less importance for radio wave capture into the duct and departure from it, is also nonlinear interaction of radio waves at different frequencies (Gurevich 1978). Let us consider, for example, two axisymmetric beams at frequencies ω_1 and ω_2 propagating in one direction through a defocusing medium [Eq. (7.1)]. We assume that the wave $\mathscr{E}_2$ is weak, so that its effect on the plasma can be ignored. Then the equations for energy densities w_1, w_2 and ray directions u_1, u_2 take, according to Eqs. (7.5) and (7.6), the following form

$$\begin{aligned}&\frac{1}{R_0}\frac{\partial w_1}{\partial\theta}+\frac{1}{\rho}\frac{\partial}{\partial\rho}(\rho u_1 w_1)=0\,,\\&\frac{1}{R_0}\frac{\partial u_1}{\partial\theta}+u_1\frac{\partial u_1}{\partial\rho}+\gamma_2\frac{\partial w_1}{\partial\rho}=0\,;\end{aligned}\qquad(7.29)$$

$$\begin{aligned}&\frac{1}{R_0}\frac{\partial w_2}{\partial\theta}+\frac{1}{\rho}\frac{\partial}{\partial\rho}(\rho u_2 w_2)=0\,,\\&\frac{1}{R_0}\frac{\partial u_2}{\partial\theta}+u_2\frac{\partial u_2}{\partial\rho}+\varkappa\gamma_2\frac{\partial w_1}{\partial\rho}=0\,;\qquad \varkappa=\frac{\omega_1^2}{\omega_2^2}\,.\end{aligned}\qquad(7.30)$$

Here, as before, we have neglected a linear refraction ($\nabla_\perp u=0$). Nonlinear refraction of the weak ray, w_2, as is seen from Eq. (7.30), is determined only by the strong wave, w_1.

We assume that at the boundary of nonlinear region, $\theta=\theta_0$, both waves have a parabolic intensity distribution [Eq. (7.8)]. Then the parabolic distribution remains also inside the plasma. In this case, insofar as the weak wave $\mathscr{E}_2$ have no effect on the strong wave $\mathscr{E}_1$, the distribution of intensities (w_1) and ray directions (u_1) in the strong wave is described as before by Eqs. (7.10) – (7.14). Substituting these equations into Eq. (7.30) and assuming the weak wave distribution of intensity to be parabolic

$$w_2=\frac{w_{20}}{f_2^2}\begin{cases}1-\dfrac{\rho^2}{a^2f_2^2}\,, & \rho\leqslant af_2\,,\\ 0\,, & \rho\geqslant af_2\,,\end{cases}\qquad u_2=\frac{\rho}{f_2R_0}\,\frac{df_2}{d\theta}\,,\qquad(7.31)$$

we obtain that f_2 obeys the equation

$$\frac{d^2f_2}{dt^2}=\varkappa\frac{f_2}{[(1+\alpha_1t)^2+t^2]^2}\,,\qquad t=\frac{\sqrt{2\gamma_2w_0}R_0}{\alpha_1}(\theta-\theta_0)\,,$$

$$\alpha_1=\frac{a}{R_1\sqrt{2\gamma_2w_0}}\,,\qquad(7.32)$$

with boundary conditions

$$f_2|_{t=0} = 1$$

$$\left.\frac{df_2}{dt}\right|_{t=0} = \alpha_1 .$$

Note that Eq. (7.32) is valid at any ρ which are less than $\rho < af_1$, i.e., if $f_2 \leqslant f_1$. This condition is always satisfied when $\varkappa \leqslant 1$.

Let us summarize here the results of the solution of Eq. (7.32). Figure 7.4 shows f_2 as a function of t at various values of the parameter $\varkappa$. Initial beams $\mathscr{E}_1$ and $\mathscr{E}_2$ were assumed in this case to be parallel ($\alpha_1 = 0$). The function f_2 is seen to increase with t. As this takes place, at great t, $f_2 \sim t$. In other words, due to nonlinearity, the beam $\mathscr{E}_2$ now becomes divergent. In this case, for $\varkappa \leqslant 1$ we have

$$f_2 \to \cos\left(\frac{\pi}{2}\sqrt{1-\varkappa}\right)\sqrt{2\gamma_2 w_0}\,\frac{R_0}{a}(\theta-\theta_0)$$

or

$$\chi_{n2} \to \cos\left(\frac{\pi}{2}\sqrt{1-\pi}\right)\sqrt{2\gamma_2 w_0} . \tag{7.33}$$

Hence, it is seen that the divergence angle of the beam $\mathscr{E}_2$, when $\varkappa \leqslant 1$, is somewhat less than that of the major beam, $\mathscr{E}_1$ [Eq. (7.18)]. When $\varkappa > 1$, the weak wave is more vigorously pushed out towards the boundaries of the beam $\mathscr{E}_1$. Its maximum divergence angle does not exceed, however, that of the strong wave [Eq. (7.18)].

Influence of Nonlinear Refraction on Radio Wave Capture into a Duct. When analyzing this problem, it is important, in general, to take into account nonlinear effects and a regular ionospheric inhomogeneity simultaneously, i.e., to consider both Eqs. (7.5) and (7.6). Here, we restrict ourselves to a simplified approach which enables the effect to be evaluated, based upon the simple solutions for a homogeneous medium (Gurevich 1971). The plot of the parameter $\gamma_2(\omega/10^8)^4$ (in $V^{-2}\,m^2$) vs the altitude z is given in Fig. 7.5 for the daytime and nighttime models of midlatitude ionosphere, according to Eqs. (7.5) and (7.6) (solid curves). The dashed curves are plotted by formulas

day:

$$\gamma_2\left(\frac{\omega}{10^8}\right)^4 = \begin{cases} \gamma_0 = 2.5\cdot 10^{-2} & \text{when} \quad z_1 \leqslant z \leqslant z_2 , \\ 0 & \text{when} \quad z \leqslant z_1 , \quad z > z_2 , \end{cases}$$

night: (7.34)

$$\gamma_2\left(\frac{\omega}{10^8}\right)^4 = \begin{cases} \gamma_0 = 3\cdot 10^{-4} & \text{when} \quad z_1 \leqslant z \leqslant z_2 , \\ 0 & \text{when} \quad z < z_1 , \quad z > z_2 , \end{cases}$$

where $z_1 = 100$ km, $z_2 = 180-200$ km. Here, ω is in rad/s, γ_2 and γ_0 in $(V/m)^{-2}$.

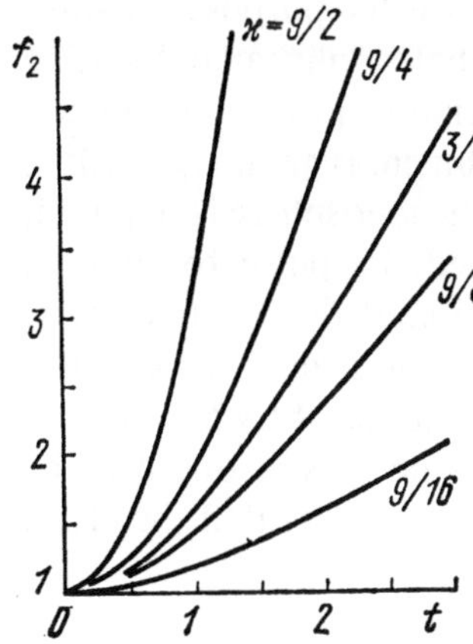

Fig. 7.4. Broadening of the weak beam

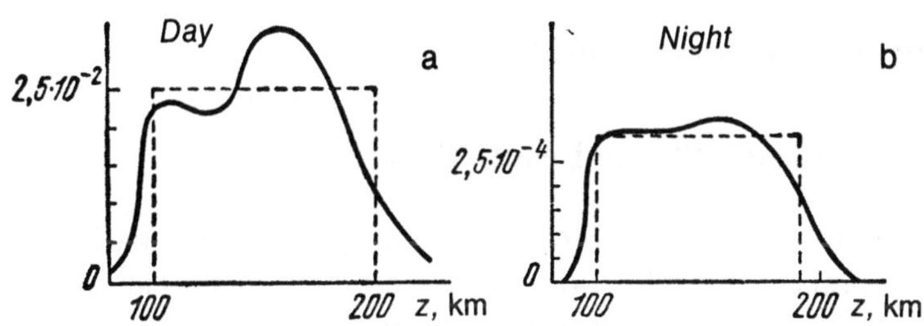

Fig. 7.5 a, b. Height profiles of the parameter $\gamma_2(\omega/10^8)^4$ (V/m)$^{-2}$ in the ionosphere

Thus, in the rough approximation [Eq. (7.34)], the ionospheric layer between the altitudes z_1 and z_2 can be treated as a homogeneous defocusing medium, i.e., the results obtained above may be used here.

As is seen from Eqs. (7.17), (7.28), and (7.33), the maximum angle, through which rays of the beam due to nonlinearity can turn, is given by

$$\chi_n = 2\sqrt{\gamma_2 w_0}\,.$$

Substituting here Eqs. (7.34) and (7.6) for γ_2 and w_0, we obtain

$$\chi_n \approx C\left(\frac{10^8}{\omega}\right)^2 \mathscr{E}_0\,, \quad C\approx 0.05 \quad (\text{day})\,, \quad C\approx 0.005 \quad (\text{night})\,. \tag{7.35}$$

Here, $\mathscr{E}_0$ is the wave field amplitude (in V/m) at the boundary of the nonlinearity region, i.e., in the ionosphere at the altitude, $z_1 \approx 100$ km. Comparing the angle χ_n with the minimum turning angles needed for radio wave capture into a duct, γ_{FE} and γ_F (Sec. 4.1), we find that the nonlinearity is important for $\mathscr{E}_0 > 1$ V/m. Note that the nonlinearity effect on radio wave capture has a threshold type: when $\chi_n < \gamma_{FE}$, γ_F, the effect is absent, when $\chi_n > \gamma_{FE}$, γ_F, it immediately becomes strong. We emphasize that since the wave field amplitude, $\mathscr{E}_0$, increases in a caustic region, nonlinear effects are intensified here (Gurevich 1978).

A possible use of mutual radio wave defocusing for the purposes of wave capture should be especially noted. The point is that nonlinear defocusing effects are connected with the electron density variation in the ionosphere. Inasmuch as a characteristic time of density establishment is great, $\tau_N \sim 10^2$ s, a rather prolonged action of the powerful wave on the ionosphere is required, $\Delta t > \tau_N$. As far as the weak wave $\mathscr{E}_2$ is concerned, its pulse for mutual defocusing can be short. Varying the time of radiation of the $\mathscr{E}_2$-pulse with respect to the modifying $\mathscr{E}_1$-pulse, one can realize the process of electron density variation. In addition, bearing in mind the fact that the wave interaction is intensified as the frequency of the wave $\mathscr{E}_2$ decreases, one can, by changing the frequency, find the optimum conditions in order that the effect be maximum.

7.2 Nonlinear Radio Wave Scattering

7.2.1 Character of Irregular Plasma Density Perturbations

Under the action of high-power radio waves, the ionospheric plasma in the F-region readily becomes unstable. Both large-scale and small-scale ionospheric plasma striations originate, waves of various types are generated. These phenomena at present are intensely examined. We outline here the principal processes.

Large-Scale Striations. The phenomenon of large-scale striations in the F-region has been discovered by Utlaut et al. when analyzing the results of ionospheric sounding in the region disturbed by a powerful transmitter (Utlaut and Cohen 1971a, b; Utlaut 1970, Utlaut et al. 1974). Later on, this phenomenon was thoroughly examined by different radio sounding techniques (Belikovich et al. 1975, Utlaut 1975, Rufenach 1973, Wright 1973, Thome and Blood 1974). It has been shown that under the action of intense field, $\mathscr{E}_0 \sim \mathscr{E}_p$ [Eq. (7.2)], a quasistationary irregular structure develops in the ionospheric F-region. The irregularities are elongated along the Earth's magnetic field. Their spectrum of transverse scales $a_\perp$ is very wide, 50 – 100 m through 1 – 10 km. A typical irregularity size $a_\perp$ exceeds the modifying wave length

$$k_0 a_\perp > 1\,, \qquad k_0 = \frac{\omega}{c}\sqrt{\varepsilon} \sim (1-3)\cdot 10^{-2}\,\mathrm{m}^{-1}\,. \tag{7.36}$$

Because of this, these irregularities are called large-scale irregularities. The density perturbations, ΔN_e, in the large-scale irregularities rapidly grow with modifying wave field amplitude, reaching at $\mathscr{E}_0 \sim \mathscr{E}_p$ the magnitudes, $(\Delta N_e/N_{e0}) \approx (1-3)\cdot 10^{-2}$. Further, the growth of ΔN_e with $\mathscr{E}_0$ slows down. The values of $(\Delta N_e/N_{e0})$ decrease rather gradually with decreasing irregularity size $a_\perp$.

Large-scale irregularities arise in 0.1 – 1 s and evolve during the time of the order of 10 s after the modifying transmitter is turned off. Their development is completed during a few minutes. The relaxation time is usually also of the order of 10 min, occasionally it can range to hours.

The theory associates the phenomenon of large-scale ionospheric scattering with a self-focusing (modulation) instability of intense radio waves in the F-region (Vas'kov and Gurevich 1974, 1979; Perkins and Valeo 1974, Abramovich 1976). Most rapidly this instability evolves in the radio wave reflection region. The complete time of the irregularity development and relaxation is of the order of the electron lifetime, τ_N.

Small-Scale Striations. The phenomenon of small-scale striations in the ionosphere under the action of intense radio waves has been found by Fialer et al. when studying HF and UHF radio wave scattering (Fialer 1974, Minkoff et al. 1974, Carpenter 1974, Barry 1974, Erukhimov et al. 1979). It was found that in the reflection region of a powerful ordinary radio wave vertically incident on the ionosphere, a quasistationary small-scale irregular structure develops. The irregularities are strongly elongated along the Earth's magnetic field. Their longitudinal length, $a_\parallel$, is of the order of hundreds of meters or even a few kilometers, and their transverse size varies from 20 – 30 cm to 10 – 30 m. For these irregularities, the condition inverse to Eq. (7.36) is satisfied

$$k_0 a_\perp < 1 . \tag{7.37}$$

The density perturbations in the irregularities increase with the modifying wave power up to the magnitudes, $\Delta N_e/N_{e0} \sim 10^{-2}$, the latter being reached when $\mathscr{E}_0 \sim \mathscr{E}_p$. After this, their subsequent growth slows down. The spectrum of $\Delta N_e/N_{e0}$ is essentially constant in the range $0.5\,\mathrm{m} \lesssim a_\perp \lesssim 20\,\mathrm{m}$, it drops sharply at smaller $a_\perp$ with decreasing $a_\perp$ (Fig. 7.6). The total region where the small-scale striations occur amounts to 10 – 20 km in the vertical and 120 – 150 km in the horizontal direction. The complete times of irregularity development or relaxation are of the order of minutes or seconds, being decreased with increasing $a_\perp$.

Together with the small-scale irregular striations, an intense (anomalous) absorption of modifying radio wave as well as of other ordinary radio waves (in the broad frequency band, $\Delta f \sim 1$ MHz) propagating through the dis-

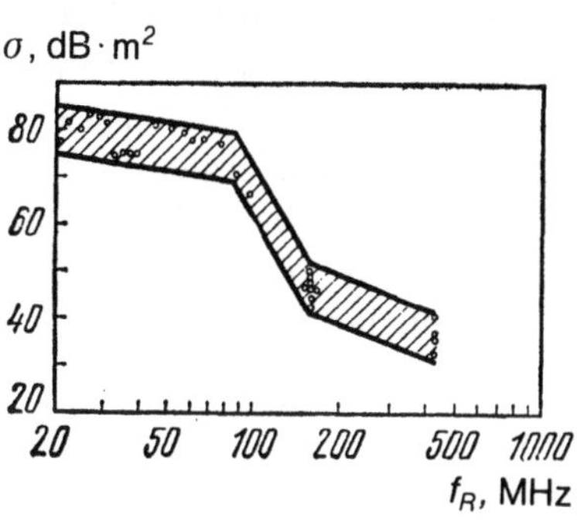

Fig. 7.6. Scattering cross-section vs the radar frequency (Rao and Thome 1974), $\sigma = (\Delta N_e/N_{e0})^2$

turbed region evolve. Small-scale irregularities are also the cause of intense scattering of HF and UHF radio waves.

The theory associates small-scale striations in the ionosphere with the nonlinear resonant instability developed as a result of intense generation of plasma oscillations in the irregularities under the action of modifying radio waves (Vas'kov and Gurevich 1975, 1977, 1979, Vas'kov 1977). Essential can also be the drift instability (Borisov et al. 1976, 1977) and dissipative parametric instability (Perkins 1974, Grach and Trakhtengerts 1975, Grach et al. 1979).

Generation of Natural Plasma Oscillations. Under the action of an intense radio wave, a generation of plasma, ion-sound, cyclotron, and other waves is possible in the *F*-region (Fejer and Leer 1972, Varshavskii and Dimant 1976, Dimant 1977, Perkins and Kaw 1971, Perkins et al. 1974, Mityakov et al. 1974, Fejer 1979, Dubois and Goldman 1979). A number of phenomena connected with the generation of plasma oscillations have been observed experimentally (Utlaut 1975, Gordon and Carlson 1974, Showen et al. 1978, Fejer et al. 1978). The plasma density perturbations in these cases turned out to be much smaller than in the case of the quasistationary striations considered above.

7.2.2 Influence of Nonlinear Scattering on Radio Wave Capture into IWD and Departure from the Duct

Irregularities with characteristic scales, $a \sim \lambda$, make the main contribution into scattering of radio waves with the length λ. For a long-distance radio wave propagation, radio waves with frequency $f \sim 10-30$ MHz are of importance. Proper optimum irregularity scales are $\sim 3-10$ m. Such irregularities, as indicated above, evolve efficiently in the reflection region of a high-power ordinary radio wave vertically incident on the ionosphere (small-scale striations). Their main feature lies in their strong elongation along the Earth's magnetic field. The ratio of the transverse irregularity scale to the longitudinal one is small, $\sim 10^{-3}$. This essentially influences the character of radio wave scattering.

Radio Wave Scattering by Anisotropic Irregularities. The Cone of Mirror Scattering. Let us consider radio wave scattering by irregularities strongly elongated along the magnetic field (Rao and Thome 1974, Booker 1956, Chapman 1953, Filipp 1974, Uspenskii and Sverdlov 1973, Uspenskii and Pudovkin 1974, Gurevich and Tsedilina 1975a, b; Gurevich et al. 1977, Kim 1977). The scattering cross-section, as before, is given by Eq. (6.8). The irregularity correlation function is assumed to be Gaussian

$$R_\varepsilon(\xi) = \exp\left\{-\left(\frac{x'^2+y'^2}{a_\perp^2}+\frac{z'^2}{a_\parallel^2}\right)\right\}, \quad \xi=(x',y',z'), \quad \frac{a_\perp}{a_\parallel}=q. \quad (7.38)$$

Here, z' is the coordinate along the Earth's magnetic field, $\boldsymbol{H}$; x', y' are coordinates in the plane orthogonal to $\boldsymbol{H}$ [Fig. (7.7)]. By Eq. (6.10), we have

$$\rho_\varepsilon(\boldsymbol{k}) = \frac{\overline{(\Delta\varepsilon)^2}}{8\pi^{3/2}} a_\perp^2 a_\parallel \exp\left\{-\frac{1}{4}[k_{x'}^2 a_\perp^2 + k_{y'}^2 a_\perp^2 + k_{z'}^2 a_\parallel^2]\right\}. \tag{7.39}$$

Here, $k_{x'}$, $k_{y'}$, $k_{z'}$ are the components of the vector $\boldsymbol{k} = \boldsymbol{k}_{sc} - \boldsymbol{k}_0$ in the coordinate system x', y', z'; as usual, $\boldsymbol{k}_0$ and $\boldsymbol{k}_{sc}$ are the wave vectors of the incident and scattered wave, $k_0 = k_z$.

Now we consider a coordinate system x, y, z, where the axis z is directed along the vertical, and the axis x is orthogonal to z in the magnetic meridian plane $z\boldsymbol{H}$. Then y and y' coincide, and the axes x, z prove to be rotated with respect to axes x', y' through an angle γ_H, where γ_H is the angle between the Earth's magnetic field and the vertical. Let the ray direction, $\boldsymbol{k}_0$, in the coordinate system x, y, z, be specified by the vertical and azimuthal angles α and φ_0, and $\boldsymbol{k}_{sc}$, by the angles β^α, φ. Then, the vectors $\boldsymbol{k}_0$, $\boldsymbol{k}_{sc}$ in the system x, y, z are

$$\begin{aligned} &k_0\,\{\cos\alpha\cdot\cos\varphi_0, \quad \cos\alpha\cdot\sin\varphi_0, \quad \sin\alpha\},\\ &k_{sc}\{\cos\beta^\alpha\cdot\cos\varphi, \quad \cos\beta^\alpha\cdot\sin\varphi, \quad \sin\beta^\alpha\}. \end{aligned} \tag{7.40}$$

Accordingly,

$$\begin{aligned} &\boldsymbol{k} = \boldsymbol{k}_{sc} - \boldsymbol{k}_0 = k_0\{(\cos\beta^\alpha\cdot\cos\varphi - \cos\alpha\cdot\cos\varphi_0),\\ &(\cos\beta^\alpha\cdot\sin\varphi - \cos\alpha\cdot\sin\varphi_0), \quad (\sin\beta^\alpha - \sin\alpha)\}. \end{aligned} \tag{7.41}$$

In order to go over to the system x', y', z', a turn in the plane xz through the angle γ_H is needed. The component of $\boldsymbol{k}$ on the axis y undergoes no changes, and the components on the axes x, z are transformed by the formulas

$$k_{x'} = k_x \cos\gamma_H + k_z \sin\gamma_H, \qquad k_{z'} = k_z\cos\gamma_H - k_x \sin\gamma_H. \tag{7.42}$$

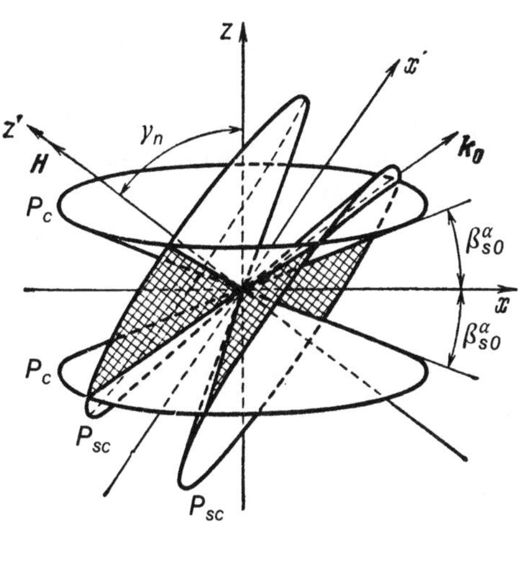

Fig. 7.7. The mirror scattering of incident wave energy

Substituting Eqs. (7.41) and (7.42) into Eq. (7.39), we obtain the following equation for the spectral density of dielectric constant ε fluctuations

$$\rho_\varepsilon(\boldsymbol{k}) = \frac{\overline{(\Delta\varepsilon)^2}}{8\pi^{3/2}} a_\perp^2 a_\parallel \exp\{-\Psi_0\},$$

$$\Psi_0 = \left(\frac{k_0}{2}\right)^2 \{a_\perp^2[(\cos\alpha\cdot\cos\varphi_0 - \cos\beta^\alpha\cdot\cos\varphi)\cos\gamma_H$$
$$+(\sin\alpha - \sin\beta^\alpha)\sin\gamma_H]^2 + a_\perp^2[\cos\alpha\cdot\cos\varphi_0 - \cos\beta^\alpha\cdot\sin\varphi]^2$$
$$+ a_\parallel^2[(\cos\beta^\alpha\cdot\cos\varphi - \cos\alpha\cdot\sin\varphi_0)\sin\gamma_H + (\sin\alpha - \sin\beta^\alpha)\cos\gamma_H]^2\}\,. \tag{7.43}$$

Here, $\overline{\Delta\varepsilon}$ are perturbations of ε caused by the powerful transmitter in the ionosphere. They are local and can be represented in the form

$$\overline{\Delta\varepsilon^2} = \left(\frac{4\pi e^2}{m\omega^2}\right)^2 (\overline{\Delta N_e^2}) = \overline{(\Delta\varepsilon)_0^2} f^0(x,y,z)\,, \quad \overline{\Delta N_e^2} = \overline{(\Delta N_e)_0^2} f^0(x,y,z)\,, \tag{7.44}$$

where $\overline{(\Delta N_e)_0^2} = \Delta N_0^2$ characterizes the maximum perturbation of N_e. The function f^0 characterizes a spatial distribution of $\overline{\Delta N_e^2}$ in the disturbed region. The distribution is determined by the wave frequency, directivity diagram of the transmitter antenna, etc. It can be put, for example, Gaussian

$$f^0 = \exp\left\{-\frac{(x-x_m)^2 + (y-y_m)^2}{B^2} - \frac{(z-z_m)^2}{A^2}\right\}, \tag{7.45}$$

where x_m, y_m, z_m are the points of maximum perturbations in the ionosphere, and A, B are the half-widths of the disturbed region in the vertical and horizontal directions, respectively. For the experiments in Boulder (Fialer 1974), $A \approx 7.5$ km, $B \approx 75$ km, $a_\perp \approx 0.5$ m, $a_\parallel \gtrsim 300-500$ m, $\Delta N_e^2/N_0^2 \approx (1-3)\cdot 10^{-4}$, where $N_0 \approx N_e(z_m)$. In Eq. (7.43) we have taken into account the fact that when the wave propagates in the ionospheric layer, the angles α, β^α change with height z by Snell's law [Eq. (6.1)], consequently, in Eq. (7.43), $\Psi_0 = \Psi_0(z)$. In the case of small angles, the functions $\alpha(z)$, $\beta^\alpha(z)$ are defined by Eq. (6.20).

The main peculiarity of scattering by strongly elongated irregularities is the formation of a so-called mirror scattering cone. Indeed, let us assume that $a_\parallel \to \infty$. Then, as is clear from Eq. (7.43), all the scattered rays β^α, φ are concentrated in the neighborhood of the surface

$$\cos\beta^\alpha\cdot\cos\varphi\cdot\sin\gamma_H - \cos\alpha\cdot\cos\varphi_0\cdot\sin\gamma_H + \sin\alpha\cdot\cos\gamma_H$$
$$-\sin\beta^\alpha\cdot\cos\gamma_H = 0\,. \tag{7.46}$$

Bearing in mind the fact that the magnetic field vector has in the coordinate system x, y, z the components $\{-H\sin\gamma_H,\ 0,\ H\cos\gamma_H\}$, we obtain from Eq. (7.40) that the cosines of angles between the incidient or scattered rays $\boldsymbol{k}_0$, $\boldsymbol{k}_{sc}$ and magnetic field $\boldsymbol{H}$ are given by, respectively

$$\frac{\boldsymbol{k}_0\boldsymbol{H}}{k_0H} = -\cos\alpha\cdot\cos\varphi_0\cdot\sin\gamma_H + \sin\alpha\cdot\cos\gamma_H\,,$$

$$\frac{\boldsymbol{k}_{sc}\boldsymbol{H}}{k_{sc}H} = -\cos\beta^\alpha\cdot\cos\varphi\cdot\sin\gamma_H + \sin\beta^\alpha\cdot\cos\gamma_H\,.$$

Hence, Eq. (7.46) can be rewritten in the form

$$\boldsymbol{k}_0\boldsymbol{H} = \boldsymbol{k}_{sc}\boldsymbol{H}\,, \qquad k_{sc} = k_0\,. \tag{7.47}$$

The scattered vectors, k_{sc}, by Eq. (7.47), form a conic surface with the axis along H and angle between the axis and generatrix equal to the angle between k_0 and H. This cone is the mirror scattering cone (see Fig. 7.7, the cones P_{sc}). In the ionospheric conditions of interest, when $a_\perp/a_\parallel \sim 10^{-3}$, the cone is quite pronounced.

Radio Wave Capture into IWD Induced by Scattering by Strongly Elongated Irregularities. The Capture Coefficient. For capture into a duct of scattered radio waves, another conic surface is of importance: $\beta^\alpha = \beta^\alpha_{s0}$. Its axis is directed along the vertical, and the angle of the generatrix is equal to $(\pi/2 - \beta^\alpha_{s0})$ (see Fig. 7.7, the cones P_c). This surface cuts off on the scattering cone two sectors (shaded), where the condition for capture into the duct, $\beta^\alpha < \beta^\alpha_{s0}$, is valid. Rays scattered into these sectors prove to be trapped into the duct. Putting in Eq. (7.46) $\beta^\alpha = \beta^\alpha_{s0}$, we obtain the lines of intersection of the cones

$$\cos\varphi = \frac{(\sin\beta^\alpha_{s0} - \sin\alpha)}{\cos\beta^\alpha_{s0}}\cot\gamma_H + \frac{\cos\alpha}{\cos\beta^\alpha_{s0}}\cos\varphi_0\,, \tag{7.48}$$

which determine the capture region boundaries.

Figure 7.7 shows that the straight line $\beta^\alpha = 0$ can be situated on the scattering cone. As is clear from Eq. (7.46), this is true if $|\cos\alpha\cdot\cos\varphi_0 - \sin\alpha\cdot\cot\gamma_H| \leqslant 1$. If, in addition, $z = z_s$, then the trapped scattered ray turns out to be sliding, $\beta^\alpha = 0$. Hence, when scattering by strongly elongated irregularities takes place, radio waves can be trapped immediately on the deep duct levels which are distinguished by a small absorption (Sec. 4.2) and the best confinement in the duct for radio waves travelling by long distances. On the other hand, insofar as a given altitude z_m is related to a well defined sliding frequency z_s ($z_m \approx z_s$; see Sec. 4.1), it is clear that the capture into the deep duct levels for irregularity produced on the height z is efficient only for frequencies f, close to f_s.

Let us obtain now the capture coefficients. To this end, the results of Sec. 6.1 can be used directly. The scattering cross-section, σ, is defined by Eqs. (6.8) and (7.43). The differential capture coefficient [Eq. (6.15)] is given by

$$G(\alpha, \beta_s^\alpha, \varphi_0, \varphi) = \frac{\partial^2 G_0}{\partial \beta_s^\alpha \partial \varphi} = \int_{z'(\beta_s^\alpha)}^{z''(\beta_s^\alpha)} \sigma \frac{\cos \beta^\alpha}{\sin \alpha} \left(\frac{d\beta^\alpha}{d\beta_s^\alpha} \right) dz ,$$

$$\sigma = \frac{\pi}{2} k_0^4 \sin^2 \Psi \rho_\varepsilon(\boldsymbol{k}) . \tag{7.49}$$

Here, $z'(\beta_s^\alpha)$, $z''(\beta_s^\alpha)$ are the boundary heights for the scattered ray trajectory

$$\sin \alpha(z) \approx \alpha(z) , \quad \cos \beta^\alpha \approx 1 , \quad \frac{d\beta^\alpha}{d\beta_s^\alpha} \approx \frac{\beta_s^\alpha}{\beta^\alpha} . \tag{7.50}$$

The total coefficient of capture [Eq. (6.16)]

$$G_0(\alpha_0, \varphi_0) = \int_{-\pi}^{\pi} d\varphi \int_{-\beta_{s0}^\alpha}^{\beta_{s0}^\alpha} d\beta_s^\alpha G \tag{7.51}$$

determines the trapped fraction of energy of the wave α_0, φ_0. The total trapped energy is given by the integration of G_0 over emission angles α_0, φ_0 [Eq. (6.17)].

When calculating the capture coefficient, G_0, it is convenient to integrate in Eq. (7.51) first with respect to φ, i.e., with respect to scattering azimuths. We recognize that the integrand in Eq. (7.51), when $(k_0 a_\parallel)^2 \gg 1$, has a sharp peak near the mirror scattering cone [Eq. (7.46)]. As a consequence, in all the quantities, excluding the factor before $(k_0 a_\parallel)^2$, the equation $\varphi = \varphi^*$ can be put and all the quantities can be taken out of the integrand. Here, φ^* is the azimuth of the mirror scattering cone defined by Eq. (7.48)

$$\cos \varphi^* = \frac{1}{\cos \beta^\alpha} [\cos \alpha \cdot \cos \varphi_0 - (\sin \alpha - \sin \beta^\alpha) \cot \gamma_H] . \tag{7.52}$$

In the factor at $(k_0 a_\parallel)^2$ we expand $\cos \varphi$ in a series near φ^*

$$\cos \varphi = \cos \varphi^* - (\varphi - \varphi^*) \sin \varphi^* + \dots \tag{7.53}$$

Substituting Eq. (7.52) into Eq. (7.53) and integrating with respect to φ, we obtain

$$G_1 = \int G d\varphi = \frac{\overline{\Delta \varepsilon_0^2} a_\perp^2 k_0^3}{8 \sin \gamma_H} \int_{z'(\beta_s^\alpha)}^{z''(\beta_s^\alpha)} \frac{f^0 \sin^2 \Psi \beta_s^\alpha}{\beta^\alpha \alpha \sin \varphi^*} dz . \tag{7.54}$$

Here, we have taken into account that a typical transverse scale of small-scale irregularities produced by intense radio waves, $a_\perp \sim 0.5$ m, is much less than the length of scattered waves in the frequency range, $10\,\text{MHz} \lesssim f \lesssim 30-40$ MHz, which we are interested in

$$(k_0 a_\perp)^2 \ll 1 . \tag{7.55}$$

Next, $\overline{\Delta \varepsilon_0^2}$ is the perturbation of the dielectric constant in the irregularity crest [Eq. (7.44)], f^0 is the function of spatial distribution of $\overline{\Delta N_e^2}$ [Eq. (7.45)].

Equation (7.50) has also been taken into account. The equation obtained for G_1 is convenient to apply for evaluation of the capture coefficient

$$G_0 = \int_{-\beta_{s0}^{\alpha}}^{\beta_{s0}^{\alpha}} G_1 \, d\beta_s^{\alpha} \sim \left(\frac{\Delta N_0}{N_{e0}}\right)^2 \frac{\omega_0^4 a_\perp^2 A}{\omega c^3 \alpha_s} \frac{\bar{\beta}_{s0}^{\alpha}}{\sin\gamma_H} f_0(x, y) . \tag{7.56}$$

Here, A is the vertical size of the disturbed region, $\Delta N_0/N_{e0}$ is the maximum electron density perturbation [Eq. (7.44)]. It follows from Eq. (7.56) that for the experiments in Boulder, the quantity $G_0 \sim 10^{-3} - 10^{-4}$. We emphasize that by Eq. (7.56), the capture coefficient rather slightly depends on the scattered wave frequency.

Azimuthal Characteristics of a Capture. The azimuthal angle φ_0, which specifies the propagation direction of a wave incident on the scattering volume, is unambiguously determined by the positions of the transmitter and of the disturbed region in the ionosphere. Of importance is, however, the fact that the azimuth of waves trapped after scattering can differ substantially from that of the incident wave. This is directly seen from Eq. (7.52), where the azimuths on the mirror scattering cone are defined. Since β^{α} varies between $-\beta_{s0}^{\alpha}$ and β_{s0}^{α}, the azimuth of trapped rays on the cone, φ^*, varies in the allowable range in accordance with Eq. (7.52). Note that $\cot\gamma_H$ for midlatitudes is usually great (for instance, for Moscow, $\gamma_H = 27°$, $\cot\gamma_H \approx 2.75$). Due to this, the azimuth φ^* can be significantly different from φ_0 at even rather small α.

Of importance is also the fact that as is seen from Eq. (7.52), for any φ_0 there are two different fields of φ^* – variation: $\varphi^* > 0$ and $\varphi^* < 0$. This is seen also from Fig. 7.8, where the results of numerical calculation of trapped ray azimuths are shown (Gurevich et al. 1977). Two radio wave beams prove to be trapped, the azimuthal width of each being $\sim 70°$.

Regions of Radio Wave Capture and Departure. Let us suppose that in some ionospheric region an artificial irregular structure has been produced. Radio wave scattering from it is of a clear-cut anisotropic type. This results in an appearance of regions for a transmitter – receiver disposition which are most preferable for the purposes of scattering-induced energy injection into a duct and output from the duct. Let us determine the structure of these regions. Note first the fact that both the azimuthal angle, φ_0, and the vertical angle, α, of waves incident on the scattering volume are unambiguously determined by the mutual disposition of the transmitter and disturbed region. Indeed, recognizing that the transmitter is situated on the Earth, we obtain from Eq. (6.1)

$$\cos\alpha = \frac{\cos\alpha_0}{\sqrt{\varepsilon'(z)}}, \quad S = R_0\theta = \int_0^{z_m} \frac{dz}{\tan[\alpha(z)]} = \int_0^{z_m} \frac{\cos\alpha_0 dz}{\sqrt{\varepsilon'(z) - \cos^2\alpha_0}} \tag{7.57}$$

or in the small angle case [Eq. (6.20)]

$$\alpha(z) = \sqrt{\alpha_0^2 - U(z) - 1}\,, \quad S = \int_0^{z_m} dz/\sqrt{\alpha_0^2 - U(z) - 1}\,. \tag{7.58}$$

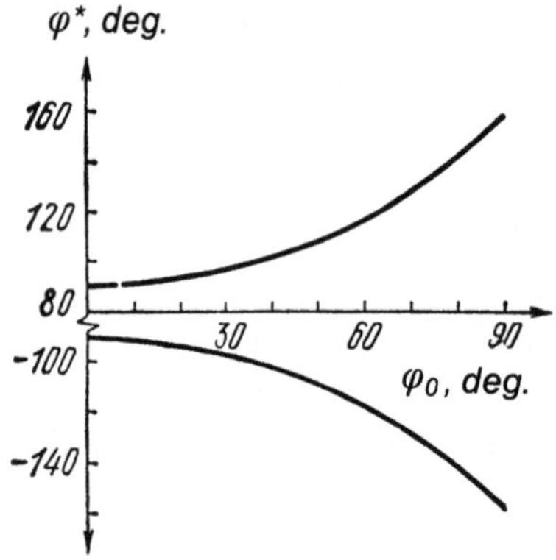

Fig. 7.8. Azimuths of the trapped rays; $\beta^\alpha = 0$, $\gamma_H = 20°$, $\alpha_0 = 20°$

Here, S is the projection on the Earth's surface of the distance between the transmitter and disturbed region, z_m is the height in the ionosphere of the disturbed region center, α_0 is the emission angle on the Earth. Naturally, $\alpha_0 \geqslant 0$. The condition $\alpha_0 = 0$ gives the maximum permissible distance between the transmitter and disturbed region

$$S_{\max} = \int_0^{z_m} \frac{dz}{\sqrt{-1-U(z)}} . \tag{7.59}$$

Equations (7.57) and (7.58) at a given $N_e(z)$ profile determine the angle α as a function of S. The plots of functions $\alpha_{z_m}(S) = \alpha(z = z_m)$, $\alpha_0(S)$ for $f = 13$ MHz, $z_m = 250$ km, and $N_e(z, \theta = 60°, t = 0^{\mathrm{h}})$ from the model EMI-72, are shown in Fig. 7.9. The dashed curve in Fig. 7.9a and the crosses in Fig. 7.9b are plotted in neglect of an ionospheric refraction. In this case

$$S_{\max} = R_0 \arccos \frac{1}{1 + z_m/R_0} \approx \sqrt{2R_0 z_m} .$$

This is true for waves of high frequency, $f \gtrsim 40$ MHz. At $f \lesssim 30-40$ MHz, the refraction is important. In particular, it results in an increase of the maximum distance, $S_{\max}$. The refraction is most important for frequencies $f \lesssim f_{\text{MUF-}F}$, since in this case the wave can reach the scattering volume not only in a direct way, but also after the reflection from the ionosphere as well.

To conceive a structure of regions of energy capture and escape on the Earth, let us consider first a simple model. We make a number of assumptions, namely: the disturbed region occupies a small volume in the neighborhood of z_m, the scattering is mirror ($a_\| \to \infty$), the wave frequency is close in value to that of the sliding mode, $f \approx f_s(z_m)$, being greater than MUF-F, so that a radio wave refraction in the ionosphere can be ignored. In addition, we restrict ourselves to the consideration of a capture of modes close to the sliding one, $\beta^\alpha \approx \beta_s^\alpha \approx 0$. Then the azimuth of the trapped wave, φ^*, is unambiguously related to the incident wave azimuth, φ_0, α

$$\cos \varphi^* = \cos \varphi_0 \cdot \cos \alpha - \sin \alpha \cdot \cot \gamma_H , \tag{7.60}$$

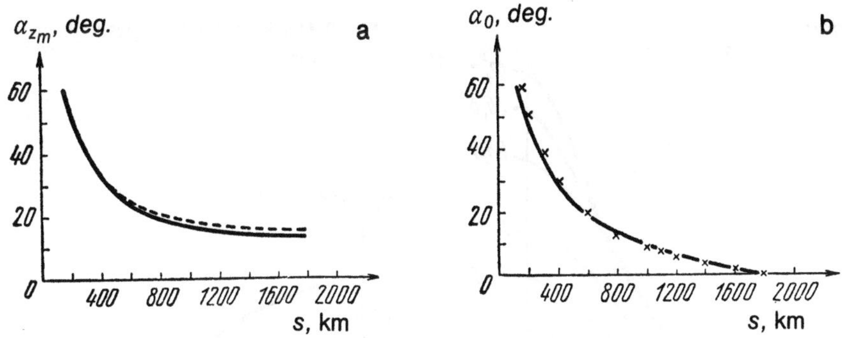

Fig. 7.9a, b. Angle α as a function of distance

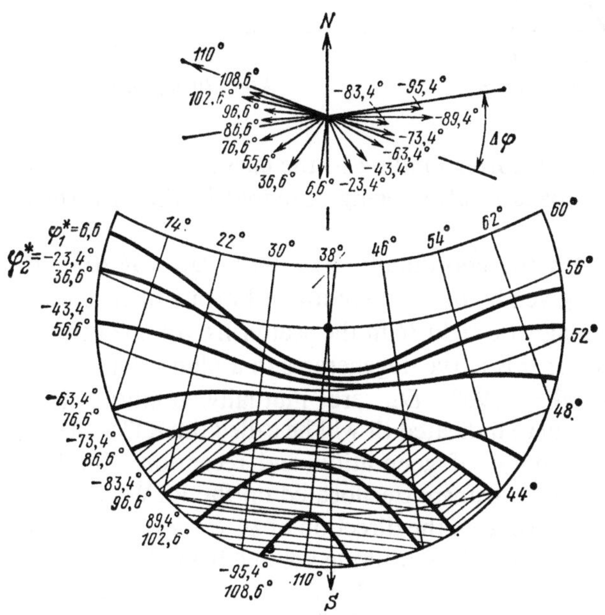

Fig. 7.10. Curves of the scattering cone intersections with the Earth for the modifier with coordinates $\varphi = 38°$, $\theta = 56°$. The magnetic meridian is shown which passes through the modifier location. The capture angles φ_1^*, φ_2^* are shown on the diagram

and, thus, to the position of the trasmitter on the Earth. Thus, specifying the trapped wave azimuth, one can find the transmitter position on the Earth.

Appropriate curves are plotted in Fig. 7.10. The scattering volume is assumed to be on the altitude, $z = 250$ km, near Moscow (the longitude, $\varphi = 38°$, latitude, $\theta = 56°$), where the magnetic dip angle is $I_H = 69.85°$, and the declination is $D_H = 6.6°$ (the magnetic field has been computed according to Guseva and Tyurmina (1976) and Cain et al. (1967) for the conditions of 1965). If the transmitter is placed on the curves shown in the figure and radiates at an angle $\alpha_0(\varphi)$ (see Fig. 7.11a), then the scattered energy will be sent horizontally (i.e., trapped on the sliding modes of the duct) in two directions of φ^*. The angle φ^* is read off from the southern direction (S): to the west, $\varphi_1^* > 0$, to the east $\varphi_2^* < 0$. Clearly if, on the contrary, the receiver is

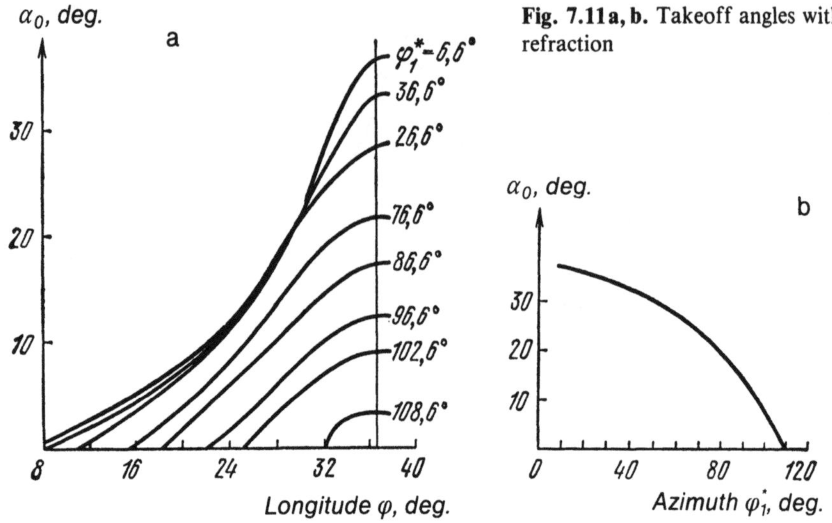

Fig. 7.11a, b. Takeoff angles with no refraction

placed on the Earth on locations corresponding to the curves in Fig. 7.10, it will receive signals which propagate, after being scattered in the disturbed region, near the sliding mode.

The curve in Fig. 7.10 with the coordinates $\varphi_1^* = 90° + D_H = 96.6°$ and $\varphi_2^* = -90° + D_H = -83.4°$ is given by the intersection of Earth's surface with the plane orthogonal to the magnetic field $\boldsymbol{H}$ in the scattering volume center, z_m. The backscattering cone for this curve degenerates to a plane. The directions, $\varphi_1^* = 90° + D_H$ and $\varphi_2^* = -90° + D_H$ are mutually opposite (see Fig. 7.10), because of this the curve indicated (it will be called the RTW curve) is distinguished from the viewpoint of generation and reception of RTW signals. Indeed, being emitted by a transmitter located on the RTW curve, a signal will propagate, after scattering, in the direction of φ_1^* (or φ_2^*) and, after round-the-world travelling, will reach the scattering volume from the opposite direction, $-\varphi_1^* + 2D_H$ (or, respectively, $-\varphi_2^* + 2D_H$). After a secondary scattering it again will appear on the Earth on the RTW curve. In the case of a RTW signal generation, the transmitter and receiver can be either superposed in one point on the RTW curve or separated. Note that the horizontal capture directions, $\varphi_{1,2}^* = \pm 90° + D_H$, for the RTW curve correspond to midlatitude path directions which are most preferable from the viewpoint of both the RTW signal occurrence (Sec. 4.1) and the minimum of absorption (Sec. 4.2).

As is seen from Fig. 7.10, for a transmitter located north from the RTW curve, the capture (or departure) occurs in southern directions. In this case, as follows from the diagram of Fig. 7.10, due to considerable dip angle, $I_H \approx 70°$, there is a finite angle range where, in the absence of refraction, no signal capture or descent to the Earth can be obtained: the quantity $|\cos\varphi^*| = |\cos\varphi_0 - \sin\alpha_{z_m} \cot\gamma_H|$ [Eq. (7.52)] proves to be greater than unity. In this angular range, the cone of reradiation is directed upward from the Earth with no intersections with its surface. Hence, in particular, it follows that in directions close to the meridional ones, for the above indicated way of energy input into a channel, in the absence of refraction, no RTW signals can be produced.

On the other hand, emission from regions in the vicinity of the RTW curve (proper directions are inside the angle $\Delta\varphi$ shown on the diagram of Fig. 7.10) has to lead to RTW signal occurrence. However, the transmitter and receiver for these directions must be separated. Appropriate regions for the transmitter and receiver positions are shown in different manners of shading. They are located in different sides with respect to the RTW curve. Transmitting energy in the region marked by the slanting hatches, one can detect RTW signals in the receiver placed in the region marked by the horizontal hatches, and vice versa. Vertical angles of emission on the Earth's surface, α_0, are unambiguously related to the distance between the transmitter and disturbed region.

In Fig. 7.11a the angles α_0 are shown vs the longitudes of emission points, φ, for various capture directions, φ_1^*, shown at the curves (see Fig. 7.10). They are shown also vs the capture azimuths, φ_1^*, on the magnetic meridian passing through the emission point (Fig. 7.11b). The emission angles are seen from the figure to be maximum in the plane of the magnetic meridian and near the scattering volume. These directions are advantageous from the energetic viewpoint.

Ray refractions were ignored above, and capture was considered only to trajectories close to the sliding one. More general numerical results for the nighttime ionospheric profile (Kim 1977) are shown in Figs. 7.12 and 7.13. The capture curves ($\beta_s^\alpha = 0$) are shown in Fig. 7.12. The capture regions are shown in Fig. 7.13 ($\beta^\alpha \leqslant \beta_{s0}^\alpha$, $z_m = 250$ km, $G \geqslant 10^{-4}$, $A = 7.5$ km). The arrow S indicates the geomagnetic meridian passing through the modifying transmitter (the fat dot in the figures). The distance in km is plotted on the corresponding straight line. The capture curves calculated with allowance for

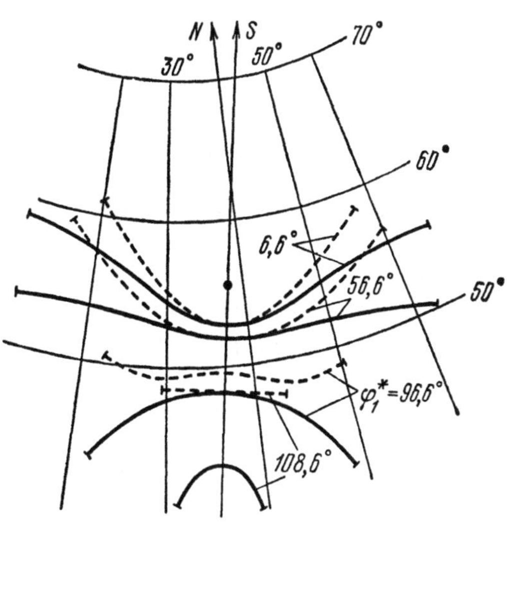

Fig. 7.12. Capture angle curves with allowance for refraction, for $f = 13$ MHz

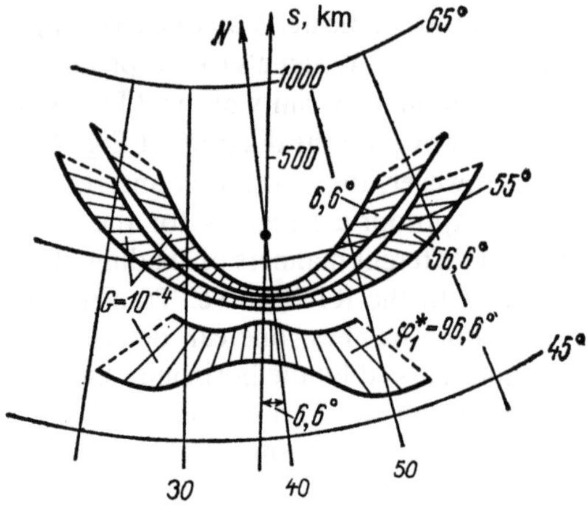

Fig. 7.13. Capture regions for $f = 13$ MHz, $a_{\parallel} = 0.3$ km, $q = 2 \cdot 10^{-3}$, $\overline{(\Delta N_0/N_{e0})^2} = 10^{-3}$

refraction (dashed line at Fig. 7.12) are seen to coincide with the curves calculated with neglect of refraction (solid lines, which are in accordance with Fig. 7.10) only at small distances S in the directions, $\varphi_1^* \approx 0$. At greater distances S, at small angles α, the discrepancy in distances S between the calculations with allowance for refraction and with neglect of refraction becomes great (see Fig. 7.9). Therefore, the proper capture curves diverge considerably. The allowance for a finite spectrum of capture angles, β_{s0}^{α}, and the vertical size of the disturbed region results in a smearing of the capture curves, i.e., in their transformation into bands with the widths $\sim 100 - 200$ km (see Fig. 7.13). In addition, the curves are distorted due to refraction, the distortion being especially strong at $f < f_{\text{MUF-}F}$.

The Output Coefficient. The energy output from a duct due to scattering by artificial irregularities can be characterized, as well as in the case of capture, by the output coefficient which is equal to the ratio of energy of scattered waves reaching the Earth's surface to the total energy in the ducts of waves incident on the irregularity. The differential output coefficient for waves propagating in the azimuth φ^* and descending to the Earth at angles α_1, φ_0 is given by

$$Q(\varphi^*, \alpha_1, \varphi_0) = \int_{-\beta_{s0}^{\alpha}}^{\beta_{s0}^{\alpha}} d\beta_s^{\alpha} \int_{z'(\beta_s^{\alpha})}^{z''(\beta_s^{\alpha})} \sigma(z, \alpha_1, \varphi_0, \varphi^*, \beta_s^{\alpha}) \frac{\beta_s^{\alpha} \sin \alpha_1}{(\beta^{\alpha})^2 \sin \alpha(z)} dz \,. \quad (7.61)$$

Here, as before, σ is the scattering cross-section [Eq. (6.8)], the angles β^{α}, φ^* characterize the wave in the duct, whereas α_1, φ_0 characterize a scattered wave descending to the Earth. The formula for the output coefficient is similar to that for the capture coefficient [Eq. (7.49)], being only slightly different in the integrand from the latter. The total output coefficient is given by the integration of Eq. (7.61) over α_1, φ_0.

A numerical calculation result for the differential output coefficient vs angles φ_0, α_1 is shown in Fig. 7.14. The calculation has been carried out for the daytime ionospheric model (Herbert 1967) for $f = 17$ MHz. It is seen that at small α_1, which correspond to long distances S, the value of $Q(\alpha_1, \varphi_0)$ is

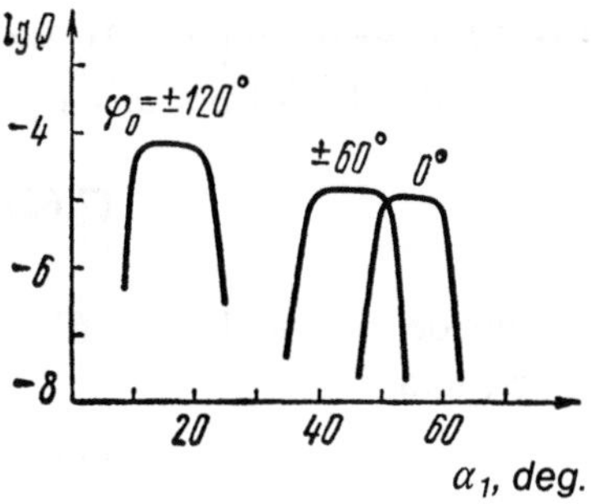

Fig. 7.14. Output coefficient for $f = 17$ MHz, $z_m = 210$ km, $\gamma_H = 27°$, $a_\parallel = 0.3$ km, $q = 10^{-2}$, $(\Delta N_0/N_{e0})^2 = 10^{-5}$

somewhat greater than that for greater angles α_1, which correspond to shorter distances. However, if we take into account a divergence of radio waves escaping from IWD, then it turns out that from the energetic point of view it is more advantageous to detect radio wave departure from ducts at great angles α_1, which correspond to a region on the Earth's surface near the projection of the scattering volume (see Fig. 7.11).

Let us draw our attention to a qualitative difference between the wave departure from the duct in the two cases: $f > f_{MUF}$ and $f < f_{MUF}$. In the first case, there is only one output region: the region of a direct ray departure and descent to the Earth after a scattering (region I in Fig. 7.15). In the second case, a ray can reach the Earth's surface after a reflection from the ionosphere as well (region II) (Kim 1977).

Large-Scale Stratification. We considered above a capture and escape induced by scattering only by small-scale irregularities formed in the ionosphere under the action of intense radio waves.

In the end of this section we also consider an influence of large-scale ionospheric striations. They must lead to intensification of multiple scattering considered in Sec. 6.2. To describe the effect of striations on radio wave

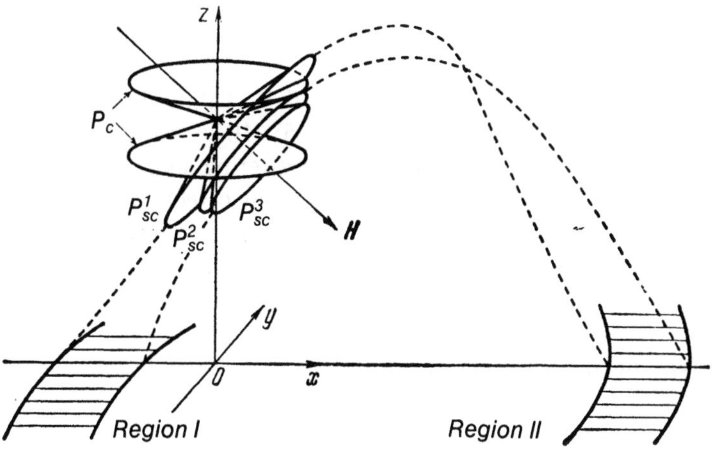

Fig. 7.15. Formation of two departure regions. Here the capture cone P_c and scattering cones P^1_{sc}, P^2_{sc}, P^3_{sc} are shown as well

capture, one can directly use Eqs. (6.77) and (6.77a). From Eq. (6.77), we obtain for trapped energy in the case of a duct without cut-off ($\omega \leqslant \omega_{\text{MUF-}E}$)

$$\Delta W = \frac{c}{\sqrt{\pi}} f_0 \sqrt{I_m \tau_1}\,, \tag{7.62}$$

where f_0 is the wave energy distribution function outside the duct [Eq. (6.67)], I_m is the maximum adiabatic invariant for the duct, and

$$\tau_1 \approx 2 R_0 \Theta_0 \int D_1(\theta')\, d\theta' \,. \tag{7.63}$$

Here, $D_1(\theta')$ is the coefficient of a diffusion caused by large-scale nonlinear stratification of the ionosphere. Roughly [Eq. (6.42)]

$$D_1 \approx \frac{\sqrt{\pi}}{4a} \overline{(\Delta\varepsilon)^2}\,, \tag{7.64}$$

where a is a characteristic striation scale. For ionospheric conditions, assuming $a \sim 1$ km, $\Delta N_0/N_{e0} \sim (1-2) \cdot 10^{-2}$, $\Delta\theta \approx (10^2\,\text{km}/R_0) \approx 0.02$, $2\Theta_0 \sim 1$, we have $\tau_1 \approx (1-5) \cdot 10^{-5}$. The capture coefficient in this case is $\sim 10^{-4}$. However, the energy here is trapped only to upper duct levels.

If the duct is cut off ($\omega > \omega_{\text{MUF-}E}$), then the trapped energy decreases sharply with increasing cut off angle. To estimate the capture, it is convenient to use the diffusion equation with respect to angles [Eqs. (6.34), (6.39), and (6.73)]. We have

$$f(\beta^\alpha) \approx \frac{f_0}{2}\left[1 - \Phi\left(\frac{\alpha-\beta^\alpha}{4\sqrt{\tau_1^*}}\right)\right] \approx \frac{\sqrt{\tau_1^*}}{(\alpha-\beta^\alpha)\sqrt{\pi}} f_0 \exp\left(-\frac{(\alpha-\beta^\alpha)^2}{4\tau_1^*}\right).$$

Here, $(\alpha-\beta^\alpha)$ is the cut-off angle (Sec. 6.1)

$$\tau_1^* = R_0 \int D_1(\theta)\, d\theta\,.$$

The capture coefficient under conditions $(\alpha-\beta^\alpha)^2 \gtrsim 4\tau_1^*$, is given by

$$G \approx C \exp\left\{-\frac{(\alpha_s-\beta_{s0}^\alpha)^2}{4\tau_1^*}\right\}, \tag{7.65}$$

where $(\alpha_s-\beta_{s0}^\alpha)$ is the minimum cut off-angle (Sec. 6.1). Taking into account that under ionospheric conditions, $2\tau_1^* \approx \tau_1 \approx (1-5) \cdot 10^{-5}$, we obtain that the coefficient of a nonlinear capture at multiple scattering sharply drops at cut-off angles as small as $\alpha_s - \beta_{s0}^\alpha \gtrsim 0.5-1°$. Note that effects of the same order of magnitude can also be produced by a curvature of ray trajectory on an isolated large-scale irregularity (Baranov et al. 1977) (see Sec. 3.4).

7.2.3 Effect of Nonlinear Scattering on Long-Distance Radio Wave Propagation

Self-Focusing Instability. In the case of long-distance propagation, the radio waves emitted at small angles to the horizon are of importance [Eq. (1.5)].

Irregular structure which is intensified in the F-region under the action of the field of such radio waves, if their power is great enough, can affect essentially both themselves and other radio waves propagating in the disturbed region.

Qualitatively the process of the growth of irregularities evolved due to self-focusing instability is as follows. In a plasma region where the radio wave field amplitude increases due to irregularity, simultaneously a stronger plasma heating takes place. It induces a further electron and ion density drop in the irregularity due to the plasma expulsion from the heated region. The density decrease leads to the growth of the refractive index, $n = \sqrt{\varepsilon} = (1 - 4\pi e^2 N_e/m\omega^2)^{1/2}$. This results in a curvature of wave trajectories, they converge, focusing in the region of increased n. Consequently, in this region the wave field also increases, the heating here intensifies, the plasma expulsion and focusing become still stronger and so on.

To obtain an idea of the theory of irregularity intensification induced by the self-focusing instability we consider first its simplest form, in the framework of the geometrical optics. To this end we use Eqs. (7.5) and (7.6) in a homogeneous focusing medium

$$\begin{aligned} &\left(\frac{1}{R_0}\frac{\partial}{\partial\theta} + \boldsymbol{u}\nabla_\perp\right)\boldsymbol{u} - \gamma\nabla_\perp w = 0 \\ &\frac{1}{R_0}\frac{\partial w}{\partial\theta} + \nabla_\perp(\boldsymbol{u}\,w) = 0\,. \end{aligned} \tag{7.66}$$

Here, we have taken into consideration the fact that unlike the lower ionosphere [Eq. (7.1)], the F-layer is a focusing medium: the electron density here decreases under the action of intense radio waves, $\Delta N_e/N_{e0} = -\gamma_3\,\mathscr{E}_0^2/\mathscr{E}_p^2$, $\gamma = 4\pi\omega_0^2\gamma_3/\omega^2\,\mathscr{E}_p^2$.

Let us consider now small perturbations δw, $\delta\boldsymbol{u}$ in the ray intensity, w, and directions, $\boldsymbol{u}$

$$w = w_0 + \delta w\,, \qquad \boldsymbol{u} = \boldsymbol{u}_0 + \delta\boldsymbol{u}\,. \tag{7.67}$$

Substituting Eq. (7.67) in Eq. (7.66) and linearizing the latter with respect to δw, $\delta\boldsymbol{u}$, we obtain

$$\begin{aligned} &\frac{1}{R_0}\frac{\partial\delta w}{\partial\theta} + u_0\frac{\partial\delta w}{\partial\rho} + w_0\frac{\partial\delta\boldsymbol{u}}{\partial\rho} = 0 \\ &\frac{1}{R_0}\frac{\partial\delta u}{\partial\theta} + u_0\frac{\partial\delta u}{\partial\rho} - \gamma\frac{\partial\delta w}{\partial\rho} = 0\,. \end{aligned} \tag{7.68}$$

The solution of Eq. (7.68) is sought as usual by the Fourier expansion

$$\begin{aligned} \delta w &= \int \delta w_q \exp[i(q_\theta\theta R_0 + q_\rho\rho)]\,d\boldsymbol{q} \\ \delta\boldsymbol{u} &= \int \delta\boldsymbol{u}_q \exp[i(q_\theta\theta R_0 + q_\rho\rho)]\,d\boldsymbol{q}\,. \end{aligned} \tag{7.69}$$

Substituting Eq. (7.69) in Eq. (7.68), we obtain a dispersion equation

$$(q_\theta+q_p u_0)^2+\gamma q_p^2 w_0=0\,. \tag{7.70}$$

Hence,

$$q_\theta=-q_p u_0\pm i q_p\sqrt{\gamma w_0}\,. \tag{7.71}$$

Substituting Eq. (7.71) in Eq. (7.70), we make sure that small perturbations in the ray intensity and directions increase exponentially rapidly as the wave penetrates into the interior of the plasma

$$\delta w\sim\exp[q_p\sqrt{\gamma w_0}R_0(\theta-\theta_0)]\,. \tag{7.72}$$

Thus, a smooth initial distribution of wave intensity in a focusing medium, in the presence of weak initial irregularities, becomes unstable, i.e., the beam gradually laminates into narrow focused filaments. Simultaneously, medium irregularities evolve and intensify.

Under actual ionospheric conditions, the character of irregularities is strongly influenced by a longitudinal diffusion, heat conductivity, wave effects, i.e., the finite wave length and smooth regular ionospheric inhomogeneity are of importance here. The analysis of self-focusing instability evolution with allowance for all these factors has been carried out by Borisov et al. (1978) and Vas'kov and Gurevich (1979). It was shown that the instability develops when the field amplitude exceeds some threshold value which under ionospheric conditions is rather small, $\mathscr{E}_0\lesssim 0.1$ V/m. In this case, the instability, as well as before [Eq. (7.71)], is of a convective type, i.e., small initial perturbations in the plasma and wave field increase as the wave propagates in the medium. This increase is characterized by the relation

$$\delta N=\delta N_e\exp(R_0\textstyle\int\varkappa d\theta)\,. \tag{7.73}$$

Here, $\varkappa$ is a spatial growth rate for the self-focusing instability. It increases with the field amplitude and plasma density. The integration in Eq. (7.73) is carried out over the whole instability region, where $\varkappa\geqslant 0$. The most intense perturbation increase occurs on the upper trajectory part, since it is the region where the character of wave propagation is especially close to that in a homogeneous medium. By the same reason, the effect is intensified, as the wave frequency becomes close to the MUF-*F*.

Estimations show that for radio waves with the frequency close to the MUF-*F*, the exponent in Eq. (7.73) is of the order of unity at the wave amplitude $\mathscr{E}_0\sim 0.1$ V/m. Thus, radio waves when $\mathscr{E}_0\gtrsim 0.1$ V/m can strongly affect the irregular ionospheric structure. The typical transverse scale (i.e., the scale transversal to the Earth's magnetic field) for irregularities, which grow most rapidly and for which the growth rate, $\varkappa$, has the optimum, is of the order of 200 – 300 m. Their longitudinal scale is much greater.

Effects of nonlinear scattering were observed experimentally in the ionosphere in the case of interaction between the intense and weak (sounding) signal. Both waves possessed close frequencies and they were emitted at a small angle with respect to the horizon in either the same direction (Bochkarev

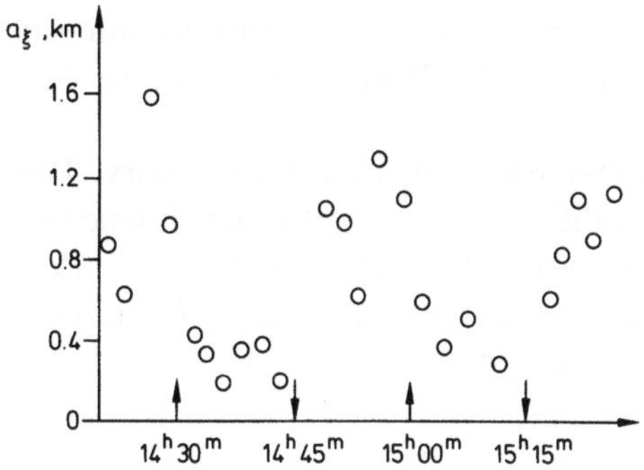

Fig. 7.16. Variation of the effective size *a* of ionospheric irregularities under the periodic action of the modifier. The moments of time when the modifier was turned on or turned off are indicated by *arrows*

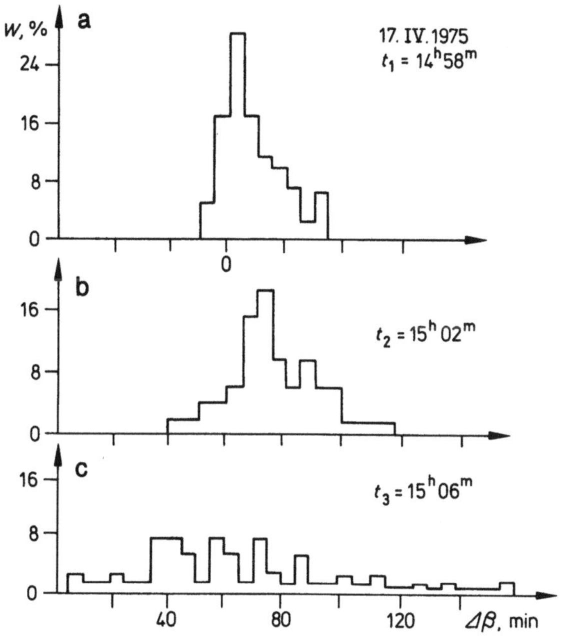

Fig. 7.17a–c. Variation of the arrival angle fluctuations of the sounding signal under the action of the modifier; $\Delta\beta$ is the arrival angle deviation (relative to a certain average value of β) in the horizontal plane; W is the probability of occurrence of $\Delta\beta$ value in corresponding angular range (5′). **a** Before the modifier is turned on; **b** 2 min after the modifier is turned on; **c** 6 min after the modifier is turned on

et al. 1979) or opposite to one another (Novozhilov and Savel'ev 1978). Figure 7.16 shows the variations in effective irregularity scales when the transmitter was turned on and turned off periodically within the period of 15 min. The effective irregularity scale is seen to vary under the action of the powerful radio wave, reducing from 0.6 – 1.5 km to 200 – 300 m.

Figure 7.17 presents histograms of variations of sounding signal angles of arrival, i.e., fluctuations of bearing. It is seen that in 6 min after the modifying turn-on, the angular spectrum of signals was expanded roughly by a factor of 3. Comparing these results with the theory [Eqs. (6.132) and (6.134)], one can determine the coefficient of diffusion with respect to horizontal angles induced by wave scattering by artificial irregularities. It proves to be $D_{\beta\beta} = 5 \cdot 10^{-12}\ \mathrm{cm}^{-1}$. Using the approximate expression for the diffusion coefficient [Eq. (7.64)], one can obtain the magnitude of density perturbations in irregularities, $\overline{(\Delta N_e/N_{e0})^2} \approx 1 \cdot 10^{-4}$. The typical irregularity scale, a, has been assumed to be equal to 200 – 300 m in accordance with Fig. 7.16.

The experimental results are in reasonable agreement with the theory. This indicates that high-power modifying facilities do exert a strong influence on irregular ionospheric structure. Furthermore, it is possible that a marked fraction of existing large-scale irregularities in the ionospheric *F*-region is produced by the process considered, i.e., has an artificial origin.

8 Radio Wave Propagation in the Magnetosphere

The electron density in the magnetosphere is comparatively small, $N_e < 3 \cdot 10^3\,\mathrm{cm}^{-3}$, so that the magnetospheric plasma, in general, affects rather slightly HF radio wave propagation: when $\omega_0^2 \ll \omega^2$, $\varepsilon_0 \approx 1$. However, the magnetospheric plasma, owing to its strong rarity, diffuses rather slowly transverse to the Earth's magnetic lines of force and transports fast along the lines. Therefore, the irregularities elongated along $\boldsymbol{H}$ are readily formed here. They evolve especially readily in sunrise – sunset periods. In this time, they could be lengthened along the whole line of force between the magnetic – conjugate points in the ionosphere. Such irregularities form waveguides for HF propagation. In spite of electron density perturbations being relatively small (dielectric constant perturbations, $\Delta\varepsilon \ll \varepsilon_0$), the irregularities can confine radio waves travelling in the neighborhood of the waveguide axis.

The present chapter is devoted to the study of wave propagation in magnetospheric waveguides.

8.1 Electromagnetic Wave Guiding by Weak Waveguides

Let us consider an electromagnetic wave propagating through a homogeneous or slightly inhomogeneous medium in the presence of irregularities which produce slight dielectric constant perturbations

$$\Delta\varepsilon \ll \varepsilon_0 . \tag{8.1}$$

If the irregularities are strongly elongated (Fig. 8.1)

$$R \gg a \gg \lambda , \tag{8.2}$$

then they can form ducts for electromagnetic wave guiding. In Eq. (8.2), λ is the wave length, a is the effective irregularity width, R is a characteristic longitudinal length of irregularity variation caused by its bending or variations of the form, dimensions, and magnitude of the dielectric constant perturbation, $\Delta\varepsilon$.

Waveguide channels formed by irregularities under conditions given by Eqs. (8.1) and (8.2) are called slight waveguides (ducts). They can confine only the waves which travel in directions close to the waveguide axis. This enables one to simplify the consideration and to analyze general properties of wave

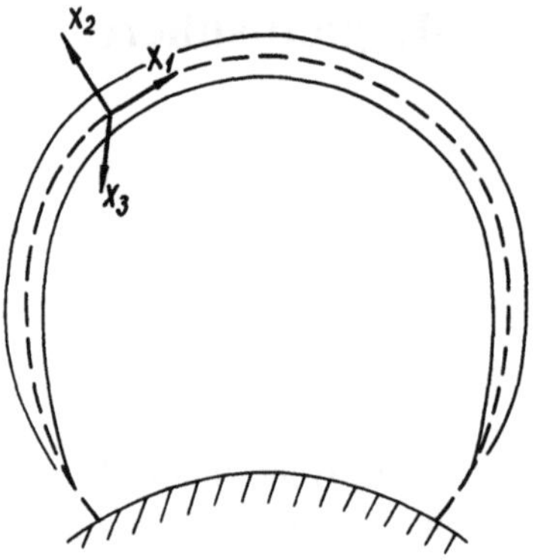

Fig. 8.1. Curvilinear coordinates in a weak waveguide. *Dashed line* shows the waveguide axis

propagation in slight waveguides. This is the subject of the present section (Gurevich 1979).

Under magnetospheric conditions, the longitudinal length R in Eq. (8.2) is rather great: of the order of the Earth's radius. The transverse scales, a, are usually much less than R (see Sec. 1.1). Thus, Eqs. (8.1) and (8.2) are satisfied in a broad range of magnetospheric irregularity parameters. Guiding becomes possible of electromagnetic waves of various types: HF, VLF (whistlers), magnetohydrodynamic, and other low frequency types of radiation.

We consider now electromagnetic wave propagation in a slight waveguide which obeys Eqs. (8.1) and (8.2). Let x_1, x_2, x_3 be orthogonal curvilinear coordinates: the coordinate x_1 is directed along the duct; x_2, x_3 are transverse coordinates (Fig. 8.1). The geometrical – optics Eq. (3.3) is

$$\boldsymbol{k}^2=\frac{\omega^2}{c^2}\varepsilon\,. \tag{8.3}$$

The wave vector $\boldsymbol{k}$ gives at any point $\boldsymbol{r}$ the direction of ray propagation, i.e., it is tangent to the ray trajectory. Analogously to Sec. 3.1, from Eq. (8.3), we derive the ray trajectory equations in the curvilinear coordinate system, x_1, x_2, x_3. We take into account that

$$\boldsymbol{k}=\nabla S=\left\{\frac{1}{h_1}\frac{\partial S}{\partial x_1},\ \frac{1}{h_2}\frac{\partial S}{\partial x_2},\ \frac{1}{h_3}\frac{\partial S}{\partial x_3}\right\}\quad k_\alpha=\frac{1}{h_\alpha}\frac{\partial S}{\partial x_\alpha}\,. \tag{8.4}$$

Here, S is the eikonal [Eq. (3.1)], and h_1, h_2, h_3 are Lame coefficients for the coordinate system x_1, x_2, x_3. Substituting Eq. (8.4) in Eq. (8.3) and differentiating successively with respect to dx_α, we obtain

$$\frac{\partial}{\partial x_\alpha}\left[\frac{1}{h_1^2}\left(\frac{\partial S}{\partial x_1}\right)^2+\frac{1}{h_2^2}\left(\frac{\partial S}{\partial x_2}\right)^2+\frac{1}{h_3^2}\left(\frac{\partial S}{\partial x_3}\right)^2\right]=\frac{\omega^2}{c^2}\frac{\partial\varepsilon}{\partial x_\alpha}$$

$$\alpha=1,2,3\,. \tag{8.5}$$

Let us transform the left-hand side of the first Eq. (8.5) ($\alpha=1$)

$$\frac{\partial}{\partial x_1}\left[\frac{1}{h_1^2}\left(\frac{\partial S}{\partial x_1}\right)^2+\frac{1}{h_2^2}\left(\frac{\partial S}{\partial x_2}\right)^2+\frac{1}{h_3^2}\left(\frac{\partial S}{\partial x_3}\right)^2\right]$$

$$=2\left[\frac{1}{h_1}\frac{\partial S}{\partial x_1}\frac{\partial}{\partial x_1}\left(\frac{1}{h_1}\frac{\partial S}{\partial x_1}\right)-\frac{1}{h_2}\frac{\partial h_2}{\partial x_1}\left(\frac{1}{h_2}\frac{\partial S}{\partial x_2}\right)^2\right.$$

$$+\frac{1}{h_2}\frac{\partial S}{\partial x_2}\left(\frac{1}{h_2}\frac{\partial^2 S}{\partial x_2\partial x_1}\right)-\frac{1}{h_3}\frac{\partial h_3}{\partial x_1}\left(\frac{1}{h_3}\frac{\partial S}{\partial x_3}\right)^2$$

$$\left.+\frac{1}{h_3}\frac{\partial S}{\partial x_3}\left(\frac{1}{h_3}\frac{\partial^2 S}{\partial x_3\partial x_1}\right)\right]$$

$$=2\left(k_1\frac{\partial k_1}{\partial x_1}-\frac{1}{h_2}\frac{\partial h_2}{\partial x_1}\cdot k_2^2+\frac{h_1}{h_2}k_2\frac{\partial k_1}{\partial x_2}+\frac{1}{h_2}\frac{\partial h_1}{\partial x_2}k_1k_2\right.$$

$$\left.-\frac{1}{h_3}\frac{\partial h_3}{\partial x_1}k_3^2+\frac{h_1}{h_3}k_3\frac{\partial k_1}{\partial x_3}+\frac{1}{h_3}\frac{\partial h_1}{\partial x_3}k_1k_3\right). \tag{8.6}$$

Substituting now Eq. (8.6) in Eq. (8.5), we write Eq. (8.5) for $\alpha=1$ in the form

$$h_1\left(\frac{k_1}{h_1}\frac{\partial k_1}{\partial x_1}+\frac{k_2}{h_2}\frac{\partial k_1}{\partial x_2}+\frac{k_3}{h_3}\frac{\partial k_1}{\partial x_3}\right)=F_1$$

$$F_1=\frac{\omega^2}{2c^2}\frac{\partial\varepsilon}{\partial x_1}+\frac{k_2^2}{h_2}\frac{\partial h_2}{\partial x_1}+\frac{k_3^2}{h_3}\frac{\partial h_3}{\partial x_1}-\frac{k_1k_2}{h_2}\frac{\partial h_1}{\partial x_2}-\frac{k_1k_3}{h_3}\frac{\partial h_1}{\partial x_3}. \tag{8.7}$$

The two remaining equations [Eq. (8.5)], $\alpha=2$, $\alpha=3$ are transformed in a similar way, yielding

$$h_2\left(\frac{k_1}{h_1}\frac{\partial k_2}{\partial x_1}+\frac{k_2}{h_2}\frac{\partial k_2}{\partial x_2}+\frac{k_3}{h_3}\frac{\partial k_2}{\partial x_3}\right)=F_2$$

$$h_3\left(\frac{k_1}{h_1}\frac{\partial k_3}{\partial x_1}+\frac{k_2}{h_2}\frac{\partial k_3}{\partial x_2}+\frac{k_3}{h_3}\frac{\partial k_3}{\partial x_3}\right)=F_3 \tag{8.8}$$

$$F_2=\frac{\omega^2}{2c^2h_1^2}\frac{\partial}{\partial x_2}(h_1^2\varepsilon)-\frac{k_2^2+k_3^2}{h_1}\frac{\partial h_1}{\partial x_2}+\frac{k_3^2}{h_3}\frac{\partial h_3}{\partial x_2}-\frac{k_1k_2}{h_1}\frac{\partial h_2}{\partial x_1}$$

$$-\frac{k_2k_3}{h_3}\frac{\partial h_2}{\partial x_3}$$

$$F_3=\frac{\omega^2}{2c^2h_1^2}\frac{\partial}{\partial x_3}(h_1^2\varepsilon)-\frac{k_2^2+k_3^2}{h_1}\frac{\partial h_1}{\partial x_3}+\frac{k_2^2}{h_2}\frac{\partial h_2}{\partial x_3}-\frac{k_1k_3}{h_1}\frac{\partial h_3}{\partial x_1}$$

$$-\frac{k_2k_3}{h_2}\frac{\partial h_3}{\partial x_2}. \tag{8.9}$$

In the expressions for F_2, F_3 [Eq. (8.9)] the quantity k_1^2 has been eliminated through Eq. (8.3)

$$k_1^2 = \frac{\omega^2}{c^2}\varepsilon - k_2^2 - k_3^2 .$$

For the spherical coordinates

$$x_1 = \theta , \quad x_2 = r , \quad x_3 = \varphi$$

the Lame coefficients are

$$h_1 = r , \quad h_2 = 1 , \quad h_3 = r\sin\theta .$$

In this case, one can easily recognize that Eqs. (8.7) – (8.9) coincide with Eqs. (3.4a) – (3.4c) ($k_1 = k_\theta$, $k_2 = k_r$, $k_3 = k_\varphi$). The characteristic equations for the system of Eqs. (8.7) – (8.9) are

$$\frac{h_1}{k_1}dx_1 = \frac{h_2}{k_2}dx_2 = \frac{h_3}{k_3}dx_3 = \frac{h_1}{F_1}dk_1 = \frac{h_2}{F_2}dk_2 = \frac{h_3}{F_3}dk_3 . \tag{8.10}$$

Hence, we have

$$\frac{dx_2}{dx_1} = \frac{h_1 k_2}{h_2 k_1} , \quad \frac{dx_3}{dx_1} = \frac{h_1 k_3}{h_3 k_1} . \tag{8.11}$$

Differentiating now Eq. (8.11) with respect to dx_1 and allowing for the fact that by Eq. (8.10)

$$\frac{dk_1}{dx_1} = \frac{F_1}{k_1} , \quad \frac{dk_2}{dx_1} = \frac{h_1}{h_2}\frac{F_2}{k_1} , \quad \frac{dk_3}{dx_1} = \frac{h_1}{h_3}\frac{F_3}{k_1}$$

we represent the characteristic equations, i.e., the equations for ray trajectories, in the form

$$\frac{d^2x_2}{dx_1^2} = \frac{h_1^2}{k_1^2 h_2^2}F_2 + \frac{k_2}{k_1}\left[\frac{d}{dx_1}\left(\frac{h_1}{h_2}\right) - \frac{h_1 F_1}{h_2 k_1^2}\right]$$

$$\frac{d^2x_3}{dx_1^2} = \frac{h_1^2}{k_1^2 h_3^2}F_3 + \frac{k_3}{k_1}\left[\frac{d}{dx_1}\left(\frac{h_1}{h_3}\right) - \frac{h_1 F_1}{h_3 k_1^2}\right] . \tag{8.12}$$

For spherical coordinates, Eq. (8.12) coincides with Eqs. (3.6) and (3.7).

Now let us simplify Eq. (8.12) for waves trapped in slight waveguides. Bearing in mind that according to Eq. (8.1), a wave can propagate in such a duct only in the direction close to the waveguide axis

$$k_2^2 , \quad k_3^2 \ll k_1^2 \tag{8.13}$$

we obtain

$$k_1^2 \approx \frac{\omega^2}{c^2}\varepsilon_0^2 = k_0^2 . \tag{8.14}$$

Equation (8.14) follows from Eqs. (8.3) and (8.13). Here ε_0 is the dielectric constant in the absence of irregularities. $\varepsilon = \varepsilon_0 + \Delta\varepsilon$; k_0 is the absolute value of the wave vector. By Eq. (8.13) we have

$$\frac{k_2}{k_0} \ll 1\,, \qquad \frac{k_3}{k_0} \ll 1\,.$$

It can be easily seen now that in the right-hand sides of Eq. (8.12) all the terms, excluding the first ones, are small in the order of k_2/k_0 and k_3/k_0 and, consequently, may be neglected. In this approximation, all the terms, excluding the first ones, should be omitted in Eq. (8.9) for F_2 and F_3. Equation (8.12) takes then a simple form

$$\frac{d^2x_2}{dx_1^2} = \frac{1}{2\varepsilon_0 h_2^2}\,\frac{\partial}{\partial x_2}(h_1^2\varepsilon)\,, \qquad \frac{d^2x_3}{dx_1^2} = \frac{1}{2\varepsilon_0 h_3^2}\,\frac{\partial}{\partial x_3}(h_1^2\varepsilon)\,. \tag{8.15}$$

Next, we take into account that by Eq. (8.2), the characteristic transverse size of the waveguide duct

$$a \sim |\Delta\varepsilon|/|\mathrm{grad}\,\Delta\varepsilon| \tag{8.16}$$

is much less than its longitudinal size which is determined by the smooth medium inhomogeneity

$$R \sim \frac{h_i}{|\mathrm{grad}\,h_i|}\,, \qquad \frac{\varepsilon_0}{|\mathrm{grad}\,\varepsilon_0|}\,. \tag{8.17}$$

Hence, the parameters h_i, ε_0 in directions transverse to the duct axis vary weakly. Therefore, it is convenient to single out the duct axis, i.e., the locus of the maximum perturbations $\Delta\varepsilon$ in the planes x_2, x_3 (i.e., in the surfaces $x_1 = \mathrm{const}$, see Fig. 8.1). Let the duct axis be defined by the relations

$$x_2 = x_{20}\,, \qquad x_3 = x_{30}\,. \tag{8.18}$$

In its neighborhood we represent the Lame coefficients, h_2, h_3, and undisturbed dielectric constant, ε_0, in the form of a series

$$\begin{aligned} h_2(x_1,x_2,x_3) &= h_2(x_1,x_{20},x_{30}) + \frac{\partial h_2}{\partial x_2}(x_1,x_{20},x_{30})(x_2-x_{20}) \\ &\quad + \frac{\partial h_2}{\partial x_3}(x_1,x_{20},x_{30})(x_3-x_{30}) + \dots\,. \end{aligned} \tag{8.19}$$

According to Eqs. (8.2), (8.16), and (8.17), within the duct $|x_2-x_{20}| \lesssim a$, $|x_3-x_{30}| \lesssim a$, the ratios of the second and third terms of the series [Eq. (8.19)] to the first one are small: namely, of the order of a/R. Hence, by neglecting small terms, one can put

$$h_2 \approx h_2(x_1,x_{20},x_{30}) = h_{20}(x_1)\,.$$

Analogously

$$h_3 = h_3(x_1, x_{20}, x_{30}) = h_{30}(x_1)$$
$$\varepsilon_0 = \varepsilon_0(x_1, x_{20}, x_{30}) = \varepsilon_0(x_1)\,. \tag{8.20}$$

In the same approximation, the coefficient h_1 should be taken with more accuracy by reason of the fact that in Eq. (8.15) it is differentiated with respect to transverse coordinates. We have

$$h_1^2(x_1, x_2, x_3) = h_{10}^2(x_1, x_{20}, x_{30}) + \frac{\partial h_1^2}{\partial x_2}(x_1, x_{20}, x_{30}) \cdot (x_2 - x_{20})$$
$$+ \frac{\partial h_1^2}{\partial x_3}(x_1, x_{20}, x_{30}) \cdot (x_3 - x_{30})\,. \tag{8.21}$$

Substituting now Eqs. (8.19) – (8.21) in Eq. (8.15), taking into account Eqs. (8.1) and (8.2) and going over to new variables, t, y, z, defined by

$$t(x_1) = \int h_{10}(x_1)\,dx_1\,, \quad x_1 = x_1(t)\,,$$
$$y = h_{20}(t)(x_2 - x_{20})\,, \tag{8.22}$$
$$z = h_{30}(t)(x_3 - x_{30})\,,$$

instead of Eq. (8.15), we obtain the equations

$$\frac{d^2y}{dt^2} = -\frac{1}{2}\frac{\partial U}{\partial y}\,, \quad \frac{d^2z}{dt^2} = -\frac{1}{2}\frac{\partial U}{\partial z}\,;$$
$$U(y, z, t) = -\frac{\Delta\varepsilon(y, z, t)}{\varepsilon_0(t)} - \frac{2y}{r_y(t)} - \frac{2z}{r_z(t)}\,, \tag{8.23a}$$

$$r_y = \frac{2h_{10}^2 h_{20}}{(\partial h_1^2/\partial x_2)_{x_1, x_{20}, x_{30}}}\,; \quad r_z = \frac{2h_{10}^2 h_{30}}{(\partial h_1^2/\partial x_3)_{x_1, x_{20}, x_{30}}}\,. \tag{8.23b}$$

Here U is the effective potential (Sec. 2.1), r_y, r_z are the radii of curvature of irregularity bending in the planes x_1, x_2 and x_1, x_3, respectively.

Equations (8.23a) and (8.23b) are finite. They have undergone all the possible simplifications. They have a simple physical analog: they describe a two-dimensional motion of the particle in the potential field, $U(y, z, t)$. The longitudinal coordinate t plays a role of time. In the one-dimensional case, Eqs. (8.23a, b) are identical to Eqs. (3.12) and (3.13).

If the potential U has a minimum with respect to y, z, then the wave can be guided by the irregularity. The trapped ray propagates in the direction of the duct axis, t, and oscillates in the transverse directions, y, z, similarly to a particle in the potential well, $U(y, z, t)$. Near the duct axis, i.e., in the vicinity of its minimum, $U_m(y_m, z_m)$, the potential U has a parabolic form

$$U(y, z, t) = U_m + \alpha(y - y_m)^2 + \beta(z - z_m)^2 + \gamma(y - y_m)(z - z_m)$$
$$\alpha = \frac{1}{2}\left(\frac{\partial^2 U}{\partial y^2}\right)_m, \quad \beta = \frac{1}{2}\left(\frac{\partial^2 U}{\partial z^2}\right)_m, \quad \gamma = \left(\frac{\partial^2 U}{\partial y\,\partial z}\right)_m \tag{8.24}$$

where U_m, α, β, γ, in general, depend on t. Going over in Eq. (8.24) to the normal coordinates by a rotation of the axes $z' = z - z_m$, $y' = y - y_m$

$$z^* = (z'\cos\theta + y'\sin\theta)\,, \qquad y^* = (-z'\sin\theta + y'\cos\theta)$$

we diagonalize U

$$U(y^*, z^*, t) = U_m + \omega_y^2 (y^*)^2 + \omega_z^2 (z^*)^2\,. \tag{8.25}$$

The rotation angle is $\theta = \frac{1}{2}\arctan[\gamma/(\beta-\alpha)]$, effective frequencies are given by

$$\omega_y^2 = \frac{\alpha+\beta+\sqrt{(\alpha-\beta)^2+\gamma^2}}{2}$$
$$\omega_z^2 = \frac{\alpha+\beta-\sqrt{(\alpha-\beta)^2+\gamma^2}}{2}\,. \tag{8.26}$$

Equations (8.23 a, b) in the new variables take a simple form of the linear oscillator equations

$$\frac{d^2y^*}{dt^2} + \omega_y^2 y^* = 0\,, \qquad \frac{d^2z^*}{dt^2} + \omega_z^2 z^* = 0\,. \tag{8.27}$$

Note that the oscillation frequencies, ω_y, ω_z, by the order of magnitude are equal to $\sqrt{U_m}/a h_{i0}$, because of the quantities α, β, γ, being, by Eqs. (8.2), (8.24) and (8.22), of the order of $U_m/a^2 h_{i0}^2$. Hence, the oscillation period in the duct is given by

$$T = \frac{2\pi}{\omega} \sim \frac{2\pi a h_{i0}}{\sqrt{U_m}}\,. \tag{8.28}$$

Of importance is the fact that the frequencies ω_y and ω_z depend on t. The characteristic length of variation of the frequencies, as well as of the quantities h_{i0}, is R [Eq. (8.17)]. If the condition for adiabaticity [Eq. (3.15)] holds

$$\frac{T}{2\pi}\,\frac{1}{\omega}\,\frac{d\omega}{dt} \approx \frac{a h_{i0}}{R\sqrt{U_m}} \ll 1 \tag{8.29}$$

then, to exponential accuracy, Eq. (8.27) has a simple solution

$$y^* = y^*(t_0)\sqrt{\frac{\omega_y(t_0)}{\omega_y(t)}}\exp\left(i\int_{t_0}^{t}\omega_y dt\right)$$
$$z^* = z^*(t_0)\sqrt{\frac{\omega_z(t_0)}{\omega_z(t)}}\exp\left(i\int_{t_0}^{t}\omega_z dt\right)\,. \tag{8.30}$$

Equation (8.27) has approximate integrals, namely, adiabatic invariants

$$I_y = \pi\omega_y (y_0^*)^2\,, \qquad I_z = \pi\omega_z (z_0^*)^2\,. \tag{8.31}$$

Here, y_0^* and z_0^* are the oscillation amplitudes. From the conservation of the invariants I_y, I_z it follows that the oscillation amplitudes change with t inversely proportional to $\sqrt{\omega(t)}$, in accordance with Eq. (8.30).

As follows from Eqs. (8.2), (8.23), and (8.24), the adiabaticity condition [Eq. (8.29)] is always satisfied for sufficiently deep and narrow ducts, when the perturbations are fairly great. Adiabatic invariants in these cases exist, of course, not only for parabolic ducts. In order that adiabatic invariants can exist in the general case, in addition to Eq. (8.29), the condition is needed for the possibility of separation of variables in the action function which describes the two-dimensional motion of a particle in the potential field U [Eq. (8.23b)] (Landau and Lifshitz 1976, Sec. 5.2). For a one-dimensional potential, $U(y, t)$, this condition is always true. By using the adiabatic invariants, a general approximate solution for the ray trajectory can be constructed which is analogous to that discussed in detail earlier (see Sec. 3.2).

We note that the same irregularities can serve for guiding of both HF and other relatively low frequency modes in the magnetosphere. The theory developed describes, in principle, the guiding by slight waveguides of all types of electromagnetic waves. However, when deriving the basic equations, the dielectric constant ε was assumed (for simplification of the formulas) to depend on the coordinates $\boldsymbol{r}$ and wave frequency ω and be independent of the wave vector $\boldsymbol{k}$. For HF and VLF waves in the magnetosphere this is true. For other waves, the dependence ε on $\boldsymbol{k}$ can be important. In this case, when differentiating with respect to the coordinates in the right-hand side of Eq. (8.5), in addition to the derivatives $\partial\varepsilon/\partial x_\alpha$ the derivatives $\partial\varepsilon/\partial k_i \cdot \partial k_i/\partial x_\alpha$ should be taken into account. This results in some change of the equations.

8.2 HF Radio Waves in the Magnetospheric Ducts

Magnetospheric Wave Ducts. Magnetospheric plasma irregularities are elongated along the Earth's magnetic field lines of force. Accordingly, the curvilinear coordinate system x_1, x_2, x_3 should be chosen in such a way that the axis x_1 be directed along the line of force, and the axes x_2, x_3 be directed orthogonal to it

$$x_1 = s\,, \quad x_2 = L\,, \quad x_3 = \varphi\,. \tag{8.32}$$

Here, s is the coordinate along the line of force, L is the McIlwain parameter, and φ is the magnetic longitude. Assuming the magnetic field to be dipole, one can express the coordinates s, L, φ in terms of standard spherical coordinates r, θ, φ by the following relations

$$s = \frac{r^2}{R_0^2 \cos\theta}\,, \quad L = \frac{r}{R_0 \sin^2\theta}\,, \quad \varphi \equiv \varphi\,. \tag{8.33}$$

Here, R_0 is the Earth's radius, θ is the geomagnetic co-latitude, i.e., the angle read off from the magnetic dipole axis.

The Lame coefficients for the coordinate system [Eq. (8.32)] can be obtained by the following way. From Eq. (8.33) we have

$$dr = \frac{R_0 \sin^4\theta}{1+3\cos^2\theta} dL + \frac{2R_0^2}{r} \frac{\cos^3\theta}{1+3\cos^2\theta} ds$$

$$d\theta = -\frac{2R_0 \sin^3\theta\cos\theta}{r(1+3\cos^2\theta)} dL + \frac{R_0^2 \sin\theta\cos^2\theta}{r^2(1+3\cos^2\theta)} ds\,.$$

Substituting these relations in the expression for the square of arc length in the spherical coordinate system, $dl^2 = dr^2 + r^2 d\theta^2 + r^2 \sin^2\theta d\varphi^2$, we obtain

$$dl^2 = \frac{R_0^4 \cos^4\theta}{r^2(1+3\cos^2\theta)} ds^2 + \frac{R_0^2 \sin^6\theta}{1+3\cos^2\theta} dL^2 + r^2 \sin^2\theta d\varphi^2\,.$$

Hence, using the definition of Lame coefficients, $dl^2 = h_1^2 ds^2 + h_2^2 dL^2 + h_3^2 d\varphi^2$, and eliminating r through Eq. (8.33), we obtain

$$h_1 = \frac{R_0 \cos^2\theta}{L \sin^2\theta \sqrt{1+3\cos^2\theta}}, \quad h_2 = \frac{R_0 \sin^3\theta}{\sqrt{1+3\cos^2\theta}}, \quad h_3 = R_0 L \sin^3\theta\,. \tag{8.34}$$

The angle θ relates to s and L by

$$\frac{s}{L^2} = \frac{\sin^4\theta}{\cos\theta}\,. \tag{8.35}$$

Let the line of force, L_0, φ_0, be the irregularity axis. Then by Eq. (8.22) we have

$$\begin{aligned} dt = h_{10} ds &= \frac{R_0 \cos^2\theta}{L_0 \sin^2\theta \sqrt{1+3\cos^2\theta}} \left(\frac{\partial s}{\partial\theta}\right)_{L_0,\varphi_0} d\theta \\ &= R_0 L_0 \sin\theta \sqrt{1+3\cos^2\theta}\, d\theta\,. \end{aligned} \tag{8.36}$$

Here, $(\partial s/\partial\theta)_{L_0,\varphi_0}$ was determined from Eq. (8.35)

$$\left(\frac{\partial s}{\partial\theta}\right)_{L_0,\varphi_0} = L_0^2 \frac{d}{d\theta}\left(\frac{\sin^4\theta}{\cos\theta}\right) = L_0^2 \frac{\sin^3\theta(1+3\cos^2\theta)}{\cos^2\theta}\,.$$

Integrating Eq. (8.36), we obtain

$$t = \frac{R_0 L_0}{2\sqrt{3}} [\sqrt{3}\cos\theta \sqrt{1+3\cos^2\theta} + \ln(\sqrt{3}\cos\theta + \sqrt{1+3\cos^2\theta})]\,. \tag{8.37}$$

In the latter equation, t is read off from the equatorial plane, $\theta = \pi/2$.

To determine the curvature radii of the duct [Eq. (8.23b)], the derivatives $(\partial h_1^2/\partial L)_{s_0}$, $(\partial h_1^2/\partial\varphi)_{s_0}$ should be found. Using Eq. (8.35), we have

$$h_1^2 = \frac{R_0^2 \cos^4\theta}{L^2 \sin^4\theta(1+3\cos^2\theta)} = \frac{R_0^2 \cos^3\theta}{s(1+3\cos^2\theta)}\,. \tag{8.38}$$

Hence,

$$\left(\frac{\partial h_1^2}{\partial L}\right)_{s_0} = \frac{R_0^2}{s_0}\,\frac{d}{d\cos\theta}\left[\frac{\cos^3\theta}{1+3\cos^2\theta}\right]\left(\frac{\partial\cos\theta}{\partial L}\right)_{s_0}. \tag{8.39}$$

To determine $(\partial\cos\theta/\partial L)_{s_0}$, we use Eq. (8.35) once more. We have

$$L^2 = s\,y(\cos\theta)\,, \qquad y(x) = \frac{x}{(1-x^2)^2}\,.$$

Hence,

$$\left(\frac{\partial\cos\theta}{\partial L}\right)_{s_0} = \frac{2L}{s_0}\,\frac{1}{dy/dx} = \frac{2L}{s_0}\,\frac{\sin^6\theta}{1+3\cos^2\theta}\,. \tag{8.40}$$

Now, substituting Eq. (8.40) in Eq. (8.39), we obtain

$$\left(\frac{\partial h_1^2}{\partial L}\right)_{s_0} = \frac{6LR_0^2}{s_0^2}\,\frac{\cos^2\theta\sin^6\theta(1+\cos^2\theta)}{(1+3\cos^2\theta)^3}\,. \tag{8.41}$$

From Eq. (8.23b) with allowance for Eqs. (8.34) and (8.35), we obtain finally

$$r_y = \frac{R_0L_0}{3}\,\frac{\sin\theta(1+3\cos^2\theta)^{3/2}}{1+\cos^2\theta}\,, \qquad r_z = \infty\,. \tag{8.42}$$

The second curvature radius turns to infinity owing to the fact that $\partial h_1^2/\partial\varphi = 0$. This must be so, because the lines of force of a dipole field do not go out from the meridional planes. The curvature radius r_y on a given line of force has its minimum at the equatorial point, $\theta = \pi/2$

$$r_{ym} = \frac{R_0L_0}{3}\,. \tag{8.43}$$

The undisturbed magnetospheric dielectric constant for high-frequency waves is given by

$$\varepsilon(t, L_0, \varphi_0) = 1 - \frac{4\pi e^2 N_e(t, L_0, \varphi_0)}{m[\omega \pm \omega_H(t, L_0, \varphi_0)]\,\omega}\,. \tag{8.44}$$

It varies slow with t due to variations in the magnetospheric plasma density $N_e(t)$ and the magnitude of the magnetic field $H(t)$ (through the gyrofrequency, $\omega_H = eH/mc$). The characteristic scales of $N_e(t)$ and $H(t)$ variations are of the order of R_0L_0. In Eq. (8.44) we have taken into account that radio waves propagate in the direction close to the irregularity axis, i.e., to the magnetic field direction. The sign $(+)$ refers to ordinary waves, and the $(-)$ refers to extraordinary waves. For short radio waves we have $\omega > \omega_H$, so that the magnitude of ε_0 is close to unity [Eq. (1.3)].

The perturbation of the dielectric constant is given by

$$\Delta\varepsilon(y,z;t)=\frac{-4\pi e^2\Delta N_e(y,z,t)}{m(\omega\pm\omega_H)\,\omega}. \tag{8.45}$$

Here, ΔN_e is the plasma density perturbation in the irregularity elongated along the magnetic field in the vicinity of a given line of force L_0, φ_0. If the perturbation $\Delta N_e(y,z,t)$ together with specific expressions for $\Delta\varepsilon$, r_y, r_z, t [Eqs. (8.45), (8.42), (8.36), and (8.37)] are given, then Eqs. (8.23 a, b) describe completely the radio wave propagation in a given magnetospheric duct.

Note that for VLF waves at $\omega \ll \omega_H$, the expressions for dielectric constant ε, $\Delta\varepsilon$, under conditions of a slight waveguide [Eqs. (8.1) and (8.13)], take the form (Semionova and Trakhtengerts 1980, Helliwell 1965)

$$\varepsilon=1+\frac{4\pi e^2 N}{m\omega\omega_H},\qquad \Delta\varepsilon\approx\frac{4\pi e^2\Delta N_e}{m\omega\omega_H}. \tag{8.46}$$

Conditions for Duct Existence. Now we proceed to the determination of conditions for waveguide existence in the magnetosphere. Within the framework of the geometrical – optics approximation, a duct exists along a given line of force in the range of longitudinal coordinate t where the potential $U(y,z,t)$ [Eq. (8.23)] has a local minimum at least with respect to one variable, y. If wave effects are taken into account, then this condition is insufficient.

At least one eigenmode should exist which corresponds to a sufficiently deep duct level, $E=E_1$ (Fig. 8.2). At a shallower level, the trapped wave damps strongly due to wave tunneling through the barrier (Secs. 2.2 and 2.3). Therefore, it is necessary that the level not only exist, but be sufficiently deep as well.

First, let us consider a one-dimensional perturbation, $\Delta N_e=\Delta N_e(y,t)$. Assuming the density distribution to be parabolic within the irregularity

$$\Delta N_e=\begin{cases}-\Delta N_{e0}\left(1-\dfrac{y^2}{a^2}\right), & |y|\leqslant a\\ 0, & |y|\geqslant a\end{cases} \tag{8.47}$$

we obtain from Eqs. (8.45) and (8.23b)

$$\begin{aligned}U(y,t)&=-\delta_0\left(1-\frac{y^2}{a^2}\right)-\frac{2y}{r_y}\\ \delta_0&=\frac{4\pi e^2\Delta N_{e0}}{m(\omega\pm\omega_H)^2\varepsilon_0}.\end{aligned} \tag{8.48}$$

At positive δ_0, i.e., at negative plasma density perturbations, the potential $U(y,t)$ is bound to have a minimum. In this case, the minimum, $U_{\min}$, and maximum, $U_{\max}$, values of the potential $U(y,t)$ are given by

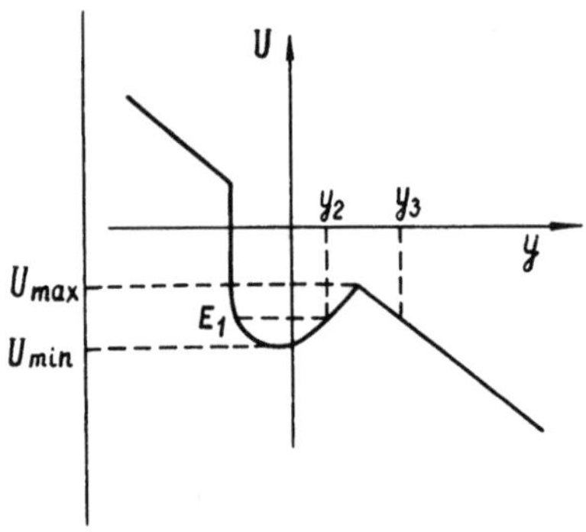

Fig. 8.2. Potential U for a parabolic duct [Eq. (8.48)]. E_1 is the first level [Eq. (8.55)], y_2 and y_3 are the boundaries of the tunneling region

$$U_{\min} = U(y_{\min}) = -\delta_0 - \frac{a^2}{\delta_0 r_y^2}\,,$$
$$U_{\max} = U(y_{\max}) = -2a/r_y\,. \tag{8.49}$$

Here, $y_{\min}, y_{\max}$ are the minimum and maximum points, respectively

$$y_{\min} = \frac{a^2}{\delta_0 r_y}\,, \qquad y_{\max} = a\,.$$

The potential difference is given by

$$U_{\max} - U_{\min} = \delta_0 - \frac{2a}{r_y} + \frac{a^2}{\delta_0 r_y^2} = \delta_0\left(1 - \frac{a}{\delta_0 r_y}\right)^2. \tag{8.50}$$

It decreases with decreasing curvature radius, r_y, and increasing duct width, a. At $r_y = a/\delta_0$ the minimum and maximum merge and disappear. Thus, the minimum of the potential U exists only at negative density perturbations, $\Delta N_e \sim -\delta_0$, for sufficiently narrow ducts

$$a \leqslant r_y \delta_0\,, \qquad -\delta_0 < 0\,. \tag{8.51}$$

This is the condition for duct existence in the geometrical-optics approximation.

Now let us take into account the wave effects. In the vicinity of its minimum, the potential is given by

$$U_y = U_{\min} + \frac{\delta_0}{a^2}(y - y_{\min})^2\,. \tag{8.52}$$

The wave equation takes here the form of Eq. (2.10) with

$$U'' = \frac{2\delta_0}{a^2}\,. \tag{8.53}$$

According to Eq. (2.10), the wave field is localized within the region

$$|y - y_{\min}| \lesssim \rho_s\,, \qquad \rho_s = \left(\frac{a^2}{\delta_0 k_0^2}\right)^{1/4} \tag{8.54}$$

$$k_0 = \omega\sqrt{\varepsilon_0}/c\,.$$

Hence, it is clear that the confinement in the duct is fairly good provided $\rho_s \ll a$, i.e., $ak_0\delta_0^{1/2} \gg 1$. This condition is additional to Eq. (8.51), both being necessary for duct existence.

Let us proceed to more accurate evaluation of the wave tunneling. The first level in the parabolic duct, by Eq. (2.11), is given by

$$E_1 = U_{\min} + \frac{\delta_0^{1/2}}{ak_0}. \tag{8.55}$$

Energy losses caused by the tunneling of the first mode, i.e., of the wave trapped onto the first level, E_1 (Fig. 8.2), are determined by the barrier penetration factor (A1.16, Appendix 1)

$$p = 2k_0 \int_{y_2}^{y_3} \sqrt{U(y) - E_1}\,dy, \qquad U(y_2) = U(y_3) = E_1. \tag{8.56}$$

Separating the integral in Eq. (8.56) into two parts and taking into account Eqs. (8.47) – (8.49), and (8.55), we have

$$\mathscr{I} = \int_{y_2}^{y_3} \sqrt{U(y) - E_1}\,dy = \int_{y_2}^{a} \left(\frac{y^2\delta_0}{a^2} - \frac{2y}{r_y} + \frac{a^2}{\delta_0 r_y^2} - \frac{\delta_0^{1/2}}{ak_0}\right)^{1/2} dy$$

$$+ \int_{a}^{y_3} \left(-\frac{2y}{r_y} + \delta_0 + \frac{a^2}{\delta_0 r_y^2} - \frac{\delta_0^{1/2}}{ak_0}\right)^{1/2} dy.$$

The second integral is equal to

$$\mathscr{I}_2 = \frac{r_y}{3}\left(\delta_0 + \frac{a^2}{\delta_0 r_y^2} - \frac{2a}{r_y} - \frac{\delta_0^{1/2}}{ak_0}\right)^{3/2}.$$

It usually exceeds the first integral. Putting, therefore, $\mathscr{I} \approx \mathscr{I}_2$, we obtain

$$p \approx \frac{2k_0 r_y}{3}\left(\delta_0 + \frac{a^2}{\delta_0 r_y^2} - \frac{2a}{r_y} - \frac{\delta_0^{1/2}}{ak_0}\right)^{3/2}. \tag{8.57}$$

The dissipation of the wave energy due to wave tunneling through the barrier, during the propagation over the distance of the order of one oscillation period T [Eq. (8.28)], is given by the quantity $\exp(-p)$ [Eq. (A1.16), Appendix 1]. Consequently, the condition for wave damping to be slight along the whole line of force in the magnetosphere, is given by

$$p \gtrsim k_m, \quad k_m \sim 1-4. \tag{8.58}$$

It is this condition that together with Eq. (8.51) determines the possibility for radio wave confinement in the magnetospheric wave duct.

Introducing dimensionless parameters

$$x = \frac{a}{r_y\delta_0}, \quad \tau = k_0 r_y \delta_0^{3/2}, \tag{8.59}$$

we rewrite Eqs. (8.58) and (8.51) with allowance for Eq. (8.57) in the form

$$\frac{2}{3}\tau\left[(x-1)^2-\frac{1}{\tau x}\right]^{3/2} \geqslant k_m, \quad x\leqslant 1\,. \tag{8.60}$$

At $k_m=0$ we obtain

$$\tau=\frac{1}{x(x-1)^2}\,. \tag{8.61}$$

The plot of τ as a function of x defined by Eq. (8.61) is shown in Fig. 8.3. It is seen that there is a minimum

$$x_m=\frac{1}{3}, \quad \tau_m=\frac{27}{4}, \quad \text{at} \quad k_m=0\,. \tag{8.62a}$$

Figure 8.3 also shows plots of $\tau(x)$ at other k_m. The region of duct existence is above the curves. It is seen that for every value of k_m there is a rather sharp minimum of $\tau(x)$, which gives the optimum conditions for duct existence. In particular,

$$\begin{aligned} &x_m\approx 0.22\,, \quad \tau_m\approx 12.5 \quad \text{at} \quad k_m=1\\ &x_m\approx 0.165\,, \quad \tau_m\approx 22 \quad \text{at} \quad k_m=4\,. \end{aligned} \tag{8.62b}$$

Eqs. (8.62 a, b) together with Eqs. (8.59) and (8.48) give the optimum magnitudes of the density perturbations, ΔN_{e0m} and duct width, a_m. At $k_m=0$

$$\Delta N_{e0m}\approx 16(ff_\pm)^{2/3}\left(\frac{L_0}{3}\right)^{-2/3}, \quad a_m\approx 2.9(ff_\pm)^{-1/3}\left(\frac{L_0}{3}\right)^{1/3}, \tag{8.63a}$$

at $k_m=4$

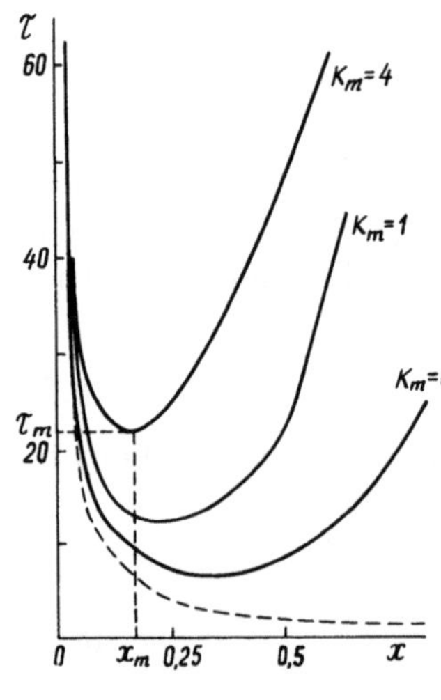

Fig. 8.3. Dependence of τ on x at various values of k_m. The region of the duct existence is situated above the curves. For $k_m=4$ the optimum values of τ_m and x_m are indicated. *Dashed curve* shows the dependence of τ on x in the case of one-dimensional perturbations, $\Delta N=\Delta N(z)$

$$\Delta N_{e0m} \approx 36(ff_\pm)^{2/3}\left(\frac{L_0}{3}\right)^{-2/3}, \quad a_m \approx 3.2(ff_\pm)^{-1/3}\left(\frac{L_0}{3}\right)^{1/3}. \quad (8.63b)$$

Here, ΔN_{e0m} is in cm^{-3}, f is the frequency, and $f_\pm = f \pm f_H$ are effective wave frequencies in MHz, a_m is the duct width in the equatorial plane (in km). The values of ΔN_{e0m} and a_m for $L_0 = 3$ at different wave frequencies are listed in Table 8.1. The effective duct width is given here both in the equatorial plane ($a_m = a_{em}$) and near the Earth at a height $z = 1000$ km (a_{0m}). It is seen that for HF radio waves the duct is rather narrow and the perturbations ΔN_{e0m} required for wave confinement must be great, approaching the undisturbed value of plasma density in the magnetosphere. Equations (8.63 a, b) show, in addition, that the extraordinary wave is confined in the duct up to greater frequencies than the ordinary wave. The difference between the limiting confinement frequencies of the extraordinary and ordinary waves for just the same duct is given by

$$\Delta f_m = f_{x\max} - f_{0\max} = f_H, \quad f_H = eH/2\pi mc.$$

Bearing in mind that confinement is determined mainly by the equatorial plane (since r_y has the minimum and a has the maximum at $\theta \to \frac{\pi}{2}$), we obtain in the approximation of the dipole Earth's magnetic field

$$\Delta f_m \approx \frac{0.9}{L_0^3}\,(\text{MHz}). \quad (8.64)$$

Hence, at $L_0 \approx 1.3 - 1.6$, we have $\Delta f_m \approx 0.25 - 0.5$ MHz, in accordance with the experimental results (Muldrew 1963, Muldrew 1967) (see Sec. 1.1.3).

We considered above only one-dimensional density perturbations with respect to y (i.e., with respect to L). In the case of one-dimensional perturbations with respect to z (i.e., with respect to the longitude φ) the picture proves to be more simple. Insofar as now the curvature radius, $r_y = \infty$, the potential $U(z, t)$ is determined by the density perturbation, $\Delta N_e(z, t)$ alone. Putting the density distribution in the irregularity to be parabolic [Eq. (8.47)], we obtain

Table 8.1. Electron density perturbation in ducts, δN_{e0m}, and the duct widths both at the geomagnetic equator, a_{em}, and near the Earth, a_{0m}, under the optimum conditions for ordinary radio wave ($L_0 = 3$)

f, MHz	$k_m = 0$			$k_m = 4$		
	δN_{e0m}, cm^{-3}	a_{em}, km	a_{0m}, km	δN_{e0m}, cm^{-3}	a_{em}, km	a_{0m}, km
1	$0.2 \cdot 10^2$	2.4	0.33	$0.4 \cdot 10^2$	3.3	0.45
3	$0.8 \cdot 10^2$	1.8	0.26	$1.6 \cdot 10^2$	2.2	0.31
5	$1.5 \cdot 10^2$	1.5	0.21	$3.1 \cdot 10^2$	1.8	0.25
10	$3.6 \cdot 10^2$	1.3	0.20	$7.8 \cdot 10^2$	1.5	0.21

from Eq. (8.55) the condition for duct existence, namely, the condition for existence of the first level, E_1, in the form

$$a k_0 \delta_0^{1/2} < 1$$

or in terms of variables defined by Eq. (8.59)

$$\tau > 1/x . \tag{8.65}$$

The curve $\tau = 1/x$ is shown in Fig. 8.3 by a dashed line, the area of duct existence being situated above the curve. It is seen that for $x \ll 1$, i.e., for narrow ducts, the conditions for their existence are close to those considered above for $N_e = N_e(y, t)$. In this case, however, broad ducts, $x > 1$, with a small density perturbation, ΔN_{e0m}, can occur as well. Of importance is, however, the fact that one-dimensional ducts with respect to z provide no channeling along the magnetic line of force and, thus, do not transfer the signal to the magnetic – conjugate point.

The picture of VLF wave confinement in magnetospheric waveguides is, on the whole, analogous. The difference lies, first, in the fact that as is clear from Eq. (8.46), VLF waves at $\omega \ll \omega_H$ are confined in irregularities with increased plasma density, $\Delta N_e > 0$. In addition, the group velocity of a VLF signal at the frequency ω is always much less than the velocity of light (Helliwell 1965)

$$\frac{v_{gr}}{c} = 2\frac{(\omega\omega_H)^{1/2}}{\omega_0}, \qquad \omega_0 = \left(\frac{4\pi e^2 N_{e0}}{m}\right)^{1/2} .$$

Hence, it is seen that under magnetospheric conditions, at $L_0 \approx 2.5-3$, $N_e \approx 10^3\,\mathrm{cm}^{-3}$, $\omega \approx 10^5\,\mathrm{s}^{-1}$ (i.e., $f = 16$ kHz), the ratio $v_{gr}/c \approx 0.1$ in accordance with the experimental results (see Sec. 1.1, Fig. 1.14).

Certainly, the particular case considered above of one-dimensional density distributions over y and the approximations made [namely, the use of the solution, Eq. (2.10), for a cut-off parabolic well] enable one only to estimate with good reliability the conditions for wave propagation in ducts. For a more accurate analysis, the solution of a two-dimensional wave equation for an arbitrary potential $U(y, z)$ is needed. Note that the case of a cylindrically – symmetric potential was considered in (Perekhvatov 1975, 1976).

Excitation of a Waveguide. Magnetospheric Resonator. Magnetospheric waveguides can trap radio waves propagating along the Earth's magnetic field. The high barrier in the vicinity of the ionospheric maximum is a hindrance at $\omega > \omega_H$. When radio waves are emitted from a satellite situated far above the ionospheric maximum, waveguide excitation undergoes essentially no obstacles. However, it is hindered to observe trapped waves on board the satellite for the following reason. A typical duct width near the Earth is rather small (about 200 – 500 m, see Table 8.1). Consequently, the duct is traversed by a satellite, travelling usually at heights $z \lesssim 1000$ km, during very short time, $\Delta t \sim 0.03 - 0.05$ s. During this time, the radio signal covers a distance ~8000 – 15000 km which is as a rule less than the distance to the magnetic – conjugate point and backwards at $L > 1.6$ [Eq. (8.37)]. This hinders the possibility for registration of duct excitation on board the satellite. The conditions for observations are bettered as the satellite trajectory approaches the geomagnetic equator, $L \rightarrow 1$, (in this case the distance covered by the signal decreases), and the wave frequency decreases (in this case the characteristic duct width increases, see Table 8.1). Both the effects are clearly seen on the experimental curves (Muldrew 1967) (see Sec. 1.1.3, Fig. 1.13).

When waves are emitted from the Earth, the ionospheric barrier proves to be an essential obstacle to excitation of magnetospheric ducts. At the polar and subpolar latitudes, i.e., outside the plasmasphere, the magnetospheric plasma density is usually insufficient to trap waves in ducts effectively. At low and midlatitudes, the magnetic field has a noticeable inclination with respect to the vertical, so that radio wave refraction in the ionosphere is here of importance. To be trapped in a magnetospheric duct, the ray must not only penetrate through the ionospheric maximum, but become directed along the magnetic field in the neighborhood of the upper ionospheric boundary as well. Hence, it is clear that only waves emitted from the Earth at a certain takeoff angle, α_0, can be trapped in the duct, the point of the capture being strongly shifted horizontally from the transmitter. This fact should be taken into consideration when observing in the magnetic – conjugate region HF signals propagating in the magnetosphere. Figure 8.4 shows the angle of inclination with respect to the horizon, α, at a height 1500 km (conventionally accepted as the ionospheric boundary) vs the horizontal distance from the transmitter, x (Ben'kova et al. 1980). The calculation has been carried out for the midlatitude ionosphere in the morning and evening hours under typical conditions of high and low solar activity. This figure also shows the dip angle of the magnetic field, I_H, for the same height, 1500 km. The intersection of the curves, i.e., the location of capture, is seen to be 500 – 1000 km apart from

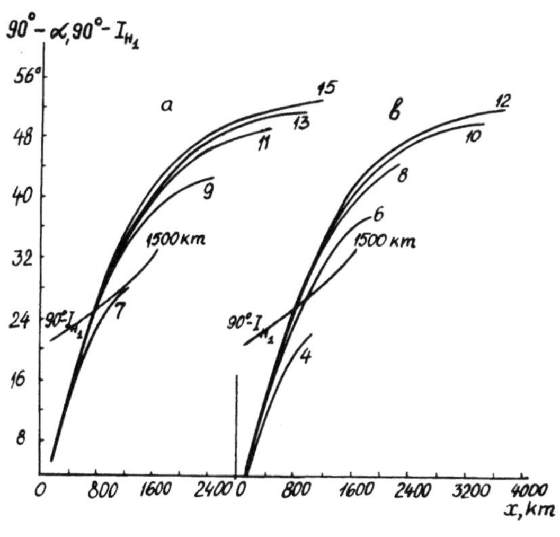

Fig. 8.4a, b. Angle of ray elevation at the height of 1500 km as a function of horizontal distance from the transmitter x. The frequency f in MHz is indicated at the curves. I_{H_1} is the angle between the magnetic field and horizon at the height of 1500. **a** High solar activity; **b** low solar activity

the transmitter. In a low solar activity period, the capture becomes possible for frequencies $f \geqslant f_0 \approx 5$ MHz, and in a high solar activity period, the capture is possible for frequencies $f \geqslant f_0 \approx 7$ MHz, in accordance with the experimental data (Sec. 1.1.3).

The waves whose frequencies are less than the critical ionospheric frequencies in the magnetic – conjugate points, $f < f_0 = \left(\frac{e^2 M_{em}}{\pi m}\right)^{1/2} \Big/ \sin I_H$ (where N_{em} is the electron density in the F-region maximum) cannot leave the magnetosphere for the Earth. For these waves, magnetospheric waveguides form cut-off resonators. The absorption in such a resonator is small. Figure 8.5 shows the wave lifetime in the resonator, i.e., the time during which the power damps by a factor of e, the damping being caused only by collisional absorption in the magnetosphere and in the region of wave reflection in the ionosphere. If the latter is predominant, the magnitude of absorption depends strongly on the structure of the reflection layer in the ionosphere. The figure demonstrates that the signal lifetime in the magnetospheric resonator can be great.

In principle, this makes the existence of a so-called Long Delay Echo possible (Al'pert 1974, Villard et al. 1969, Muldrew 1979). Indeed, let us imagine that at the instant of a signal passing through the ionospheric maximum, the plasma density has been decreased here, enabling the wave to penetrate into the duct. During the whole time of wave propagation in the duct to the magnetic – conjugate point and backwards, $\Delta t \sim 0.2 - 0.4$ s, the plasma density in the maximum can increase, resulting in the wave reflection from the maximum and, thus, in capture into the magnetospheric resonator. After a few successive reflections, the wave can leave the resonator due to the same cause as when trapping. In this case, as is clear from Fig. 8.5, the total signal time delay can amount to a few seconds, or even 10 s. Certainly, insofar as for

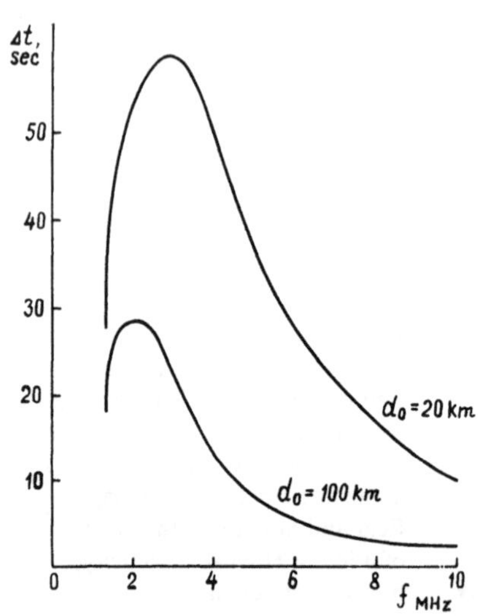

Fig. 8.5. Frequency dependence of the time interval Δt, for which the power of an ordinary wave trapped in the magnetospheric resonator is attenuated by a factor of e. The effective thickness of the reflecting ionospheric layer is indicated at the curves: $N_e \approx N_{e0}(z_0 - z)/d_0$, $N_{e0} = \frac{m(\omega + \omega_H)}{4\pi e^2}$, z_0 is the wave reflection point

these phenomena special conditions are required, these phenomena must be fairly rare. Besides, in addition to collisional absorption, some other mechanisms of wave transformation and tunneling are possible, resulting in losses of the trapped wave energy (Budden 1961).

Let us mention another possible cause for the occurrence of signals with great time delays, which is discussed in the literature (Muldrew 1979, Crawford et al. 1970). The cause lies in the transformation of a part of radio wave power on ionospheric irregularities into natural plasma oscillations, in particular, into an extraordinary wave trapped in the ionosphere near the plasma resonance point, $\omega \approx \omega_0$ (a so-called *z*-mode). This wave has small propagation velocity. Transforming after a time of about a few seconds or tens of seconds again into the ordinary wave, it can be received on the Earth. We have made here an essential assumption concerning the long lifetime of *z*-mode in the ionosphere. Under common equilibrium (or nearly equilibrium) ionospheric conditions, natural oscillations, in particular, *z*-mode, have a lifetime $\sim 10^{-3} - 10^{-2}$ s. In order that the *z*-mode lifetime be increased by three-four orders of magnitude, it is necessary that the ionosphere should be in a nonequilibrium state which is strongly excited and unstable with respect to the *z*-mode. The ionospheric excitation can be produced, for example, by fast electrons precipitating from the magnetosphere in the auroral regions.

Appendix 1. Wave Equation. Radio Wave Field Distribution in Connected Waveguide Ducts

A1.1 Wave Equation

A decomposition of the vectorial wave equation [Eqs. (2.1) and (2.4)] into scalar components in a spherically-symmetric medium has been carried out in the monograph (Fok 1970, Chap. 14) for various field sources: vertical dipole, horizontal dipole, and vertical and horizontal magnetic dipoles. Here, we shall carry out the decomposition in the general form for arbitrarily distributed sources (Borisov 1975).

The field sources, according to Eq. (2.1), are external electric currents $\boldsymbol{j}$ and charges, $\rho = \frac{i}{\omega} \operatorname{div} \boldsymbol{j}$. Let us consider a spherical coordinate system and separate the vertical (radial) and horizontal components of the source currents

$$\boldsymbol{j} = \boldsymbol{j}_1 + \boldsymbol{j}_2 + \boldsymbol{j}_3 , \quad \rho = \frac{i}{\omega} \operatorname{div} \boldsymbol{j} = \rho_1 + \rho_2 + \rho_3 ,$$

where $\boldsymbol{j}_1$ is the radial component

$$\boldsymbol{j}_1 = \boldsymbol{j}_r , \quad \rho_1 = \frac{i}{\omega} \operatorname{div} \boldsymbol{j}_1 ,$$

and horizontal components $\boldsymbol{j}_2$ and $\boldsymbol{j}_3$ are chosen so that

$$\operatorname{rot} \boldsymbol{j}_2 \neq 0 , \quad \operatorname{div} \boldsymbol{j}_2 = 0 , \quad \rho_2 = 0 , \tag{A1.1}$$

$$\operatorname{rot} \boldsymbol{j}_3 = 0 , \quad \operatorname{div} \boldsymbol{j}_3 \neq 0 , \quad \rho_3 = \frac{i}{\omega} \operatorname{div} \boldsymbol{j}_3 . \tag{A1.2}$$

An arbitrary current $\boldsymbol{j}$ can be represented as the sum of the components $\boldsymbol{j}_1, \boldsymbol{j}_2, \boldsymbol{j}_3$.

First, let us derive the equation for the magnetic field $\boldsymbol{H}_1$ generated by the radial current $\boldsymbol{j}_1$. From Eq. (2.1) we obtain

$$\tilde{\varepsilon} \operatorname{rot} \left(\frac{1}{\tilde{\varepsilon}} \operatorname{rot} \boldsymbol{H}_1 \right) - \tilde{\varepsilon} k_0^2 \boldsymbol{H}_1 = \frac{4\pi}{c} \operatorname{rot} \boldsymbol{j}_1 , \quad k_0 = \omega / c . \tag{A1.3}$$

We substitute $\boldsymbol{H}_1$ in the form $\boldsymbol{H}_1 = \operatorname{rot} \boldsymbol{e}_r A_1$, into this equation, where $\boldsymbol{e}_r$ is the unit vector directed along the radius. For A_1, taking into account Eq. (A1.1), we obtain the following scalar equation

$$\tilde{\varepsilon}\frac{\partial}{\partial r}\left(\frac{1}{\tilde{\varepsilon}}\frac{\partial A_1}{\partial r}\right)+\frac{1}{r^2\sin\theta}\frac{\partial}{\partial\theta}\left(\sin\theta\frac{\partial A_1}{\partial\theta}\right)+\frac{1}{r^2\sin^2\theta}\frac{\partial^2 A_1}{\partial\varphi^2}+\tilde{\varepsilon}k_0^2 A_1$$

$$=-\frac{4\pi}{c}j_1\,. \tag{A1.4}$$

Here, (r, θ, φ) is the spherical coordinate system. Equation (A1.4) coincides with those obtained by Krasnushkin (1947, 1948). The electric and magnetic field components are expressed in terms of the potential A_1, for example

$$\mathscr{E}_{1r}=\frac{1}{ik_0}\left[\frac{\partial}{\partial r}\left(\frac{1}{\tilde{\varepsilon}}\frac{\partial A_1}{\partial r}\right)+k_0^2A_1\right],\quad \mathscr{E}_{1\theta}=\frac{1}{ik_0\tilde{\varepsilon}r}\frac{\partial^2 A_1}{\partial r\,\partial\theta},$$

$$\mathscr{E}_{1\varphi}=\frac{1}{ik_0\tilde{\varepsilon}r\sin\theta}\frac{\partial^2 A_1}{\partial r\,\partial\varphi}.$$

Now let us derive the equation for the field $\mathscr{E}_2$ (or $\boldsymbol{H}_2$) generated by the horizontal current $\boldsymbol{j}_2$. From Eq. (2.1) we have

$$\operatorname{rot}\operatorname{rot}\mathscr{E}_2-\tilde{\varepsilon}k_0^2\,\mathscr{E}_2=-ik_0\frac{4\pi}{c}\boldsymbol{j}_2\,.$$

According to Eq. (A1.1), we put: $\boldsymbol{j}_2=\operatorname{rot}\boldsymbol{e}_r J$, $\mathscr{E}_r=\operatorname{rot}\boldsymbol{e}_r A_2$. Substituting the quantities $\boldsymbol{j}_2$ and $\mathscr{E}_r$ in this form in Eq. (A1.3), we obtain the following equation for the potential A_2

$$\frac{\partial^2 A_2}{\partial r^2}+\frac{1}{r^2\sin\theta}\frac{\partial}{\partial\theta}\left(\sin\theta\frac{\partial A_2}{\partial\theta}\right)+\frac{1}{r^2\sin^2\theta}\frac{\partial^2 A_2}{\partial\varphi^2}+\tilde{\varepsilon}k_0^2A_2=ik_0\frac{4\pi}{c}J\,. \tag{A1.5}$$

The scalar equation for the fields generated by the third current component, $\boldsymbol{j}_3$, can be derived in the same way. To this end, taking into account Eq. (A1.2), we put $\boldsymbol{j}_3=\nabla G$. Then the equation for $\boldsymbol{H}_3$ is in the same form as Eq. (A1.3), but in the right-hand side the term $\frac{4\pi}{c}\operatorname{rot}\boldsymbol{g}$, is present, where $\boldsymbol{g}=G\nabla\ln\tilde{\varepsilon}$. Introducing the potential A_3 so that $\boldsymbol{H}_3=\operatorname{rot}\boldsymbol{e}_r A_3$, we obtain for A_3 Eq. (A1.4), where j_1 is replaced by g.

Thus, the solution of scalar Eqs. (A1.4) and (A1.5) enables one to determine the field in a spherically-symmetric layer at arbitrary source distribution. The solution is naturally to be sought by separation of variables

$$A=Z(r)\,Y(\theta,\varphi)\,.$$

The angular part Y is described by spherical functions. For the radial part in the propagation region, i.e., in the region with no sources, we obtain in the case of Eq. (A1.4)

$$\frac{d^2\Psi}{d\zeta^2}+[C+W(\zeta)]\,\Psi=0\,,\quad Z(r)=\sqrt{\tilde{\varepsilon}}\,r\,\Psi\,,$$
$$\zeta=\ln r\,,\quad W=k_0^2\tilde{\varepsilon}r^2-\sqrt{\tilde{\varepsilon}}\,r\frac{d^2}{d\zeta^2}\left(\frac{1}{\sqrt{\tilde{\varepsilon}}\,r}\right); \tag{A1.6}$$

and in the case of Eq. (A1.5)

$$\frac{d^2\Psi}{dr^2}+k_0^2\left[\tilde{\varepsilon}(r)+\frac{C}{r^2k_0^2}\right]\Psi=0\,,\quad \Psi=Z(r)\,. \tag{A1.7}$$

Here, C is the separation constant for the spherical function Y of the p-th order, $C=-p(p+1)$.

Thus, the solution of the vectorial Eqs. (2.1) and (2.4) reduces to the solution of two one-dimensional scalar Eqs. (A1.6) and (A1.7). Next, bearing in mind that the ionosphere is a thin spherical layer [Eq. (1.7)], i.e., assuming

$$\Psi=\Psi(z)\,,\quad z=r-R_0\,,\quad z\ll R_0$$

(z is the altitude, R_0 is the Earth's radius), we represent them in the form

$$\frac{d^2\Psi}{dz^2}+k_0^2(E-U(z))\,\Psi=0$$
$$E=\frac{C}{R_0^2k_0^2}=-\frac{p(p+1)}{R_0^2k_0^2}\,. \tag{A1.8}$$

Here, E is a constant, and in the case of Eq. (A1.6)

$$U(z)=-\tilde{\varepsilon}(z)+\frac{2z}{R_0}E\,, \tag{A1.9}$$

and in the case of Eq. (A1.7)

$$U(z)=-\tilde{\varepsilon}(z)\left(1+\frac{2z}{R_0}\right)-\left[\frac{3}{4\tilde{\varepsilon}^2k_0^2}\left(\frac{d\tilde{\varepsilon}}{dz}\right)^2-\frac{1}{2\tilde{\varepsilon}k_0^2}\frac{d^2\tilde{\varepsilon}}{dz^2}\right]. \tag{A1.10}$$

Next, we take into account that by Eqs. (1.2), (1.6), and (1.8), under conditions of interest, $\tilde{\varepsilon}$ is always close to unity [Eq. (1.3)]

$$\tilde{\varepsilon}\approx-E\approx1\,,$$

and the term in the square brackets in Eq. (A1.10) is negligibly small. Therefore, the equation for both the field components [Eqs. (A1.6) and (A1.7)] with the potentials given in Eqs. (A1.9) and (A1.10) in the first approximation are of the identical form [Eq. (A1.8)] with the potential [Eq. (2.7)]

$$U(z)=-\varepsilon(z)-\frac{2z}{R_0}\approx-1+\frac{4\pi e^2N_e(z)}{m\omega^2}-\frac{2z}{R_0}\,. \tag{A1.10a}$$

In this equation we have also ignored, by Eqs. (1.6) and (1.1), the imaginary part of ε, as well as the influence on ε of the Earth's magnetic field. Thus, vectorial Eq. (2.4) in the approximation considered is reduced under ionospheric conditions to one scalar Eq. (A1.8) with the potential given in Eq. (2.7) i.e., Eq. (A1.10a). All the possible simplifications have been made here.

A1.2 Radio Wave Field Distribution in Connected Waveguide Ducts

In Sec. 2.2 we have considered the radio wave field distribution in an isolated waveguide duct. Real structure of wave ducts in the ionosphere is, however, of a more complicated character: commonly, two or even more mutually connected ducts exist (see Figs. 2.1 and 2.2); wave penetration from one duct to another, as well as from the ducts to the Earth, become possible. The analysis of these effects is the subject of the present section (Borisov and Gurevich 1975).

Quasiclassical Approximation. We consider a system of mutually connected ducts and study it, first, in the quasiclassical approximation, assuming the interduct barrier to be slightly penetrable. We consider successively three characteristic cases:

I. $f < f_{\text{MUF-}E}$.

Let the wave frequency be less than the maximum usable frequency of the E-region, MUF-E. The plot of $U(z)$ is shown in this case in Fig. A1.1a. Let us consider a wave emitted at an angle of α_0 and reflected from the E-layer. The level $E = -\cos^2\alpha_0$ [Eq. (2.8)] for this wave is shown by a dashed line; $z = 0$ is the Earth's surface. Taking into account that at the point $z = z_1 = 0$ an infinitely high barrier exists, we obtain, by Eqs. (2.12) – (2.14) (Landau and Lipshitz 1977, Sec. 47) that in the ground-based duct I the field is given by

$$\Psi_{\mathrm{I}} = \frac{a}{\sqrt{k(z)}}\left\{\exp\left[i\int_{z_1}^{z} k\,dz - i\frac{\pi}{4}\right] - \exp\left[-i\int_{z_1}^{z} k\,dz - i\frac{\pi}{4}\right]\right\}. \qquad \text{(A1.11)}$$

Within the barrier (region III), the wave field is given by

$$\Psi_{\mathrm{III}} = \frac{b_1}{\sqrt{|k|}}\exp\left[\int_{z_2}^{z}|k|\,dz\right] + \frac{b_2}{\sqrt{|k|}}\exp\left[-\int_{z_2}^{z}|k|\,dz\right]. \qquad \text{(A1.12)}$$

Finally, in the duct II (ionospheric duct FE, see Sec. 4.1) we shall seek the function Ψ so that only a damping wave is present when $z > z_4$

$$\Psi_{\mathrm{II}} = \frac{C}{\sqrt{k(z)}}\left\{\exp\left[i\int_{z}^{z_4} k\,dz - i\frac{\pi}{4}\right] + \exp\left[-i\int_{z}^{z_4} k\,dz + i\frac{\pi}{4}\right]\right\}. \qquad \text{(A1.13)}$$

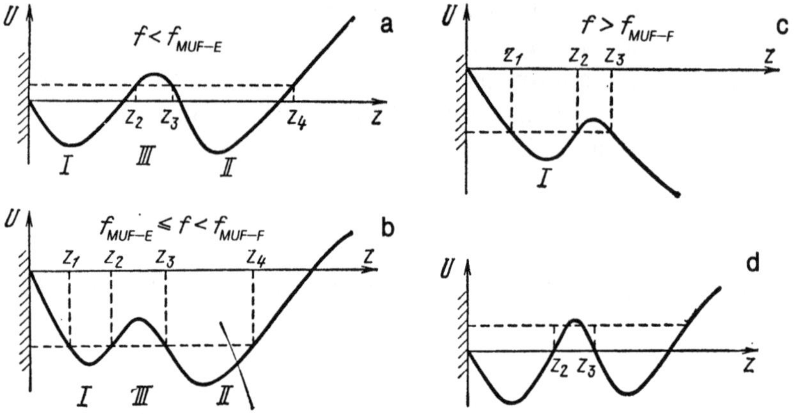

Fig. A1.1a – d. Various forms of the potential $U(z)$

Using matching rules for quasiclassical functions at the point $z = z_2$, we obtain

$$b_1 = 2ae^{i\frac{\pi}{4}} \cos\left[\int_{z_1}^{z_2} k\,dz - \frac{\pi}{4}\right],$$
$$b_2 = ae^{i\frac{\pi}{4}} \sin\left[\int_{z_1}^{z_2} k\,dz - \frac{\pi}{4}\right]. \tag{A1.14}$$

Analogously for $z = z_3$ we have

$$b_1 = C\exp\left[-\int_{z_2}^{z_3} |k|\,dz\right] \sin\int_{z_3}^{z_4} k\,dz\,,$$
$$b_2 = 2C\exp\left[\int_{z_2}^{z_3} |k|\,dz\right] \cos\int_{z_3}^{z_4} k\,dz\,. \tag{A1.15}$$

Eliminating in Eqs. (A1.14) and (A1.15) the constants a and C, we obtain an equation for eigenlevels E_n (or for eigenwave numbers) of the system

$$\cot\left(\int_{z_1}^{z_2} k\,dz - \frac{\pi}{4}\right) \cot\left(\int_{z_3}^{z_4} k\,dz\right) = \frac{p}{4},$$
$$p = \exp\left[-2\int_{z_2}^{z_3} |k|\,dz\right], \tag{A1.16}$$

where p is the barrier penetrance. Here, we consider the case of the slightly penetrable barrier, $p \ll 1$.

In the absence of a resonance, i.e., when the levels E_n and E_m differ markedly, Eq. (A1.16) defines two sets of eigennumbers. For one of them we have

$$\int_{z_1}^{z_2} k_n \, dz \approx \frac{3\pi}{4} + \pi n + \delta_1 , \quad k_n = k_0 \sqrt{E_n - U(z)} , \tag{A1.17}$$

and for the other

$$\int_{z_3}^{z_4} k_m \, dz \approx \frac{\pi}{2} + \pi m + \delta_2 , \quad k_m = k_0 \sqrt{E_m - U(z)} . \tag{A1.18}$$

Here, δ_1, δ_2 are small quantities

$$\delta_1 = -\frac{p}{4} \tan \left[\int_{z_3}^{z_4} k_n \, dz \right] , \quad \delta_2 = -\frac{p}{4} \tan \left[\int_{z_1}^{z_2} k_m \, dz - \frac{\pi}{4} \right] , \tag{A1.19}$$

and n, m are integer numbers corresponding to the wave mode numbers for the first and second ducts. Naturally, if the barrier penetrance tends to zero, $p \to 0$, then the Eqs. (A1.18) and (A1.13) coincide with Eqs. (2.13) and (2.15).

The expressions obtained enable one to determine, under stationary conditions, the intensity of field penetration through the barrier from the ground-based duct I to the ionospheric duct II and vice versa. Indeed, let us assume that there is a stationary field distribution with the eigenvalue E_n in the duct I, i.e., the n-th duct eigenmode is excited here. Then $k_n(z)$ satisfies Eq. (A1.17), and from Eqs. (A1.13)–(A1.15) we obtain

$$C = \frac{a p^{1/2}}{2} e^{i\frac{\pi}{4}} \frac{\sin \left[\int_{z_1}^{z_2} k_n \, dz - \frac{\pi}{4} \right]}{\cos \left[\int_{z_3}^{z_4} k_n \, dz \right]} . \tag{A1.20}$$

Here, a is the field amplitude in the duct I, C is that in the duct II, p is the barrier penetrance [Eq. (A1.16)]. Hence, $|C/a| \sim p^{1/2}$, i.e., the field is localized mainly in the duct I, and the field in the duct II is $p^{-1/2}$ times weaker.

On the contrary, if the ionospheric duct II is excited, i.e., if $k_m(z)$ satisfies Eq. (A1.18), then we obtain for the field amplitude in the ground-based duct I

$$a = \frac{C}{2} p^{1/2} e^{-i\frac{\pi}{4}} \frac{\sin \left[\int_{z_3}^{z_4} k_m \, dz \right]}{\cos \left[\int_{z_1}^{z_2} k_m \, dz - \frac{\pi}{4} \right]} .$$

Now $|a/C| \sim p^{1/2}$.

Equations obtained [Eqs. (A1.17) and (A1.18)] are true only far from the resonance. In the resonance, the eigenlevels in both ducts nearly coincide. In this peculiar case, the argument of the tangent is close to $\pi n+\frac{\pi}{2}$, so corrections δ are not small. The resonant mode field in the absence of absorption is distributed roughly equal between both ducts ($a \sim C$). However, under real ionospheric conditions the absorption is significant. Therefore, the resonance case will be considered below with allowance for the absorption.

II. $f_{\text{MUF-}E}<f<f_{\text{MUF-}F}$.

The plot of $U(z)$ is shown in Fig. A1.1b. Unlike the previous case, the field in the duct I will be sought based on the condition that when $z<z_1$, only the damping wave exists. In this case

$$\Psi_{\mathrm{I}}=\frac{a}{\sqrt{k(z)}}\left\{\exp\left[i\int_{z_1}^{z} k\,dz-i\frac{\pi}{4}\right]+\exp\left[-i\int_{z_1}^{z} k\,dz+i\frac{\pi}{4}\right]\right\}. \tag{A1.21}$$

The matching rule in the point $z=z_2$ yields

$$b_1=2a\cos\int_{z_1}^{z_2} k\,dz\,,\quad b_2=a\sin\int_{z_1}^{z_2} k\,dz\,. \tag{A1.22}$$

The field in the duct II is given by Eq. (A1.13), and at the point $z=z_3$ the former result, Eq. (A1.15) is valid. The equation for eigenwave numbers has now the form

$$\cot\left[\int_{z_1}^{z_2} k\,dz\right]\cot\left(\int_{z_3}^{z_4} k\,dz\right)=\frac{p}{4}. \tag{A1.23}$$

From Eq. (A1.23) follows, as before, that far from a resonance two sets of eigenwave numbers exist, for which

$$\int_{z_1}^{z_2} k_n\,dz=\frac{\pi}{2}+\pi n+\delta_1\,,\quad \int_{z_3}^{z_4} k_m dz=\frac{\pi}{2}+\pi m+\delta_2\,,$$

where δ_1, δ_2 are small corrections analogous to those in Eq. (A1.19).

If in the second waveguide one of the eigenmodes is excited, then the field amplitude in the first waveguide is given by

$$a=\frac{1}{2}Cp^{1/2}\frac{\sin\int_{z_3}^{z_4} k\,dz}{\cos\int_{z_1}^{z_2} k\,dz}. \tag{A1.24}$$

In the opposite case, i.e., if the first waveguide is excited, then

$$C=\frac{1}{2}ap^{1/2}\frac{\sin\int_{z_1}^{z_2} k\,dz}{\cos\int_{z_3}^{z_4} k\,dz}. \tag{A1.25}$$

Finally, the field on the Earth's surface can be formed by matching the solution in the duct I with that in the region $z < z_1$

$$\Psi_0 = \frac{a}{\sqrt{|k|}} \exp\left[-\int_0^{z_1} |k| dz\right]. \tag{A1.26}$$

Taking into account that by Eq. (A1.10a), at small z, $U(z) = -1 - \frac{2z}{R_0}$, we obtain from Eq. (A1.26) an additional factor of field attenuation on the Earth ($z = 0$) for a wave with $f > f_{\text{MUF-}E}$ propagated in the ground-based duct

$$p = \exp\left[-2\int_0^{z_1} |k| dz\right] = \exp\left[-\frac{8\sqrt{2}\pi z_1^{3/2}}{3\lambda R_0^{1/2}}\right], \tag{A1.27}$$

where $\lambda = c/f$ is the wave length.

III. $f > f_{\text{MUF-}F}$.

The plot of $U(z)$ in this case is shown in Fig. A1.1c. Here the only duct exists. If we neglect wave tunneling through the barrier, we have

$$\cot\int_{z_1}^{z_2} k\, dz \approx 0, \quad \int_{z_1}^{z_2} k\, dz = \frac{\pi}{2} + \pi n. \tag{A1.28}$$

The tunneling results in energy losses. With allowance for wave penetration, we obtain instead of Eq. (A1.28)

$$\cot\int_{z_1}^{z_2} k\, dz = i\frac{p}{4}, \quad \delta k = p\Big/\left(4k_1\int_{z_1}^{z}\frac{dz}{k_1}\right).$$

Here, $k = k_1 + i\delta k$, where k_1 is defined as before by Eq. (A1.18), and $\int_{z_1}^{z_2} \delta k\, dz = p/4$. Wave tunneling plays here a role of absorption. As before, the field on the Earth is related to the field in the duct by Eq. (A1.26).

In conclusion, we note that if one of the waveguides is excited, the field in the other one can be expanded in terms of the first waveguide eigenfunctions. In this case, the expansion coefficients decrease exponentially rapid with increasing difference between the wave numbers of the excited modes of the first and second waveguides (see Landau and Lifshitz 1977). In other words, these are modes whose levels E_m are close to the levels of modes excited in the first waveguide, E_n, that are excited in the second waveguide.

Influence of Absorption. Equations (A1.16) and (A1.23) enable the stationary field distribution in ducts to be investigated also when absorption takes place. We consider this case in more detail by way of example of Eq. (A1.23). We assume that absorption takes place in the duct I, so that here the wave number k is complex. We represent k in the form $k = k_1 + ik_2$, where, by Eq. (2.6),

$$k_1 \approx k_0 (E - \operatorname{Re} U(z))^{1/2}, \quad k_2 = \frac{1}{2} k_0 \frac{\operatorname{Im} \varepsilon(z)}{(E - \operatorname{Re} U(z))^{1/2}}, \quad \left|\frac{k_2}{k_1}\right| \ll 1.$$

Here, $U(z)$ is given as before by Eq. (2.7), and $\mathrm{Im}\,\varepsilon(z) \approx \nu_e \omega_0^2/\omega^3$. In this case, we have from Eq. (A1.22)

$$\frac{b_1}{b_2} = 2\,\frac{\cot\left(\int\limits_{z_1}^{z_2} k_1\,dz\right) - i\tanh\left[\int\limits_{z_1}^{z_2} k_2 dz\right]}{1 + i\cot\left[\int\limits_{z_1}^{z_2} k_1\,dz\right]\tanh\left(\int\limits_{z_1}^{z_2} k_2 dz\right)}\,.$$

If $\left|\cot\left(\int\limits_{z_1}^{z_2} k_1\,dz\right)\tanh\left(\int\limits_{z_1}^{z_2} k_2\,dz\right)\right| \ll 1$, then we have the following equation for eigenwave numbers

$$\cot\left(\int\limits_{z_3}^{z_4} k_1\,dz\right)\left\{\cot\left(\int\limits_{z_1}^{z_2} k_1\,dz\right)\left[1 - \tanh^2\int\limits_{z_1}^{z_2} k_2\,dz\right]\right.$$
$$\left. - i\tanh\int\limits_{z_1}^{z_2} k_2 dz\left[1 + \cot^2\int\limits_{z_1}^{z_2} k_1 dz\right]\right\} = \frac{p}{4}\,. \tag{A1.29}$$

In the reverse case, if $\left|\cot\left(\int\limits_{z_1}^{z_2} k_1\,dz\right)\tanh\left(\int\limits_{z_1}^{z_2} k_2\,dz\right)\right| \gg 1$, we have

$$\cot\int\limits_{z_3}^{z_4} k_1\,dz = i\frac{p}{4}\tanh\int\limits_{z_1}^{z_2} k_2\,dz\,. \tag{A1.30}$$

Let us analyze Eq. (A1.29) in the resonance case, i.e., when both the quantities $\cot\int\limits_{z_1}^{z_2} k\,dz$ and $\cot\int\limits_{z_3}^{z_4} k\,dz$ are simultaneously close to zero. Putting $\int\limits_{z_1}^{z_2} k_1 dz = \frac{\pi}{2} + \pi n + \delta$, we rewrite this equation in the presence of slight absorption, $\tanh\int\limits_{z_1}^{z_2} k_2\,dz \ll 1$, in the form

$$\delta^2 + i\tanh\int\limits_{z_1}^{z_2} k_2\,dz \cdot \delta - \frac{p}{4h} = 0\,. \tag{A1.31}$$

Here

$$h = \cot\int\limits_{z_3}^{z_4} k_1\,dz/\cot\int\limits_{z_1}^{z_2} k_1\,dz\,.$$

Solving Eq. (A1.31)

$$\delta_\pm = -\frac{i}{2}\tanh\int\limits_{z_1}^{z_2} k_2\,dz \pm \frac{1}{2}\sqrt{\frac{p}{h} - \tanh^2\int\limits_{z_1}^{z_2} k_2\,dz}\,, \tag{A1.32}$$

we see that when $p \gg h \cdot \tanh^2 \int_{z_1}^{z_2} k_2 dz$, the waves corresponding to δ_+ or δ_- are localized in neither of the ducts. If $p \ll h \cdot \tanh^2 \int_{z_1}^{z_2} k_2 dz$, then $\delta_+ \approx -\frac{ip}{4h} \cdot \coth \int_{z_1}^{z_2} k_2 dz$, and the wave is localized in the first waveguide. When $\delta = \delta_- \approx -i \tanh \int_{z_1}^{z_2} k_2 dz$, the wave is localized in the second waveguide.

Usually the quantity p is small, so that the latter condition is satisfied. In other words, in the resonance case in the presence of absorption the wave is also localized in the first or in the second duct. The only exception is represented by the levels close to the duct boundary.

Finally, let us take into account slight absorption in the vicinity of the barrier. We put $k_0^2(E - U(z)) = -\varkappa^2 + i\sigma$. Restricting ourselves to the case when $\int_{z_2}^{z_3} \frac{\sigma}{\varkappa} dz \ll 1$, we see that also this absorption can be taken into account in all the preceding equations if we make a replacement

$$\exp\left[-\int_{z_2}^{z_3} |k| dz\right] \to \exp\left[-\int_{z_2}^{z_3} \varkappa dz\right] \times \left\{\cos \frac{1}{2} \int_{z_2}^{z_3} \frac{\sigma}{\varkappa} dz + i \sin \frac{1}{2} \int_{z_2}^{z_3} \frac{\sigma}{\varkappa} dz\right\}.$$

Wave Tunneling through the Barrier. All the previous considerations pertain to the case when the barrier is slightly penetrable (Fig. A1.1a), allowing the quasiclassical approximation to be applied. At the same time, for the analysis of the problem of wave capture into or departure from the duct, of most importance is the region of sufficiently great penetrance, when the exact solution of Eq. (A1.8) is needed.

Insofar as the radio wave field length λ is in the ionosphere much less than the typical barrier width, a, a considerable deviation from the quasiclassics would be expected only in the case when the wave propagates in the close vicinity of the barrier maximum (Fig. A1.1d). But in this region, the barrier may be approximated by a parabola. Thus, it is enough to consider the exact solution of Eq. (A1.8) for the parabolic barrier and then match it to the quasiclassical solution at some points z_2, z_3 (Fig. A1.1d) where both the solutions are valid.

Putting in the region $z_2 < z < z_3$

$$U(z) = U_m\left[1 + \left(\frac{z - z_m}{a}\right)^2\right], \quad U_m < 0, \tag{A1.33}$$

where $U_m = U(z_m)$ is the maximum barrier "height", we obtain instead of Eq. (2.10)

$$\frac{d^2\Psi}{d\xi^2} + [\xi^2 - \varepsilon_1]\Psi = 0. \tag{A1.34}$$

Here

$$\xi = (z - z_m) \sqrt[4]{\frac{-k_0^2}{a^2} U_m}, \quad \varepsilon_1 = -(-U_m + E) \frac{k_0 a}{\sqrt{-U_m}}. \tag{A1.35}$$

The general solution of Eq. (A1.34) has the form [Whittaker and Watson 1946]

$$\Psi(\xi) = b_1 D_{-\frac{1+i\varepsilon_1}{2}} [(1+i)\xi] + b_2 D_{-\frac{1+i\varepsilon_1}{2}} [-(1+i)\xi], \tag{A1.36}$$

where $D_p(z)$ is the parabolic cylinder function. Matching the exact and quasi-classical solutions, as well as their derivatives, at the points z_2, z_3, one can obtain both the equation for eigenwave numbers and the ratio of the field amplitudes in the ducts I, II. We choose z_2 and z_3 so that

$$z_m = \frac{z_2 + z_3}{2}, \quad 1 \ll |\xi^2(z_2)| = |\xi^2(z_3)| \gg \varepsilon_1^2. \tag{A1.37}$$

To find the ratio of the field amplitudes in the ducts I, II, it is convenient to take as independent solutions of Eq. (A1.34) linear combinations of the D-functions rather than the functions themselves (the asymptotics of D-functions are used)

$$\begin{aligned} \varphi_1(\xi) &= v_1 + \frac{1}{2} \exp\left[-\frac{\pi\varepsilon_1}{2} - i\frac{\pi}{2}\right] v_2, \\ \varphi_2(\xi) &= v_2 + \frac{1}{2} \exp\left[-\frac{\pi\varepsilon_1}{2} - i\frac{\pi}{2}\right] v_1, \end{aligned} \tag{A1.38}$$

where

$$v_2(\xi > 0) = v_1(\xi < 0) = \exp(-i\chi) + \exp\left(-i\frac{\pi}{2}\right) \frac{f(\varepsilon_1)}{1 + e^{-\pi\varepsilon_1}} e^{i\chi},$$

$$v_2(\xi < 0) = v_1(\xi > 0) = \exp\left(-\frac{\pi\varepsilon_1}{2}\right) \frac{f(\varepsilon_1)}{1 + e^{-\pi\varepsilon_1}} e^{i\chi}.$$

Here

$$f(\varepsilon) = \frac{\sqrt{2\pi}}{\Gamma\left(\frac{1 - i\varepsilon_1}{2}\right)} e^{-\frac{\pi\varepsilon_1}{2}} e^{i\frac{\varepsilon_1}{2}} \left(\frac{\varepsilon_1}{2}\right)^{-i\frac{\varepsilon_1}{2}}, \quad e^{i\chi} = e^{i\frac{\xi^2}{2}} |\xi|^{i\frac{\varepsilon_1}{2}} e^{-i\frac{\varepsilon_1}{4}} \varepsilon_1^{i\frac{\varepsilon_1}{4}} 2^{-i\frac{\varepsilon_1}{2}}.$$

At $\varepsilon_1 = 0$, the function $f(\varepsilon_1) = \sqrt{2}$, and at $\varepsilon_1 = \infty$, $f(\varepsilon_1) = 1$. The functions φ_1 and φ_2 when $\varepsilon_1 \gg 1$ describe standing waves in the first and second waveguides, respectively. The factor $\exp[-\pi\varepsilon_1/2]$ gives the penetrance through the parabolic barrier

$$p^{1/2} = \exp\left[-\int_{-\sqrt{\varepsilon_1}}^{\sqrt{\varepsilon_1}} \sqrt{\varepsilon_1 - \xi^2}\, d\xi\right] = \exp\left[-\frac{\pi\varepsilon_1}{2}\right]. \qquad \text{(A1.39)}$$

It characterizes the energy capture into the duct. Finally, the factor $e^{i\chi}$ describes the wave phase increment from the turning point $\xi = \sqrt{\varepsilon_1}$ to some point ξ: $\exp[i\chi] = \exp\left[i\int_{\sqrt{\varepsilon_1}}^{\xi} \sqrt{\xi^2 - \varepsilon_1}\, d\xi\right]$. For high barrier ($\varepsilon_1 \gg 1$) in the case when the first waveguide is excited, by using Eq. (A1.38), we obtain the ratio of field amplitudes in the ducts I and II [see Eq. (A1.20)]. The result obtained on the basis of the solution of Eq. (A1.34) shows that Eq. (A1.20) is approximately valid down to $\varepsilon_1 \approx 1$. If $\varepsilon_1 \ll 1$, the field amplitudes in both the ducts are of the same order of magnitude, $|a|/|C| \sim 1$.

Appendix 2. Analytical Models of Ionospheric Parameters

For investigation of long-distance propagation regularities and for development of radio path calculation techniques, the analytical models of ionospheric parameters are needed. Such models should represent with sufficient reliability the main regularities of ionospheric variations. Let us formulate the general requirements the models should satisfy.

1. Since the models are to be used for the calculation of radio wave trajectories and their characteristics, they should include the functions which everywhere with respect to the coordinates are continuous with the first derivatives.
2. The models should cover the height range of 50 – 500 km, i.e., they should describe the *D, E,* and *F* layers, as well as the valley between the *E* and *F* layers. The description of ionospheric parameters in the lower ionosphere is necessary to calculate the radio wave absorption. A consideration of the interlayer ionization is important for the investigation of the interlayer duct.
3. It is necessary that the models should represent latitudinal – longitudinal and sunspot cycle variations of ionospheric parameters known from the experiments. The consideration of such dependences in the models allows to investigate the global properties of long-range propagation.
4. According to points 1 – 3, the analytical models should not be too complicated (i.e., they should be described by such a set of functions, of which computation does not take too much time), because the model should be used as a constituent in the more complex programs.

Below, the model of a three-dimensional electron density distribution and electron collision frequency is given which is constructed in Gurevich et al. (1973), Gayvoronskaya et al. (1974), Gurevich et al. (1981), Israetel (1982). Let us note that there are some other three-dimensional analytical models of electron density distribution (Jones and Stewart 1970, Nisbet 1971, Ching and Chiu 1973, Myung 1973, Polyakov et al. 1978, Rawer et al. 1978, Llewellyn and Bent 1973, Kohnlein 1977). However, they do not always satisfy the above requirements to a full extent.

A2.1 Three-Dimensional Equinoctial Model of Electron Density in the Ionosphere

The electron density model, satisfying points 1 – 4, was constructed in Gurevich et al. (1972, 1973) (see also Gurevich and Tsedilina 1979a) for the equinoctial conditions during solar activity (SA) minimum. Later on a number of modifications were made in the model (Israetel and Tsedilina 1982, Gurevich et al. 1981). In Tushentsova and Fishchuk 1975, the latitudinal variations of model (Gurevich et al. 1972, 1973) coefficients were revised and refined. In Israetel and Tsedilina (1982), Kadukhin and Soboleva (1978) the SA dependence and some additional modifications were introduced.

Here the complete description of the analytical equinoctial model of the ionosphere EMI-81 is given. The model is constructed for magnetically quiet conditions (Gurevich et al. 1981). The model (Gurevich et al. 1972, 1973) is cited in the text as the model EMI-72. Mainly, it corresponds to the EMI-81 for the low SA. However, the EMI-72 does not completely describe the latitudinal variations of electron density (see Tushentsova and Fishchuk 1975, Gurevich and Tsedilina 1979a).

Description of the Model. The EMI is constructed in the spherical coordinates z, θ, φ ($z = R - R_0$, R is a radial distance read off from the Earth's center), R_0 is the Earth's radius, θ is the latitude, φ is the longitude, θ is read off from the equator towards the poles. $\theta = 0 - \pi/2$ is the northern hemisphere, $\theta = 0 - (-\pi/2)$ is the southern hemisphere, φ is read off from the noon: $\varphi = 0 - \pi$ is the eastern hemisphere, $\varphi = 0 - -\pi$ is the western hemisphere. Solar activity dependence is taken into account by specifying the parameter W (the Wolf number or a monthly mean smoothed sunspot number). At present, in scientific publications, in addition to W, some other solar activity indices are used, namely, $F_{10.7}$ which is the solar radio flux in units of $10^{-22}\,\mathrm{W\,m^{-2}\,MHz^{-1}}$ at wavelength 10.7 cm, I_{F2} and I_E which are the ionospheric indices of SA. A relationship among the indices is discussed, for instance, in Minnis and Bazzard (1959), Maeda (1972).

Parameters θ, φ, and W are the key ones for determination of electron density at a given height $z = z_0$ or for determination of the height profile $N_e(z, \theta = \theta_0, \varphi = \varphi_0, W = W_0)$.

Height Dependence of N_e must describe its exponential variations known from the experiments. The following principal parameters describing the height variations of N_e were chosen: the heights of peak density in the layers E, $z_{\max E} = z_1$, and F, $z_{\max F} = z_2$, the height of minimum density between E and F layers $z_{\min F} = z_V$ and, of course, the values of density at these characteristic points N_1, N_2, N_V. The whole height range is divided into four regions:

1. the region below z_1;
2. the region from z_1 to z_V;
3. the region from z_V to z_2;
4. the region above z_2.

The electron density within each region is described by the following functions:

$$N_e = N_1 \exp\left\{ -\frac{\alpha_1 (z-z_1)^2}{1+\beta_1 (z_1 - z)} \left[1 + \gamma \frac{\frac{z-z_{st}}{a}}{1+\left(\frac{z-z_{st}}{a}\right)^4} \right] \right\}, \quad z < z_1, \tag{A2.1}$$

$$N_e = N_V \exp A(z), \quad z_m \geqslant z \geqslant z_1, \tag{A2.2}$$

$$N_e = N_V \exp B(z), \quad z_2 \geqslant z > z_m, \tag{A2.3}$$

$$N_e = N_2 \exp(-M), \quad M = \frac{\alpha_2 (z-z_2)^2}{1+\beta_2 (z-z_2)^{3/2}}, \quad z > z_2, \tag{A2.4}$$

$$A(z) = \ln\left(\frac{N_1}{N_V}\right) \frac{(z_V - z)}{(z_V - z_1)^3} [z_V^2 + z z_V + 3 z z_1 - 3 z_V z_1 - 2z^2],$$

$$B(z) = \ln\left(\frac{N_2}{N_V}\right) \frac{(z - z_V)}{(z_2 - z_V)^3} [z_V^2 + z z_V + 3 z z_2 - 3 z_V z_2 - 2z^2].$$

Functions $A(z)$ and $B(z)$ are the cubic curves. They are obtained by integration of the expressions

$$A(z), \quad B(z) = k \int_{z_V}^{z} (z - z_{1,2})(z_V - z)\, dz \tag{A2.5}$$

where the constant k is found from the following conditions: $N_e(z_1) = N_1$, $N_e(z_2) = N_2$. A typical property of these curves [Eq. (A2.5)] is that at their middle point $z_{01,02} = \frac{1}{2}(z_V + z_{1,2})$ the logarithm of density is equal to a half sum of logarithms at the terminal points

$$\ln N_e(z_{02}) = \frac{1}{2}(\ln N_2 + \ln N_V), \quad \ln N_e(z_{01}) = \frac{1}{2}(\ln N_1 + \ln N_V),$$

which mainly represent the height dependence of electron density in the regions (2) and (3). Thus, Eqs. (A2.2) and (A2.3) define the height dependence of density between the maxima of F and E layers without introducing any additional parameters characterizing the half-thickness of the F layer. In Eq. (A2.1) the factor in brackets, including parameters a, z_{st}, γ differs from unity only at heights $z \approx z_{st} \pm a$. It describes a "step" in the D region at heights of 60 – 80 km. The parameter z_{st} defines the step height, while a and γ define the step width and height.

Besides the principal parameters, N_1, N_V, N_2, z_1, z_V, z_2, the Eqs. (A2.1)–(A2.4) involve parameters α_1, α_2, β_1, β_2. These parameters describe the form of the layers above z_{maxF} and below z_{maxE}. The parameters α_1 and α_2 characterize the square law dependence of N_e on z below z_{maxE} and above z_{maxF}, respectively. Numerical values of α_1 and α_2 were determined by the empirical profiles (Soboleva 1973) from the electron density variations near peaks. In this case only the square factor in Eqs. (A2.1) and (A2.4) was taken into account. The parameter α_1 was determined at the height which is less than z_1 approximately by 10 km, whereas the parameter α_2 was determined at the height which is greater than z_2 by 20 – 50 km depending on the shape of the empirical curves. The parameters β_1 and β_2 characterize the height dependence $N_e(z)$ far from the layer peaks. That is why the numerical values of β_1, β_2 should be determined on the basis of the empirical profiles (Soboleva 1973) by asymptotic behavior of Eq. (A2.1) and Eq. (A2.4) far from the layer peak in the region where $\beta_1(z-z_1) \geqslant 1$ and $\beta_2(z-z_2)^{3/2} \gg 1$. Figure A2.1 a,b shows the averaged empirical profiles within the height range of 100 – 1000 km (solid lines) (Soboleva 1973, Kadukhin and Soboleva 1978). For comparison, in the same figure the height profiles (broken lines) calculated by Eqs. (A2.1) – (A2.4) are shown (see also Figs. A2.12 and A2.13). The agreement between the calculated and empirical profiles is seen to be quite good.

Solar Activity Dependence of N_e is defined by the dependence of the height profile parameters $F \equiv N_1, N_V, N_2, z_1, z_V, z_2, \alpha_1, \alpha_2, \beta_1, \beta_2, a, z_{st}, \gamma$ on W. The SA dependence for the parameters $F \equiv N_1, N_V, N_2$ is accepted as follows

$$F(W) = F_1\left(\frac{1+\mu W}{1+\mu W_1}\right)^2 \tag{A2.6}$$

where F_1 are the values of parameters F at $W = W_1 = 10$ (SA minimum Gurevich et al. 1972, 1973; Tushentsova and Fishchuk 1975). Equation (A2.6) cor-

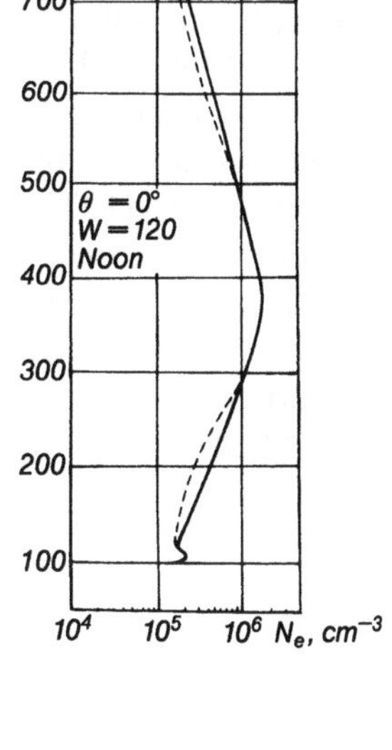

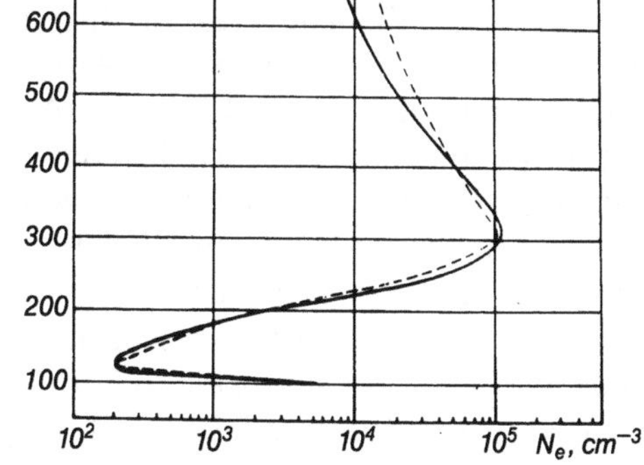

Fig. A2.1 a,b. Comparison of the EMI-81 height profiles of electron density (*broken curve*) with the empirical profiles (*solid curve*): **a** noon, $\theta = 0°$, $W = 120$; **b** midnight, $\theta = 50°$, $W = 10$

responds to linear dependence of the monthly mean values of critical frequencies for E and F_2 layers on the monthly mean smoothed values of W (Al'pert 1974, Minnis and Bazzard 1959, Appleton 1959). The numerical values of coefficients $\mu \equiv \mu_1, \mu_V, \mu_2$ at different values of θ (see below) are determined by the values of parameters at $W = W_1 = 10$ and $W = W_2 = 120$ as follows

$$\mu = \frac{\sqrt{F_2} - \sqrt{F_1}}{\sqrt{F_1}\, W_2 - \sqrt{F_2}\, W_1} \tag{A2.7}$$

where $F_1 = F(W_1)$, $F_2 = F(W_2)$. In Maeda and Fukao (1972) and Maeda (1972) another SA dependence of N_1 is suggested

$$N_1(W) = (N_1)_0 \sqrt{1 + \rho W} \tag{A2.8}$$

where $(N_1)_0 = N_1(W = 0)$. Figure A2.2a demonstrates the curves $N_1(W)$ plotted according to Eq. (A2.6) (curve 1) and Eq. (A2.8) (curve 2) for the equinoctial conditions at noon for the latitude $\theta = 50°$. Here, the coefficients $(N_1)_0$ and ρ are calculated in the same way as the coefficients μ by the values of N_1 at $W = 10$ (Soboleva 1973) and $W = 120$ (Kadukhin and Soboleva 1978a). These values of N_1, as well as the value of N_1 at $W = 190$ from Kadukhin and Soboleva (1978a), are shown in Fig. A2.2a by crosses. One can see that at $W = 190$ the differences among the curves 1, 2 and N_1 from Kadukhin and Soboleva (1978a) are less than 6%. In turn, the curves 1 and 2 in the range of $W = 0 - 220$ differ from each other by less than 7.5%. It can be explained by the smallness of coefficients μ_1 and ρ ($\mu_1 \sim 2 \cdot 10^{-3}$, $\rho \sim 0.01$).

The dependence of the F_2 layer critical frequency on W in a number of cases can decline from the linear one. In this case both the limitation and intensification of the growth of f_0F_2 at high SA ($W \approx 140$) (Kerblay and Lyakhova 1969, Serafimov and Lilov 1975) may occur. Figure A2.2b shows the curve $N_2(W)$ plotted according to Eq. (A2.6) in the same way as the curve 1, shown in Fig. A2.2a. In Fig. A2.2b, the little circles correspond to the values of N_2 obtained by recalculation of f_0F_2 from Kerblay and Lyakhova (1969) (p 16, Fig. 1.18, IV). The greatest difference between the data of Kerblay and Lyakhova (1969) and the curve $N_2(W)$ [Eq. (A2.6)] is about 50%, but mainly it is no more than 15 – 20%. At $W = 190$, the value of N_2 calculated according to Eq. (A2.6) is different from N_2 (Kadukhin and Soboleva 1978a) (crosses) by less than 6%. In Kadukhin and Soboleva (1978a) it is noted that with increasing SA from $W_1 = 10$ to $W_2 = 120$, the parameter z_1 decreases by 2 – 3 km. However, such variations are within the measurement accuracy of z_1. Therefore, let us accept that z_1 does not depend on SA. For the parameters $F \equiv z_V, z_2$, according to Kadukhin and Soboleva (1978b), Kerblay and Lyakhova (1969), Anufrieva and Shapiro (1976), the following SA dependence is accepted

$$F(W) = F_1 \frac{1 + \eta W}{1 + \eta W_1} \tag{A2.9}$$

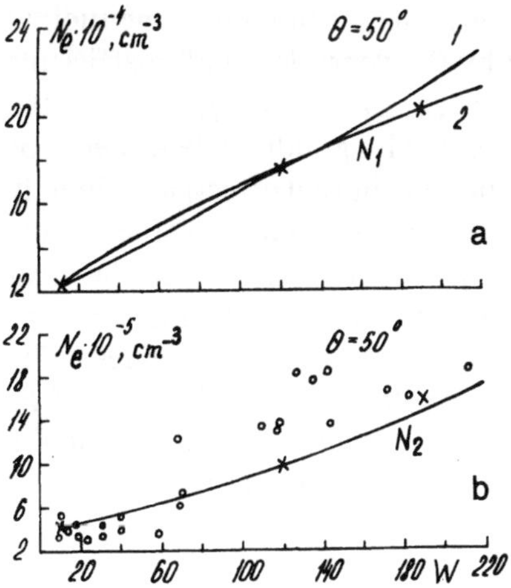

Fig. A2.2 a, b. Dependence of N_1 and N_2 on the sunspot number W, $\theta = 50°$

where F_1 are the values of parameters F at $W = W_1 = 10$. The numerical values of coefficients $\eta \equiv \eta_V$, η_2 are determined in the same way as the coefficients μ, according to the following equation

$$\eta = \frac{F_2 - F_1}{F_1 W_2 - F_2 W_1} . \tag{A2.10}$$

The height variations of electron density $N_e(z)$ when $z > z_2$ (above the F layer maximum) for $W_1 = 10$ and $W_2 = 120$ are similar with an accuracy of $\sim 20-30\%$ (Soboleva 1973, Fatkullin et al. 1981, Kadukhin and Soboleva 1978a, b). But the measurement accuracy of $N_e(z)$ in the F region is about 20% (Kadukhin and Soboleva 1977a, b), so let us assume that when $z > z_2$, the profile $N_e(z)$ shape does not depend on SA, i.e., the coefficients α_2 and β_2 do not depend on W.

When $z < z_1$, the midday profiles $N_e(z)$ (D region) for $W_1 = 10$ (Soboleva 1973) and $W_2 = 120$ (Kadukhin and Soboleva 1977a) are also similar with an accuracy of $\sim 30-40\%$. At the same time, the measurement accuracy of $N_e(z)$ in the D region is about 40% (Kadukhin and Soboleva 1978a). So, one can assume that α_1 and the additional coefficients γ, z_{st}, and a, characterizing the step in the D region, do not depend on SA. At midnight at the heights of about 70–80 km the similarity is violated, because at $W = 120$ the electron density at these heights decreases much slower than at $W = 10$ (Soboleva 1973, Kadukhin and Soboleva 1978a) (see also Figs. A2.12 and A2.13). Thus, it proved to be worthwhile to introduce the SA dependence only in the coefficient β_1 analogously to Eq. (A2.9)

$$\beta_1(W) = \beta_1 \frac{1 + \xi W}{1 + \xi W_1} . \tag{A2.11}$$

The numerical values of the coefficient ξ are determined through the values of β_1 at $W = W_1 = 10$ (Soboleva 1973) and $W = W_2 = 120$ (Kadukhin and Soboleva 1978a) analogously to the coefficients η [Eq. (A2.10)].

Latitudinal – Longitudinal Dependence of N_e. The latitudinal – longitudinal distribution of electron density is specified by the dependence of coefficients $F \equiv N_1, N_V, N_2, z_1, z_V, z_2, \alpha_1, \alpha_2, \beta_1, \beta_2, \gamma, z_{st}, a, \mu_1, \mu_V, \mu_2, \eta_V, \eta_2, \xi$ on the latitude θ and longitude φ. Generally speaking, the longitudinal dependence of coefficients F is determined both by solar time (the diurnal effect) and directly by the dependence on the longitude φ. Here, we shall consider only the diurnal effect. In this case, naturally, the daytime $F_d(\theta)$ (corresponding to the local noon $\varphi = 0$) and nighttime $F_n(\theta)$ (corresponding to the midnight $\varphi = \pm\pi$) values of all the parameters F prove to be the reference values. The coefficients F_d and F_n are assumed to vary only with latitude θ. In terms of F_d and F_n, it is convenient to approximate the longitudinal variations of the parameters $F(\theta, \varphi) = N_1, N_V, N_2$ by means of the exponential functions

$$F(\theta, \varphi) = \exp\{f(\theta, \varphi)\},$$

$$f(\theta,\varphi) = \begin{cases} f_n(\theta) + \left(\dfrac{f_d(\theta) - f_n(\theta)}{2}\right)\left[1 + \dfrac{(\pi - |\varphi_m|)(|\varphi| - |\varphi_m|)}{\pi - |\varphi|}\right]^{-\lambda}, \\ \qquad \varphi_m \geqslant \varphi \geqslant -\pi, \\ f_d - \left(\dfrac{f_d - f_n}{2}\right)\left[1 + \dfrac{\varphi_m(|\varphi_m| - |\varphi|)}{\varphi}\right]^{-\lambda}, \quad 0 \geqslant \varphi \geqslant \varphi_m, \\ f_d - \left(\dfrac{f_d - f_n}{2}\right)\left[1 + \dfrac{\varphi_e(\varphi_e - \varphi)}{\varphi}\right]^{-\lambda}, \quad \varphi_e > \varphi \geqslant 0, \\ f_n + \left(\dfrac{f_d - f_n}{2}\right)\left[1 + \dfrac{(\pi - \varphi_e)(\varphi - \varphi_e)}{\pi - \varphi}\right]^{-\lambda}, \quad \pi \geqslant \varphi \geqslant \varphi_e, \end{cases} \tag{A2.12}$$

where $f_n = \ln F_n(\theta)$, $f_d = \ln F_d(\theta)$. The latitudinal approximation of F_d and F_n is given below.

The parameters φ_m and φ_e define an asymmetry of the morning and evening values of F with respect to the noon. Generally speaking, these parameters may depend on the latitude. However, in the EMI-81 they are assumed to be constant. The parameters φ_m and φ_e define the longitude of the morning and evening values of F [according to Eq. (A2.12), $F = \sqrt{F_d F_n}$ at $\varphi = \varphi_m, \varphi_e$].

According to Eq. (A2.6), the values of F_d and F_n for the parameters N_1, N_V, N_2 are equal to

$$\begin{aligned} F_d(W, \theta) &= F_{1d}(\theta)\left[\frac{1 + \mu_d(\theta)\, W}{1 + \mu_d(\theta)\, W_1}\right]^2 \\ F_n(W, \theta) &= F_{1n}(\theta)\left[\frac{1 + \mu_n(\theta)\, W}{1 + \mu_n(\theta)\, W_1}\right]^2. \end{aligned} \tag{A2.13}$$

Table A2.1. Values of φ_m and φ_e for the parameters $F \equiv N_1, N_V, N_2$

Parameter	φ_m	φ_e
N_1	$-0.5\ \pi$	$0.536\ \pi$
N_V	$-0.49\ \pi$	$0.53\ \pi$
N_2	$-0.425\ \pi$	$0.622\ \pi$

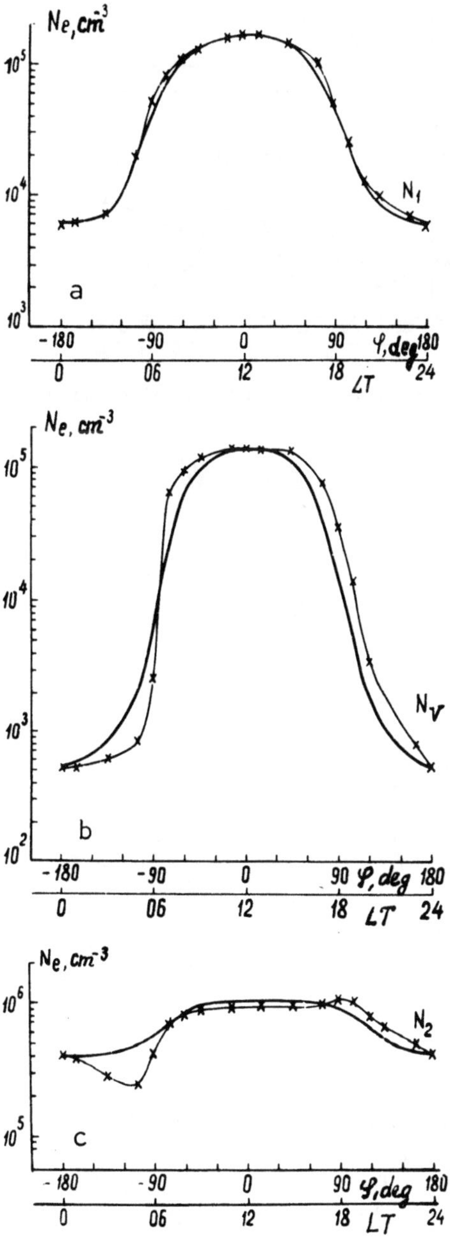

Fig. A2.3 a–c. Comparison between the diurnal variation of the parameters (**a**) N_1; (**b**) N_V; (**c**) N_2 calculated by EMI-81, and the empirical one (*crosses*) at $\theta = 50°$, $W = 120$

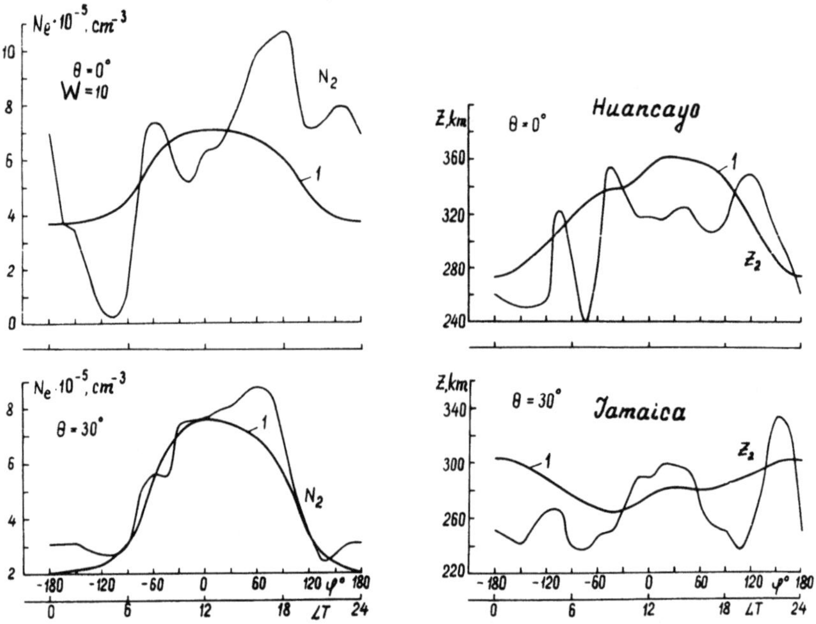

Fig. A2.4. Comparison of the diurnal variation of N_2 and z_2 calculated by EMI-81 (*curves 1*) with the empirical one at different latitudes

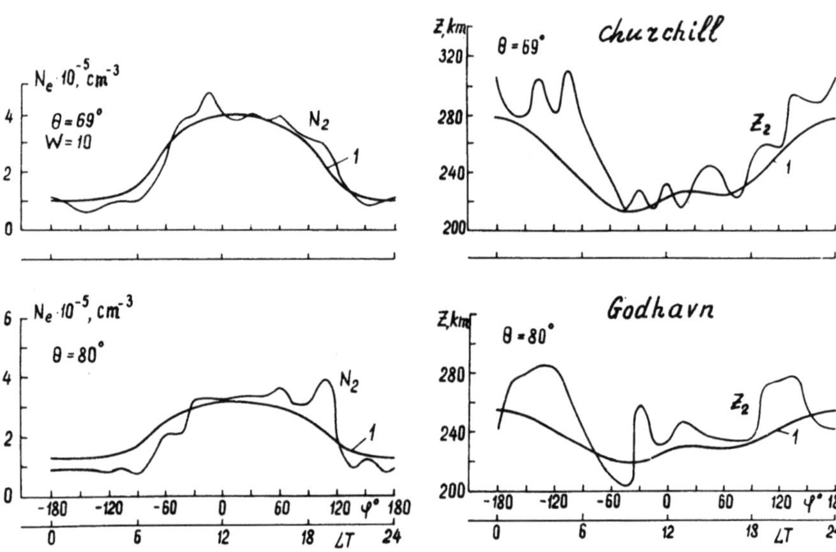

Fig. A2.5. The same as in Fig. A2.4

Let us note that all the parameters defining the longitudinal dependence of N_e in the EMI-81 are constructed on the basis of the diurnal variations of the empirical profiles $N_e(z)$ (Soboleva 1972) for March equinox 1964–1965, $\theta = 50°$, $W = 10$. In this case, the fraction in brackets in [Eq. (A2.13)] is equal to 1, hence, $F_d(W=10, \theta) = F_{1d}(\theta)$, $F_n(W=10, \theta) = F_{1n}(\theta)$. The values of φ_m and φ_e, calculated by the data of Soboleva (1972), for the parameters N_1, N_V, N_2 are given in Table A2.1. The parameter λ in Eq. (A2.12) is constant. Its value was chosen to be equal to 3 for the parameter N_V, and equal to 2 for the parameters N_1 and N_2. Figure A2.3 a,b,c shows the diurnal variations of the parameters N_1, N_V, N_2 (Kadukhin and Soboleva 1978b) for high SA (small crosses) and their approximation, according to Eqs. (A2.12) and (A2.13) (solid curves). One can see that the differences are mainly less than 10%. Maximum discrepancies (roughly a factor of 2) take place for the parameter N_V at predawn and postevening hours. This also takes place at low SA.

Note that the approximation, Eq. (A2.12), represents with high accuracy the diurnal variation of the parameters N_1, N_2, N_V both at various SA and at various latitudes. This is seen from Figs. A2.4 and A2.5, where for $W=10$ the comparison is given between the model curves 1 [Eq. (A2.12)] and empirical ones (Kadukhin and Soboleva 1978c). The latter are obtained by means of calculation of $N_e(z)$ – profiles for the 18th of March 1965.

The latitudinal dependence of F_{1d}, F_{1n}, μ_d, μ_n is approximated as follows

$$\begin{aligned}
N_{1d} &= [1.27 + 0.33(1-x)^8 - 0.75x^7] \cdot 10^5\,,\\
N_{1n} &= \{1.5 + 3.53 \cdot 2^8 \cdot x^4(1-x)^2[10(1-2x)^4 + 2^8 \cdot x^4(1-x)^6]\\
&\quad + 0.159/(1.00315 - x) + 0.016/(0.0025 + x^2)\} \cdot 10^3\,,\\
\mu_{1d} &= [1.8 - 0.206(1-x)^{18} + 2.2x^{15}] \cdot 10^{-3}\,,\\
\mu_{1n} &= \{5.3 - 1.3x^{18} - [100 + 220(1-x)^4]x^3(1-x)^2 - 4.216(1-x)^2\} \cdot 10^{-3}\,.\\
N_{Vd} &= [1.15 - 0.73x^8 + 0.14(1-x)] \cdot 10^5\,,\\
N_{Vn} &= \{0.42/(1.001 - x) + 0.073/(0.0143 + x) - 14.176x(1-x)^3\\
&\quad + 2^8x^5(1-x)^2[327(1-2x)^4 + 2^{28}x^{13}(1-x)^{16}]\} \cdot 10^2\,,\\
\mu_{Vd} &= (1.1 + 3.8x^{17}) \cdot 10^{-3}\,, \qquad \text{(A2.14)}\\
\mu_{Vn} &= \{4.9 + [6.6 - 326x^2(1-x)^6 - 1.5(1-x)^4](1-x) - 27.5x(1-x)^3\\
&\quad - 6.1(1-x)^6\} \cdot 10^{-3}\,,\\
N_{2d} &= [4.8 + 20.4 \cdot 2^8 \cdot x^2(1-x)^8(1-2x)^4 + 0.48/(0.185 + x)\\
&\quad - 0.316/(1.121 - x)] \cdot 10^5\,,\\
N_{2n} &= [0.9 + 2.8(1-x)^3 + 0.0136/(1.008 - x)] \cdot 10^5\,,\\
\mu_{2d} &= \{5.67 - [137 + 445x - 150(1-x)^{19}]x(1-x)^{12} - 40x(1-x)^{32}\\
&\quad + [25 - 400x(1-x)^{12}]x(1-x)^3 - 35x^3(1-x)^2 - 6x^3 + 4.13x^{10}\} \cdot 10^{-3}\,,\\
\mu_{2n} &= \{3.8 + [22 + 340x^3(1-x)^2](1-x)^2 - (19.6 + 40x)(1-x)^5\\
&\quad + [40 + 92(1-x)^8]x(1-x)^{12}\} \cdot 10^{-3}\,.
\end{aligned}$$

Here, $x = \sin^2\theta$. Eqs. (A2.14) are obtained by the analytical approximation of the reference points which correspond to the empirical data from Soboleva (1973), Kadukhin and Soboleva (1978a). In order to satisfy the continuity condition mentioned above, the latitudinal variation of all the model parameters [Eqs. (A2.14), (A2.17), (A2.20), and (A2.21)] was approximated by trigonometrical functions ($\sin\theta$, $\cos\theta$) in even powers. In Figs. A2.6, A2.7 the comparison of analytical dependences (solid curves) with corresponding empirical ones (small crosses) is illustrated. The approximation inaccuracy [Eq. (A2.14)] at reference points is mainly less than 5%, and in the whole range of θ, the inaccuracy is no more than 10%.

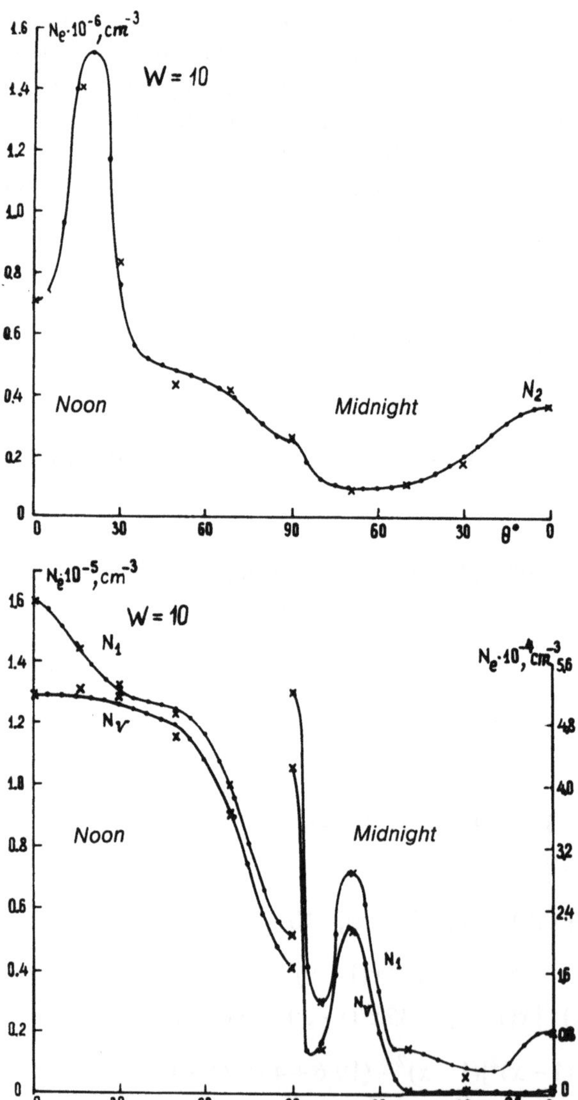

Fig. A2.6. Comparison of the latitudinal variation of the EMI-81 parameters N_1, N_2, N_V (*dots*) with the empirical one (*crosses*) at $W = 10$

For longitudinal variation of the parameters $F \equiv z_1, z_V, z_2$, the approximation analogous to Eqs. (A2.2) and (A2.3) is of convenience. The whole range of φ from $-\pi$ to π is divided into four regions. In each region the value of the parameters $F \equiv z_1, z_V, z_2$ is equal to

$$F(\varphi,\theta) = F_1 + \frac{(F_2 - F_1)(\varphi - \varphi_1)}{(\varphi_2 - \varphi_1)^3}\,[\varphi_1^2 + \varphi_1\varphi + 3\,\varphi\varphi_2 - 3\,\varphi_1\varphi_2 - 2\,\varphi^2]\,. \quad \text{(A2.15)}$$

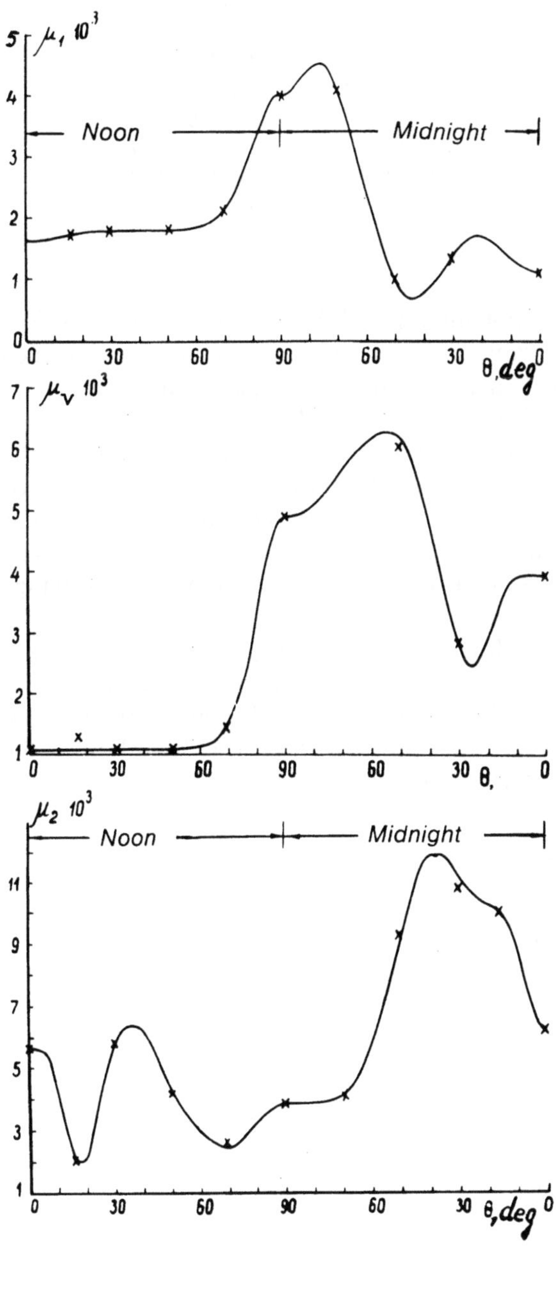

Fig. A2.7. Comparison between the analytical approximation of latitudinal variation of the parameters $\mu = \mu_1, \mu_2, \mu_V$, and the reference values of μ (*crosses*) calculated according to Eq. (A2.7) by the empirical curves N_1, N_2, N_V at $W = 10$ and at $W = 120$

Table A2.2. Values of φ_1, φ_2, F_1, F_2 for the parameters z_1, z_V, z_2

Parameter	φ_1	φ_2	F_1	F_2
z_1	π	$-0.39\ \pi$	F_n	$(1+0.15b)\ F_d^*$
	$-0.39\ \pi$	$0.0833\ \pi$	$(1+0.15b)\ F_d^*$	F_d
	$0.0833\ \pi$	$0.54\ \pi$	F_d	$(1+0.15b)\ F_d^*$
	$0.54\ \pi$	π	$(1+0.15b)\ F_d^*$	F_n
z_V	$-\pi$	$-0.416\ \pi$	F_n	$(1+0.3b)\ F_d$
	$-0.416\ \pi$	0	$(1+0.3b)\ F_d$	F_d
	0	$0.483\ \pi$	F_d	$(1+0.25b)\ F_d$
	$0.483\ \pi$	π	$(1+0.25b)\ F_d$	F_n
z_2	$-\pi$	$-0.208\ \pi$	F_n	$(1-0.05b)\ F_d$
	$-0.208\ \pi$	$0.125\ \pi$	$(1-0.05b)\ F_d$	$(1+0.016b)\ F_d$
	$0.125\ \pi$	$0.333\ \pi$	$(1+0.016b)\ F_d$	$(1+0.008b)\ F_d$
	$0.333\ \pi$	π	$(1+0.008b)\ F_d$	F_n

Equation (A2.15) is derived by applying Eq. (A2.5) in the interval $\varphi_1-\varphi_2$. The quantities φ_1 and φ_2 are constant, and determine the boundaries of each region, see Table A2.2; F_1 and F_2 are the values of the parameters $F \equiv z_1$, z_V, z_2 at $\varphi=\varphi_1$ and $\varphi=\varphi_2$, respectively, i.e., $F(\varphi_1)=F_1$, $F(\varphi_2)=F_2$. The values of F_1 and F_2 are defined in terms of the noon F_d and midnight F_n values of the parameters $F \equiv z_1$, z_V, z_2, see Table A2.2. In this table, the factor $b=1-0.01/(\cos^2\theta+0.01)$ is introduced to satisfy the continuity conditions at the poles; at $\theta=\pm\pi/2$, $b=0$, hence, $F_1=F_2=F_d=F_n$ (see below). The expressions marked by an asterisk in Table A2.2, are recommended to change for $(F_d+F_n)/2$.

Figure A2.8 shows the longitudinal variation of the parameters $F \equiv z_1$, z_V, z_2 from Soboleva (1972) and its analytical approximation, according to Eq. (A2.15). As is seen from the figure, the greatest differences are observed for the parameter z_V. However, they do not exceed 15%. Mainly, the differences lie within a few percent. Equation (A2.15) represents with high

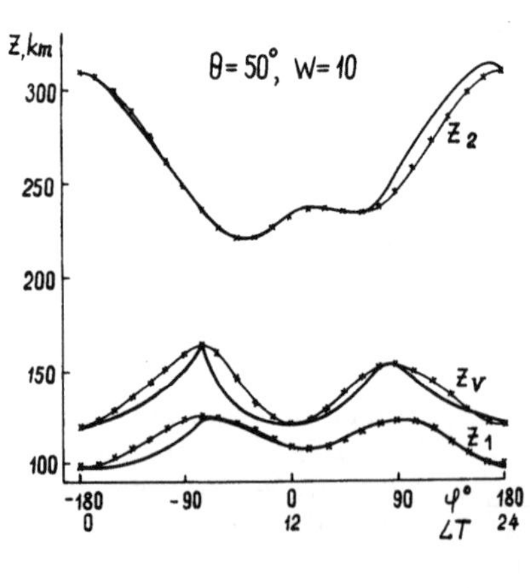

Fig. A2.8. Comparison of the diurnal variation of the EMI-81 parameters z_1, z_2, z_V (*crosses*) with the empirical one (*solid curves*) for $\theta=50°$, $W=10$

accuracy the diurnal variation of the parameters z_1, z_V, z_2 both at various SA and at various latitudes (see Figs. A2.4 and A2.5). For the noon, F_d, and midnight, F_n, values of the parameters $F \equiv z_1$, z_V, z_2 we have from Eq. (A2.9)

$$F_d(W,\theta) = F_{1d}(\theta)\frac{1+\eta_d(\theta)\,W}{1+\eta_d(\theta)\,W_1}$$
$$F_n(W,\theta) = F_{1n}(\theta)\frac{1+\eta_n(\theta)\,W}{1+\eta_n(\theta)\,W_1}\,. \tag{A2.16}$$

The latitudinal variation of the parameters $F_{1d} = z_{1d}$, z_{Vd}, z_{2d}; $F_{1n} = z_{1n}$, z_{Vn}, z_{2n}; η_d and η_n is approximated as follows

$$z_{1d} = 114.2 - 10.2x - 0.088/(0.0086+x) + 0.013/[0.001625+(1-x)^2]\,,$$
$$z_{1n} = 109 - 9x + 0.0129/[0.001075+(1-x)^2] - 0.1875/(0.0208+x^2)\,,$$
$$\eta_{1d} = \eta_{1n} = 0\,,$$
$$z_{Vd} = 124 - 8x^2 - 0.225/(0.025+x) + 0.149/(1.0212-x)\,, \tag{A2.17}$$
$$z_{Vn} = 125 - 11x + 0.144/(1.016-x)\,,$$
$$\eta_{Vd} = 2.965\cdot x^{16}\cdot 10^{-4}\,,$$
$$\eta_{Vn} = [16 - 10\cdot 2^{12}\cdot x^6(1-x)^6 - 13.035x^{13} - 14.543(1-x)^{11}]\cdot 10^{-4}\,,$$
$$z_{2d} = 221 + 135(1-x)^3 + 0.266/(1.00834-x)\,.$$

Equations (A2.17) are obtained analogously to Eqs. (A2.14) with the same accuracy. This is seen from Fig. A2.9, where the analytical dependence of z_{1d}, z_{1n}, z_{Vd}, z_{Vn}, z_{2d}, z_{2n} on θ [Eqs. (A2.17)] is compared with the empirical one at $W = 10$ (Soboleva 1973).

For the longitudinal variation of the parameters $F \equiv \alpha_1$, α_2, β_1, β_2, the following approximation is assumed, allowing a smooth transition from the daytime to nighttime values

$$F = F_n + (F_d - F_n)\cos^2\varphi/2\,. \tag{A2.18}$$

The possibility for using Eq. (A2.18) for the parameters α_2 and β_2 is illustrated by Fig. A2.10, where the diurnal dependence of N_e from Soboleva (1972) at the height of $z = 1000$ km is demonstrated. In this figure, the dependence calculated by the analytical model according to Eqs. (A2.4), (A2.12), and (A2.18) is also shown. One can see that the differences mainly do not exceed 10%.

As was pointed out above, the coefficients α_{1d}, α_{2d}, β_{1d}, β_{2d}, α_{1n}, α_{2n}, β_{2n} do not depend on W. Coefficient β_{1n} depends on W as follows

$$\beta_{1n}(W,\theta) = \beta_{1n}(\theta)\frac{1+\xi(\theta)\,W}{1+\xi(\theta)\,W_1}\,. \tag{A2.19}$$

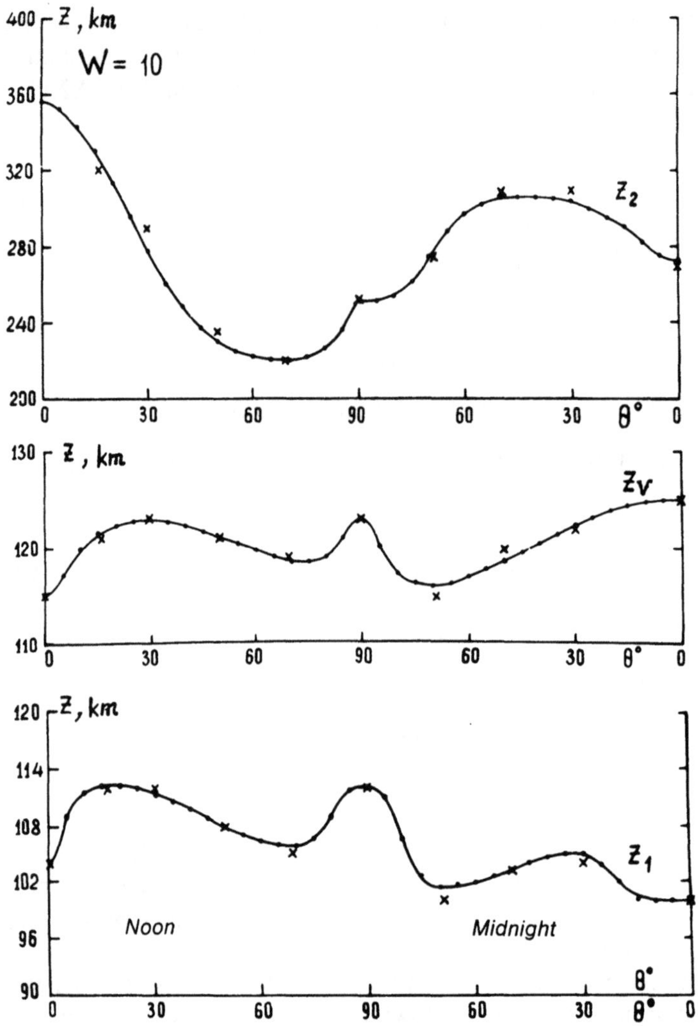

Fig. A2.9. Comparison of the latitudinal variation of the EMI-81 parameters z_1, z_2, z_V (*dots*) with the empirical data (*crosses*) at $W = 10$

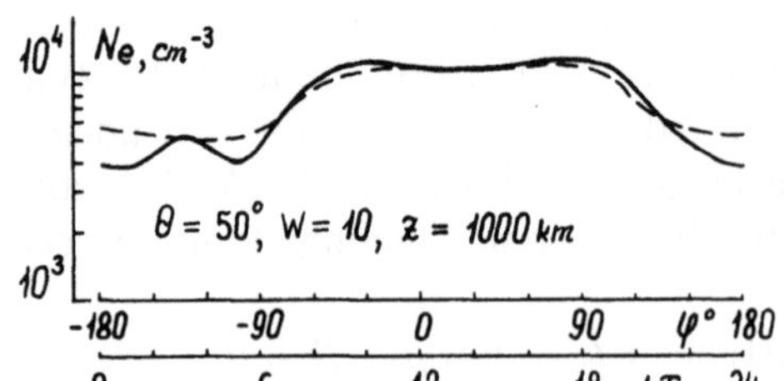

Fig. A2.10. Comparison between the diurnal variation of electron density calculated by the EMI-81 (*broken curve*) at $z = 1000$ km, and the empirical data (*solid curve*)

The latitudinal dependence of the coefficients α_{1d}, α_{1n}, β_{1d}, β_{1n}, α_{2d}, α_{2n}, β_{2d}, β_{2n}, ξ, obtained analogously to Eqs. (A2.14) and (A2.17), is given by the equations

$$\begin{aligned}
\alpha_{1d} &= [2.43(1-x)^{11}+2.64x-0.0109/(1.0143-x)]\cdot 10^{-2},\\
\alpha_{1n} &= [7x-5.2x^4+8.51(1-x)^9]\cdot 10^{-2},\\
\beta_{1d} &= [7.92(1-x)^8+12.6x^2-6.92x^{11}]\cdot 10^{-2},\\
\beta_{1n} &= [5.68x^4+14.28(1-x)^7+3\cdot 2^{17}\cdot x^8(1-x)^8]\cdot 10^{-2},\\
\alpha_{2d} &= [6.5-3.5x-5.5(1-x)^6]\cdot 10^{-4},\\
\alpha_{2n} &= [3.034-0.035/(x+0.035)]\cdot 10^{-4},\\
\beta_{2d} &= [2.4+3.3(1-x)-5.07(1-x)^4+2.25x(1-x)^3\\
&\quad -15x^3(1-x)^2]\cdot 10^{-3},\\
\beta_{2n} &= [2.6-0.2x^5-1.22(1-x)^5]\cdot 10^{-3},\\
\xi &= \{28-28x^{17}-27.5\cdot 2^{12}\cdot x^6(1-x)^6-[55-37\\
&\quad \cdot(1-x)^{17}](1-x)^{13}\}\cdot 10^{-3}.
\end{aligned} \tag{A2.20}$$

Comparison of the analytical dependences of the parameters α_1 and β_1 with the empirical reference points, obtained according to the data from

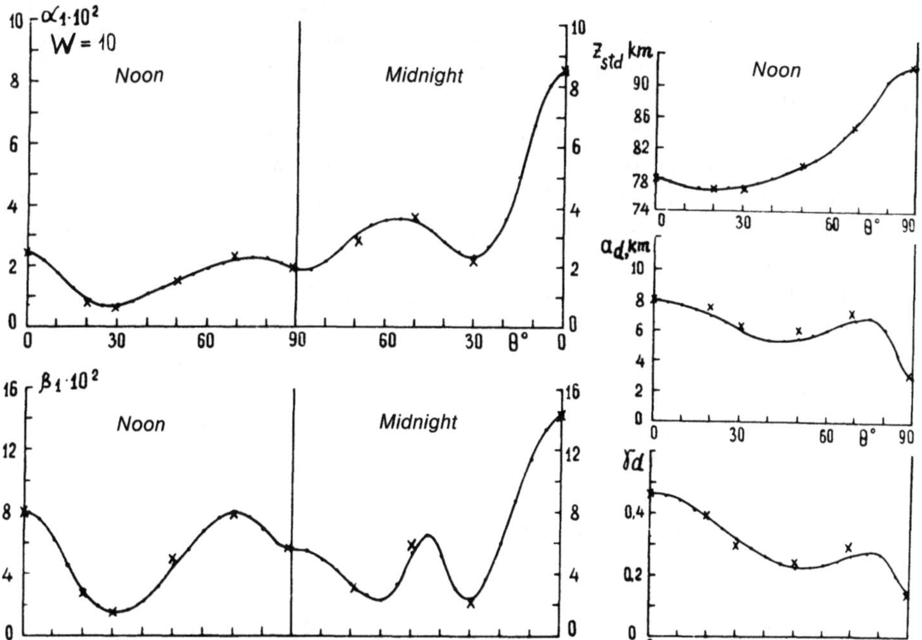

Fig. A2.11. Comparison between the analytical approximation of latitudinal variation of the coefficients α_1, β_1 [Eq. (A2.20)], z_{std}, a_d, γ_d [Eq. (A2.21)], and the empirical data (*crosses*)

Soboleva (1973) at $W = 10$, is illustrated in Fig. A2.11. The difference between the analytical approximation [Eq. (A2.20)] and reference points is generally less than 5% and no more than 10%.

Latitudinal – longitudinal variation of the parameters $F \equiv \gamma$, z_{st}, a is approximated by the following equations

$$\text{when} \quad -\frac{\pi}{2} \leqslant \varphi \leqslant \frac{\pi}{2} \qquad a = a_d(\theta),$$
$$\gamma = \gamma_d(\theta) \cdot b(\varphi) \cdot b(\theta),$$
$$z_{st} = z_{std}(\theta) + 5\, b(\theta) \cos^2 \varphi,$$

$$\text{when} \quad \frac{\pi}{2} < |\varphi| \leqslant \pi \quad \gamma = 0,$$

$$\text{where} \quad b(\varphi) = 1 - 0.01/(0.01 + \cos^2 \varphi),$$

$$a_d = 3.0 + 5.0(1-x)^2 + 5.83x^2 - \frac{0.1166}{1.02 - x}, \tag{A2.21}$$

$$\gamma_d = 0.226x^5 + \frac{0.2640}{0.5666 + x} - \frac{0.005}{1.02 - x},$$

$$z_{std} = 68 + \frac{4.1322}{1.2066 - x} + \frac{0.03}{0.019 + x}.$$

Here, $x = \sin^2 \theta$, the same as before. Latitudinal dependences of the parameters a_d, γ_d, z_{std} [Eq. (A2.21)] are obtained by means of analytical approximation of numerical values calculated by the empirical profiles (Soboleva 1973) for $W = 10$ (see Fig. A2.11).

In Fig. A2.12 the EMI-81 height profiles of electron density [Eqs. (A2.1) and (A2.12) – (A2.21)] in the D region are compared with the empirical ones (Soboleva 1973). One can see that in some cases the analytical and empirical profiles differ by one order, but mainly, the differences are far less, being about a factor of 2.

Since the reliability of experimental data for the D region is poor, and their dispersion is great (Fatkullin et al. 1981), it does not prove to be worthwhile to achieve higher accuracy, all the more so, since this would result in a more complicated analytical model.

Calculation of the height profiles with different values of φ and θ within the range of $z_2 \leqslant z \leqslant 1000$ km has shown that, in general, they differ from the corresponding empirical ones by no more than a factor of 1.5 – 2. In the valley region, the differences mainly do not exceed 20 – 30% and in several cases when the valley between the E and F layers manifests itself most strongly, the differences within the height ranges of $z_V < z < z_2$ can reach a factor of 2. Of course, at the heights of $z \approx z_1$, z_V, z_2 the differences are minimal.

Thus, Eqs. (A2.1) – (A2.4) and (A2.12) – (A2.21) and Tables A2.1 and A2.2 represent the EMI-81. Figure A2.13 illustrates the comparison of the EMI-81 profiles with the empirical ones within the range of $z \geqslant 100$ km.

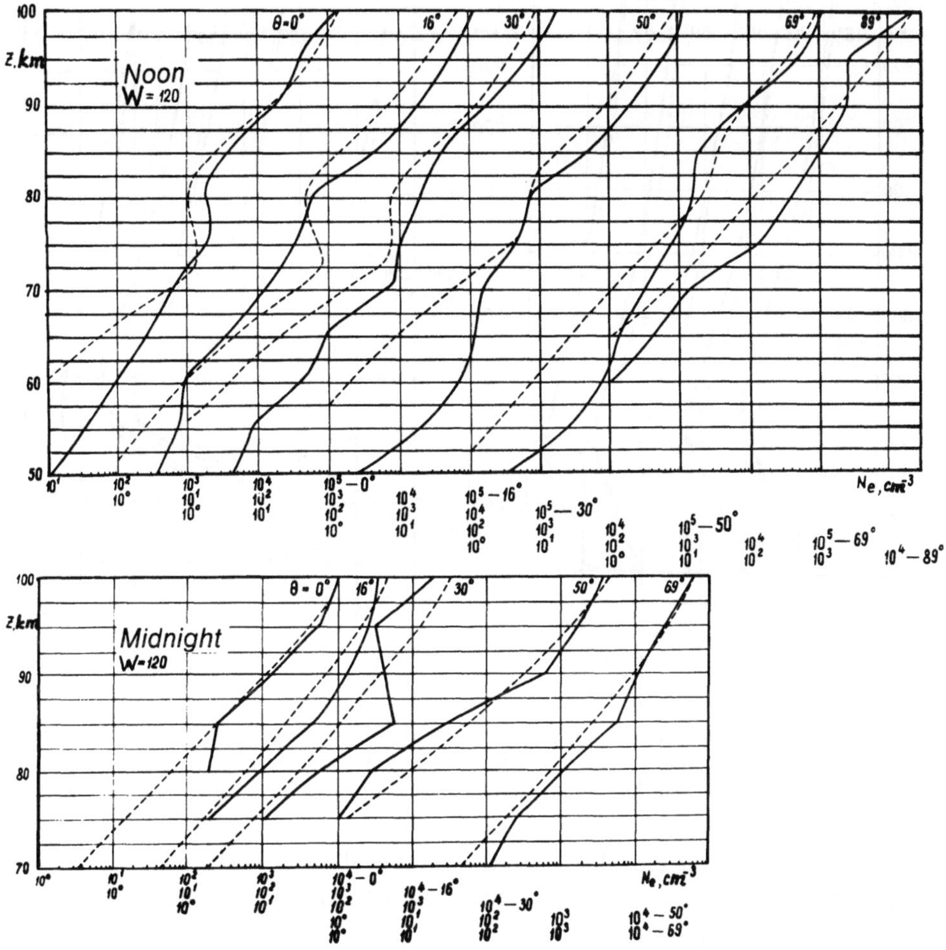

Fig. A2.12. Comparison between the EMI-81 height profiles of electron density (*broken curves*), and the empirical data below $z_{\max E}$ for $\varphi = 0$ (noon) and $\varphi = \pm\pi$ (midnight) at different latitudes, $W = 120$. The latitude is indicated near the profiles

Note that the principle parameters for the EMI-81 height profile are the extreme values, a number of which are determined by the data of the worldwide network of the vertical sounding stations. That is why the model can be easily corrected. It is most convenient to make the correction of the model (using the data of vertical sounding) when using it in the calculations of the adiabatic invariant method (Chaps. 4 and 5). In this case, it is necessary to define the $N_e(z)$ profiles only at the fixed points spaced 1000 – 1500 km over the path. When the critical frequency of the F_2 layer (f_0F_2) is used to correct the EMI-81, only the parameters $N_{2d}(W, \theta)$, $N_{2n}(W, \theta)$ are substituted as follows

$$N_{2d,2n} = \frac{f^2_{0d,0n}}{8.06 \cdot 10^7}.$$

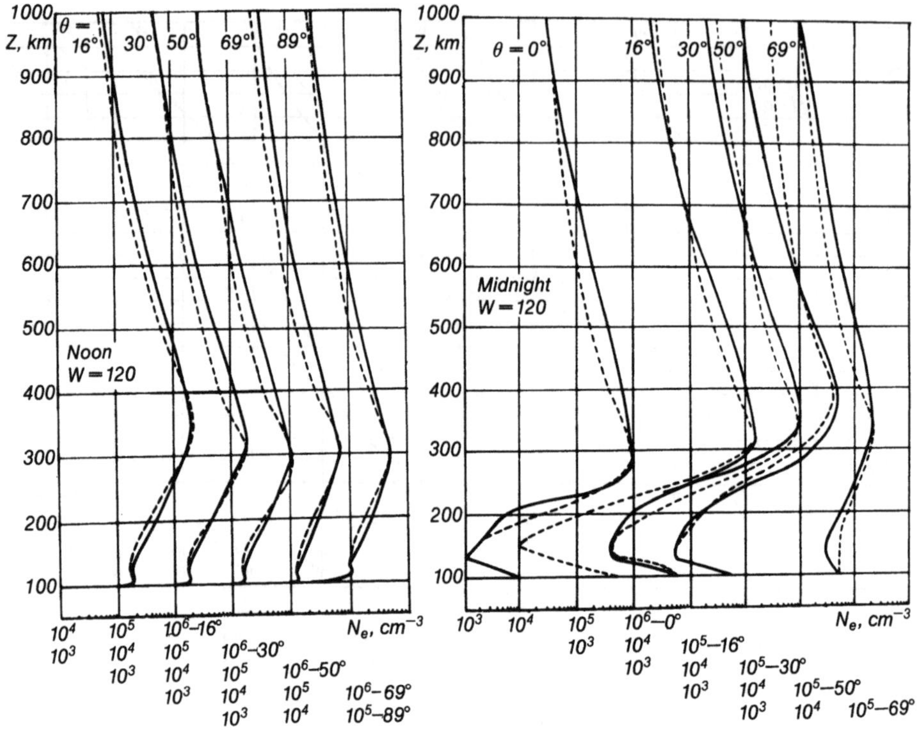

Fig. A2.13. Comparison between the EMI-81 height profiles of electron density (*broken curves*) at different latitudes, and the empirical data above $z = 100$ km for $\varphi = 0$ (noon) and $\varphi = \pm\pi$ (midnight) at $W = 120$. The latitude is indicated near the profiles

For the ray-tracing calculations, the numerical values of f_{0n}, f_{0d} or corresponding values of N_{2n}, N_{2d} defined at the fixed points θ_i along the path are necessary to interpolate with respect to θ.

In Gurevich et al. (1981) the derivatives of $N_e(z, \theta, \varphi)$ with respect to the coordinates z, θ, φ, and also the computer program for N_e in ALGOL-60 are given.

Note that the EMI-81 parameters N_2 [Eqs. (A2.12) – (A2.14)] and z_2 [Eqs. (A2.15) – (A2.17)] can be used for the global description of the parameters (N_m and z_m) of the simple one-layer parabolic model [Sec. 5.1, Eq. (5.2)]. In this case, the parameter y_m can be expressed in terms of Eq. (A2.3). Expanding Eq. (A2.3) in the vicinity of the F_2 layer maximum z_2 in a Tailor series in powers of $(z - z_2)$, we obtain the following equation for the parameter y_m [Eq. (5.2)]

$$y_m = \frac{z_2 - z_V}{\sqrt{3\ln(N_2/N_V)}} .$$

Thus, the parameters N_2, z_2, N_V, z_V of the present model define the latitudinal – longitudinal and solar activity dependences for the parabolic layer parameters N_m, z_m, y_m under the equinoctial conditions.

A2.2 Three-Dimensional Equinoctial Model of Electron Effective Collision Frequency

The distribution model of effective collision frequency ν_e (Gayvoronskaya et al. 1974, 1973, Israetel 1982) is constructed analogously to the model of N_e on the basis of latitudinal – longitudinal variations of height profiles within the range of 50 – 700 km (Fatkullin et al. 1981, Gayvoronskaya et al. 1973, Novikova and Soboleva 1978).

The analytical model of ν_e is constructed in spherical coordinates z, θ, φ; $z = R - R_0$. The coordinate reading is the same as in the N_e model. The dependence on SA is taken into account by specifying the parameter W analogously to the N_e model (see Sec. A2.1).

Height Dependence of ν_e. In Fig. A2.14 a, b the height profiles of ν_e (solid lines) are plotted to the semilogarithmic scale for the noon and midnight at the latitudes 0°, 30°, 50°, 75°. The profiles are calculated on the basis of the empirical height profiles of N_e for equinox during 1964 – 1965 under magnetically quiet conditions (Soboleva 1972). For convenience, the curves corre-

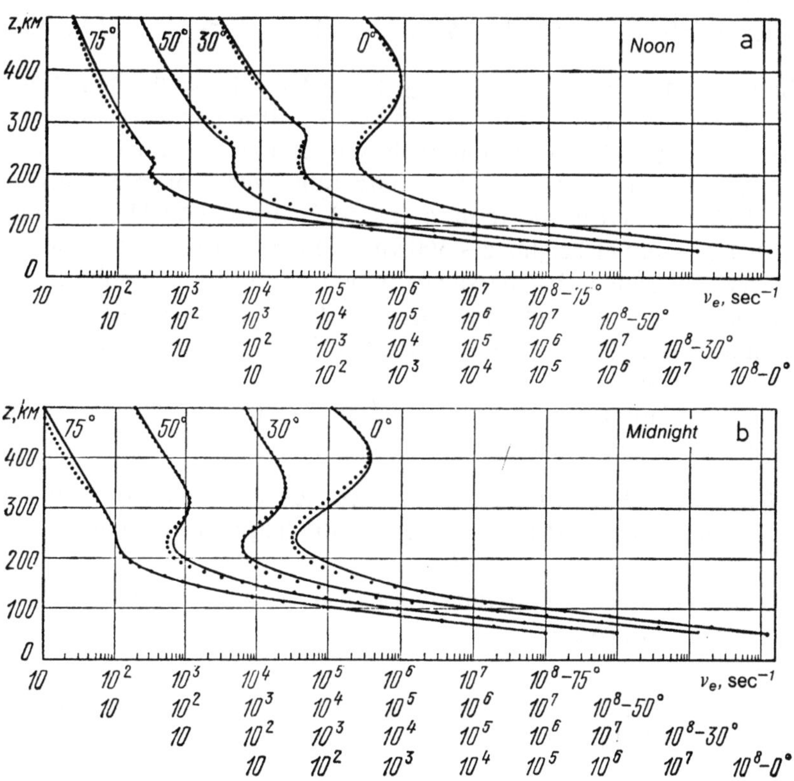

Fig. A2.14. Comparison between the height profiles calculated by Eqs. (A2.22) (*dots*), and the empirical data (Fatkullin et al. 1981), $W = 10$

sponding to different latitudes are displaced from one another along the abscissa, so each curve has its own scale of ν_e. All the profiles of ν_e are characterized by the presence of one maximum and one minimum which at $\theta = 75°$ at night degenerate into one point of inflexion.

For analytical description of these profiles, the exponential functions were used analogously to the N_e model (see Sec. A2.1). The whole height range is divided into three regions: (1) the region below the minimum of ν_e; (2) the region between the minimum and maximum of ν_e; (3) the region above the maximum of ν_e. Accordingly, the parameters are used characterizing the height variation of ν_e in each region, namely, z_1 is the height of ν_e minimum, ν_1 is the minimum value of ν_e, z_2 is the height of ν_e maximum, ν_2 is the maximum value of ν_e. The dependence of ν_e on z in each region is described as follows

$$\nu_e(z) = \nu_1 \cdot \exp\{\alpha_1 (z_1 - z)^{5/2}/[1 + \beta_1 (z_1 - z)^{3/2}]\}, \quad z < z_1, \tag{A2.22a}$$

$$\nu_e(z) = \nu_2 \cdot \exp\left\{\left[\ln \frac{\nu_1}{\nu_2}\right]\left[\frac{(z_2 - z)(z_2^2 + z_2 z + 3 z_1 z - 3 z_2 z_1 - 2 z^2)}{(z_2 - z_1)^3}\right]\right\}, \tag{A2.22b}$$

$$z_1 \leqslant z \leqslant z_2$$

$$\nu_e(z) = \nu_2 \cdot \exp\{-\alpha_2 (z - z_2)^2/[1 + \beta_2 (z - z_2)^{3/2}]\}, \quad z > z_2 . \tag{A2.22c}$$

Besides the main parameters ν_1, ν_2, z_1, z_2, Eqs. (A.2.22) include the parameters α_1, β_1 defining variation of the curves below z_1 [Eq. (A2.22a)], whereas the parameters α_2, β_2 define variation of the curves above z_2 [Eq. (A2.22c)]. In Eq. (A2.22a) the term $\beta_1 (z_1 - z)^{3/2}$ is comparable with 1 or more only at heights $z \sim 50$ km, whereas the term $\beta_2 (z - z_2)^{3/2}$ in Eq. (A2.22c) is comparable with 1 or more at heights $z \sim 500$ km. Consequently, the parameters α_1 and α_2 were determined by the values of ν_e (Fig. A2.14 a, b, solid lines) when $z \lesssim z_1$ and $z \gtrsim z_2$, assuming that β_1, $\beta_2 = 0$. The parameters β_1 and β_2 were determined at $\beta_1 (z_1 - z)^{3/2} \sim 1$ when $z \gtrsim 50$ km, and $\beta_2 (z - z_2)^{3/2} > 1$ when $z \approx 500$ km, respectively.

The height variations, $\nu_e(z)$, calculated by Eq. (A2.22) are shown in Fig. A2.14 a, b by the dotted curves.

Solar Activity Dependence of ν_e. The SA dependence of electron collison frequency is determined by the dependence of the height profile parameters $F \equiv \nu_1$, ν_2, z_1, z_2, α_1, α_2, β_1, β_2 on W.

The SA dependence of the parameters $F \equiv \nu_1$, ν_2 has the form which is accepted similar to Eq. (A2.6)

$$F(W) = F_1 \left(\frac{1 + \mu W}{1 + \mu W_1} \right)^2 \tag{A2.23}$$

where F_1 are the values of parameters $F \equiv \nu_1$, ν_2 at $W = W_1 = 10$ (SA minimum). Numerical values of the coefficients $\mu \equiv \mu_1$, μ_2 are determined by the values of the parameters F for different θ at $W = 10$ and $W = 120$ from Fatkullin et al. (1981), Gayvoronskaya et al. (1973), and Novikova and Soboleva (1978) analogously to Eq. (A2.7).

For the parameters $F \equiv z_1, z_2$, the SA dependence has the form which is accepted to be analogous to Eq. (A2.9)

$$F(W) = F_1 \cdot \frac{1 + \eta W}{1 + \eta W_1}. \tag{A2.24}$$

Numerical values of the coefficients $\eta = \eta_1, \eta_2$ are determined in the same manner as the coefficients μ analogously to Eq. (A2.10).

The height variations of ν_e within the range of $700 \text{ km} \geqslant z \geqslant z_1$ at low SA are similar to the variation of $\nu_e(z)$ at high SA. That is why we can assume that the coefficients α_2 and β_2 are independent of SA. In the lower ionosphere within the height range of $150 \text{ km} \gtrsim z > 50 \text{ km}$ the similarity is violated. According to Fatkullin et al. (1981), Gayvoronskaya et al. (1973), and Novikova and Soboleva (1978), at high SA ($W = W_2 = 120$), the collision frequency $\nu_e(z)$ increases with decreasing height much slower than at low SA ($W = W_1 = 10$). Consequently, it is worthwhile to introduce the SA dependence into the coefficient β_1 in the form analogous to Eq. (A2.11)

$$\beta_1(W) = \beta_1 \frac{1 + \xi W}{1 + \xi W_1}. \tag{A2.25}$$

The numerical values of $\xi(\theta)$ are determined by the values of $\beta_1(W)$ at $W = 10$ and $W = 120$ analogously to the coefficients η, see Eq. (A2.10).

Latitudinal – Longitudinal Dependence of ν_e. The latitudinal – longitudinal distribution of electron collision frequency is defined by the dependence of the parameters $F \equiv \nu_1, \nu_2, z_1, z_2, \alpha_1, \alpha_2, \beta_1, \beta_2$ on the latitude θ and longitude φ. In the general case, the longitudinal dependence of the parameters F is defined by the time of day and directly by the longitudinal effect. However, we shall consider below only the diurnal variations of ν_e analogously to the EMI-81. In this case, it proves to be convenient to express the diurnal dependence of the parameters F in terms of the noon $F_d(\varphi = 0)$ and midnight $F_n(\varphi = \pm\pi)$ values. These values are assumed to depend only on the latitude θ (see below). To approximate the parameters $F(\theta, \varphi)$, it proved to be possible to use the dependence on φ in the form analogous to longitudinal dependence of the EMI-81 (see Sec. A2.1).

Longitudinal dependence of the parameters ν_1 and ν_2 is accepted as follows

$$F(\theta, \varphi) = \exp\{f(\theta, \varphi)\}, \quad f(\theta, \varphi) = \left\{ \begin{array}{l} f_n(\theta) + \dfrac{f_d(\theta) - f_n(\theta)}{2} \\ \times \left[1 + \dfrac{(\pi - |\varphi_m|)(|\varphi| - |\varphi_m|)}{\pi - |\varphi|}\right]^{-\lambda}, \\ \qquad \varphi_m \geqslant \varphi \geqslant -\pi, \end{array} \right.$$

$$f(\theta,\varphi)=\begin{cases} f_d(\theta)-\dfrac{f_d(\theta)-f_n(\theta)}{2}\left[1+\dfrac{\varphi_m(|\varphi_m|-|\varphi|)}{\varphi}\right]^{-\lambda}, & 0\geqslant\varphi>\varphi_m\,, \\ f_d(\theta)-\dfrac{f_d(\theta)-f_n(\theta)}{2}\left[1+\dfrac{\varphi_e(\varphi_e-\varphi)}{\varphi}\right]^{-\lambda}, & \varphi_e>\varphi\geqslant 0\,, \\ f_n(\theta)+\dfrac{f_d(\theta)-f_n(\theta)}{2}\left[1+\dfrac{(\pi-\varphi_e)(\varphi-\varphi_e)}{\pi-\varphi}\right]^{-\lambda}, & \pi\geqslant\varphi\geqslant\varphi_e\,, \end{cases} \tag{A2.26}$$

where $f_n(\theta)=\ln F_n(\theta)$, $f_d(\theta)=\ln F_d(\theta)$, the parameters φ_m and φ_e define an asymmetry of $F(\varphi)$ curves with respect to the noon. These parameters define the value of F equal to $\sqrt{F_dF_n}$ (see Sec. A2.1).

According to Eq. (A2.23), the values of F_d and F_n for parameters v_1 and v_2 are equal to

$$\begin{aligned} F_d(W,\theta)&=F_{1d}(\theta)\left[\frac{1+\mu_d(\theta)\,W}{1+\mu_d(\theta)\,W_1}\right]^2 \\ F_n(W,\theta)&=F_{1n}(\theta)\left[\frac{1+\mu_n(\theta)\,W}{1+\mu_n(\theta)\,W_1}\right]^2 . \end{aligned} \tag{A2.27}$$

To specify the values of the parameters λ, φ_m, φ_e in Eq. (A2.26), the diurnal variation of the height profile $v_e(z)$ was calculated at the latitude $\theta=50°$ (Gayvoronskaya et al. 1973). In this case, the collision frequency $v_e=v_{en}+v_{ei}=\sum\limits_k v_{en_k}+v_{ei}$ was calculated by the following equations from Banks (1966) and Gurevich (1978)

$$\begin{aligned} v_{eN_2}&=2.5\cdot10^{-11}N_{N_2}[1+0.0092\,T_e^{1/2}]^{-1}T_e\,, \\ v_{eO_2}&=1.82\cdot10^{-10}N_{O_2}[1+0.036\,T_e^{1/2}]\,T_e^{1/2}\,, \\ v_{eO}&=2.8\cdot10^{-10}N_O T_e^{1/2}\,, \\ v_{eHe}&=4.6\cdot10^{-10}N_{He}T_e^{1/2}\,, \\ v_{ei}&=\frac{12.65\,N_e}{T_e^{3/2}}\lg\left(\frac{220\,T_e}{N_e^{1/3}}\right), \end{aligned} \tag{A2.28}$$

where v_{eN_2}, v_{eO_2}, v_{eHe}, v_{eO} are the collision frequencies of the electrons with the molecules of nitrogen N_2, oxygen O_2, helium He, and with the atoms of oxygen O; N_{N_2}; N_{O_2}, N_{He}, N_O are concentrations of the neutral components in cm^{-3}, T_e is the electron temperature in absolute degrees.

When calculating the height profiles of ν_e, the model of neutral composition of the atmosphere was used at the values of $T_e = 700°$ ($21^h - 03^h$ LT), $T_e = 800°$ ($10^h - 15^h$ LT) and $T_e = 750°$ at the rest time. In this case, at heights $z \sim 100 - 120$ km a correction of percentage of atomic oxygen has been carried out according to the model of Jacchia (1971). The values of T_e were chosen according to Evans (1970). For the time periods $10^h - 15^h$ and $21^h - 03^h$ LT the T_e curves used were obtained at Millstone – Hill in April 1965 in the daytime and at night, respectively. For the other times, the values of T_e were specified as arithmetic mean between the midnight value and noon one.

The diurnal variations of electron density N_e were taken from Soboleva (1972). They correspond to the average empirical model of N_e constructed for magnetically quiet conditions of the equinox and solar activity minimum at $\theta = 50°$. The calculation of ν_e was performed also using the equation for the total collision frequency between the electrons and neutral molecules

$$\nu_{en} = 3.6 \cdot 10^{-10} N_n T_e^{1/2}, \tag{A2.29}$$

where N_n is the total concentration of the neutral molecules.

The difference between the values of ν_e calculated by Eqs. (A2.28) and (A2.29) is about 10 – 15% (always less than 30%) and lies within the accuracy of specifying an effective cross-section for electron collisions with various molecules. Therefore, the calculation of ν_e for determination of radio wave absorption appears to be carried out by Eq. (A2.29) rather than by the complicated Eq. (A2.28).

Figure A2.15 shows the diurnal variations of the parameters $\nu_1 = \nu_{e\min}$, $\nu_2 = \nu_{e\max}$, $z_1 = z_{\min}$ and $z_2 = z_{\max}$. It is seen that on the middle latitudes the height of $\nu_e(z)$ minimum varies during 24 h from 230 – 220 km at night to 180 – 170 km by day, whereas the value of ν_e remains constant in the daytime between 8^h and 16^h at heights of 190 – 210 km [the profiles $\nu_e(z)$ have an inflection in this height range].

On the basis of Fig. A2.15, the values of parameters φ_m, φ_e (which are constant for all the latitudes) were chosen (see Table A2.3). As a result of preliminary calculations, the parameter λ was chosen to be equal to 2. Figure A2.15a shows the values of ν_1 and ν_2 calculated by Eqs. (A2.26) for φ_m and φ_e indicated in Table A2.3 (the values of ν_d and ν_n were taken from the figure). The maximum differences between ν_1 and ν_2 values obtained by Eqs. (A2.26), and empirical data take place at predown and postsunset hours. But these differences do not exceed 20 – 25%.

The latitudinal dependence of the parameters $F_{1d} \equiv \nu_{1d}$, ν_{2d}, μ_{1d}, μ_{2d}; $F_{1n} \equiv \nu_{1n}$, ν_{2n}, μ_{1n}, μ_{2n} is obtained by means of the analytical approximation of their numerical values at various latitudes. The dependence is as follows

$$\nu_{1n} = 5.4 \cdot 10^1 - \frac{2.8 \cdot 10^1}{1 + 4.0 \cdot 10^2 x} + \frac{1.46 \cdot 10^2}{1 + 3.5 \cdot 10^1 (1-x)},$$

$$\nu_{1d} = 4.18 \cdot 10^2 - \frac{1.86 \cdot 10^2}{1 + 2.3 \cdot 10^1 x} - \frac{2.07 \cdot 10^2}{1 + 1.4 \cdot 10^1 (1-x)},$$

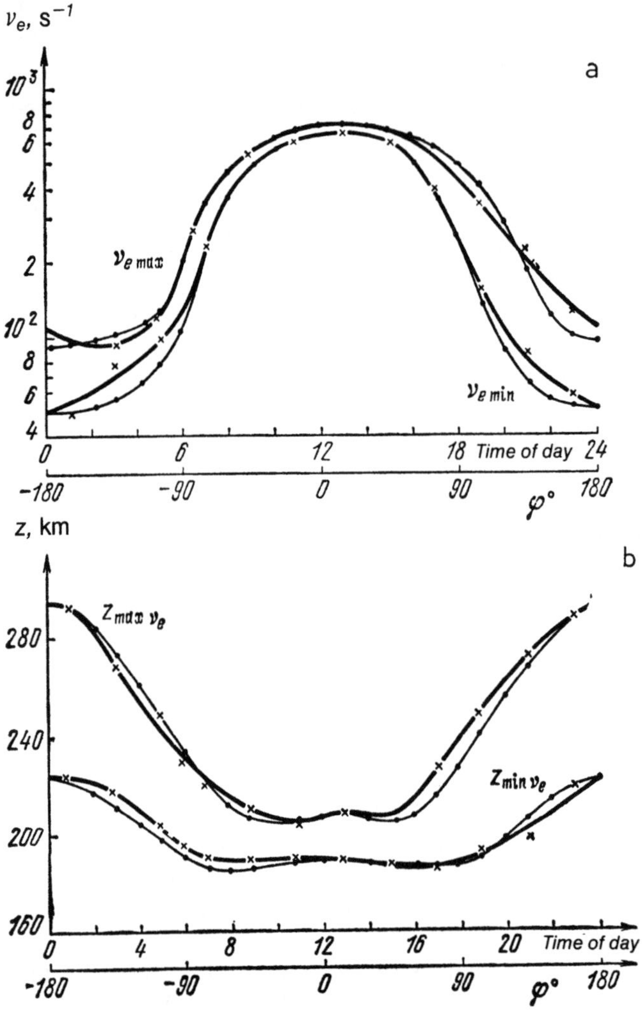

Fig. A2.15 a, b. Comparison between the diurnal variation of the parameters ν_1, ν_2 (**a**) and z_1, z_2 (**b**) calculated by Eqs. (A2.26) and (A2.31) (*dots*), and the reference points (*crosses*), $W = 10$

$$\nu_{2n} = 8.0 \cdot 10^1 + 2.6 \cdot 10^2(1-x)^2 + \frac{1.28 \cdot 10^2}{1 + 8.03 \cdot 10^1(1-x)},$$

$$\nu_{2d} = 4.32 \cdot 10^2 + \frac{4.8 \cdot 10^2}{1 + 4.0 \cdot 10^2 x} - \frac{2.24 \cdot 10^2}{1 + 1.64(1-x) \cdot 10^1},$$

$$\mu_{1d} = [5(1-x)^{12} + 11x(1-x)^3 + 600x^6(1-x)^4] \cdot 10^{-3},$$

$$\mu_{1n} = [7.42(1-x)^{18} + 52x^7(1-x) - 428.4x^6(1-x)^4 - 250x(1-x)^{32}] \cdot 10^{-3},$$

$$\mu_{2d} = \{6.22 + 6.17 \cdot 10^2 x(1-x)^{32} + [220 + 1800x(1-x)^{12}]x(1-x)^{12} - 5.308x^2 - 20x^3(1-x)^2 - 40x^7(1-x)\} \cdot 10^{-3},$$

$$\mu_{2n} = [21.5 - 20.588x - 13.96(1-x)^3 + 2 \cdot 10^3 x^6(1-x^4)] \cdot 10^{-3}. \quad \text{(A2.30)}$$

where $x = \sin^2\theta$.

Table A2.3. Values of φ_m and φ_e for the parameters ν_1 and ν_2

Parameter	φ_m	φ_e
ν_1	$-0.429\ \pi$	$0.546\ \pi$
ν_2	$-0.467\ \pi$	$0.680\ \pi$

To describe the longitudinal dependence of the parameters z_1, z_2, the following equations is convenient. The whole range of φ from $-\pi$ to π is divided into four regions. In each region the value of the parameter is equal to

$$F(\varphi,\theta)=F_1+\frac{(F_2-F_1)(\varphi-\varphi_1)}{(\varphi_2-\varphi_1)^3}\,[\varphi_1^2+\varphi_1\varphi+3\,\varphi\varphi_2-3\,\varphi_1\varphi_2-2\,\varphi^2]\,. \quad (A2.31)$$

Here, the quantities φ_1 and φ_2 define the boundaries of the regions, F_1 and F_2 are the values of F at $\varphi=\varphi_1$, φ_2, respectively. The quantities φ_1 and φ_2 are taken to be constant and independent of θ (see Table A2.4). When using the parameters from Table A2.4 to obtain better agreement of the analytical model with the empirical one, it is recommended for higher latitudes $\theta \geqslant 60°$ to substitute the values φ_1 and φ_2 (marked by one and two asterisks) for $-\pi/2$, $\pi/2$ and the values F_1 and F_2 (marked by one asterisk) for $(F_d+F_n)/2$, respectively.

The values of F_1 and F_2 are determined in terms of the noon, F_d, and midnight, F_n, values of the parameters and are also given in Table A2.4. The coefficient $b=1-0.01/(\cos^2\theta+0.01)$ is needed to satisfy the continuity conditions for the parameters F at the poles (at $\theta=\pm\pi/2$, $b=0$, hence $F=F_1=F_2=F_n=F_d$).

In Fig. A2.15b the approximation of z_1 and z_2, according to Eq. (A2.31), is shown. Here the numerical values of F_n and F_d from Fatkullin et al. (1981) (Table A2.4) were used. The maximum differences between the analytical curves [Eq. (A2.31)] and empirical ones are observed for z_2, however, these differences do not exceed 5%.

Table A2.4. Values of φ_1, φ_2, F_1, F_2 for the parameters z_1 and z_2

Parameter	φ_1	φ_2	F_1	F_2
z_1	$-\pi$	$-0.333\ \pi^*$	F_n	$(1-0.025\,b)\ F_d^*$
	$-0.333\ \pi^*$	0	$(1-0.025\,b)\ F_d^*$	F_d
	0	$0.416\ \pi^{**}$	F_d	$(1-0.025\,b)\ F_d^*$
	$0.416\ \pi^{**}$	π	$(1-0.025\,b)\ F_d^*$	F_n
z_2	$-\pi$	$-0.2\ \pi$	F_n	$(1-0.01\,b)\ F_d$
	$-0.2\ \pi$	$0.333\ \pi$	$(1-0.01\,b)\ F_d$	$(1+0.005\,b)\ F_d$
	$0.0833\ \pi$	$0.25\ \pi$	$(1+0.005\,b)\ F_d$	$(1-0.015\,b)\ F_d$
	$0.25\ \pi$	π	$(1-0.015\,b)\ F_d$	F_n

According to Eq. (A2.25), we can express the noon, F_d, and midnight, F_n, values of the parameters $F \equiv z_1, z_2$ as follows

$$F_d(W,\theta) = F_{1d}(\theta) \cdot \frac{1+\eta_d(\theta)\,W}{1+\eta_d(\theta)\,W_1}$$
$$F_n(W,\theta) = F_{1n}(\theta) \cdot \frac{1+\eta_n(\theta)\,W}{1+\eta_n(\theta)\,W_1}. \qquad \text{(A2.32)}$$

For latitudinal variation of the parameters $F_{1d}(\theta) \equiv z_{1d}, z_{2d}, \eta_d$; $F_{1n} \equiv z_{1n}, z_{2n}, \eta_n$, the following equations are obtained

$$z_{1n} = 2.18 \cdot 10^2 + 2.47 \cdot 10^1 x^2 + \frac{2.16 \cdot 10^1}{1+4.0 \cdot 10^2 x} - \frac{2.33 \cdot 10^1}{1+1.49 \cdot 10^3 (1-x)},$$

$$z_{1d} = 2.20 \cdot 10^2 - 3.46 \cdot 10^1 x^8 + \frac{2.02 \cdot 10^1}{1+4.0 \cdot 10^2 x} + \frac{3.46 \cdot 10^1}{1+1.49 \cdot 10^3 (1-x)},$$

$$z_{2n} = 3.2 \cdot 10^2 - 1.04 \cdot 10^2 x^4 + 8.0 \cdot 10^1 (1-x)^7,$$

$$z_{2d} = 2.71 \cdot 10^2 - 5.9 \cdot 10^1 x + \frac{1.04 \cdot 10^2}{1+2.45 \cdot 10^1 x},$$

$$\eta_{1d} = \left[25x^9 - \frac{16.39 \cdot 10^{-3}}{1.001-x} + 1700x^6(1-x)^4 - \frac{3.86 \cdot 10^{-3}}{x+0.001}\right] \cdot 10^{-4},$$

$$\eta_{1n} = \left[\frac{8.609 \cdot 10^{-3}}{1.001-x} + 60(1-x) - 56(1-x)^2 - 8270x^6(1-x)^4 \right.$$
$$\left. - \frac{11.39 \cdot 10^{-3}}{0.001+x}\right] \cdot 10^{-4},$$

$$\eta_{2d} = [1.7 + 3.34x^6 - 1.09(1-x)^{18}] \cdot 10^{-3},$$
$$\eta_{2n} = [5.04x^8 + 60x^3(1-x)^2 - 2.44(1-x)^{18}] \cdot 10^{-3}. \qquad \text{(A2.33)}$$

The longitudinal variation of parameters $\alpha_1, \alpha_2, \beta_1, \beta_2$ is approximated as follows

$$F = F_n + (F_d - F_n)\cos^4(\varphi/2). \qquad \text{(A2.34)}$$

Equation (A2.34) provides a smooth transition from the noon values to midnight ones.

As shown above, the coefficients $\alpha_{1d}, \alpha_{2d}, \beta_{2d}, \alpha_{1n}, \alpha_{2n}, \beta_{2n}$ are independent of W. The coefficients β_{1n}, β_{1d} depend on W as follows

$$\beta_{1d}(\theta, W) = \beta_{1d}(\theta)\left(\frac{1+\xi_d(\theta)\,W}{1+\xi_d(\theta)\,W_1}\right)^2$$
$$\beta_{1n}(\theta, W) = \beta_{1n}(\theta)\left(\frac{1+\xi_n(\theta)\,W}{1+\xi_n(\theta)\,W_1}\right)^2. \qquad \text{(A2.35)}$$

The latitudinal dependence of parameters α_{1d}, α_{2d}, β_{1d}, β_{2d}, α_{1n}, α_{2n}, β_{1n}, β_{2n}, ξ_d, ξ_n has the form

$$\alpha_{1n}=\left[6.8-3.8x^5+2^{11}x^6(1-x)^6-\frac{3.4}{1+2.2x}+\frac{3.8}{1+1.49\cdot 10^3(1-x)}\right]\cdot 10^{-5},$$

$$\alpha_{1d}=\left[3.5+7.8x^6-\frac{0.7}{1+4.0\cdot 10^2x}-\frac{5.5}{1+1.49\cdot 10^3(1-x)}\right]\cdot 10^{-5},$$

$$\alpha_{2n}=\left[2.1-2.1(1-x)^2+\frac{1.6}{1+4.0\cdot 10^2x}+\frac{2.6}{1+1.49\cdot 10^3(1-x)}\right]\cdot 10^{-4},$$

$$\alpha_{2d}=\left[8.0-2^{12}x^5(1-x)^5-\frac{6.75}{1+4.0\cdot 10^2x}-\frac{3.3}{1+5.53\cdot 10^1(1-x)}\right]\cdot 10^{-4},$$

$$\beta_{1n}=[2.5x^{35}+0.382(1-x)^5+2.6x(1-x)^3+1.95\cdot 10^3x^6(1-x)^4]\cdot 10^{-4},$$

$$\beta_{1d}=\left[0.2+70x^6-2^{10}x^6(1-x)^6-\frac{0.1}{1+4.0\cdot 10^2x}\right.$$
$$\left.-\frac{4.7}{1+1.49\cdot 10^3(1-x)}\right]\cdot 10^{-4},$$

$$\beta_{2n}=\left[2.8x^{13}+1.35\cdot 2^{10}x^5(1-x)^5+\frac{0.5}{1+1.6\cdot 10^1x}\right]\cdot 10^{-3},$$

$$\beta_{2d}=\left[5.0-3.0\cdot 2^4x^2(1-x)^2-\frac{4.55}{1+4.0\cdot 10^2x}\right.$$
$$\left.-\frac{2.2}{1+9.4\cdot 10^1(1-x)}\right]\cdot 10^{-3}, \qquad \text{(A2.36)}$$

$$\xi_d=[6.858x^2+6\cdot 10^3x^6(1-x)^4+19.5x(1-x)^3+30.5x(1-x)^{12}]\cdot 10^{-3},$$

$$\xi_n=[6.858x^{16}+1.216\cdot 10^4x^{14}(1-x)^2$$
$$+2.05\cdot 10^3x^2(1-x)^6+65.5x(1-x)^{12}]\cdot 10^{-3}.$$

Equation (A2.36), as well as Eqs. (A2.30) and (A2.33), are obtained by means of approximation of latitudinal variation of the parameters F. The approximation has been performed on the basis of the numerical values of parameters F known at a number of points $\theta=\theta_i$ for the noon and midnight at high and low SA.

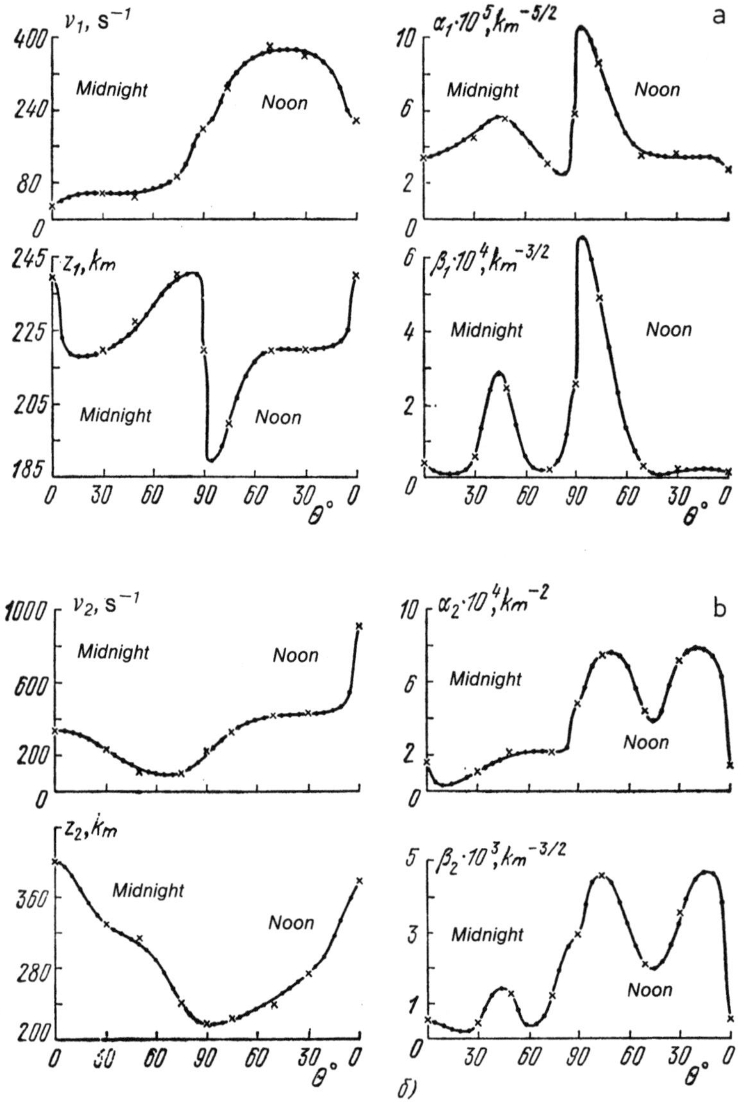

Fig. A2.16 a,b. Comparison between the analytical approximation of latitudinal variation of the parameters ν_1, z_1, α_1, β_1 **(a)** and ν_2, z_2, α_2, β_2 **(b)** (*dots*), and the reference points (*crosses*) at $W = 10$

The values of F at the point $\theta = 90°$ were specified as arithmetic mean between the noon and midnight values for $\theta = 75°$. In Fig. A2.16 the latitudinal variations of parameters $F \equiv \nu_1$, ν_2, z_1, z_2, α_1, α_2, β_1, β_2 are shown [Eqs. (A2.30), (A2.33), and (A2.35)]. The crosses show the reference points from Fatkullin et al. (1981). It is seen that there is quite good agreement of the approximated values with the reference ones. The differences are mainly no more than 15% and in most cases they are less than 5%.

Thus, Eqs. (A2.22), (A2.26), (A2.27), and (A2.30) – (A2.36) and Tables A2.3 and A2.4 give the three-dimensional analytical model of effective collision frequency of the electrons. The model of ν_e, as well as the model of N_e (Sec. A2.1), is constructed for equinoctial conditions.

Appendix 3. Wave Theory of HF Radio Wave Propagation in a Horizontally – Inhomogeneous Ionosphere

N. D. BORISOV and A. V. GUREVICH

Radio wave propagation in a horizontally – inhomogeneous ionosphere was considered in Chap. 3 in the geometrical – optics approximation. In the present Appendix, exact solutions of the wave equation are constructed and analyzed. This enables one to refine the results obtained in Chap. 3 and to determine their field of applicability more precisely, as well as to study wave effects which are ignored in the geometrical – optics approximation (Borisov and Gurevich 1975, 1976a, b, 1977a, b; Kravtsov 1965, Borisov 1978, Borisov and Gurevich 1977c, Popov 1976, Kinber et al. 1978).

A3.1 Wave Equation in a Slightly Inhomogeneous Ionosphere

Electromagnetic wave propagation is described in the general case by the vectorial wave Eq. (2.4).

In the spherically – symmetric ionosphere a wave field can be decomposed into the vector components. In so doing, the initial equation is separated into independent scalar equations for separate components (see Appendix 1). In the general case, when $\tilde{\varepsilon}$ depends on all the coordinates, it is impossible to obtain a system of independent scalar equations. However, taking into account the smallness of the horizontal ionospheric inhomogeneity, $\delta \ll 1$ [Eq. (1.9)], one can obtain such a decomposition in the first-order approximation with respect to the parameter δ.

Let us introduce for wave Eq. (2.4) a vector potential $\boldsymbol{A}_2$ so that the electric field $\mathscr{E}$ be expressed in terms of $\boldsymbol{A}_2$ in the following way: $\mathscr{E} = \frac{1}{\tilde{\varepsilon}} \operatorname{rot} \tilde{\varepsilon} \boldsymbol{A}_2$. In this case, in a space free of sources, the relation $\operatorname{div} \tilde{\varepsilon} \mathscr{E} = 0$ is satisfied identically. Substituting the expression for $\mathscr{E}$ in wave Eq. (2.4) and using a vectorial analysis formula $\frac{1}{\tilde{\varepsilon}} \operatorname{rot} \tilde{\varepsilon} \boldsymbol{A}_2 = \operatorname{rot} \boldsymbol{A}_2 + [\nabla \ln \tilde{\varepsilon}\,, \boldsymbol{A}_2]$, we obtain for the potential $\boldsymbol{A}_2$

$$\Delta \boldsymbol{A}_2 - \operatorname{rot}[\nabla \ln \tilde{\varepsilon}\,, \boldsymbol{A}_2] + \tilde{\varepsilon} k_0^2 \boldsymbol{A}_2 = \nabla \Phi\,, \quad k_0 = \frac{\omega}{c}\,, \tag{A3.1}$$

where Φ is an arbitrary scalar function ($\nabla \Phi$ appears after elimination of the operator rot). In the particular case of the spherical symmetry, $\tilde{\varepsilon} = \tilde{\varepsilon}(r)$, putting the vector $\boldsymbol{A}_2$ to be radial, $\boldsymbol{A}_2 = \boldsymbol{e}_r A$, and $\Phi = 2A/r$, we obtain Eq. (A1.5). The equation gives the field of the wave generated by the eddy current localized on the surface $r = \text{const}$.

The second independent solution may be sought as in Appendix 1 by introducing a potential $\boldsymbol{A}_1$ so that $\boldsymbol{H} = \text{rot}\boldsymbol{A}_1$. Substituting the latter equation in wave Eq. (A1.3), we obtain for the potential $\boldsymbol{A}_1$

$$\Delta \boldsymbol{A}_1 - (\nabla \ln \tilde{\varepsilon})\,\text{div}\boldsymbol{A}_1 + \tilde{\varepsilon} k_0^2 \boldsymbol{A}_1 = \tilde{\varepsilon} \nabla \Phi\,, \tag{A3.2}$$

where $\Phi(\boldsymbol{r})$ is as before an arbitrary scalar function. In the particular case of the spherical symmetry, $\tilde{\varepsilon} = \tilde{\varepsilon}(r)$, putting $\boldsymbol{A}_1 = \boldsymbol{e}_r A$, $\Phi = 2A/\tilde{\varepsilon} r$, we obtain Eq. (A1.4). It gives the field of the wave generated by the radial current.

In the general case, $\tilde{\varepsilon} = \tilde{\varepsilon}(\boldsymbol{r})$, vectorial Eqs. (A3.1) and (A3.2) are rather complicated. In our conditions, they can be simplified, if taking into account the smallness of the horizontal ionospheric inhomogeneity [Eq. (1.9)]. Let us solve Eqs. (A3.1) and (A3.2) in the form of a series in the powers of the small parameter $\delta \sim \dfrac{\partial \tilde{\varepsilon}}{\partial x}$ (x is a coordinate along the Earth's surface)

$$\boldsymbol{A}_1 = \boldsymbol{A}_{10} + \sum_{k=1} \boldsymbol{A}_{1k} \delta^k\,, \quad \boldsymbol{A}_2 = \boldsymbol{A}_{20} + \sum_{k=1} \boldsymbol{A}_{2k} \delta^k\,,$$

where $\boldsymbol{A}_{10} = \boldsymbol{e}_r A_{10}$, $\boldsymbol{A}_{20} = \boldsymbol{e}_r A_{20}$. (Further, when evaluating the terms $A_{11}\delta$ and $A_{21}\delta$, the exact expression for the parameter δ will be written.)

As a lower approximation with respect to δ, we have the following equations for A_{20}, A_{10}

$$\begin{aligned} &\Delta A_{20} + \tilde{\varepsilon} k_0^2 A_{20} = \frac{2}{r}\,\frac{\partial A_{20}}{\partial r}\,, \\ &\Delta A_{10} - \frac{1}{\tilde{\varepsilon} r^2}\,\frac{\partial \tilde{\varepsilon}}{\partial r}\,\frac{\partial}{\partial r}(r^2 A_{10}) + \tilde{\varepsilon} k_0^2 A_{10} = 2\tilde{\varepsilon}\frac{\partial}{\partial r}\left(\frac{A_{10}}{r\tilde{\varepsilon}}\right). \end{aligned} \tag{A3.3}$$

Here, the functions Φ_{10} and Φ_{20} are chosen in the same form as in the spherically – symmetric case.

A further simplification of Eqs. (A3.3) can be made bearing in mind the fact that the ionosphere is a thin spherical layer, $z = r - R_0 \ll R_0$ [Eq. (1.7)]. Let us introduce, instead of A_{10}, A_{20}, new potentials A_2^0, A_1^0, defining them by relations: $A_2^0 = A_{20}$, $A_1^0 = \dfrac{A_{10}}{\sqrt{\tilde{\varepsilon}}}$. Bearing in mind that $\tilde{\varepsilon} \approx 1$ [Eq. (1.3)] and $\left|\dfrac{1}{\tilde{\varepsilon} k_0}\,\dfrac{\partial \tilde{\varepsilon}}{\partial z}\right| \ll 1$ [Eq. (1.8)], we obtain for A_1^0, A_2^0 a common equation

$$\frac{\partial^2 A}{\partial z^2} + \Delta_{\theta,\varphi} A - U k_0^2 A = 0\,, \quad U \approx -\left(\tilde{\varepsilon} + \frac{2z}{R_0}\right). \tag{A3.4}$$

Here,

$$\Delta_{\theta,\varphi} = \frac{1}{R_0^2}\left(\frac{1}{\sin\theta}\frac{\partial}{\partial\theta}\sin\theta\frac{\partial}{\partial\theta} + \frac{1}{\sin^2\theta}\frac{\partial^2}{\partial\varphi^2}\right)$$

is the angular part of the Laplacian. In Eq. (A3.4), the dielectric constant $\tilde{\varepsilon}$ depends, in general, on all the variables z, θ, φ. Therefore, no separation of variables analogous to that performed in Appendix 1 is possible in the general case. However, we recognize that in the geometrical optics, the ray behavior is managed mainly by $\tilde{\varepsilon}$ variations in the vertical direction z, where the gradients $\partial\tilde{\varepsilon}/\partial z$ are rather great as well as those along the ray path. Let the latter direction be the axis of φ, so that $\tilde{\varepsilon} = \tilde{\varepsilon}(z, \varphi)$, then the potential A may be sought as a function of two coordinates z, φ only. This corresponds to a wave field generated by a current which is independent of the coordinate θ. In this case, Eq. (A3.4) takes the form

$$\frac{\partial^2 A}{\partial z^2} + \frac{\partial^2 A}{\partial x^2} - k_0^2 U(z,x) A = 0\,, \quad x = R_0\varphi\,, \quad U = -\varepsilon - \frac{2z}{R_0}\,. \tag{A3.5}$$

(From now on, no radio wave absorption will be taken into account, therefore, here $\tilde{\varepsilon}$ is replaced by its real part ε.) In Eq. (A3.5), the fact has been taken into account that the wave propagates along a great circle arc, i.e., $\theta \approx \pi/2$, $\sin\theta \approx 1$, z is the vertical direction, x is the horizontal direction along the wave path. The quantity $U(z,x)$ has the same meaning of the potential as before. The maxima of U along the vertical z separate the ionospheric wave ducts.

A3.2 Quasispherical Approximation

In a horizontally-homogeneous, i.e., spherically-symmetric ionosphere, $U = U(z)$, solutions of Eq. (A3.5) are represented in the form of expansion in terms of eigenfunctions $\Psi_n(z)$

$$A = \sum_n C_n \Psi_n(z) e^{ik_n x}\,. \tag{A3.6}$$

The eigenfunctions obey Eq. (A1.8)

$$\frac{\partial^2 \Psi_n}{\partial z^2} - (k_0^2 U + k_n^2)\,\Psi_n = 0\,. \tag{A3.7}$$

Here, $k_n^2 = -k_0^2 E_n$ are the squares of eigenwave numbers. The functions $\Psi_n(z)$ form a complete set of eigenmodes for a spherically-symmetric waveguide (see Secs. 2.2 and 2.3).

In our case, the ionosphere is inhomogeneous, $U = U(z,x)$, but its horizontal inhomogeneity is slight. Therefore, it is also natural in this case to make an attempt at solving Eq. (A3.5) by expansion into eigenfunctions of a spherically – symmetric waveguide [Eq. (A3.7)]

$$A(z,x) = \sum_n \frac{C_n}{\sqrt{k_n(x)}} \Psi_n e^{i\int k_n dx} . \tag{A3.8}$$

Here we have taken into account the fact that due to the dependence of U on the horizontal coordinate x, the eigenwave vectors k_n and eigenfunctions Ψ_n vary slowly with x. In the expansion of Eq. (A3.8) it is natural for it to be termed the quasispherical approximation. A similar approach in the waveguide theory is called the method of transversal cross-section (Katsenelenbaum 1961). This approximation is closely related to the adiabatic invariant approximation considered in Chaps. 3 and 4. Indeed, for great number modes, the quantity $\int_{z_1}^{z_2} dz \sqrt{-(k_0^2 U + k_n^2)}$ is constant

$$\int_{z_1}^{z_2} dz \sqrt{-(k_0^2 U + k_n^2)} = \pi \left(n + \frac{1}{2} \right) \tag{A3.9}$$

where z_1, z_2 are the turning points. This constant with an accuracy to a numerical factor coincides with the adiabatic invariant I [see Eq. (3.17), $\bar{E} = -k_n^2/k_0^2$)].

It is important that the mode excitation coefficients, C_n, in the quasispherical approximation do not conserve along the path, but vary with $C_n = C_n(x)$. This variation depends on the excitation coefficients for the adjacent modes. Thus, quasispherical modes, unlike normal modes of a spherically – symmetric waveguide (Sec. 2.1), are not independent: they tangle and mix as the wave propagates. The rate of mode mixing depends on the horizontal ionospheric inhomogeneity, i.e., on the explicit form of $U(z,x)$. Wave propagation in a duct is conveniently described in terms of quasispherical modes Ψ_n, only in the case when mode mixing along the path is weak. In the following section we estimate the mixing for specific real ionospheric models.

A3.3 Interaction of Quasispherical Modes in the Real Ionosphere

First, we derive the equation for the coefficients C_n, i.e., the coefficients of potential A expansion into quasispherical modes. To this end, we substitute Eq. (A3.8) in wave Eq. (A3.5), using the orthonormalization property of functions Ψ_n. Note that in a waveguide slightly inhomogeneous in the horizontal x-direction, the coefficients C_n vary rather slowly with x:

$|d\ln C_n/dx| \ll k_0$. As a result we obtain a set of connected equations for C_n (see Landau and Lifshitz 1977, Sec. 40)

$$\frac{dC_n}{dx} + \sum_m \sqrt{\frac{k_m}{k_n}}\, H_{nm} \exp[i \textstyle\int (k_m - k_n)\, dx]\, C_m = 0 . \tag{A3.10}$$

Here, $H_{nm} = \int dz\, \Psi_n^* \dfrac{\partial \Psi_m}{\partial x}$ is an operator characterizing the interaction between different modes. (Note that the diagonal elements of H_{nm} are equal to zero due to conservation of normalizations of Ψ_n.) In the spherically-symmetric ionosphere $\partial \Psi_m/\partial x = 0$, and any connection between modes disappears.

To estimate the interaction H_{nm} in the real ionosphere, we consider a region in the neighborhood of the sliding point, $z = z_s$, where the potential $U(z, x)$ has a local minimum

$$U(z_s, x) = U_m(x) .$$

Here, the effective dielectric constant, $-\varepsilon - \dfrac{2z}{R_0} = U(z, x)$, can be thought to be parabolic [Eq. (2.9)]: $U = U_m + \dfrac{1}{2} U''(z - z_s)^2$. Let us go over to dimensionless variables ζ, R_n, ρ_s [see Eq. (2.10)]. Equation (A3.7) in these variables

$$\frac{d^2 \Psi_n}{d\zeta^2} + (R_n - \zeta^2)\, \Psi_n = 0 \tag{A3.11}$$

has solutions in terms of Hermite polynomials [Eq. (2.11)]

$$\Psi_n = S_n e^{-\zeta^2/2} H_n(\zeta) , \quad R_n = 2n + 1 ,$$

where $S_n = (\sqrt{\pi}\, n!\, 2^n)^{-1/2}$.

Eigenwave vectors k_n for the waveguide model under consideration, $k_n = k_0 \sqrt{-E_n}$ [Eq. (2.16)] are given by

$$k_n = k_0 \sqrt{-U_m} - \frac{1}{2} \left(-\frac{U''}{2U_m} \right)^{1/2} (2n + 1) . \tag{A3.12}$$

They form a discrete spectrum with the spacings

$$\Delta k \equiv |k_n - k_{n-1}| = \left(-\frac{U''}{2U_m} \right)^{1/2} = \frac{1}{\rho_s^2 k_0} . \tag{A3.13}$$

Using the recurrence relations for Hermite polynomials, we represent the derivative $\partial \Psi_m / \partial x$, involved in H_{nm}, in the form

$$\frac{\partial \Psi_m}{\partial x} = S_m e^{-\zeta^2/2}\left(mH_{m-1} - \frac{1}{2}H_{m+1}\right) a(x)\,, \qquad \text{(A3.14)}$$

where $a(x) \approx -\dfrac{1}{\rho_s}\dfrac{dz_s}{dx}$. Taking into account Eq. (A3.14) and the definition of H_{nm}, we obtain

$$H_{nm} = a(x)\left\{\sqrt{\frac{m}{2}}\,\delta_{n,m-1} - \sqrt{\frac{m+1}{2}}\,\delta_{n,m+1}\right\}. \qquad \text{(A3.15)}$$

Here, $\delta_{n,m}$ is the Kronecker symbol.

Now let us evaluate the quantity $a(x)$ for the interlayer duct. A characteristic property of this duct lies in the fact that at the night side it rises to substantially greatly altitude than at the day side, so that the difference between the sliding altitudes at the night side, $z_{s\max}$ and at the day side, $z_{s\min}$, reaches $z_{s\max} - z_{s\min} \sim 10^2$ km. Let us put, for estimation, $z_s(x) = z_s^{(0)} + \delta z_s \cos\dfrac{x}{R_0}$, where $z_s^{(0)}$ is the mean altitude, and δz_s is the deviation from the mean value. In this case, we obtain $a(x) = \dfrac{\delta z_s}{R_0 \rho_s}\sin\dfrac{x}{R_0}$.

Now let us give the estimation results for $a(x)$ in two specific cases. Using the ionospheric model considered in Sec. 2.2, Table 2.1, for the frequency $\omega = 1.5 \cdot 10^8\,\mathrm{s}^{-1}$, we obtain $a(x) = 10^{-2}\sin\dfrac{x}{R_0}\,\mathrm{km}^{-1}$. Using the experimental profiles of the electron density at the equator, taken from Soboleva (1973) (Appendix 2) for the frequency $\omega = 1.25 \cdot 10^8\,\mathrm{s}^{-1}$, we obtain $a(x) = 3 \cdot 10^{-3} \sin(x/R_0)\,\mathrm{km}^{-1}$.

Thus, for HF waves in the case of great number modes ($n \gg 1$), the interaction H_{nm} between the adjacent modes proves to be rather strong, $|H_{n,n-1}| \sim \sqrt{\dfrac{n}{2}} \cdot a \sim 10^{-2} - 10^{-1}\,\mathrm{km}^{-1}$. The coefficients C_n vary rather rapidly as the wave travels in the duct, and the modes are mixed. For the perturbation theory to be applicable, the parameter $\left|\dfrac{H_{n,n-1}}{\Delta k}\right|$ is required to be small, $\left|\dfrac{H_{n,n-1}}{\Delta k}\right| \ll 1$. This is not satisfied under ionospheric conditions. The perturbation theory fails over the distance of the order of $|H_{n,n-1}|^{-1} \sim 10^2$ km, i.e., much less than that of one hop. The number of effectively interacting modes, $\sim \left|\dfrac{H_{n,n-1}}{\Delta k}\right|$, can amount to a few tens. This means that because of real ionospheric inhomogeneity, the quasispherical modes are unusable for description of long-distance HF wave propagation in ionospheric ducts. Therefore, it is necessary to construct other eigensolutions, which conserve rather long as the wave propagates in the horizontally – inhomogeneous ionosphere.

A3.4 Adiabatic Modes of Ionospheric Waveguides

It was indicated above that the quasispherical approximation proves to be insufficient under real ionospheric conditions. Therefore, the system of eigenmodes of an ionospheric waveguide should be constructed simultaneously involving the horizontal ionospheric inhomogeneity.

Let us consider the wave equation in an inhomogeneous medium [Eq. (A3.5)]. Taking into account Eq. (1.8), we shall seek the solution in the quasiclassical form [Eq. (A3.8)]

$$A(z,x) = \tilde{\Psi}(z,x)\frac{1}{\sqrt{k}}e^{i\int k\,dx}. \tag{A3.16}$$

Here, $k = k(x)$ is the horizontal wave number, which is always close to $k_0 = \omega/c$ owing to small ionospheric waveguide depth [Eq. (1.7)]. The function $\tilde{\Psi}$ varies with x slower than the rapidly oscillating exponential factor. Substituting Eq. (A3.16) in Eq. (A3.5), we obtain the following parabolic equation for the function $\tilde{\Psi}(z,x)$

$$\frac{\partial^2\tilde{\Psi}}{\partial z^2} - (k_0^2 U + k^2)\,\tilde{\Psi} + i2k\frac{\partial\tilde{\Psi}}{\partial x} = 0. \tag{A3.17}$$

The second derivative, $\partial^2\tilde{\Psi}/\partial x^2$, is neglected here due to small horizontal inhomogeneity of the ionosphere: $|1/k \cdot \partial U/\partial x| \sim 1/kR_0 \ll 1$. If, in addition, the term $i2k\,(\partial\tilde{\Psi}/\partial x)$ is neglected in Eq. (A3.17), then the equation of quasispherical approximation [Eq. (A3.7)] would be obtained. The latter, however, as shown above, fails in describing a horizontal inhomogeneity, so the term $i2k\,(\partial\tilde{\Psi}/\partial x)$ is of importance.

The system of independent eigensolutions of Eq. (A3.17), as will be shown below, is the desired system of eigenmodes of a horizontally – inhomogeneous ionospheric waveguide. From now on we shall call them the adiabatic modes, since to find them, a graduality of ionospheric variation in the horizontal direction is used. The adiabatic modes almost do not mix in the case of HF radio wave propagation over not only a one-hop path but over long-distance and RTW paths either. Thus, these modes play a role of the normal modes of a spherically – symmetric waveguide. In other words, if the system of adiabatic modes has been constructed, then the field distribution would have to be easily found as superposition of these modes at arbitrary distances from the transmitter.

Below, a special method will be given for the construction of adiabatic modes in horizontally – inhomogeneous ionospheric waveguides.

Parabolic Duct. First, we consider the parabolic waveguide model [Eq. (2.9)]

$$U(z,x) = U(z_s) + \frac{1}{2}U''(z_s)(z - z_s)^2. \tag{A3.18}$$

Unlike the spherically-symmetric case (Sec. A3.3), now the sliding mode height, z_s, and the duct width, U'', depend on the longitudinal coordinate x. The model [Eq. (A3.18)] is always valid in the real horizontally-inhomogeneous ionosphere for low modes, since near the bottom the duct has a parabolic form.

In the case of Eq. (A3.18), going over to dimensionless variables ζ, ρ_s, R [Eq. (2.10)], we obtain instead of Eq. (A3.17) the equation

$$\frac{\partial^2 \Psi_1}{\partial \zeta^2} + (R - \zeta^2)\, \Psi_1 + i 2 k \rho_s^2 \frac{\partial \Psi_1}{\partial x} = 0 . \tag{A3.19}$$

This equation has the exact solution for arbitrary functions $z_s(x)$, $U(z_s)$, $U''(z_s)$ (Baz' et al. 1971). However, we are interested now in the case when the parameters adiabatically slowly vary along x

$$\frac{dz_s}{dx} \ll \frac{z_s}{R_0 \Theta}, \quad \frac{d\rho_s}{dx} \ll \frac{\rho_s}{R_0 \Theta} . \tag{A3.20}$$

Note that for regular ionospheric variations connected, for example, with the transitions from the day side to the night one or from the polar region to midlatitude and equatorial ones, Eq. (A3.20) is usually well satisfied. The typical gradients are $\frac{dz_s}{dx} \sim 10^{-2}$, $\frac{d\rho_s}{dx} \sim 10^{-5}$ (Gurevich 1971). Since the spatial ray oscillation period in the parabolic duct is $\Theta R_0 = 2\pi k \rho_s^2$, we see that under condition of Eq. (A3.20), the last term in Eq. (A3.19) is small compared with the two other terms.

In accordance with Baz' et al. (1971), we shall seek independent particular solutions for $\tilde{\Psi}$ [Eq. (A3.17)] and Ψ_1 [Eq. (A3.19)] in the form

$$\tilde{\Psi}(z, x) = \frac{a_0}{\rho_s^{1/2}} \Psi_{1n}(\zeta, x) , \quad \Psi_{1n} = \Psi_n(\tilde{\zeta}) \exp(i\mu(x)\zeta + i\nu(x)\zeta^2) \tag{A3.21}$$

where $\tilde{\zeta} = \zeta + \delta\zeta(x)$, $a_0 = \text{const}$.

Substituting Ψ_{1n} in Eq. (A3.19) and taking into account the slowness of medium variations in the horizontal direction x, we obtain approximately

$$\mu = k \rho_s \frac{dz_s}{dx} , \quad \nu = k \rho_s \frac{1}{2} \frac{d\rho_s}{dx} , \quad \delta\zeta = k \rho_s^2 \frac{d\mu}{dx} . \tag{A3.22}$$

The functions $\Psi_n(\tilde{\zeta})$ obey the equation

$$\frac{d^2 \Psi_n}{d\tilde{\zeta}^2} + (R_n + \mu^2 - \tilde{\zeta}^2)\, \Psi_n = 0 \tag{A3.23}$$

and are expressed in terms of Hermite polynomials of the argument $\tilde{\zeta}$

$$\Psi_n = \frac{1}{\sqrt{\pi n! 2^n}} e^{-\tilde{\zeta}^2/2} H_n(\tilde{\zeta}) .$$

At greater n the functions are rather well described by the geometrical – optics approximation [Eq. (2.16)].

Now let us compare the solution obtained with an analogous quasispherical solution (Sec. A3.2). The comparison of Eqs. (A3.11) and (A3.12) with Eqs. (A3.21) – (A3.23) shows that longitudinal wave numbers for quasispherical and adiabatic modes are somewhat different in magnitude so that the differences are $\delta k = \mu^2/2k\rho_s^2$. Such relatively small variations in k lead to significant discrepancies in the wave phase at long distances. So $\Delta\varphi = 10^1 - 10^2$ when $\Delta x \sim 10^3$ km, i.e., at the distance as small as the one-hop path, the phase change is rather great. Another distinction in both the solutions consists in the appearance of the factor $\rho_s^{-1/2}$ in Eq. (A3.21) involved by requirement of the normalization to be constant [the requirement follows from the initial Eq. (A3.17)].

Finally, the principal discrepancy between the solutions consists in the appearance of the phase factor $\exp[i\mu(x)\zeta + i\nu(x)\zeta^2]$ and shift in the argument of the function Ψ_n by $\delta\zeta(x)$. It is easily seen that due to the discrepancies the adiabatic modes represent packets of quasispherical modes, $\Psi_{1n} = \sum_m C_m \Psi_m(\zeta)$. Let us estimate the packet widths at various magnitudes of the longitudinal inhomogeneity. In the case of vary small gradients, $|\mu(x)| \ll 1/\sqrt{n}$, $|\delta\zeta| \ll 1/\sqrt{n}$, making in Eq. (A3.21) the expansion into the Taylor series, one can recognize that the coefficients C_m decrease by a power law when moving away from the central mode.

A different situation arises when the absolute values of the parameters μ, ζ become greater than $1/\sqrt{n}$. Consider the case in more detail. For simplicity, we shall assume that $\nu(x) = 0$. Introducing the operator $\hat{L}_n = \partial^2/\partial\tilde{\zeta}^2 + 2n + 1 - \tilde{\zeta}^2$, and taking into account that $\hat{L}_n \Psi_n(\tilde{\zeta}) = 0$, we obtain for the coefficients desired, $C_m = \int \Psi_m(\zeta)\, \Psi_{1n} d\zeta$, the equation

$$i2\mu \sqrt{\frac{m}{2}}\,[C_{m-1} - C_{m+1}] - 2\delta\zeta \sqrt{\frac{m}{2}}\,[C_{m-1} + C_{m+1}] + [2(n-m) - (\delta\zeta)^2 - \mu^2]\, C_m = 0\,. \tag{A3.24}$$

We put $C_m = f(m) \exp[i\varkappa(n-m)]$, where $\tan\varkappa = -\delta\xi/\mu$. Then for the function $f(m)$ we obtain

$$i2A_m[f(m-1) - f(m+1)] + [2(n-m) - (\delta\zeta)^2 - \mu^2]\, f(m) = 0\,. \tag{A3.25}$$

Here, $A_m = \sqrt{\frac{m}{2}}\,[\mu\cos\varkappa - \delta\zeta\sin\varkappa]$, $|A_m| = \sqrt{\frac{m}{2}(\mu^2 + (\delta\zeta)^2)}$. Assuming $f(m)$ to be a slow changing function of m, we obtain from Eq. (A3.25)

$$f(m) \approx f(n) \exp\left[i\frac{(n-m)^2}{4A_n}\right]. \tag{A3.26}$$

This solution is true, if $|(n-m)/2A_n| < 1$. Conditions $n \gg 1$, $m \gg 1$ are also assumed to be satisfied.

Let the inverse inequality, $|(n-m)/2A_n|>1$, be satisfied. Putting for definiteness $A_n>0$, $m>n$, and introducing $C(m)=f(m)\exp[\varkappa(n-m)]$, where $\tan\varkappa=\mu/\delta\zeta$, we obtain approximately

$$2A_n[f(m-1)+f(m+1)]+2(n-m)f(m)=0\,. \tag{A3.27}$$

This equation has at great m the solution

$$f(m)\sim f(n+2A_n)/\Gamma\left(\frac{m-n}{A_n}+1\right). \tag{A3.28}$$

The coefficients $f(m)$ are seen to decrease rapidly in absolute value with increasing m.

Thus, in the presence of inhomogeneities strong enough, $|A_n|\gg 1$, eigenmodes represent packets of undisturbed modes with the effective width, $m_{max}-m_{min}\approx 4|A_n|$. For typical ionospheric gradients, $dz_s/dx\sim 10^{-2}$, $\mu(x)\sim 1$, the packet half-width is by the order of magnitude, $|m-n|\sim\sqrt{n}k\rho_s\cdot(dz_s/dx)\sim\sqrt{n}$. This agrees with the expression for the number of efficiently interacting quasispherical modes, $|m-n|\sim|H_{n,n-1}/\Delta k|\sim k\rho_s|dz_s/dx|\sqrt{n}$.

The general solution of the initial wave equation is a superposition of particular solutions

$$A=\sum_n C_n\frac{\Psi_n(\tilde{\zeta})}{\sqrt{k_n\rho_s}}\exp[i(\int k_n dx+\mu(x)\zeta+\nu(x)\zeta^2)]\,. \tag{A3.29}$$

Taking into consideration the accuracy of the solutions constructed, it can be shown that under ionospheric conditions the coefficients C_n remain essentially constant up to the distances of the order of RTW. To make sure, one should substitute Eq. (A3.29) in initial Eq. (A3.5), and take into account the corrections which are proportional to the second derivatives with respect to the longitudinal coordinate.

Duct of Arbitrary Form. The solution considered above for an arbitrary waveguide can be applied only near the bottom where the duct is roughly parabolic. To construct modes with arbitrary numbers, we consider Eq. (A3.17) in common coordinates z, x.

We shall seek particular solutions of the equation in the form

$$\tilde{\Psi}_n=a(x)[\Psi_n(z,x)e^{i\varphi}+i\delta\Psi_n(z,x)]\,. \tag{A3.30}$$

If the term $\delta\Psi_n$ in Eq. (A3.30) is neglected, then the general solution of Eq. (A3.30) is analogous to that considered above. The coefficient $a(x)$ in this case is determined by the requirement for the normalization to be conserved, the major function Ψ_n is associated with the standing wave in the duct. The phase factor $\exp(i\varphi)$ and the small term $\delta\Psi_n$ yield corrections connected with the longitudinal medium inhomogeneity. Substituting Eq. (A3.30) in Eq. (A3.17), we obtain a system of two coupled equations

$$\frac{\partial^2 \Psi_n}{\partial z^2} + \left[k_0^2 (E_n - U(z,x)) - \left(\frac{\partial \varphi}{\partial z}\right)^2 - 2k \frac{\partial \varphi}{\partial x} \right] \Psi_n$$

$$- 2k e^{-i\varphi} \delta \Psi_n \frac{d \ln a}{dx} = 0 \,, \tag{A3.31}$$

$$\Psi_n \left(\frac{\partial^2 \varphi}{\partial z^2} + 2k \frac{d \ln a}{dx} \right) + 2 \frac{\partial \varphi}{\partial z} \frac{\partial \Psi_n}{\partial z} + 2k \frac{\partial \Psi_n}{\partial x}$$

$$+ e^{-i\varphi} \left\{ \frac{\partial^2 \delta \Psi_n}{\partial z^2} + k_0^2 (E_n - U(z,x)) \, \delta \Psi_n + i2k \frac{\partial \delta \Psi_n}{\partial x} \right\} = 0 \,. \tag{A3.32}$$

Equation (A3.31) serves to find the functions Ψ_n. Equation (A3.32) we shall break into two equations in order that one of them yield the phase φ

$$\frac{\partial \varphi}{\partial z} \frac{\partial \Psi_n}{\partial z} + k \frac{\partial \Psi_n}{\partial x} = 0 \,, \tag{A3.33}$$

whereas the other yields the functions $\delta \Psi_n$

$$\frac{\partial^2 \delta \Psi_n}{\partial z^2} + k_0^2 (E_n - U(z,x)) \, \delta \Psi_n + i2k \frac{\partial \delta \Psi_n}{\partial x}$$

$$+ e^{i\varphi} \Psi_n \left[\frac{\partial^2 \varphi}{\partial z^2} + 2k \frac{d \ln a}{dx} \right] = 0 \,. \tag{A3.34}$$

The smallness of the longitudinal inhomogeneity gives a simple way for the solution of the system of equations presented above. We assume that individual terms in Eqs. (A3.31) and (A3.32) are of the following orders of magnitude

$$k_0 \frac{\partial \tilde{\Psi}}{\partial x} \sim \delta k_\perp^2 \tilde{\Psi} \,, \quad \left(\frac{\partial \varphi}{\partial z}\right)^2 \sim k_0 \frac{\partial \varphi}{\partial x} \sim \delta^2 k_\perp^2 \tilde{\Psi} \,,$$

$$k_0 \frac{d \ln a}{dx} \sim \delta^2 k_\perp^2 \,. \tag{A3.35}$$

Here, $k_\perp$ is the characteristic magnitude of the transverse wave number in the duct, δ is the small parameter of the expansion. The relations in Eq. (A3.35) are chosen in such a way that they can agree with actual ionospheric conditions. This can be shown if considering again the parabolic waveguide model studied above. The parameter δ is then $\delta = dz_s/dx$.

Let us proceed to the solution of Eqs. (A3.31)–(A3.34). In the lowest approximation, a function $\Psi_n(0)$ is sought which obeys the equation

$$\frac{\partial^2 \Psi_n^{(0)}}{\partial z^2} + k_0^2 (E_n - U(z,x)) \, \Psi_n^{(0)} = 0 \,. \tag{A3.36}$$

For modes with great numbers, $n \gg 1$, the WKB approximations may be used

$$\Psi_n^{(0)} = \frac{a(x)}{[k_0^2(E_n - U)]^{1/4}} \cos\left(k_0 \int_{z_1}^{z} (E_n - U)^{1/2} dz - \frac{\pi}{4}\right), \tag{A3.37}$$

where z_1 is the turning point.

Substituting Eq. (A3.37) in Eq. (A3.33), we obtain the phase

$$\varphi = -k \int_{z_*}^{z} \frac{dz'}{(E_n - U)^{1/2}} \frac{\partial}{\partial x} \int_{z_1}^{z'} (E_n - U)^{1/2} dz''. \tag{A3.38}$$

Here, z_* is an arbitrary initial coordinate. The solution of Eq. (A3.34) in neglect of small corrections, $k\dfrac{\partial \delta \Psi_n}{\partial x}$, will be sought in the form

$$\delta \Psi_n = \eta(z, x) \frac{\partial \Psi_n}{\partial z}.$$

In so doing we obtain for $\eta(z, x)$ the following approximate equation

$$k_0^2\left[2(E_n - U)\frac{\partial \eta}{\partial z} - \frac{\partial U}{\partial z}\eta\right] - e^{i\varphi}\left[\frac{\partial^2 \varphi}{\partial z^2} + 2k\frac{d \ln a}{dx}\right] = 0, \tag{A3.39}$$

which has the following solution

$$\eta(z, x) = \frac{1}{2k_0^2(E_n - U)^{1/2}} \int_{z_1}^{z} \frac{dz}{(E_n - U)^{1/2}} e^{i\varphi}\left[\frac{\partial^2 \varphi}{\partial z^2} + 2k\frac{d \ln a}{dx}\right]. \tag{A3.40}$$

Now let us return to the equation for Ψ_n, which with an accuracy of δ^2 has the form

$$\frac{\partial^2 \Psi_n}{\partial z^2} + \left\{k_0^2[E_n - U(z, x)] - \left(\frac{\partial \varphi}{\partial z}\right)^2 - 2k\frac{\partial \varphi}{\partial x}\right\} \Psi_n = 0. \tag{A3.41}$$

Comparing Eq. (A3.41) with Eqs. (A3.7) and (A3.17), we see that allowance for small terms results in some variation of the potential, this in turn results in a small shift of the eigenwave numbers. In particular, the condition for standing wave existence means now the conservation of the integral

$$\int_{z_1}^{z_2} dz \left\{k_0^2[E_n - U(z, x)] - \left(\frac{\partial \varphi}{\partial z}\right)^2 - 2k\frac{\partial \varphi}{\partial x}\right\}^{1/2} = \pi\left(n + \frac{1}{2}\right), \tag{A3.41a}$$

taken between the turning points [see Eq. (A3.9)]. The arbitrariness in the choice of the initial coordinate z_* in Eq. (A3.38), as can be easily seen, does not influence the final results. Indeed, if z_* varies, then an extra phase factor $\exp\{i\delta\varphi(x)\}$ appears, but the combination $k_n + d\delta\varphi/dx$ in this case suffers no changes.

Note finally that the correction connected with the function $\delta\Psi_n$ can be represented in the form of superposition of adiabatic modes

$$\delta\Psi_n = \sum_m C_m(x)\exp(i\int \Delta k_m dx)\,\tilde{\Psi}_m . \tag{A3.42}$$

Insertion of Eq. (A3.42) into Eq. (A3.34) enables one to find the system of equations determining the dependence of the coefficients C_m on the longitudinal coordinate x. In the general case (if $m \neq n$) these coefficients have the order of magnitude $\sim\delta^2$, i.e., under ionospheric conditions they are negligible.

Thus, approximate solutions of the parabolic equation for the great number modes have the form

$$\tilde{\Psi}_n(z,x) = \frac{a(x)}{(k_0 k_n)^{1/2}(E_n - U)^{1/4}}\cos\left\{\int_{z_1}^{z}\left[k_0(E_n - U(z,x)) - \left(\frac{\partial\varphi}{\partial z}\right)^2 - 2k\frac{\partial\varphi}{\partial x}\right]^{1/2} dz - \frac{\pi}{4}\right\}\exp\left[-ik\int_{z_*}^{z}\frac{dz'}{(E_n - U)^{1/2}}\frac{\partial}{\partial x}\cdot\int_{z_1}^{z'}(E_n - U)^{1/2}dz''\right] . \tag{A3.43}$$

These solutions represent packets of quasispherical modes whose widths are determined by local horizontal gradients.

In regions where the longitudinal inhomogeneity becomes very small the functions in Eq. (A3.43) convert into standard quasispherical modes.

Thus, for modes with great numbers ($n \gg 1$) one succeeds in construction of the quasiclassical solution [Eq. (A3.43)]. For low modes, when the WKB approximation fails, the duct profile can always be regarded as parabolic, and the exact solution [Eq. (A3.21)] can be applied. For intermediate n, both approaches give the same result, i.e., Eq. (A3.21) and Eq. (A3.43) overlap.

Connection Between the Vector Potential Components. Now let us find out what corrections appear when the first-order terms with respect to δ are taken into account in the initial Eqs. (A3.1) and (A3.2). The corrections, in general, could be comparable with those taken into account. Let us write down the system of two coupled equations following from Eq. (A3.2) in the coordinates z, x,

$$\begin{aligned} \Delta A_z - k_0^2 U A_z &= \frac{\partial \ln U}{\partial z}\frac{\partial A_z}{\partial z} + \frac{\partial \ln U}{\partial z}\frac{\partial A_x}{\partial x} , \\ \Delta A_x - k_0^2 U A_x &= \frac{\partial \ln U}{\partial x}\frac{\partial A_x}{\partial x} + \frac{\partial \ln U}{\partial x}\frac{\partial A_z}{\partial z} . \end{aligned} \tag{A3.44}$$

In Borisov and Gurevich (1976a) it has been shown that the terms $\sim(\partial \ln U/\partial x)(\partial A/\partial x)$ in the right-hand sides of Eq. (A3.44) result in only rather small corrections in the solutions. Therefore, below we restrict ourselves to the corrections describing the interaction between the potentials A_z and A_x

$$\begin{aligned} \varDelta A_z - k_0^2 U A_z &= \frac{\partial \ln U}{\partial z}\,\frac{\partial A_x}{\partial x}\,, \\ \varDelta A_x - k_0^2 U A_x &= \frac{\partial \ln U}{\partial x}\,\frac{\partial A_z}{\partial z}\,. \end{aligned} \tag{A3.45}$$

We shall seek the solution of Eq. (A3.45) in the form of a series with respect to adiabatic modes

$$\begin{aligned} A_z &= \sum_n C_n \tilde{\Psi}_n \frac{\exp(i\int k_n dx)}{\sqrt{k_n}}\,, \\ A_x &= \sum_n B_n \tilde{\Psi}_n \frac{\exp(i\int k_n dx)}{\sqrt{k_n}}\,. \end{aligned} \tag{A3.46}$$

Substituting Eq. (A3.46) in the second equation, we obtain the following approximation chain of equations for B_n

$$\begin{aligned} \frac{dB_n}{dx} = -\frac{i}{2k_0}\Bigg[\sum_{m\neq n}\left(\int \tilde{\Psi}_n^* \frac{\partial \ln U}{\partial x}\,\frac{\partial \tilde{\Psi}_m}{\partial z}\,dz\right)\exp\left\{i\int (k_m - k_n)\,dx\right\} C_m \\ +\left(\int \tilde{\Psi}_n^* \frac{\partial \ln U}{\partial x}\,\frac{\partial \tilde{\Psi}_n}{\partial z}\,dz\right) C_n\Bigg]\,. \end{aligned} \tag{A3.47}$$

Let the initial conditions be specified in the form

$$C_m(x_0) = \delta_{ml}\,, \quad B_m(x_0) = 0\,. \tag{A3.48}$$

Returning to the first Eq. (A3.45), with allowance for Eqs. (A3.47) and (A3.48), we obtain

$$\begin{aligned} \frac{dC_l}{dx} = -\frac{i}{4k_0}\Bigg[\sum_{n\neq l}\exp\left[i\int_{x_0}^{x}(k_n - k_l)\,dx\right]\left(\int \tilde{\Psi}_l^* \frac{\partial \ln U}{\partial z}\,\tilde{\Psi}_n dz\right) \\ \times \int_{x_0}^{x}\left(\int \tilde{\Psi}_n^* \frac{\partial \ln U}{\partial x}\,\frac{\partial \tilde{\Psi}_l}{\partial z}\,dz\right)\exp\left[i\int_{x_0}^{x'}(k_l - k_n)\,dx''\right] C_l(x')\,dx' \\ +\left(\int \tilde{\Psi}_l^* \frac{\partial \ln U}{\partial z}\,\tilde{\Psi}_l dz\right)\int_{x_0}^{x}\left(\int \tilde{\Psi}_l^* \frac{\partial \ln U}{\partial x}\,\frac{\partial \tilde{\Psi}_l}{\partial z}\,dz\right) C_l(x')\,dx'\Bigg]\,. \end{aligned} \tag{A3.49}$$

Let us evaluate the variation of C_l along the path, retaining in the right-hand side of Eq. (A3.49) only the most important terms from all the slow varying ones. Putting $C_l = \exp\left(i \int_{x_0}^{x} \gamma(x')\,dx'\right)$, we obtain

$$\gamma^2 \sim \frac{i}{4k_0}\left(\int \tilde{\Psi}_l^* \frac{\partial \ln U}{\partial z}\,\tilde{\Psi}_l\,dz\right)\left(\int \tilde{\Psi}_l^* \frac{\partial \ln U}{\partial x}\,\frac{\partial \tilde{\Psi}_l}{\partial z}\,dz\right), \tag{A3.50}$$

which under ionospheric conditions is by the order of magnitude, $|\gamma| \lesssim 10^{-6}\,\mathrm{km}^{-1}$. The coefficient B_l, according to Eq. (A3.47), increases roughly by the linear law

$$B_l(x) \approx \frac{i}{2k_0}\int_{x_0}^{x}\left(\int \tilde{\Psi}_l^* \frac{\partial \ln U}{\partial x}\,\frac{\partial \tilde{\Psi}_l}{\partial z}\,dz\right) C_l(x')\,dx'\,. \tag{A3.51}$$

However, even on long-distance paths its absolute value is rather small, $|B_l| \lesssim 10^{-2}$.

Thus, whereas in the spherically – symmetric ionosphere the vector potential breaks into independent components, the horizontal ionospheric irregularity leads to an interaction between the components resulting in the generation of another wave whose polarization is different from the original wave.

Accuracy of the Adiabatic Approximation. The solutions obtained above enable one to answer the question on the accuracy of the conservation of the adiabatic invariant I considered in Chaps. 3 and 4. The simple adiabatic invariant [Eq. (3.17)], by Eq. (A3.9), is equivalent to the quasispherical mode number. As shown in Sec. A3.3, the expansion of the initial field into the quasispherical modes is not conserved. Effective mixing of neighboring modes takes place. This means that in the real ionosphere, strictly speaking, a redistribution of energy between different adiabatic invariants I occurs as well.

But to what extent is the redistribution significant? The answer is given by the analysis of the adiabatic modes. The adiabatic functions $\tilde{\Psi}_n$, being expanded into quasispherical modes, represent packets with the effective width with respect to n, $\Delta n \sim (1-3)\cdot 10$. The adiabatic modes are conserved along long-distance and RTW paths. This means that wave packets of quasispherical modes with the width Δn are conserved. In other words, the packets are conserved which have the width with respect to adiabatic invariants I, $\Delta I \sim (\Delta n/n) I$. Note, however, that the total number of the modes in the IWD is fairly great, $n_m \sim 10^2 - 10^3$. Therefore, $\Delta n/n_m \sim 10^{-1} - 10^{-2}$, and, consequently, $\Delta I/I_m \sim 10^{-1} - 10^{-2}$, where I_m is the maximum adiabatic invariant of the duct. This means that at sufficiently smooth invariant distribution of wave energy in a slightly inhomogeneous duct, energy redistribution, as the wave propagates, is insignificant.

Note that as follows from Eq. (A3.41a), a more precise adiabatic invariant can be introduced which directly takes into account the horizontal ionospheric inhomogeneity. Such an invariant is conserved during very long-distance propagation. It differs from the simple invariant I [Eq. (3.17)] by a small quantity, $\Delta I \sim (10^{-1} - 10^{-2})I$. We also note that we considered here the horizontal ionospheric inhomogeneities which are smooth with respect to the adiabaticity parameter, $\mu \ll 1$ [Eq. (3.15)]. If, on the contrary, inhomogeneities are rather small-scaled, so that Eq. (3.15) is not satisfied, then the adiabatic invariant is not conserved along the path. In this case, as the wave travels, the adiabatic modes $\tilde{\Psi}_n$ [Eq. (A3.43)] are tangled, so they are inconvenient for representation of a field in irregular waveguides. In this case, it is advisable to use eigenfunctions of the parabolic equation constructed in Borisov (1978, 1980). In these works, a rigorous criterium for adiabaticity of perturbations is given, and the diffusion coefficient with respect to adiabatic invariants is found for small-scale irregularities. A geometrical – optics analog of these results is represented in Chap. 6 (pp. 200 – 205).

A3.5 Radio Wave Capture into IWD

Now we consider in the framework of wave approach the problem of radio wave capture into ducts in a horizontally – inhomogeneous ionosphere. In the geometrical – optics approximation, this problem was studied in Sec. 3.4. Further (Chaps. 6 and 7), it was found out that the regular refraction is one of the most effective mechanisms for ionospheric waveguide excitation. However, these results have been obtained with allowance for no wave effects. These effects, generally speaking, are important for rays whose trajectories are close to boundaries of the capture region. Therefore, the rigorous solution of the wave equation will be applied below for the description of the capture.

A3.5.1 Neighborhood of the Separatrix

In Sec. 3.2 it was shown in the geometrical – optics approximation that for ray capture into a duct or for departure from the duct, it is necessary that the ray trajectory intersects the separatrix, namely, a distinctive trajectory separating the capture region. Near the separatrix, $\theta = \theta_s$, the condition for applicability of the adiabatic approximation [Eq. (3.15)] is violated due to the increase in the oscillation period Θ. Indeed, taking into account the fact that near the maximum the interduct barrier always can be put parabolic

$$U(z) = U_m + \frac{1}{2} U''_m (z - z_m)^2 \quad (U''_m < 0) ,$$

we represent the ray oscillation period in the duct in the form of Eq. (2.32)

$$\Theta = \frac{2}{R_0}\int_{z_1}^{z_2}\frac{dz}{\sqrt{E-U(z)}} \approx \Theta_0\left[1+\frac{1}{2\pi}\ln\frac{\Delta U}{\delta E}\right],$$

$$\Theta_0 = \frac{2\pi}{R_0}\sqrt{\frac{2}{U''(\tilde{z}_s)}} = \frac{2\pi k_0 \rho_s^2}{R_0}, \quad \delta E = E - U_m . \tag{A3.52}$$

Here, ΔU is the duct depth, $U''(\tilde{z}_s)$ is the duct width in the vicinity of the sliding point where the parabolic model [Eq. (2.9)] was used. When $\delta E \to 0$, the period Θ logarithmically diverges. Substituting now Eq. (A3.52) in the condition for adiabatic approximation applicability [Eq. (3.15)], we obtain that in the neighborhood of the separatrix, where

$$\delta E < \Delta U e^{-\frac{2\pi}{\mu_0}}, \quad \mu_0 = \frac{\Theta_0}{2\pi P_i}\frac{dP_i}{d\theta} \tag{A3.53}$$

[P_i is a parabolic duct characteristic, see Eq. (3.15)], the adiabatic approximation breaks down. But the field of adiabaticity violation near the separatrix

$$\delta\theta \sim \frac{\delta E}{|dU_m/d\theta|}$$

proves to be exponentially narrow with respect to the parameter $1/\mu_0$. Consequently, at sufficiently small μ_0, always $\delta\theta \ll \Theta$ [see Eq. (3.26)], it is this circumstance that enables the rules for transition through the separatrix to be formulated [Eq. (3.44 a – c)].

Now let us evaluate the contribution of wave effects to the separatrix neighborhood. We confine ourselves, for simplicity, to the quasispherical approximation, and assume that the barrier separates the ground-based duct E and interlayer duct FE. Matching the quasiclassical solutions of Eq. (A3.7) in both ducts with the exact solution in the neighborhood of the parabolic duct, we obtain the equation for determination of the eigenvalues

$$\cos(\Phi_1+\Phi_2) + V(R)\cos(\Phi_1-\Phi_2) = 0 . \tag{A3.54}$$

Here, $V(R) = [1+e^{\pi R}]^{-1/2}$, $\Phi_{1,2}$ are the phase integrals in the first and second ducts, respectively. Equation (A3.54) gives the generalization of the results of Appendix 1 to the case of the above barrier transmission, $R > 0$.

In the field of applicability of the geometrical optics we have

$$\Phi = k_0 \int_{z_1}^{z_2} [E-U(z)]^{-1/2} dz ,$$

where z_1, z_2 are the turning points.

If the wave is in the *FE*-duct ($\pi R \gg 1$), then from Eq. (A3.54) the standard condition follows for standing wave existence, $\cos(\Phi_1 + \Phi_2) = 0$.

In the opposite limit, $\pi R \ll -1$, the equation defines the system of modes for two independent waveguides, $\cot \Phi_1 \cdot \cot \Phi_2 = 0$. In this case, in either duct a phase integral

$$k_0 \int\limits_{z_1}^{z_2} (E - U(z,x))^{1/2} dz = \pi\left(n + \frac{1}{2}\right),$$

is conserved, which, as shown in Chap. 3, determines the refraction over the ray oscillation period. In the region $1 \lesssim \pi R \leqslant 0$, the connection between the waveguides becomes rather strong due to wave tunneling. In this case, the eigenvalues no longer satisfy the relation

$$k_0 \int\limits_{z_1}^{z_2} (E_n - U(z,x))^{1/2} dz = \text{const},$$

but are determined from the solution of the general Eq. (A3.54). Consequently, because of the wave effects near the separatrix, in the determination of eigenwave vectors, a difference appears from the geometrical optics results. Thus, in the narrow region

$$\delta E = |U_m - E| < \frac{[U_m'']^{1/2}}{k_0} \sim 10^{-4} \tag{A3.55}$$

the geometrical-optics approximation should be refined.

As follows from the comparison of Eqs. (A3.55) and (A3.53) in the case of small longitudinal inhomogeneity, $\mu \ll 1$, the wave theory in the vicinity of the separatrix leads to stronger limitations on the field of applicability for the adiabatic approximation than the ray theory.

A3.5.2 Antenna Radiation Field

Now let us consider the structure of the field generated by an antenna in the duct. Previously we analyzed the eigenmodes of ionospheric waveguides which are determined by a discrete set of wave numbers. However, an actual spectrum of generated waves is not, in general, discrete. This fact should be borne in mind when studying the process of radio wave capture into a duct.

So let on the Earth's surface an antenna be placed which emits a packet of waves with the frequency ω at a small takeoff angle. The medium where waves propagate will be regarded, as before, as a two-dimensional medium, gradually varying in the horizontal direction x [Eq. (1.9)]. The field of HF waves produced by the source $j(z,x)$ (antenna current) in such a medium obeys the wave equation

$$\frac{\partial^2 A}{\partial z^2} + \frac{\partial^2 A}{\partial x^2} - k_0^2 U(z,x) A = \frac{4\pi}{c} j(z,x). \tag{A3.56}$$

In the spherically – symmetric ionosphere, the solution of Eq. (A3.56) can be easily written in terms of the Green function

$$A = \frac{4\pi}{c} \int dx e^{ikx} \int G_k(z, z') j_k(z') dz' . \tag{A3.57}$$

Here, $j_k(z)$ is the Fourier component of the source, $j_k(z) = \frac{1}{2\pi} \int e^{ikx} j(z, x) dx$, $G_k(z, z')$ is the Green function for Eq. (A3.57). In a particular case, when the potential is parabolic, $U(z) = U(z_m) - \frac{1}{2} \mid U''(z_m) \mid (z - z_m)^2$, and the source represents a point-like dipole in the z, x-space, an explicit analytical expression for the field can be found which is similar to that obtained by Tinin (1973) for the spherical wave

$$A = A_0 \int_{-\infty}^{\infty} e^{ikx} \Gamma(-\nu) D_\nu(\eta) D_\nu(-\eta_i) , \quad \eta > \eta_i , \tag{A3.58}$$

where A_0 is a constant depending on the source power, $\Gamma(x)$ is the Gamma function, D_ν is the parabolic cylinder function,

$$\nu = -\frac{i}{2} k_0^2 \rho_s^2 (E - U(z_m))^{-1/2} , \quad \eta = \sqrt{2} e^{i\frac{\pi}{4}} \frac{z - z_m}{\rho_s} , \quad \eta_i = \sqrt{2} e^{i\frac{\pi}{4}} \frac{z_i - z_m}{\rho_s} ,$$

z_i is the height where the antenna is situated.

Now we show that in a horizontally – inhomogeneous ionosphere the field can also be represented in the form analogous to Eq. (A3.57) but with another Green function. In this case, of importance is the fact that near the Earth, where the source is situated, the potential $U(z, x)$ depends only on the coordinate z: $U(z, x) \approx -\left(1 + \frac{2z}{R_0}\right)$.

We introduce the Green function in the following way

$$G_k(z, z', x) = \begin{cases} \frac{1}{\Delta} \tilde{\Psi}_k^{(+)}(z, x) \tilde{\Psi}_k^{(-)}(z', x) , & z > z' , \\ \frac{1}{\Delta} \tilde{\Psi}_k^{(-)}(z, x) \tilde{\Psi}_k^{(+)}(z', x) , & z \leqslant z' . \end{cases} \tag{A3.59}$$

Here,

$$\Delta = \left. \frac{\partial \tilde{\Psi}_k^{(+)}}{\partial z} \tilde{\Psi}_k^{(-)} - \frac{\partial \tilde{\Psi}_k^{(-)}}{\partial z} \tilde{\Psi}_k^{(+)} \right|_{z = z'} ,$$

and $\tilde{\Psi}_k^{(\pm)}$ are the solutions of Eq. (A3.17) corresponding to waves propagating from the source ($\tilde{\Psi}^{(+)}$) or to the source ($\tilde{\Psi}^{(-)}$).

We have used the parabolic equation for the function instead of the general wave equation, using the graduality of medium variations in the

x-direction. Similarly to the way of adiabatic mode construction, one can approximately obtain

$$\tilde{\Psi}_k^{(\pm)} \approx \frac{1}{k_0^{1/2}(E-U)^{1/4}} \exp\left\{ \pm ik_0 \int_{z_i}^{z} (E-U)^{1/2} dz \right.$$
$$\left. - ik \int_{z_i}^{z} \frac{dz'}{(E-U)^{1/2}} \frac{\partial}{\partial x} \int_{z_i}^{z} (E-U)^{1/2} dz \right\}, \qquad \text{(A3.60)}$$

where z_i is the source coordinate. In the vicinity of the source, the functions $\tilde{\Psi}_k^{(\pm)}$ in Eq. (A3.60) represent solutions of the wave equation in a horizontally–inhomogeneous ionosphere, and $G_k(z, z', x)$ satisfy the standard conditions required commonly upon the Green function. Direct substitution of Eq. (A3.57) together with the Green function [Eq. (A3.59)] in the initial Eq. (A3.56), demonstrates that it is really the solution. Thus, for the electromagnetic wave field outside the source region, the representation is true

$$A = \int e^{ikx} \tilde{\Psi}_k^{(+)}(z, x) C(k) dk, \qquad \text{(A3.61)}$$

where

$$C(k) = \frac{4\pi}{c} \int j_k(z') \tilde{\Psi}_k^{(-)}(z') dz'.$$

We see that the antenna generates a continuous spectrum whose width is determined by the directivity diagram. If, however, the waveguide is sufficiently extensional (i.e., if its longitudinal length considerably exceeds the ray oscillation period in the duct, $\Delta x \gg \Theta R_0$), then due to wave interference, a discrete spectrum gradually manifests itself.

A3.5.3 Radio Wave Capture into a Duct. Role of Wave Effects

Let us consider radio wave capture into the interlayer duct after the first wave reflection from the F layer. In other words, we assume that the wave emitted by a near-ground antenna passes through the barrier and comes into the interlayer duct. Further, reflecting from the F layer, the wave once again approaches the barrier. The energy portion which is left in the interlayer duct as a result of a partial reflection from the barrier will be referred to as trapped energy. As shown below, in a horizontally–inhomogeneous ionosphere the wave is trapped in the duct if the factor of reflection from the barrier increases along the path. The increase in the reflection factor can be induced by the increase in both the barrier "height" and the effective duct "depth".

Let the transmitter antenna be situated in the duct I at the height $z = z_i$, as shown in Fig. A3.1. The barrier separating the ducts I and II can be regarded as parabolic, since actually only a small vicinity of the barrier maximum is important for the capture. So we put

$$U(z,x) = U(z_m) + \frac{1}{2} U''(z_m)\,[z - z_m(x)]^2\,, \tag{A3.62}$$

where $U''_{(z_m)} < 0$. The transmission, d, and reflection, r, factors for such a barrier have the form [see also Eq. (A1.39)]

$$d_k = \frac{e^{\pi R/4}}{[e^{-\pi R/2} + e^{\pi R/2}]^{1/2}}\,, \quad r_k = \frac{e^{-\pi R/4}}{[e^{-\pi R/2} + e^{\pi R/2}]^{1/2}} \tag{A3.63}$$

where $R = k_0^2 \rho_s^2 [E - U(z_m)]$.

For the field Fourier components with the wave number $k < k_0 |U(z_m(x_1))|^{1/2}$, the transmission factor is close to unity, for $k > k_0\,|\,U(z_m(x_1))\,|^{1/2}$ it is nearly zero. The quantity $k = k_0\,|\,U(z_m)\,|^{1/2}$ corresponds to $d_k = r_k = 1/\sqrt{2}$. In terms of the geometrical optics, this value corresponds to the ray 3, separating the transmitted, 2, and reflected, 1, wave beams. The total field trapped in the duct II can be represented approximately (namely, with no multiple reflections) in the form

$$A = \int dk\,C(k)\,d_k(x_1) \exp\left(ikx + i \int_{z_i}^{z_2} p\,dz \right) \cdot \tilde{\Psi}_k(z,x)\,. \tag{A3.64}$$

Here, $p = k_0(E - U)^{1/2}$ and z_2 is, in the case of wave tunneling, the turning point in the duct I, and in the case of the above barrier transmission, $z_2 = z_m$. The function $\tilde{\Psi}_k(z,x)$ describes the transverse wave tunneling in the interlayer duct II and obeys the parabolic equation

$$\frac{\partial^2 \tilde{\Psi}_k}{\partial z^2} + \left[-k_0^2 \frac{U''}{2} (z_m)(z - z_m)^2 - (k_0^2 U(z_m) + k^2) \right] \tilde{\Psi}_n + i2k \frac{\partial \tilde{\Psi}_k}{\partial x} = 0\,. \tag{A3.65}$$

Here, we have assumed, for simplicity, that the ground-based duct I does not change along the path.

The wave field in the interlayer duct, after the reflection at the point x_2, we represent in the form

$$\tilde{\Psi}_k = r_{\tilde{k}}(x_2)\, e^{i\Phi_k(x)}\, \tilde{\Psi}_{1k}(z,x)\,. \tag{A3.66}$$

Here, $\Phi_k(x)$ is the progressive phase as the wave travels from the source z_1 to the barrier z_2 transverse to the duct, $\Phi_k(x) \approx k_0 \int [E - U(z,x)]^{1/2} dz$, the function $\tilde{\Psi}_{1k}$ satisfies the equation

$$\frac{\partial^2 \tilde{\Psi}_{1k}}{\partial z^2} - \left[k_0^2\, U(z,x) + \left(k^2 + 2k \frac{d\Phi}{dx}\right) \right] \tilde{\Psi}_{1k} + i2k \frac{\partial \tilde{\Psi}_{1k}}{\partial x} = 0\,. \tag{A3.67}$$

The factor of reflection from the barrier, $r_{\tilde{k}}$, involves the wave number $\tilde{k} = k + \frac{2}{k} \frac{d\Phi}{dx}$. Note that the correction $\Delta k = \frac{2}{k} \frac{d\Phi}{dx}$ describes the variation along the path of the reflection factor due to refraction in the horizontally-inhomogeneous medium. The explicit expression for the solution of

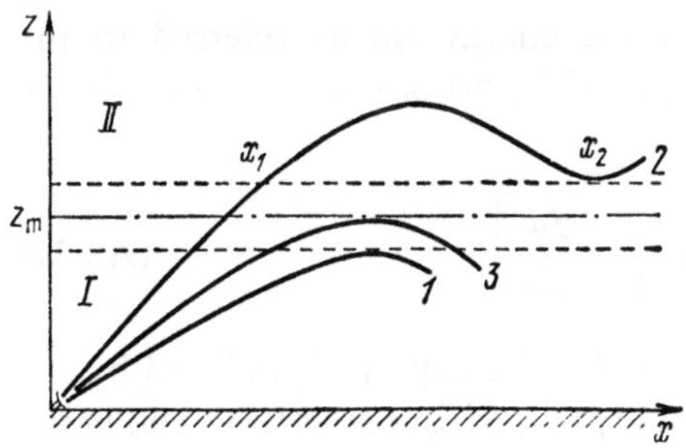

Fig. A3.1. Capture of HF radio waves into the duct II in a horizontally – inhomogeneous ionosphere. The altitude z_m corresponds to the peak of the barrier separating the ducts I and II

Eq. (A3.67) in terms of parabolic cylinder functions is given in Borisov and Gurevich (1977c). In the geometrical – optics approximation, the function $\tilde{\Psi}_{1k}(z,x)$ has the form

$$\tilde{\Psi}_{1k}(z,x) = \frac{1}{\sqrt{\tilde{p}_k}} \exp\left[i \left(\int_{z_3}^{z} \tilde{p}_k dz' - \int_{z_3}^{z} \frac{dz'}{\tilde{p}_k} \frac{\partial}{\partial x} \int_{z_3}^{z'} \tilde{p}_k dz'' \right) \right]. \tag{A3.68}$$

where $\tilde{p}_k = \left[-k_0^2 U(z,x) - \left(k^2 + 2k \frac{d\Phi}{dx} \right) \right]^{1/2}$, z_3 is the turning point in the interlayer duct (or the coordinate of the barrier crest, $z_3 = z_m$, in the case of above barrier transmission).

Using Eqs. (A3.64) and (A3.68), we obtain the field of the wave trapped in the duct II after the reflection from the barrier

$$A = \int C(k) d_k(x_1) r_{\tilde{k}}(x_2) \exp\left[i \left(kx + \int_{z_i}^{z_2} p dz + \Phi(x) \right.\right.$$

$$\left.\left. + \int_{z_3}^{z} \tilde{p}_k dz - \int_{z_3}^{z} \frac{dz'}{\tilde{p}_k} \frac{\partial}{\partial x} \int_{z_3}^{z'} \tilde{p}_k dz'' \right) \right]. \tag{A3.69}$$

Let us evaluate, by using Eq. (A3.63), the range of the wave numbers, Δk, determining for the capture. We also estimate the role of the wave effects. To this end, we separate the field A into two parts: the "geometrical – optics" part A_1 with the transmission factor $d_k(x_1) > 1/\sqrt{2}$ and the "wave" part A_2 with $d_k(x_1) < 1/\sqrt{2}$. Let us find the relation between the respective energy fluxes penetrating into the duct II. We assume that the coefficients $C(k)$ in Eq. (A3.69) which are determined by the antenna emission [Eq. (A3.61)] sufficiently slowly vary through the wave number range Δk corresponding to the capture region. Wave numbers, for which, on the one hand, the transmission factor d_k [Eq. (A3.63)] is great at the point x_1, and, on the other hand, the reflection factor r_k at the point x_2 (Fig. A3.1) is not negligibly small, contribute mainly to the capture. Hence, we obtain the estimation

$$\Delta k = \frac{k_0}{2} \left[U(z_m(x_2)) - U(z_m(x_1)) + \frac{2}{k_0} \frac{d\Phi}{dx} \right]. \tag{A3.70}$$

Accordingly, for the ratio between the energy fluxes referred to the geometrical-optics, A_1, and wave, A_2, field portions in the duct II, we obtain

$$\frac{A_1}{A_2} \sim \rho_s^2 k_0^2 \left[U(z_m(x_2)) - U(z_m(x_1)) + \frac{2}{k_0} \frac{d\Phi}{dx} \right]. \tag{A3.71}$$

Thus, according to Eqs. (A3.69) and (A3.70), the capture is induced by the increase both in the barrier height along the path, $dU_m(z_m)/dx > 0$, and in the effective duct depth $d\Phi/dx > 0$. Under real ionospheric conditions, both processes lead to the effects of the same order of magnitude. Equations (A3.69) and (A3.70) are in a good agreement with the result obtained in Sec. 3.4. The role of the wave effects becomes essential only for very small longitudinal gradients, $d\Phi/dx \lesssim k_0 \rho_s^2$. However, in the latter case the total wave energy trapped in the duct proves to be less significant. Thus, the capture coefficient into a waveguide duct, G, in a horizontally-inhomogeneous ionosphere is determined essentially by the geometrical-optics effects, see Sec. 3.4.

Principal Symbols

MUF-F or $f_{\text{MUF-}F} = \dfrac{\omega_{\text{MUF-}F}}{2\pi}$	maximum usable frequency for the F-layer
MUF-E or $f_{\text{MUF-}E} = \dfrac{\omega_{\text{MUF-}E}}{2\pi}$	maximum usable frequency for the E-layer
$f = \omega/2\pi$	wave frequency
$f_0 = \omega_0/2\pi$	plasma (Langmuir) frequency
ω_H	gyrofrequency of electrons
$f_s = \omega_s/2\pi$	frequency of sliding ray in the ionosphere
z	height (altitude)
R_0	the Earth's radius
λ	wavelength in medium
e, m	electron charge and mass
c	velocity of light
H	magnetic field intensity
$\mathscr{E}$	electric field intensity
$S(\boldsymbol{r})$	eikonal
E	parameter, characterizing the oscillation level in a duct
N_e	electron density
T_e	electron temperature
$\boldsymbol{k} = (k_z, k_\theta, \lambda_\varphi)$	wave vector
α	angle between a ray and the horizontal
α_0	angle of takeoff (emission)
α_s	angle of elevation at the sliding level
α_1	angle of arrival
r, θ, φ	spherical coordinates
ν_e	electron collision frequency
$\tilde{\varepsilon} = \varepsilon - i\,\dfrac{4\pi\sigma}{\omega}$	complex dielectric constant
ε	real part of $\tilde{\varepsilon}$
$\varepsilon' = \varepsilon + 2z/R_0$	effective dielectric constant
z_s	height of sliding
E, FE, F	designations of ionospheric wave ducts (IWD)
z_{sE} and z_{sF}	height of sliding in the E- and F-ducts
z_1 (z_{1E}, z_{1F})	reflection height of horizontal ray (in the ducts E and F)

$z_{\min}, z_{\max}$	minimum and maximum trajectory heights
Θ	period of ray oscillations
$\Theta_E, \Theta_{FE}, \Theta_F$	oscillation period of ray with the maximum invariants in the ducts *E, FE, F*
I	ray invariant
I_m	maximum invariant of a duct, or invariant of upper ray
$I_1, I_E^I, I_{FE}^I, I_F^I$	invariant of horizontal ray and invariants of rays in the ducts *E, FE, F*
$I_E, I_{FE}, I_F, I_{E+FE}$	maximum invariants of the ducts *E, FE, F,* and total invariant of the ducts *E* and *FE*
Γ	absorption of the ray over half oscillation
$\Gamma_1, \Gamma_E, \Delta_{FE}, \Gamma_F$	absorption of horizontal ray in the ducts *E, FE, F*
K	absorption averaged over oscillation
$K_s, K_1, K_E, K_{FF}, K_F$	absorption of sliding ray, horizontal ray, in the ducts *E, FE, F*
A	total absorption over very long-distance path
A_{RTW}	total absorption of around-the-world signal
$A_s, A_1, A_E, A_{FE}, A_F$	absorption of sliding ray, horizontal ray, in the ducts *E, FE, F*
$\mathscr{P}$	oscillation number of a ray over a path
n	integer number of oscillations over a path, or the mode number
F^h, E^h, F^hF^r, etc.	simple communication channels of a path (CCP) with hop (h), ricochet (r) and combined trajectories (see Sec. 5.2)
$E^hF^r, E^h(FE), E^h(FE)E^h$, etc.	complex CCP with combined trajectories (see Sec. 5.2)
$I^m = I_{\max}$	maximum invariant of CCP
$I_{\min}$	minimum invariant of CCP
α_{0n}, α_{1n}	angle of takeoff and arrival of $n-$th mode
I_{0n}	invariant of $n-$th mode
μ	adiabaticity parameter
$U(z) = -\varepsilon'$	effective potential
p	penetrance of barrier
G	capture coefficient, G_0 – total capture coefficient (with respect to angles)
Q	output coefficient
W	total trapped or transmitted power
$D(\alpha, \varphi)$	antenna directivity diagram with respect to vertical emission angles, (α) and azimuthal ones, (φ)
$f(\theta, z, \varphi;\ \alpha, \beta)$	distribution function of radio wave beam

	intensity with respect to azimuthal (β) and vertical (α) angles at a point (θ, z, φ)
α^m	maximum angle of takeoff
α_E^m, α_F^m	maximum angles of capture into the ducts E and F
α_{FE}	angle of capture into the duct FE
α_E, α_F	angles of capture into the ducts E and F
β^m	maximum angle of capture for a transmitter located in the ionosphere
β_E, β_{FE}, β_F	angles of capture β^m for the ducts E, FE, F
γ	turning angle
γ_E, γ_{FE}, γ_F	turning angles for the ducts E, FE, and F
S	distance along the Earth's surface
D_H, I_H	magnetic declination and dip angle
γ_H	angle between the magnetic field $\boldsymbol{H}$ and the vertical
D	diffusion coefficient
D_1	nonlinear diffusion coefficient
a	irregularity scale
$a_\parallel$, $a_\perp$, $q = a_\perp/a_\parallel$	longitudinal and transverse irregularity scales and anisotropy coefficients
$\overline{(\Delta\varepsilon)^2}$, $\overline{(\Delta N_e)^2}$	square of fluctuations of dielectric constant and electron density
θ_{sc}	angle of scattering
ρ_ε	spectral density of dielectric constant fluctuations
σ	scattering cross-section
$\boldsymbol{k}_{sc}$	wave vector of scattered energy
R_ε	correlation function of irregularities in a medium with dielectric constant ε
$\boldsymbol{A}$	vector potential
j	current density
$\Delta\varepsilon_n$	perturbation of ε under nonlinear action
$\mathscr{E}_p$	plasma field
Π	Poynting's vector
τ_N	density relaxation time
t_{RTW}	group time delay of around-the-world signals
Ψ	angular distance along a radio path
$\varkappa$	absorption factor
$v_{gr} = d\omega/dk$	group velocity of a signal
β^α	angle between the ray and the horizontal in a wave duct
β_{s0}^α	maximum angle of capture β^α into wave duct
φ^*	azimuth of capture

β	azimuth of transmission or arrival, read off clockwise from the north
$\beta^F_{\min}$	optimum azimuth of minimum absorption
$\Gamma(\theta, \boldsymbol{R}, \rho)$	coherence function of a random field
Ψ_n	wave functions
χ	divergence angle of beam
w	linear density of energy
Φ	ray oscillation phase
x_1, x_2, x_3	curvilinear coordinates
h_1, h_2, h_3	Lame coefficients
L	McIlwain coordinate
f_0, f_x	frequencies of ordinary and extraordinary waves
N_m, z_m, y_m	electron density in the maximum of parabolic layer, height and semithickness of the layer
z_1, z_2, z_V	altitudes of electron density maximum in the *E*- and *F*-layers, and the height of minimum electron density between the *E*- and *F*-layers
N_1, N_2, N_V	electron density at heights z_1, z_2, z_V

List of Abbreviations

adiab	adiabatic
CCP	Communication Channel of the Path
cmp	complex
EMI	Equinoctial Model of the Ionosphere
h	hop
IWD	Ionospheric Wave Duct
LDOS	Long-Distance Oblique Sounding
LOF	Lowest Observed Frequency
LT	Local Time
LUF	Lowest Usable Frequency
MOF	Maximum Observed Frequency
m.p.	middle point of the path
MUF	Maximum Usable Frequency
OS	Oblique Sounding
r	ricochet
RTW	Round-the-World
SA	Solar Activity
SFCW	Sweep Frequency Continuous Wave
smp	simple
s	sliding
TID	Travelling Ionospheric Disturbances
tr	tracing

References

Abramovich BS (1976) Spatial instability of a plane wave during nonlocal heating of magnetoactive plasma. Izv Vyssh Uchebn Zaved Radiofiz 19:329 (in Russian)

Afraimovich EL (1982) Interference methods for radio sounding of the ionosphere. Nauka, Moscow (in Russian)

Agaryshev AI (1981) Features of very long-range propagation of decameter radio waves in the high solar activity period. Geomagn Aeron 21:472 (in Russian)

Agaryshev AI, Unuchkov VE (1977) On the random and regular component of short wave azimuth variation in the case of very long-range propagation. Izv Vyssh Uchebn Zaved Radiofiz 20:1108 (in Russian)

Agaryshev AI, Unuchkov VE (1975) Application of the phase radio direction finder for the study on the arrival angles of round-the-world signals. Geomagn Aeron 15:754 (in Russian)

Agy V, Davies K, Salaman R (1959) An Atlas of oblique-incidence ionograms. NBS Techn Note, No 31

Al'pert YaL (1974) Radio wave propagation and the ionosphere. Consultants Bureau, New York

Al'pert YaL, Vitshas LN, Krasnushkina VI, Sinel'nikov VM (1971) On the results of investigation of local and integral ionospheric electron density by means of coherent radio waves transmitted from an artificial satellite, part II. Geomagn Aeron 11:595

Anufrieva TA, Shapiro BS (1976) Geometrical parameters of the ionospheric F_2 layer. Nauka, Moscow (in Russian)

Appleton EV (1959) The normal E region of the ionosphere. Proc IRE 47:155

Bailey DR (1958) The Effect of Echo on the Operation of High-Frequency Communication Circuits. IRE Trans Anten Prop, p 325

Banks PM (1966) Collision frequencies and energy transfer. Electrons. Planet Space Sci 14:1085

Banks PM (1965) Measurements of Antipodal High-Frequency Radio Signals. J Geophys Res 70:625

Banks PM (1972) Behaviour of thermal plasma in the magnetosphere and topside ionosphere. In: Critical problems of magnetospheric physics, Collection of Papers, Madrid, p 157

Baranov VA, Popov AV (1975) The perturbation method for rays in quasi-stratified medium. In: Propagation of decameter radio waves, Collection of Papers, IZMIRAN, Moscow, p 14 (in Russian)

Baranov VA, Egorov IB, Popov AV (1977) Calculation of the antipodal and around-the-world focusing in a horizontally-inhomogeneous ionospheric waveguide. In: Diffraction effects of decameter radio waves in the ionosphere, Collection of Papers, Nauka, Moscow, p 31 (in Russian)

Baranov VA, Egorov IB, Popov AV (1977) Radio wave focusing in the case of global propagation in an ionospheric waveguide. In: Theory of diffraction and radio wave propagation, Collection of Papers, Nauka, Moscow, p 196 (in Russian)

Baranov VA, Popov AV, Cherkashin YuN (1977) The effect of scattering by large-scale irregularities on radio wave capture into ionospheric waveguide channel. In: Diffraction effects of decameter radio waves in the ionosphere, Collection of Papers, Nauka, Moscow, p 43 (in Russian)

Barker JI, Grossi MD (1970) Results of the OV4-1 dual satellite experiment on guided ionospheric propagation. Radio Sci 5:983

Barry GH (1974) HF-VHF communications experiments using man-made field aligned ionospheric scatters. Radio Sci 9:1025

Bates HF, Albee PR (1966) On the strong influence of the F_1 layer on medium to high latitude HF propagation. UAGR-175, March 1966

Baz' AI, Sel'dovich YaB, Perelomov AM (1971) The scattering, reactions and decays in non-relativistic quantum mechanics. Nauka, Moscow (in Russian)

Belikovich VV, Benediktov EA, Getmantsev GG, Erukhimov LM, Zuikov NA, Komrakov GP, Korobkov YuS, Kotik DS, Mityakov NA, Rapoport VO, Sazonov YuA, Trakhtengerts VYu, Frolov VL, Cherepovitskii VA (1975) Nonlinear phenomena in upper ionosphere. Sov Phys Usp 17:615

Belikovich VV, Ben'kova NP, Bryantsev VF, Bukin GV, Matyugin SN, Protashchik AA, Cherepovitskii VA (1976) Measurements results of the HF-signal arrival angles on the paths between Moscow and the receivers located in a region of the Indian and Atlantic oceans. Geomagn Aeron 16:841

Belyaev ST (1958) Kinetic equation for rare gases in the strong fields. In: Plasma physics and a problem of controlled thermonuclear reactions, Nauka, Moscow, vol 3, p 50 (in Russian)

Benediktov EA, Ben'kova NP, Bukin GV, Getmantsev GG, Yefimova TV, Komrakov GP, Malyshev SK, Mityakov NA, Perekhvatov YuK, Skrebkova LA, Shilkov NP (1971) An experiment of the HF wave reception at the magnetic conjugate point. Geomagn Aeron 11:252

Benediktov EA, Ben'kova NP, Bukin GV, Matyugin SN, Mitaykov NA, Perekhvatov YuK (1974) Investigation of the anomalous propagation modes of HF pulse signals on transequatorial paths. IV. Some results of reception of the reverse and round-the-world echo-signals. In: Problems of short radio wave propagation, Collection of Papers, part II, IZMIRAN, Moscow, p 17 (in Russian)

Ben'kova NP, Bukin GV, Gotsakova LS, Yashin YuYa (1978) On the mechanisms of magnetospheric HF propagation. Geomagn Aeron 18:448

Ben'kova NP, Bukin GV, Gotsakova LS, Kolokolov LE, Shubin VN (1980) Influence of ionospheric conditions on the magnetospheric HF propagation. Geomagn Aeron 20:1014 (in Russian)

Berdeyans D, Bocharov VI, Lobachevskii LA, Martinez R, Suarez I, Tushentsova IA (1975) Types of propagation of radio waves of decameter range, according to observations by the OBS method on Cuba – Soviet Union paths. Geomagn Aeron 15:135

Bochkarev GS, Kim VYu, Lobachevskii LA, Lyannoy BE, Migulin VV, Sergeenko OS, Cherkashin YuN (1979) Interaction of decameter radio waves on the frequencies close to the MUF-F_2 at oblique propagation. Geomagn Aeron 19:830 (in Russian)

Boldovskaya IG, Egorov IB, Kiyanovskii MP (1971) Computational program of ray trajectory in the isotropic spherically-stratified parabolic ionosphere. In: Ray approximation and problems of radio wave propagation. Nauka, Moscow, p 280 (in Russian)

Booker HG (1956) A theory of scattering by nonisotropic irregularities with application to radar reflections from the aurora. J Atmos Terr Phys 8:204

Booker HG (1962) Guidance of radio and hydromagnetic waves in the magnetosphere. J Geophys Res 67:4135

Borisov ND (1975) A representation of the electromagnetic wave field in a spherically-symmetric ionosphere. Geomagn Aeron 15:1086 (in Russian)

Borisov ND (1978) The effect of large-scale irregularities on waveguide propagation of short radio waves. Geomagn Aeron 18:267 (in Russian)

Borisov ND (1980) Wave approach to the problem of short wave scattering in irregular ionospheric waveguides. Izv Vyssh Uchebn Zaved Radiofiz 23:1147 (in Russian)

Borisov ND, Gurevich AV (1975) Distribution of the short radio wave field in ionospheric wave ducts. In: Investigation of very long-range propagation of short radio waves, Collection of Papers, Nauka, Moscow, p 3 (in Russian)

Borisov ND, Gurevich AV (1975) On the theory of short radio wave propagation in the ionospheric waveguide channels. In: Proc XI All-Union Conference on Radio Wave Prop, vol 4, Kazan' University Press, Kazan', p 79 (in Russian)

Borisov ND, Gurevich AV (1976a) Adiabatic modes of short radio waves in horizontally-inhomogeneous ionosphere. Preprint, IZMIRAN, No 2 (145), Moscow (in Russian)

Borisov ND, Gurevich AV (1976b) On the theory of short radio wave propagation in horizontally-inhomogeneous ionosphere. Izv Vyssh Uchebn Zaved Radiofiz 19:1275 (in Russian)
Borisov ND, Gurevich AV (1977a) Adiabatic modes of short radio waves in horizontally-inhomogeneous ionosphere. Izv Vyssh Uchebn Zaved Radiofiz 20:1876 (in Russian)
Borisov ND, Gurevich AV (1977b) The wave theory of HF propagation in horizontally-inhomogeneous ionosphere, and the wave theory of HF capture into ionospheric ducts. In: Very long-range radio wave propagation and ionospheric models, Collection of Papers, IZMIRAN, Moscow, p 13 (in Russian)
Borisov ND, Gurevich AV (1977c) Excitation of ionospheric waveguides by ground-based sources. Izv Vyssh Uchebn Zaved Radiofiz 20:1876 (in Russian)
Borisov ND, Gurevich AV (1980) Comparison between the waveguide propagation calculation in the adiabatic approximation, and the analytical solution. In: Interaction between HF/VHF radio waves and the ionosphere, Collection of Papers, IZMIRAN, Moscow, p 76 (in Russian)
Borisov ND, Vas'kov VV, Gurevich AV (1976) Excitation of drift instability under the action of radio waves at the *F*-layer of the ionosphere. Geomagn Aeron 16:369
Borisov ND, Fishchuk DI, Tsedilina EE (1977) Influence of inhomogeneous ionospheric structure on the group time delay of around-the-world signals. In: Very long-range radio wave propagation and ionospheric models, Collection of Papers, IZMIRAN, Moscow, p 90 (in Russian)
Borisov ND, Kim VYu, Tsedilina EE (1977) Capture of radio waves emitted from the Earth into an ionospheric wave duct and their escape from the duct. In: Very long-range radio wave propagation and ionospheric models. Collection of Papers, IZMIRAN, Moscow, p 4 (in Russian)
Borisov ND, Vas'kov VV, Gurevich AV (1977) Short-waved drift-dissipative instability. Fiz Plazmy 3:168 (in Russian)
Borisov ND, Fishchuk DI, Tsedilina EE (1978) Frequency dependence of group time delay of around-the-world signals. Izv Vyssh Uchebn Zaved, Radiofiz 21:372 (in Russian)
Borisov ND, Vas'kov VV, Gurevich AV (1978) Self-focusing instabilities of radio waves at oblique sounding of the ionosphere. Geomagn Aeron 18:827 (in Russian)
Born M, Wolf E (1968) Principles of Optics. Pergamon, New York
Borowski S, Jarkowski S, Manczarski S (1956) Prace instituty LACZNOSCI, III, ZESZYT 1:52
Brekhovskikh LM (1957) Waves in stratified media. Academy of sciences of the USSR Press, Moscow (in Russian)
Bremmer H (1946) Terrestrial Radio Waves, Theory of Propagation. Elsevier, Amsterdam
Bubenik DM (1975) Some results of one year of continuous around-the-world HF soundings over a transpolar path. In: Proc of American Fall meeting of URSI, Los Angeles, Sept
Bubenik DM, Fraser-Smith AC, Sweeney LE, Villard OG Jr (1971) Observations of time structure in high-frequency radio signals propagated around-the-world. J Geophys Res 76:1088
Bubnov VA, Rumyantsev GA (1975) Propagation of round-the-world and reverse echo-signals outside of the great circle plane. Izv Vyssh Uchebn Zaved, Radiofiz 18:1383 (in Russian)
Bubnov VA, Golyan SF, Shlionskii AG (1975) Features of attenuation of the very long-distance signals. In: Ionospheric propagation of short radio waves, Collection of Papers, IZMIRAN, Moscow, p 90 (in Russian)
Budden KG (1961) Radio Waves in the Ionosphere. Cambridge University Press, Cambridge
Budden KG (1980) The theory of radio windows in the ionosphere and magnetosphere. J Atmos Terr Phys 42:287
Bukin GV (1978) Experimental investigation of the magnetospheric HF propagation on the Earth-Earth path. Geomagn Aeron 18:440 (in Russian)
Bukin GV, Molchanov OA (1976) A comparison between experimental results on magnetospheric propagation of HF, and VLF radio waves. Geomagn Aeron 16:287
Cain IC, Hendricks SI, Langel RA, Hudson WV (1967) A proposed model for international geomagnetic reference Field-1965. J Geomagn Geoelectr 19:335
Calvert W (1981) The role of wave ducting in magnetospheric measurements. Rep at the URSI congress, Washington
Carpenter GB (1974) VHF and UHF bistatic observation of a region of the ionosphere modified by a high power radio transmitter. Radio Sci 9:965

Carrara N, De Giorgio MT, Pellegrini PF (1970) Guided propagation of HF radio waves in the ionosphere. Space Sci Rev 11:555
Chapman S (1953) The geometry of radio echoes from aurora. J Atoms Terr Phys 3:1
Chernii BF (1962) Radio wave propagation. Sov Radio, Moscow (in Russian)
Chernov LP (1958) Wave propagation in the medium with random irregularities. Academy of sciences of the USSR Press, Moscow (in Russian)
Ching BK, Chiu YT (1973) A phenomenological model of global ionospheric electron density in the E, F_1 and F_2 regions. J Atoms Terr Phys 35:1615
Chiu AT (1975) An improved phenomenological model of ionospheric electron density. J Atoms Terr Phys 37:1563
Crawford FW, Sears DM, Bruce RL (1970) Possible observations and mechanism of very long delayed radio echoes. J Geophys Res 75:7326
Danilin VA (1967) Ionospheric waveguide channels for distant radio communication below the maximum of the F_2 layer. Cosmic Res 5:199
Danilin VA (1968) Averaged field in an ionospheric waveguide channel in the case of radio communication with an artificial satellite. Geomagn Aeron 8:840 (in Russian)
Davies K (1969) Ionospheric radio waves. Blaisdell Publishing Company, Massachusetts
Dimant YaS (1977) Excitation of dissipative parametric instability in ionospheric F-region. Geomagn Aeron 17:528 (in Russian)
Dolin LS (1968) Equations for correlation functions of wave beam in a randomly-inhomogeneous medium. Izv Vyssh Uchebn Zaved Radiofiz 11:840 (in Russian)
Dolukhanov MP (1972) Radio wave propagation. Svyazizdat, Moscow (in Russian)
Drobzhev VI, Kudelin GM, Nurgozhin VI, Pelenitsin GM, Rudina MP, Troitskii BV, Ynkovets AF (1975) Wave disturbances in the ionosphere. Alma-Ata (in Russian)
Dubois DF, Goldman MV (1979) Spectrum and anomalous resistivity for saturated parametric instability. Phys Rev Lett 28:218
Eckersley TL (1927) Short-Wave Wireless Telegraphy JIEE 65:600
Egorov IB, Egorova VN (1976) Numerical simulation of long-distance HF signal propagation. In: Propagation of decameter radio waves, Collection of Papers, Nauka, Moscow, p 100 (in Russian)
Egorova VN, Egorov IB, Manaenkova NI, Fishchuk DI, Tushentsova IA, Tsedilina EE (1980a) Comparison between the adiabatic calculations of long-distance HF path and the ray-tracing one. I. In: Interaction between HF/VHF radio waves and the ionosphere, Collection of Papers, IZMIRAN, Moscow, p 88 (in Russian)
Egorova VN, Egorov IB, Manaenkova NI, Fishchuck DI, Tushentsova IA, Tsedilina EE (1980b) Comparison between the adiabatic calculations of long-distance HF path and the ray-tracing one. II. In: Interaction between HF/VHF radio waves and the ionosphere, Collection of Papers, IZMIRAN, Moscow, p 100
Erukhimov LM, Mityakova EE, Myasnikov EN, Polyakov SV, Rakhlin AV, Sinel'nikov VM (1977) On the spectrum of artificial ionospheric irregularities at various heights. Izv Vyssh Uchebn Zaved Radiofiz 20:1814
Erukhimov LM, Metelev SA, Mityakova EE, Myasnikov EN, Rakhlin AV, Uryadov VP, Frolov VL (1979) Experimental investigations of the artificial ionospheric turbulence. In: Thermal nonlinear phenomena in plasmas, Collection of Papers, IPF AN SSSR, Gorki, p 7 (in Russian)
Evans JV (1970) Millstone Hill Thompson scatter results for 1965. Planet Space Sci 18:1225
Fatkullin ZN, Zelenova GI, Kozlov VK, Legen'ka AD, Soboleva TN (1981) Empirical models of midlatitude ionosphere. Nauka, Moscow (in Russian)
Fejer JA (1979) Ionospheric modifications and parametric instabilities. Rev Geophys Space Phys 17:135
Fejer JA, Leer E (1972) Excitation of parametric instabilities by radio waves in the ionosphere. Radio Sci 7:481
Fejer JA, Rinnert K, Woodman RF (1978) Detection of stimulated Brillouin scattering by the Tacamarca radar. J Geophys Res 83:2133
Fenwick RB (1963) Round-the-World High-Frequency Propagation. Techn Rep 71, Radioscience Laboratory, Stanford University, Stanford

Fenwick RB (1966) Sweep-Frequency, Spaced-Station Measurements of Round-the-World HF Propagation. Techn Rep 122, Radioscience Laboratory, Stanford University, Stanford

Fenwick RB, Villard OG Jr (1963a) A test of the importance of ionosphere-ionosphere reflections in long distance and around-the-world high frequency propagation. J Geophys Res 68:5659; A Test of the reality and practical importance of ionosphere-ionosphere reflections in long distance HF Propagation. Techn Rep 76, Stanford Electr Lab, California, Stanford

Fenwick RB, Villard OG Jr (1963b) The Effect of Magnetic Storms on Around-the-World High-Frequency Propagation. J Geophys Res 68:4683

Fenwick RB, Villard OG Jr (1963c) Measurements of the frequency dependence of Round-the-World HF pulse time delays and dispersions. Proc IEEE 51:1240

Fialer PA (1974) Field-aligned Scattering from a heated region of the ionosphere – observations at HF and VHF. Radio Sci 9:923

Filipp ND (1974) Radio wave scattering in anisotropic ionosphere. Shtiintsa Press, Kishiniov (in Russian)

Fishchuk DI (1974) Algol-program for computation of distribution of the effective electron collision frequency in the ionosphere. IZMIRAN, Moscow (in Russian)

Fishchuk DI (1975) Generalized algorithm for computations in Algol of the fundamental parameters of very long-range paths. IZMIRAN, Moscow (in Russian)

Fishchuk DI, Tsedilina EE (1975a) A new method of calculation of the very long-distance radio paths. In: Investigation of very long-range propagation of short radio waves, Collection of Papers, IZMIRAN, Moscow, p 42 (in Russian)

Fishchuk DI, Tsedilina EE (1975b) Maps of the global distribution of some parameters of ionospheric wave ducts, part II. In: Investigation of very long-range propagation of short radio waves, Collection of Papers, IZMIRAN, Moscow, p 33 (in Russian)

Fishchuk DI, Tsedilina EE (1975c) Capture angles of short radio waves propagating in ionospheric wave ducts. In: Investigation of very long-range propagation of short radio waves, Collection of Papers, IZMIRAN, Moscow, p 23 (in Russian)

Fligel' MD (1962) Geographical distribution of ionospheric absorption. Geomagn Aeron 2:1091 (in Russian)

Fligel' MD (1963) Geographical distribution of radio wave absorption in the ionosphere. Thesis, IZMIRAN, Moscow (in Russian)

Fok VA (1970) Problems of diffraction and electromagnetic wave propagation. Sov Radio, Moscow (in Russian)

Francis SH (1975) Global propagation of atmospheric gravity waves (review). J Atmos Terr Phys 37:1011

Freiman ME (1975) The adiabatic capture of radio waves into an ionospheric wave duct. In: Investigation problems of the lower ionosphere and geomagnetism, Novosibirsk University Press, Novosibirsk, p 5 (in Russian)

Furutsu K (1951) Calculation of field intensity around the antipode of the transmitting station. Rep Ionosph Res Japan 5:159

Gayvoronskaya TV, Soboleva TN, Tushentsova IA, Tsedilina EE (1973) Three-dimensional distribution model of effective electron collision frequency, and diurnal variation of midlatitude profiles in quiet ionosphere. Preprint, IZMIRAN, No 9 (in Russian)

Gayvoronskaya TV, Soboleva TN, Tushentsova IA, Tsedilina EE (1974) Three-dimensional analytical model of distribution of effective electron collision frequency in quiet ionosphere. Geomagn Aeron 14:25 (in Russian)

Gayvoronskaya TV, Egorova VN, Egorov IB, Fishchuk DI, Tsedilina EE (1975) Comparison between path calculation by the adiabatic invariant method, and standard ray-tracing calculations. In: Investigation of very long-range propagation of short radio waves, Collection of Papers, IZMIRAN, Moscow, p 54 (in Russian)

Gdalevich GL, Ozerov VD, Vsekhsvyatskaya IS, Novikova LN, Soboleva TN (1980) Investigation of ionospheric variability at the height of 500 km according to "Cosmos-900" satellite data. Geomagn Aeron 20:809 (in Russian)

Gel'fond OA, Gurevich AV, Tsedilina EE (1982) Calculation of long-distance paths by the adiabatic method with a parabolic ionospheric model. Geomagn Aeron 22:953 (in Russian)

George PL (1971) The global morphology of the quantity $\int N\nu dh$ in the *D*- and *E*-regions of the ionosphere. J Atmos Terr Phys 33:1893
Gershman BN (1974) Dynamics of ionospheric plasma. Nauka, Moscow (in Russian)
Gerson NC (1972) Ray tracing over a transequatorial path. Can J Phys 50:976
Gerson NC, Hengen IG, Pipp RM, Webster JB (1969) Radio Wave propagation to the antipode. Can J Phys 47:2143
Getmantsev GG, Erukhimov LM (1969) Radio star and satellite scintillation. In: Annals of the IQSY, Cambridge University Press, Cambridge, p 229
Getmantsev GG, Komrakov GP, Korobkov PP, Mironenko LF, Mityakov NA, Rapoport VO, Trakhtengerts, VYu, Frolov VL, Cherepovitskii VA (1973) Some results of investigations of nonlinear phenomena in the *F*-layer of the ionosphere. JETP Lett 18:364
Ginzburg VL (1970) Propagation of Electromagnetic Waves in Plasmas. Pergamon, New York
Ginzburg VL, Gurevich AV (1960) Nonlinear phenomena in a plasma located in an alternating electric field. Sov Phys Usp 3:115, 175
Golyan SF (1972) Analysis of results of the field strength measurements on HF radiolink Moscow-Molodiozhnaya (Antarctic continent). In: Proc X All-Union Conference on Radio Wave Prop, Section 1, Irkutsk, p 287 (in Russian)
Golyan SF (1973) Experimental investigations of the amplitude fluctuations of HF signals. In: Problems of short radio wave propagation, Collection of Papers, part I, Nauka, Moscow, p 78 (in Russian)
Golyan SF (1974) On the optimum conditions for long-distance propagation of short radio waves. In: Problems of short radio wave propagation, Collection of Papers, part II, Nauka, Moscow, p 30 (in Russian)
Golyan SF (1975) Optimal conditions for extra-long-range short-wave propagation. Radiophys Quant Electron 18:1044
Golyan SF, Shlionskii ShG (1971) Propagation of circumterrestrial echo signals. Geomagn Aeron 11:76
Golyan SF, Kerblai TS, Kovalevskaya EM (1975) Trajectories of the direct and return signals over the Moscow-Molodiozhnaya path. In: Ionospheric propagation of short radio waves, Collection of Papers, IZMIRAN, Moscow, p 20 (in Russian)
Golyan SF, Tushentsova IA, Tsedilina EE (1975) Absorption on the Moscow-Molodiozhnaya path. In: Investigation of very long-range propagation of short radio waves, Collection of Papers, IZMIRAN, Moscow, p 135 (in Russian)
Golyan SF, Israetel AG, Lobachevskii LA, Nekrasov BYu, Smirnov VB, Suares JB, Tushentsova IA, Fishchuk DI, Tsedilina EE, Shumilov IA (1981) An investigation of multipath HF propagation over two long-distance paths. In: Diffraction effects of short radio waves, Collection of Papers, IZMIRAN, Moscow, p 24 (in Russian)
Gordon WE, Carlson HC (1974) Arecibo heating experiments. Radio Sci 9:1041
Grach SM, Trakhtengerts VYu (1975) Parametric excitation of ionospheric irregularities extended along the magnetic field. Izv Vyssh Uchebn Zaved Radiofiz 18:1288 (in Russian)
Grach SM, Mityakov NA, Rapoport VO, Trakhtengerts VYu (1979) Thermal parametric instability in ionospheric plasma. In: Thermal nonlinear phenomena in plasmas, Collection of Papers, IPF AN USSR, Gorki, p 46 (in Russian)
Grinberg GA (1943) On the radio wave propagation above the Earth's surface with allowance for its sphericity and atmospheric inhomogeneity. Izv Akad Nauk USSR 7:99 (in Russian)
Grossi MD, Langworthy BM (1966) Geometric optics investigation of HF and VHF guided propagation in the ionospheric whispering gallery. Radio Sci 1:877
Grosskopf J, Vogt K (1941) Einfallswinkelmessungen im Kurzwellenbericht. TFT 30:19
Gurevich AV (1971) Effect of non-linearity on the generation of signals circling the Earth. Geomagn Aeron 11:810
Gurevich AV (1972) Self-action of radio waves in the ionospheric layers *E* and F_1. Geomagn Aeron 12:24 (in Russian)
Gurevich AV (1978) Nonlinear phenomena in the ionosphere. Springer, Berlin Heidelberg New York

Gurevich AV (1979) Electromagnetic wave guiding in weak waveguide channels in the ionosphere and magnetosphere. Geomagn Aeron 19:462 (in Russian)

Gurevich AV, Borisov ND (1982) Variation of adiabatic invariant due to the large-scale irregularities. In: Interaction between short radio waves and the ionosphere, Collection of Papers, IZMIRAN, Moscow, p 66 (in Russian)

Gurevich AV, Kim VYu (1976) The effect of scattering by the magnetic-field-aligned irregularities of the ionosphere on radio wave capture into an ionospheric wave duct. In: Propagation of decameter radio waves, Collection of Papers, Nauka, Moscow, p 5 (in Russian)

Gurevich AV, Tsedilina EE (1973) To the theory of very long-range propagation of short radio waves. Geomagn Aeron 13:283 (in Russian)

Gurevich AV, Tsedilina EE (1975) The distribution of energy trapped in an ionospheric wave duct. In: Propagation of decameter radio waves, Collection of Papers, IZMIRAN, Moscow, p 5 (in Russian)

Gurevich AV, Tsedilina EE (1975) Radio wave capture into an interlayer ionospheric wave duct. In: Propagation of decameter radio waves, Collection of Papers, IZMIRAN, Moscow, p10 (in Russian)

Gurevich AV, Tsedilina EE (1975a) Theoretical investigation of very long-range propagation of short radio waves. Sov Phys Usp 115:540 (in Russian)

Gurevich AV, Tsedilina EE (1975b) Capture of radio waves into an ionospheric duct during scattering by artificial irregularities. Geomagn Aeron 15:1005 (in Russian)

Gurevich AV, Tsedilina EE (1976) Scattering of radio waves in ionospheric wave ducts. Izv Vyssh Uchebn Zaved, Radiofiz 19:43 (in Russian)

Gurevich AV, Tsedilina EE (1977) Lateral deviations of radio waves over long-distance paths. In: Very long-range radio wave propagation and ionospheric models, Collection of Papers, IZMIRAN, Moscow, p 104 (in Russian)

Gurevich AV, Tsedilina EE (1979a) Very long-distance propagation of high frequency radio waves. Nauka, Moscow (in Russian)

Gurevich AV, Tsedilina EE (1979b) The effect of multiple scattering by the irregularities in the case of long-range HF propagation in wave ducts. Geomagn Aeron 19:53 (in Russian)

Gurevich AV, Tsedilina EE (1980) Radio wave capture into an ionospheric duct during multiple scattering. In: Interaction between HF/VHF radio waves and the ionosphere, Collection of Papers, IZMIRAN, Moscow, p 50 (in Russian)

Gurevich AV, Tsedilina EE (1983) Electric field amplitude of short radio waves in the inhomogeneous ionosphere. Geomagn Aeron 23:764 (in Russian)

Gurevich AV, Tsedilina EE (1985) (to be published) Fluctuations and broadenings of the short wave bearing on long-distance paths. Izv Vyssh Uchebn Zaved 28 (in Russian)

Gurevich AV, Fishchuk DI, Tsedilina EE (1972) Three-dimensional analytical model of electron concentration distribution in quiet ionosphere. Preprint, IZMIRAN, No 36 (in Russian)

Gurevich AV, Fishchuk DI, Tsedilina EE (1973) Three-dimensional analytical model of electron concentration distribution in quiet ionosphere. Geomagn Aeron 13:31 (in Russian)

Gurevich AV, Erukhimov LA, Kim VYu, Uryadov VP, Tsedilina EE (1975) The effect of scattering on radio wave capture into ionospheric wave ducts. Izv Vyssh Uchebn Zaved Radiofiz 18:1305 (in Russian)

Gurevich AV, Krylov AL, Tsedilina EE (1976) Electric fields in the Earth's Magnetosphere and Ionosphere. Space Sci Rev 19:59

Gurevich AV, Kim VYu, Tsedilina EE (1977) The effect of nonlinear scattering on radio wave capture into the interlayer ionospheric duct, and on radio wave departure from the duct. In: Very long-range radio wave propagation and ionospheric models, Collection of Papers, IZMIRAN, Moscow, p 30 (in Russian)

Gurevich AV, Israetel AG, Soboleva TN, Tushentsova IA, Fishchuk DI, Tsedilina EE (1981) Global analytical equinoctial model of electron concentration of the ionosphere (EMI-81). Preprint IZMIRAN No 49a, No 49(362)

Guseva ZG, Tyurmina LO (1976) Instruction for use of the algorithms and programs of spherical harmonic analysis. Nauka, Moscow (in Russian)

Harnischmacher E (1960) A calculation method of ionospheric propagation conditions for very high and antipode distance. In: Electromagnetic Wave Propagation, Academic, New York, p 527

Haselgrove J (1955) Ray theory and a new method for Ray-Tracing. Proc Cambridge Conf Phys Ionosphere, The Phys Society, London, p 355

Helliwell RA (1965) Whistlers and related ionospheric phenomena. Stanford University Press, Stanford

Herbert T (1967) Tables of virtual height for models of monotonic ionospheric layers. Radio Sci 2:1269

Hess HA (1948) Investigation of High-Frequency Echoes. Proc IRE 36:981

Hess HA (1949) Part II – Investigation of High-Frequency Echoes. Proc IRE 37:986

Hess HA (1952) Part III – Investigation of High-Frequency Echoes. Proc IRE 40:1065

Ionospheric predictions. US Department of Commerce 1982

Israetel AG (1982) Solar activity dependence of analytical model for effective electron collision frequency. In: Interaction between short radio waves and the ionosphere, Collection of Papers, IZMIRAN, Moscow, p 145 (in Russian)

Israetel AG, Tsedilina EE (1982) Solar activity dependence of equinoctial analytical model for ionospheric electron density. Geomagn Aeron 22:738 (in Russian)

Israetel AG, Lobachevskii LA, Manaenkova NI, Tushentsova IA, Fishchuk DI, Tsedilina EE (1982) Model investigations of HF propagation conditions over long-distance paths. In: Proc VI All-Union Simposium on Ionospheric Modelling (Tomsk, June 1982), IZMIRAN, Moscow, p 15 (in Russian)

Isted GA (1958) Round-the-World Echoes. Marconi Rev 21:173

Jacchia LG (1971) Revised static models of the thermosphere and exosphere with empirical temperature profiles. SAO Special Report, No 332

Jones WB, Stewart FG (1970) A numerical method for global mapping of plasma frequency. Radio Sci 5:773

Kadukhin GF, Soboleva TN (1978a) Latitude variations of electron density for the radio wave propagation calculations. In: Propagation of decameter radio waves, Collection of Papers, IZMIRAN, Moscow, p 140 (in Russian)

Kadukhin GF, Soboleva TN (1978b) Empirical model for height-diurnal variations of electron density in the mid-latitude ionosphere at heights of 50 – 900 km under equinoctial conditions during high solar activity. In: Radio wave propagation in the ionosphere, Collection of Papers, IZMIRAN, Moscow, p 137 (in Russian)

Kadukhin GF, Soboleva TN (1978c) Space-time variations of key parameters of electron density height profiles in the unperturbed ionosphere. In: Trajectory characteristics of short radio waves, Collection of Papers, IZMIRAN, Moscow, p 130 (in Russian)

Katsenelenbaum BZ (1961) Theory of irregular waveguides with slowly varying parameters. Nauka, Moscow (in Russian)

Kazantsev AN (1956) Development of the calculation method for electric field strength of short radio waves. Trudy IRE, Radiotekhnika, Collection of Papers No 2, Academy of Sciences of the USSR Press, Moscow, p 134 (in Russian)

Kazantsev AN, Lukin DS (1966) The field strength of short radio waves emitted from an artificial earth satellite. Kosm Issled 4:221, 238 (in Russian)

Kazantsev AN, Lukin DS, Spiridonov YuG (1967) An investigation method of radio wave propagation in inhomogeneous magnetoactive ionosphere. Kosm Issled 5:593 (in Russian)

Kelso JM (1968) Ray-tracing in the Ionosphere. Radio Sci 3:1

Kerblay TS, Kovalevskaya EM (1974) Trajectories of short radio waves in the ionosphere, Nauka, Moscow (in Russian)

Kerblay TS, Lyakhova LN (1969) (eds) Fundamentals of Long-Term Forecasting. Nauka, Moscow (in Russian)

Khodzha-Akhmetov ChL, Koshurko MS (1975) Preliminary results of oblique sounding of the ionosphere over the Moscow – Antarctic Continent path. Proc XI All-Union Conference on Radio Wave Prop, vol 4, Kazan' University Press, Kazan', p 103 (in Russian)

Kim VYu (1974a) A role of scattering in the case of excitation of an ionospheric waveguide. In: Radio wave propagation and plasma instabilities in the ionosphere and magnetosphere. Collection of Papers, IZMIRAN, Moscow, p 3 (in Russian)

Kim VYu (1974b) To the theory of propagation of pulses in an ionospheric waveguide. In: Problems of short radio wave propagation, Collection of Papers, Nauka, Moscow, p 124 (in Russian)

Kim VYu (1975a) Role of the scattering by caustic in the case of excitation of ionospheric wave duct. In: Propagation of decameter radio waves. Collection of Papers, Nauka, Moscow, p 95 (in Russian)

Kim VYu (1975b) On the multimode broadening of signals in an ionospheric waveguide. In: Investigation of very long-range propagation of short radio waves, Collection of Papers, IZMIRAN, Moscow, p 15 (in Russian)

Kim VYu (1977a) Influence of the scattering and Earth's magnetic field on radio wave capture and propagation in ionospheric wave ducts. Thesis, IZMIRAN, Moscow

Kim VYu (1977b) The radio wave capture and departure regions in the case of scattering by artificially modified ionospheric region: problem of radio wave propagation in ionospheric wave ducts. In: Very long-range radio wave propagation and ionospheric models, Collection of Papers, IZMIRAN, Moscow, p 54 (in Russian)

Kim VYu, Fishchuk DI, Tsedilina EE (1975) The effect of magnetic field on long-distance HF propagation. In: Ionospheric propagation of short waves, Collection of Papers, Nauka, Moscow, p 58 (in Russian)

Kinber BE, Komissarova NN, Kravtsov YuA (1978) Ray theory of wave propagation in the inhomogeneous refractive waveguides: drive of the wave-duct width, and mode transformation. Radiotekhn Elektron 21:1219 (in Russian)

Kinber BF, Komissarova NN, Kravtsov YuA (1979) Ray theory of wave propagation in the inhomogeneous refractive waveguides: mode transformation and drive of the wave duct width. Izv Vyssh Uchebn Zaved Radiofiz 22:414 (in Russian)

Kiyanovskii MP (ed) (1971) Ray approximation and problems of radio wave propagation. Collection of translated Papers, Nauka, Moscow (in Russian)

Kiyanovskii MP, Ostrovskii IYa (1972) Some results of observation of round-the-world signals in the eastern part of the USSR. In: Proc X All-Union Conference on Radio Wave Prop, Section 1, Irkutsk p 360 (in Russian)

Kiyanovskii MP, Ostrovskii IYa (1974) On the role of scattering in the propagation of round-the-world signals. In: Actual problems of radio wave propagation, Collection of Papers, Nauka, Moscow, p 162 (in Russian)

Klyatskin VI (1975) Statistical description of the dynamic systems with fluctuating parameters. Nauka, Moscow (in Russian)

Klyatskin VI, Tatarskii VI (1973) The diffusive random process approximation in some non-stationary statistical problems of physics. Usp Fiz Nauk 110:499 (in Russian)

Knecht RW, Van Zandt TE, Russel S (1961) First pulsed radio soundings of the topside of the ionosphere. J Geophys Res 66:3078

Köhnlein W (1977) Electron density models of the ionosphere. Paper presented at Joint Assemb Intern Assoc Geomagn Aeron, Assoc Atmos Phys, Washington Seattle, Aug

Krasnushkin PE (1947) The Normal-Wave Method Applied to Long-Range Radio Communications. Moscow University Press, Moscow (in Russian)

Krasnushkin PE (1948) On the waveguide properties of inhomogeneous media. Zh Tekhn Fiz 18:431 (in Russian)

Krasnushkin PE (1971) Qualitative study of radio wave trajectory in a stratified medium by means of the phase trajectory method. Dokl Akad Nauk USSR 200:1313 (in Russian)

Kravtsov YuA (1965) Modification of the geometrical-optics technique for a wave tunneling through caustic. Izv Vyssh Uchebn Zaved Radiofiz 8:659 (in Russian)

Kravtsov YuA, Rytov SM (1975) Statistical problems in the diffraction theory. Usp Fiz Nauk 115:239 (in Russian)

Kravtsov YuA, Tinin MV, Cherkashin YuN (1979) Possible mechanisms of excitation of ionospheric wave ducts. Geomagn Aeron 19:769 (in Russian)

Kulikova VV, Shashun'kina VM, Yudovich LA (1980) Generation of internal gravity waves in the magnetic storm period. Izv Vyssh Uchebn Zaved Radiofiz 23:1015 (in Russian)

Kunsucker RD, Bates HF (1969) Surwey of polar and auroral region effects on HF propagation. Radio Sci 4:347

Landau LD, Lifshitz EM (1955) The classical theory of field. Pergamon, New York

Landau LD, Lifshitz EM (1976) Mechanics. Pergamon, New York

Landau LD, Lifshitz EM (1977) Quantum mechanics. Pergamon, New York

Landau LD, Lifshitz EM (1984) Electrodynamics of Continuous Media. Pergamon, New York

Llewellyn SK, Bent RB (1973) Documentation and description of the Bent ionospheric model: Report SAMSO TR-73-252, Space and Missile Syst Organ, Los Angeles

Lobachevskii LA, Suares J, Tushentsova IA, Fishchuk DI, Tsedilina EE (1978) Study of propagation conditions for HF radio waves by the adiabatic invariant method on the Nikolaev-Havana path (multifrequency model experiment). Proc XII All-Union Conference on Radio Wave Prop, Nauka, Tomsk, p 147 (in Russian)

Lobachevskii LA, Suares JB, Tushentsova IA, Fishchuk DI, Tsedilina EE (1979a) Adiabatic analysis of decameter wave propagation conditions over a long-distance path at a fixed frequency, part I. Geomagn Aeron 19:474 (in Russian)

Lobachevskii LA, Suares JB, Tushentsova IA, Fishchuk DI, Tsedilina EE (1979b) Adiabatic analysis of decameter wave propagation conditions over a long-distance path at a fixed frequency, part II. Geomagn Aeron 19:834 (in Russian)

Lobachevskii LA, Suares JB, Tushentsova IA, Fishchuk DI, Tsedilina EE (1979c) Adiabatic analysis of decameter wave propagation conditions over a long-distance path at a fixed frequency, part III. Geomagn Aeron 19:1038 (in Russian)

Lobachevskii LA, Nekrasov BYu, Smirnov VB, Suares JB, Tushentsova IA, Fishchuk DI, Tsedilina EE, Shumilov IA, Israetel AG (1981) Multimode broadening of HF signals on long-distance paths. Geomagn Aeron 21:652 (in Russian)

Loftus BT, Van Zandt TE, Calvert W (1966) Observations of conjugate ducting by the fixed-frequency topside-sounder satellite. Ann Geophys 22:539

Lukin DS, Spiridonov YuG, Fominykh SI, Shkolnikov VA (1973) Calculation of the field amplitude, fading depth and group delay in the case of short wave propagation in the magneto-active ionosphere. In: Actual problems of decameter radio wave propagation, Collection of Papers, Nauka, Moscow, p 5 (in Russian)

Luscombe GW (1957) Delayed Signals in Ionospheric Forward-Scatter Communication. Nature 180:138

Maeda K (1972) *E*-region electron density profiles. Space Res XII, p 1228

Maeda K, Fukao S (1972) Study of the solar activity dependence of *E* region peak electron density and some atmospheric parameters. J Geomagn Geoelectr 24:289

Manaenkova NI (1980) On the comparison of long-distance path calculation by the adiabatic and ray-tracing techniques. In: Propagation of decameter radio waves, Collection of Papers, IZMIRAN, Moscow, p 80 (in Russian)

Manaenkova NI, Tsedilina EE (1981) Geometrical-optics effects of azimuthal deviations of short radio waves over long-distance paths. Preprint IZMIRAN, No 16 (in Russian)

Manaenkova NI, Tsedilina EE (1982) Azimuthal deviations of short radio waves on meridional long-distance paths. Izv Vyssh Uchebn Zaved Radiofiz 26:22 (in Russian)

Martines Brunet R, Popov AV (1980) A simplified model for long-distance propagation of short radio waves. In: Propagation of decameter radio waves, Collection of Papers, IZMIRAN, Moscow, p 52 (in Russian)

Minkoff J, Kugelman P, Weissman I (1974) Radio frequency scattering from a heated ionospheric volume. I. VHF/UHF field-aligned and plasma-line backscatter measurements. Radio Sci 9:941

Minnis CM, Bazzard GH (1959) Some indices of solar activity based on ionospheric and radio noise measurements. J Atmos Terr Phys 14:213

Mityakov NA, Rapoport VO, Trakhtengerts VYu (1974) Induced scattering of radio waves in *F*-layer of the ionosphere. Planet Space Sci 22:95

Monthly prediction for short radio wave propagation (1975), Nauka, Moscow (in Russian)

Muldrew DB (1963) Radio propagation along magnetic field-aligned sheets of ionization observed by the Alouette topside sounder. J Geophys Res 68:5355

Muldrew DB (1967) Medium frequency conjugate echoes observed on topside-sounder data. Can J Phys 45:3935

Muldrew DB (1979) Generation of Long-delayed Echoes. J Geophys Res 84:5199

Myung Ki Lee (1973) Propagation studies using a theoretical ionospheric model. Report 408, Pennsylvania University Press, Pennsylvania

Nevolin GN, Shlionskii ShG (1973) Antipodal radio wave propagation. In: Investigation of short wave propagation, Collection of Papers, Nauka, Moscow, p 56 (in Russian)
Neyer H, Rawer K (1951) Veränderungen des Einfallswinkels der Kurzwellen im Transatlantikverkehr unter dem Einfluß des Sonnenfleckenzyklus. Arch Elektr Übertrag 5:215
Nisbet JS (1971) On the construction and use of a simple ionospheric model. Radio Sci 6:437
Novikova LN, Soboleva TN (1978) Effective electron collision frequency for the conditions of high solar activity. In: Trajectory characteristics of short radio waves, Collection of Papers, IZMIRAN, Moscow, p 190 (in Russian)
Novozhilov VI, Savel'ev SM (1978) Irregular structure of the ionosphere in the field of strong obliquely incident radio wave. Geomagn Aeron 18:214 (in Russian)
Obayashi T (1959) A possibility of the long distance HF propagation along the exospheric field-aligned ionizations. Rep Ionosph Space Res, Japan 13:177
Ortner I (1959) Around-the-World Echoes Observed on a Transpolar Transmission Path. J Geophys Res 64:2464
Ostrovskii IYa (1972) Azimuthal characteristics of around-the-world signal in the winter period. Geomagn Aeron 12:763 (in Russian)
Ostrovskii IYa (1974) On the reciprocity of round-the-world radio paths. In: Problems of short radio wave propagation, Collection of Papers, part II, IZMIRAN, Moscow, p 61 (in Russian)
Ostrovskii IYa, Shlionskii ShG (1974) Azimuthal, time and frequency regularities of the round-the-world signal propagation. In: Problems of short radio wave propagation, Collection of Papers, part II, IZMIRAN, Moscow, p 52 (in Russian)
Ostrovskii IYa, Shlionskii ShG (1974) The effect of space-time variation in the *F*-region on the propagation of round-the-world signals. In: Problems of short radio wave propagation, Collection of Papers, part II, IZMIRAN, Moscow, p 65 (in Russian)
Ostrovskii IYa, Shlionskii AG (1975) Determination of optimum conditions for around-the-world signal propagation. In: Ionospheric propagation of short radio waves, Collection of Papers, IZMIRAN, Moscow, p 76 (in Russian)
Pedersen PO (1929) Wireless echoes of long delay. Proc IRE 17:1750
Perekhvatov YuK (1975, 1976) Theory of HF propagation in cylindrical inhomogeneous stratified plasma ducts. In: Propagation of decameter radio waves, Collection of Papers, Nauka, Moscow, p 26, p 30 (in Russian)
Perkins FW (1974) A theoretical model for short-scale field-aligned plasma density striations. Radio Sci 9:1065
Perkins FW, Kaw PK (1971) On the role of plasma instabilities in ionospheric heating by radio waves. J Geophys Res 76:282
Perkins FW, Valeo KJ (1974) Thermal self-focusing of electromagnetic waves in plasmas. Phys Rev Lett 32:1234
Perkins FW, Oberman G, Valeo EJ (1974) Parametric instabilities and ionospheric modification. J Geophys Res 79:1478
Pipp RM, Webster YB (1964) An experimental investigation of signal strength in the area around a transmitter's antipode. J NBS 68(D):333
Pitteway MLV, Cohen R (1961) A waveguide interpretation of "temperature-latitude spread F" on equatorial ionograms. J Geophys Res 66:3141
Polyakov VM, Sazhin VI, Svistunov KV, Semeney YuA, Tinin MV, Unuchkov VE (1977) Comparison of the results of experimental and theoretical investigations of return signal propagation on the Khabarovsk-Irkutsk path. Izv Vyssh Uchebn Zaved Radiofiz 20:1866 (in Russian)
Polyakov VM, Sukhodol'skaya VE, Ivel'skaya MK, Shapranova GV (1978) Semiempirical model of the ionosphere. Nauka, Moscow (in Russian)
Popov AV (1976) Eigenmodes of the irregular waveguides. Dokl Akad Nauk SSSR 230:122 (in Russian)
Quaeck E (1927) Propagation of Short Waves Around the Earth, Further Communication of the Propagation of Short Waves. Proc IRE 15:341, 1065
Quaeck E, Möegel H (1929) Double and Multiple Signals with Short Waves. Proc IRE 17:791
Rao PB, Thome GD (1974) A model of RF scattering from field-aligned heater-induced irregularities. Radio Sci 9:987

Rawer K, Ramakrishnan S, Bilitza D (1978) International Reference Ionosphere, URSI-COSPAR
Rufenach CL (1973) Radio scintillation of stellar signals during artificial ionospheric modifications. J Geophys Res 78:5611
Rufenach CL (1974) Wavelength dependence of radio scintillation: ionosphere and interplanetary irregularities. J Geophys Res 79:1562
Rytov SM (1966) Introduction to the statistical physics. Nauka, Moscow (in Russian)
Rytov SM, Kravtsov YuA, Tatarskii VI (1978) An introduction to the statistical radiophysics. The random fields. Nauka, Moscow (in Russian)
Sazhin VI, Tinin MV (1975a) Role of the sliding mode in excitation of ionospheric waveguides. Izv Vyssh Uchebn Zaved 18:1389 (in Russian)
Sazhin VI, Tinin MV (1975b) Long-distance propagation by means of the Pedersen ray. Geomagn Aeron 15:748 (in Russian)
Semionova VI, Trakhtengerts VYu (1980) Some features of waveguide VLF propagation in the magnetosphere. Geomagn Aeron 20:1021 (in Russian)
Serafimov K, Lilov A (1975) Effect of solar activity on critical frequencies in ionospheric region F (layer F_2) above Sofia. C R Acad Bulg Sci 28:1207
Shchukin AN (1940) Radio wave propagation. Svyazizdat, Moscow (in Russian)
Shlionskii ShG (1965) Attenuation of radio waves emitted from an artificial satellite and propagating on ground-detached trajectories. Geomagn Aeron 5:1061 (in Russian)
Shlionskii ShG (1974) Some trajectory characteristics of radio wave guiding in the ionosphere. Variations of the maximum of $rn(r)$-function in the ionosphere. Refractive characteristics of radio wave capture and descent from ionospheric waveguides. In: Problems of short radio wave propagation, part II, Collection of Papers, IZMIRAN, Moscow, p 77, 88, 95 (in Russian)
Shlyuger IS (1974) Experimental investigation of nonlinear effects in E and F-regions of ionosphere. JETP Lett 20:334
Showen RL, Duncan LM, Cragin BL (1978) Observations of plasma instabilities in a multiple pump ionospheric heating experiment. Geophys Rev Lett 5:187
Smirnov VB (ed) (1972) Oblique sounding of the ionosphere, Collection of Papers, Gidrometeoizdat, Leningrad (in Russian)
Soboleva TN (1972) Model profiles of electron concentration diurnal distribution in a quiet ionosphere at middle latitudes. Preprint, IZMIRAN, Moscow (in Russian)
Soboleva TN (1973) Model of latitude electron concentration distribution in quiet ionosphere. Preprint, IZMIRAN, No 16 (in Russian)
Stratton JA (1941) Electromagnetic Theory. McGraw-Hill, New York
Suares JB, Reborido J (1979) Daily regularities of the group of channels in a trace for a distance of 10000 km in 10 MHz. Resumenes de la segunda Jornada cientifica. IGA, Academia C. Cuba, Habana, p 94
Suares JB, Tushentsova IA (1980) Diurnal variations of a set of communication channels of the path (CCP) for the path length of 10000 km at the frequency $f = 10$ MHz. In: Interaction between HF/VHF radio waves and the ionosphere, Collection of Papers, IZMIRAN, Moscow, p 112 (in Russian)
Talanov VI (1965) On the self-focusing of wave beams in nonlinear media. JETP Lett 2:218 (in Russian)
Tamm IE (1966) Principles of the electricity theory. Nauka, Moscow (in Russian)
Tatarskii VI (1967) Wave propagation in a turbulent atmosphere. Nauka, Moscow (in Russian)
Tatarskii VI (1971) The effect of turbulent atmosphere on wave propagation. Jerusalem
Taylor AH, Young LG (1928) Studies of High-Frequency Radio Wave Propagation. Proc IRE 16:561
Thome GD, Blood DW (1974) First observations of HF backscatter from field-aligned irregularities produced by ionospheric heating. Radio Sci 9:917
Tinin MV (1973) Role of the Pedersen ray and directed waves connected with it in the case of wave propagation along a parabolic layer. In: Studies on geomagnetism, aeronomy and solar physics, Collection of Papers, Nauka, Moscow, vol 29, p 157 (in Russian)
Tinin MV (1976) On the waveguide propagation in smoothly-inhomogeneous, quasistratified medium (ionosphere). In: Studies on geomagnetism, aeronomy and solar physics, Collection of Papers, Nauka, Moscow, vol 39, p 166

Tsedilina EE (1974) Investigation of the global properties of ionospheric wave ducts, I. Geomagn Aeron 14:851
Tsedilina EE (1975) Radio wave propagation in ionospheric wave ducts over around-the-world paths. Geomagn Aeron 15:457 (in Russian)
Tsedilina EE (1975) Radio wave absorption on around-the-world midlatitude paths. In: Ionospheric propagation of short radio waves, Collection of Papers, IZMIRAN, Moscow, p 117 (in Russian)
Tsedilina EE (1980) Discrete structure of around-the-world signals. Geomagn Aeron 20:148 (in Russian)
Tushentsova IA, Fishchuk DI (1975) Refinement of latitudinal variation of equinoctial analytical model coefficients for three-dimensional ionospheric electron concentration distribution. In: Investigation of very long-range propagation of short radio waves, Collection of Papers, IZMIRAN, Moscow, p 143 (in Russian)
Tushentsova IA, Tsedilina EE (1976) Azimuthal regularities of RTW absorption. Geomagn Aeron 16:559
Tushentsova IA, Tsedilina EE (1977) On the problem of azimuthal regularities of RTW absorption. In: Investigation methods of radio wave propagation regularities, Collection of Papers, Nauka, Moscow, p 37 (in Russian)
Tushentsova IA, Tsedilina EE (1983) Multifrequency model analysis of propagation conditions for HF waves on the Nikolaev-Havana path. Geomagn Aeron 23:41 (in Russian)
Tushentsova IA, Fishchuk DI, Tsedilina EE (1974) The maps of global distribution of some ionospheric wave channel parameters. Preprint 5, IZMIRAN, Moscow
Tushentsova IA, Fishchuk DI, Tsedilina EE (1975a) The absorption distribution in ionospheric wave ducts, and the absorption of around-the-world echo-signals. In: Investigation of very long-range propagation of short radio waves, Collection of Papers, IZMIRAN, Moscow, p 8 (in Russian)
Tushentsova IA, Fishchuk DI, Tsedilina EE (1975b) Maps of the global absorption distribution in ionospheric wave ducts. In: Investigation of very long-range propagation of short radio waves, Collection of Papers, IZMIRAN, Moscow, p 104 (in Russian)
Tushentsova IA, Fishchuk DI, Tsedilina EE (1975c) Global absorption distribution of short radio waves in the ionosphere. Izv Vyssh Uchebn Zaved, Radiofiz 18:1279 (in Russian)
Tushentsova IA, Fishchuk DI, Tsedilina EE (1975d) Investigation of the global properties of ionospheric wave ducts II. Geomagn Aeron 15:62
Tushentsova IA, Fishchuk DI, Tsedilina EE (1977a) The refined maps of the global distribution of some ionospheric wave duct parameters for equinoctial conditions during solar activity minimum. In: Very long-range radio wave propagation and ionospheric models, Collection of Papers, IZMIRAN, Moscow, p 110 (in Russian)
Tushentsova IA, Fishchuk DI, Tsedilina EE (1977b) Some features of the global radio wave propagation for equinoctial conditions during solar activity minimum. In: Very long-range radio wave propagation and ionospheric models, Collection of Papers, IZMIRAN, Moscow, p 136 (in Russian)
Tushentsova IA, Fishchuk DI, Tsedilina EE (1977c) The refined maps of the global absorption distribution in ionospheric wave ducts for equinoctial conditions during solar activity minimum. In: Very long-range radio wave propagation and ionospheric models, Collection of Papers, IZMIRAN, Moscow, p 146 (in Russian)
Tushentsova IA, Fishchuk DI, Tsedilina EE (1983) Model ionograms of oblique sounding over the Nikolaev-Havana path. Geomagn Aeron 23:573 (in Russian)
Tushentsova IA, Fishchuk DI, Tsedilina EE, Israetel AG (1982) Calculation of the oblique sounding ionograms for the Nikolaev-Havana path on the basis of the EMI-81 model. In: Interaction between short radio waves and the ionosphere, Collection of Papers, IZMIRAN, Moscow, p 93 (in Russian)
Uspenskii MV, Pudovkin AI (1974) Ground-based waveguide propagation of radio waves in the aurora region. In: Problems of short radio wave propagation, part II, Collection of Papers, IZMIRAN, Moscow, p 101 (in Russian)

Uspenskii MV, Sverdlov YuL (1973) Forward scattering of VHF in the auroral ionosphere. In: High-latitude geophysical phenomena, Collection of Papers, Nauka, Leningrad, p 312 (in Russian)

US Standard Atmosphere Supplements issued by ESSA-NASA-USAF (1966)

Utlaut WF (1970) An ionospheric modification experiment using very high power, high frequency transmission. J Geophys Res 75:6402

Utlaut WF (1975) Ionospheric modifications. A potential for Extended Range VHF-UHF Communications and Plasma Physics Research. Proc IEEE 63:1022

Utlaut WF, Cohen R (1971a) Modifying the ionosphere with intense radio waves. Science 174:245

Utlaut WF, Cohen R (1971b) Modifying the ionosphere with intense radio waves. Sov Phys Usp 109:371 (in Russian)

Utlaut WF, Violette EJ, Melanson LL (1974) Radar cross section measurements and vertical incidence effects observed with Platteville at reduced power. Radio Sci 9:1033

Varshavskii II, Dimant YaS (1976) Parametric excitation of Langmuir oscillations in the ionosphere by powerful ordinary radio wave. Geomagn Aeron 16:57

Vas'kov VV (1969) Radio wave scattering by the trail of a body travelling in the ionosphere in the vicinity of a caustic. Geomagn Aeron 9:847 (in Russian)

Vas'kov VV (1977) Small-scale perturbations of electron density in magnetoplasma with the heating sources. In: Very long-range radio wave propagation and ionospheric models, Collection of Papers, IZMIRAN, Moscow, p 61 (in Russian)

Vas'kov VV, Gurevich AV (1974) Stratification of ionospheric plasmas in the reflection region of ordinary radio wave. JETP Lett 20:93

Vas'kov VV, Gurevich AV (1975) Nonlinear resonance instability of plasma in the reflection region of ordinary electromagnetic wave. Sov Phys JETP 42:91

Vas'kov VV, Gurevich AV (1977) Resonance instability of small-scale plasma perturbations. Sov Phys JETP 73:923 (in Russian)

Vas'kov VV, Gurevich AV (1979) Self-focusing and resonance instabilities in the ionospheric *F*-layer. In: Thermal nonlinear phenomena in plasmas, Collection of Papers, IPF AN SSSR, Gorki, p 81 (in Russian)

Vas'kov VV, Golyan SF, Gurevich AV, Gruzdev YuV, Dimant YaS, Kim VYu, Lobachevskii LA, Migulin VV, Panchenko VA, Petrov MS, Polimatidi VP, Sitnikov VI, Shoya LD, Shlyuger IS, Yurin KI (1982) Stimulated ionization of the upper ionosphere by an intense radio wave. JETP Lett 34:558

Villard OG, Graf CR, Lomasuey JM (1969) Long-delayed echoes. Radio's "flying saucer" effect. QST 53:38

Voogt AH De (1950) Les Echoes Radioélectriques Autour de la Terre. L'Onde Electrique 30:433

Vorobiov VV (1971) Scattering of the light beam in the medium with regular refraction. Izv Vyssh Uchebn Zaved Radiofiz 14:1283 (in Russian)

Vovk VYa, Lukashkin VM, Smirnov BV (1981) A measurement of HF radio wave arrival angles on the Moscow–Antarctic Continent path. Proc XIII All-Union Conference on Radio Wave Prop. Nauka, Gorki, p 213 (in Russian)

Vsekhsvyatskaya IS, Kalinin YuK (1975) Estimation of a height factor for around-the-world echoes according to experimental results. Geomagn Aeron 15:1101 (in Russian)

Vsekhsvyatskaya IS, Kalinin YuK (1979) On the interference of decameter radio waves in the vicinity of the antipode. Geomagn Aeron 19:470

Wernik AW, Liu CH (1974) Ionospheric irregularities causing scintillation of GHz frequency radio signals. J Atmos Terr Phys 36:871

Whittaker ET, Watson GN (1946) A course of modern analysis, vol 1. Cambridge University Press, Cambridge

Whittaker ET, Watson GN (1946) A course of modern analysis, vol 2, Cambridge University Press, Cambridge

Woyk E (Chvojkova) (1958) Determination of radio propagation elements due to an artificial earth satellite. Nature 181:1195

Woyk E (Chvojkova) (1959) The antipodal reception of Sputnik III. Proc IRE 3:1144
Woyk E (Chvojkova) (1974) Radio trajectory formula. 1. Theory. Round-the-world echoes and other features. Geomagn Aeron 14:57; Radio trajectory formula. II. Practical applications. Geomagn Aeron 14:466 (in Russian)
Wright JW (1973) Kinesonde observations of ionosphere modification by intense electromagnetic fields from Platteville, Colorado. J Geophys Res 78:5622

Subject Index

Physics and Chemistry in Space

Editors: **L. J. Lanzerotti, J. T. Wasson**

Volume 10
A. v. Gurevich
Nonlinear Phenomena in the Ionosphere
Translated from the Russian by J. G. Adashko
1978. 76 figures, 17 tables. X, 370 pages. ISBN 3-540-08605-6

Volume 9
A. Nishida
Geomagnetic Diagnosis of the Magnetosphere
1978. 119 figures, 1 table. VIII, 256 pages. ISBN 3-540-08297-2

Volume 8
A. Hasegawa
Plasma Instabilities and Nonlinear Effects
1975. 48 figures. XI, 217 pages. ISBN 3-540-06947-X

Volume 7
M. Schulz, L. J. Lanzerotti
Particle Diffusion in the Radiation Belts
1974. 83 figures. IX, 215 pages. ISBN 3-540-06398-6

Volume 6
S. J. Bauer
Physics of Planetary Ionospheres
1973. 89 figures. VIII, 230 pages. ISBN 3-540-06173-8

Volume 5
A. J. Hundhausen
Coronal Expansion and Solar Wind
1972. 101 figures. XII, 238 pages. ISBN 3-540-05875-3

Volume 4
A. Omholt
The Optical Aurora
1971. 54 figures. XIII, 198 pages. ISBN 3-540-05486-3

Volume 3
I. Adler, J. I. Trombka
Geochemical Exploration of the Moon and Planets
1970. 129 figures. X, 243 pages. ISBN 3-540-05228-3

Volume 2
J. G. Roederer
Dynamics of Geomagnetically Trapped Radiation
1970. 94 figures. XIV, 166 pages. ISBN 3-540-04987-8

Volume 1
J. A. Jacobs
Geomagnetic Micropulsations
1970. 81 figures. VIII, 179 pages. ISBN 3-540-04986-X

Springer-Verlag
Berlin
Heidelberg
New York
Tokyo

MIX
Papier aus verantwortungsvollen Quellen
Paper from responsible sources
FSC® C105338

If you have any concerns about our products,
you can contact us on
ProductSafety@springernature.com

In case Publisher is established outside the EU,
the EU authorized representative is:
Springer Nature Customer Service Center GmbH
Europaplatz 3, 69115 Heidelberg, Germany

Printed by Libri Plureos GmbH
in Hamburg, Germany